NATIONAL GEOGRAPHIC

STARGAZER'S ATLAS

THE ULTIMATE GUIDE TO THE NIGHT SKY

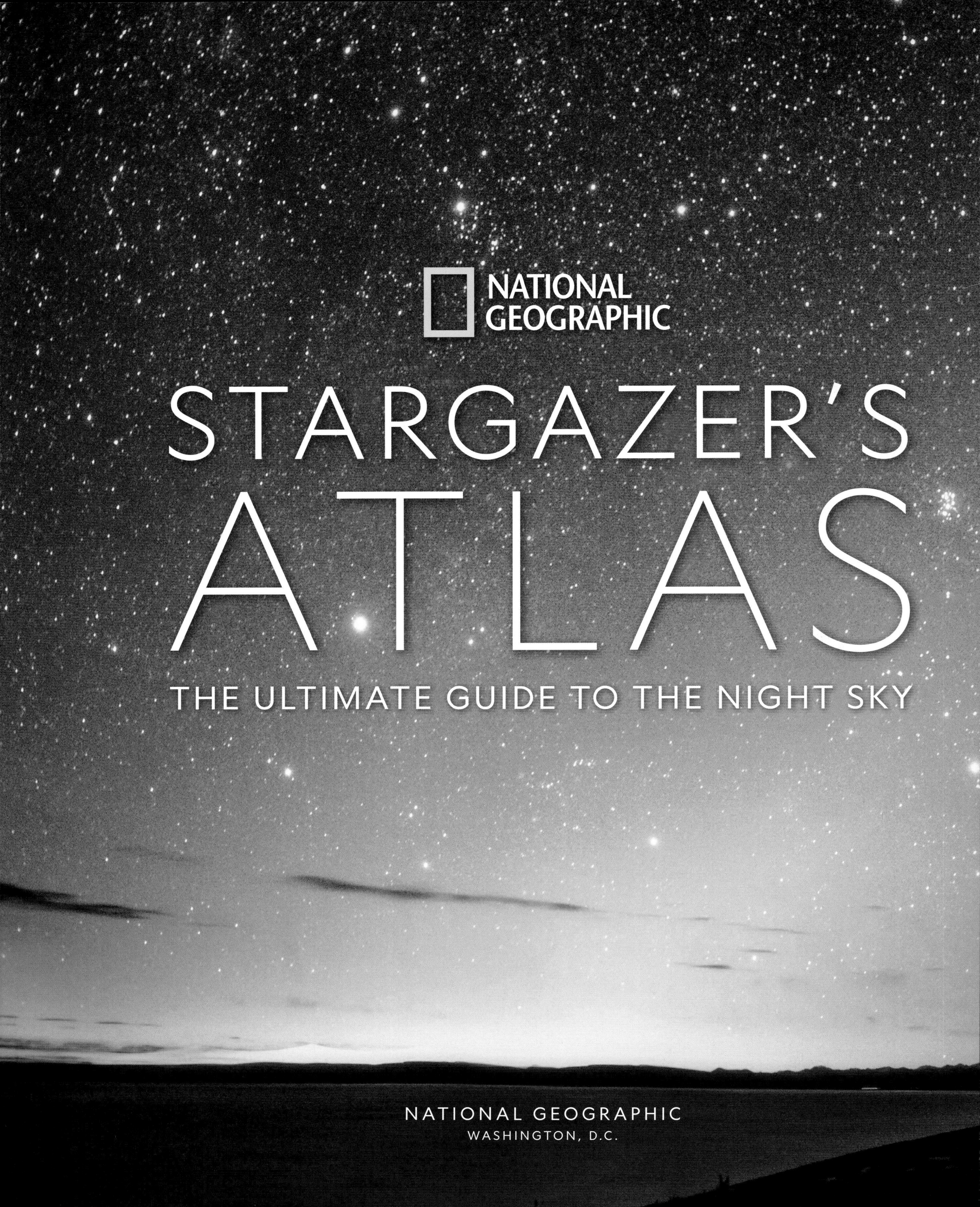

NATIONAL GEOGRAPHIC

STARGAZER'S ATLAS

THE ULTIMATE GUIDE TO THE NIGHT SKY

NATIONAL GEOGRAPHIC

WASHINGTON, D.C.

An observer views zodiacal light—sunlight reflected from dust particles in the solar system—over Lake Namtso in Tibet.

Two Hubble instruments made the observations used to create this stunning composite image of the globular star cluster Caldwell 84.

CONTENTS

Foreword: The Joy of Stargazing by Andrew Fazekas | 6

About This Book | 8

CHAPTER ONE
STARS | 12

Where Are They? | 16   What Are They? | 30
The Stellar Life Cycle | 38

CHAPTER TWO
SUN, MOON, EARTH | 52

The Sun | 56   The Moon | 68   The View From Earth | 84

CHAPTER THREE
THE SOLAR SYSTEM | 92

Origins of the Solar System | 96   Mercury | 102   Venus | 108   Earth | 114   Mars | 120   Jupiter | 128
Saturn | 136   Uranus | 144   Neptune | 150   Dwarf Planets | 154   Small Worlds | 158

CHAPTER FOUR
THE NIGHT SKY | 164

Navigating the Night Sky | 168   Using the Sky Maps | 172
Constellations | 194

CHAPTER FIVE
MORE CELESTIAL SIGHTS | 278

Meteors and Comets | 282   Diving Into the Deep Sky | 288   Deep-Sky Treasures | 292
Supernovae | 304   Passing Worlds | 308   Tracking Satellites | 310
Ghostly Sky Glows | 314

CHAPTER SIX
LET'S GO GAZING | 318

Archaeoastronomy | 322   Astrotourism Today | 340   Future Flights | 354

CHAPTER SEVEN
SEEING FARTHER | 360

Our Cosmic Neighborhood | 364   Advances in Astronomy | 368   Alien Planets | 388
Searching for Life | 398   Where Do Humans Fit In? | 404

Glossary | 416   Notable Deep-Sky Objects | 418   Skywatcher's Calendar | 420

About the Authors and Acknowledgments | 422

Illustrations Credits | 423   Index | 425

The distinctive stars of the Orion constellation (left) and the Pleiades (right) shine over an ancient palace in Iran's Fars Province.

THE JOY OF STARGAZING

Stargazing is one of the oldest pastimes in human history: looking up in wonder at the night sky, finding and naming shapes and objects of interest, observing predictable cycles through time, and marveling at the unexpected. Through the ages, people have looked to the heavens for inspiration and excitement, solace and peace.

Today, when so many of us live in cities, we forget to look up. Round-the-clock lighting obscures the night sky. Screen time dominates the view. Yet the stars and planets are always there, ready for you to observe and explore. We hope this book intrigues you, reminds you, guides you, and reconnects you to the wonders of the universe, no matter where you are.

This glorious book is an atlas—and so much more. As an atlas, it offers you maps of moons and planets, charts of all 88 constellations visible in the Northern and Southern Hemispheres, and a guide to world travel for the sake of stargazing. As a compendium of general knowledge about the cosmos, it features the basic science of astronomy today and glimpses of future explorations going ever further.

In the study of astronomy, space and time are intimately connected. With our technology, we peer out into space and back in time. That glimmer of light that hits your eyes as a star has been traveling years to get to you. Most stars visible to the naked eye are on average 75 light-years away. That means the light you see tonight took 75 years—a human lifetime!—to travel from its source to you.

We live in an epoch when our technology allows us to experience the universe as never before. Telescopes both on Earth and in the sky increase our knowledge about distant stars, galaxies, planets, and celestial systems. Even as you are reading these words, complex spacecraft are traveling through the universe, gathering information—with some even beyond the influence of our sun, out in deepest space, and still sending back precious data.

So today, as we stand outside at night and gaze up at the starry sky, we too can join that primeval sense of wonder and get a clearer understanding of just what we are looking at. Thanks to imagery generated by giant observatories and space probes, no longer is that galaxy just the cotton-ball smudge seen through a backyard telescope. When you appreciate the science behind your observations, you understand that this flicker of light you can barely see is a factory of newborn stars, or a cluster of old stars circling together, or a dying star, exploding as it expires.

You don't need much: If you get really intrigued, you can invest in binoculars or a personal telescope, but you don't even need to do that. Just find a comfortable place, unobscured by artificial lighting, where you can allow your eyes the 20 or 30 minutes they need to adjust to the dark—and take it from there. Gaze up in wonder at the night sky.

We have created this book with the philosophy that knowledge of the science of astronomy can only deepen your sense of wonder. Just imagine: On a starry night, you are basking in light that left its source long before you were born. In fact, you are gazing far back in time, thousands, even millions of years.

From your own backyard.

—ANDREW FAZEKAS
THE NIGHT SKY GUY

ABOUT THIS BOOK

Welcome to *National Geographic Stargazer's Atlas,* a book designed to provide everyone interested in the wonders of the night sky with a full compendium of maps and charts, graphics, photographs, and ample information to guide you to a new level of knowledge and wonder about our universe.

Each chapter contributes to the experience. Chapters 1, 2, and 3 introduce the basic science of stars, our nearby neighbors the sun and moon, and the planets in our solar system. Chapters 4 and 5 provide charts and pointers for finding all 88 constellations in the Northern and Southern Hemisphere, as well as other skywatching phenomena. Chapter 6 addresses the traveler in everyone, describing the world's best stargazing destinations and the ancient sites where our ancestors did the same. Finally, chapter 7 looks into the future. What questions are astronomers asking about the universe, and what spacecraft will advance our knowledge?

Photos and Fact Boxes
Today as never before, spacecraft and telescopes are bringing us detailed imagery, as well as information to inform authentic illustrations, to help picture planets and other celestial objects. Photos and art throughout are paired with recurrent fact boxes, offering the basics on objects in the night sky.

Graphics, Sidebars, and Cross-References
Key concepts of science are presented in accessible text accompanied by explanatory graphics. Sidebars throughout dig deeper into compelling stories and important ideas, and marginal cross-references connect passages throughout the book for fuller understanding.

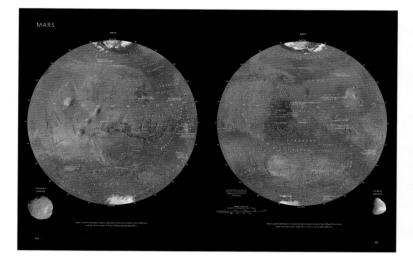

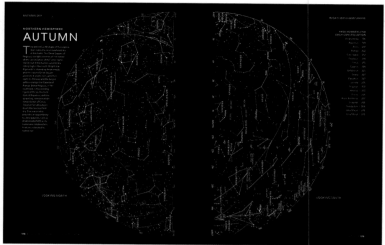

Solar System Maps

Maps of all the planets and many moons in our solar system provide topographic details and note locations such as impact craters, canyons, and exploration sites.

Seasonal Sky Charts

Sky charts for all four seasons in both the Northern and Southern Hemispheres orient the stargazer eager to identify constellations.

Constellations

A star chart is provided for every constellation recognized by the International Astronomical Union, along with a list of the main stars, deep-sky objects within the constellation, and neighboring constellations. Star-hopping charts advise you on how to use familiar constellations to find others that are harder to identify. Sidebars detail the myths behind constellation names and other noteworthy sightings visible within constellations.

Sidebars With More Information

"Skylist" sidebars enumerate more examples to amplify the main text. "See for Yourself" sidebars offer practical, personal ways to experience the science. General information sidebars tell stories and offer deeper explanations for the ideas on each page.

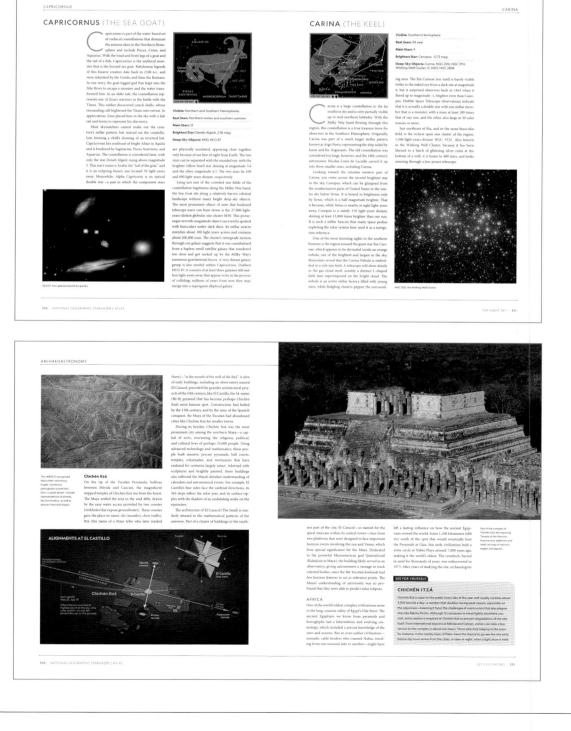

A Perseid meteor streaks across the sky over Saskatchewan, Canada, during a stargazing party.

The Milky Way as seen from Easter
Island. The famous stone moai in this
picture are facing south.

CHAPTER ONE

STARS

THE CHILDREN OF GRAVITY

Whenever you feel the warmth of the summer sun on your face, whenever you see a star twinkling in the midnight sky, you are witnessing the result of an unending battle. Rather than looking at stars as eternal and unchanging lights in the sky, modern scientists see stars as both children and victims of gravity. They are objects that are born when gravity acts on the material in interstellar clouds, and they then spend the rest of their lives devising strategies to avert total gravitational collapse.

When you think about the way people looked at stars in the past, it's important to keep one thing in mind. Until the 19th century, there were no gaslit or electric streetlights, even in major cities. For every observer, the night sky would have looked like what we can see today only if we get far away from cities (a point we will emphasize in chapter 6). The incredible display of the night sky would simply have been an everyday part of everyone's life.

Small wonder, then, that the ancients peopled the sky with gods and demons, and that they believed the nightly display contained important and immediate information on human life. Small wonder also that they thought the positions of the stars and planets needed to be consulted when important decisions were to be made.

Every night this incredible display of lights would rotate slowly across the sky. Over time, some of those lights would change their relationship to others, while others would not. The moving lights are what we recognize as planets today. The other lights—the ones that don't seem to change their relative positions—are called stars.

Consider the Milky Way, the most spectacular stellar display in a dark night sky. Today, we understand that the Milky Way looks the way it does because we are looking at its disk from the inside. The Apache of the American Southwest, however, considered the Milky Way to be the first path traveled by the sun and moon, while Hungarians believed it was the road down which the warrior Csaba (the legendary son of Attila the Hun) would ride to fight when the Hungarian nation was threatened. The builders of the Nasca lines in Peru, on the other hand, apparently concentrated on the dark spaces in the Milky Way—the places where the starlight is obscured by clouds of dust. To them, the absence of stars was more important than their presence.

A collection of stars known as a globular cluster, 7,800 light-years from Earth

WHERE ARE THEY?

The stars are out of reach. How do we know where they really are?

Traditional ways of looking at the stars changed abruptly with the development of modern science. The stars stopped being gods and demons and became ordinary physical objects—objects that could be studied and understood like everything else. In fact, we can consider the scientific view of stars to be centered around two questions:

Where are they?

What are they?

Each of these seemingly simple questions is actually a lot deeper than it appears, since each leads us to think about the nature of the universe at a profound level. This fact might not have been obvious to early astronomers, who tended to be more concerned with tasks like constructing calendars and casting horoscopes. In many societies the stars and the planets were thought to have a direct influence on human life, so many of those early astronomers were involved in what we call astrology today. (Note that astrology is universally rejected by modern astronomers. When we think about the stars now we prefer to think deeply about the two questions posed above rather than trying to get advice for our daily lives.)

Let's start with the first question. The universe may be three-dimensional, but our view of it is not. The sky looks to us like a two-dimensional bowl arching overhead, a bowl sprinkled with the lights we call stars. There is no clue in this display as to the actual distance to any of these lights. A star may appear dim, for example, because it really is dim or because it is far away. If we want to ask one of the most fundamental questions that we can ask—"How big is the universe?"—we have to find some way of adding a third dimension to what we see in

the sky. This challenge is known as the problem of constructing the "astronomical distance ladder" because, as we shall see, its solution involves piling one sort of observation on top of another, like rungs on a ladder.

The problem of determining the size of the universe has gone through several historical stages. First, it involved figuring out the size of Earth—a problem solved by the Alexandrian Greeks before the birth of Christ. Then we had to find the distances to objects in the solar system: a problem that wasn't solved until the late 1600s by a team of French astronomers. The next step involved finding

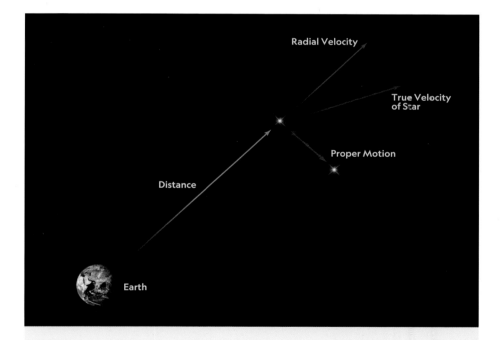

THE MOTION OF STARS

Although they appear stationary, all the stars we see in the sky are in motion. For one thing, our entire galaxy rotates, completing a turn every 250 million years or so. Moreover, stars typically have some additional motion, more or less random, on top of that rotation. For astronomical studies that involve a typical human lifetime, however, this extra motion is usually just ignored, though it does mean that millions of years from now the familiar constellations will look different because the stars that create the images will have moved.

(Opposite) A cluster of galaxies as seen by the Hubble Space Telescope

The astronomical term "parallax" comes from the Greek verb meaning "to change."

the distances to the stars, which was finally done in 1838. Today, we are struggling with the continuing problem of finding the distances to the farthest visible galaxies, a challenge that must be met if we want to know the age of the universe itself. As with all great intellectual quests, this one starts with some fairly simple ideas.

At this point, try a simple experiment. Hold your finger out and look at it with one eye closed. Now open that eye and close the other one. Do you notice how your finger seems to move against the background?

This phenomenon, as everyday and simple as it seems, gives us the tool for measuring distances—not only the distance between your eye and your finger, but also the distance between Earth and the stars. It involves some simple geometry and is called parallax.

PARALLAX

Parallax is the first rung on the astronomical distance ladder, and, for the record, modern satellites in orbit can use this technique to measure the distances to stars over 30,000 light-years away. The history of our ability to measure distances in the sky—of getting from early measurements to the abilities of those satellites—can be thought of as a kind of leapfrogging of competencies in increasing the accuracy of our instruments and increasing the baseline of our measurements.

Here's how it works: Suppose you want to measure the distance to an object but can't do it directly. You may, for example, want to know how far it is from where you're standing to an unusually shaped rock across a canyon. Suppose further that you don't have any way to get to the other side of the canyon. One way you could proceed would be to make a

Three stars are moving with high velocity (shown in boxes) in this view of the center of the Orion Nebula. These stars will eventually leave the nebula.

couple of measurements. You begin by marking out a line of known length—say a 15-meter (50-ft) tape measure—on your side of the canyon. This is known as the baseline of your measurement. At one end of the baseline you measure the angle between the tape and the line of sight to the rock, and then do the same at the other end of the tape. These measurements define a triangle whose vertices are the rock and the two ends of your tape. Basic geometry tells us that if you know one side and two angles of a triangle you can determine everything else about it, including the lengths of the other two sides. These other sides, of course, represent the distance to the rock.

In general, the farther away an object is, the longer the baseline has to be to get an accurate measurement. For example, with the 15-meter baseline we assumed and instruments capable of measuring an angle to an accuracy of a degree, we could determine the distance to the rock only if it was less than about 760 meters (about half a mile) away.

If we wanted to find the distance to the rock when it's beyond this limit, we could do one of two things: We could use more accurate instruments, so that we could distinguish the difference between the line-of-sight angles, or we could increase the baseline, so that the difference in those angles becomes larger than the limit of accuracy of our optical instruments. In the case of the rock, for example, using instruments capable of measuring to half a degree would allow us to find the distance to objects over 1,500 meters (about a mile) away from us, or, if we can't change the instruments, we could get the same effect by increasing the baseline to 30 meters (about 100 ft).

MEASURING EARTH

The first baseline available to astronomers was the surface of Earth. The idea was that by observing a distant object like the moon from two points on the planet, we could find the distance to the object. In this scheme the longest baseline at hand was the diameter of Earth itself, and as it happens, this quantity was first measured by the third-century B.C. Alexandrian scientist Eratosthenes of Cyrene.

Eratosthenes knew that on the summer solstice sunlight penetrated to the bottom of a deep well in the Egyptian town of Syene, near modern Aswan. He then measured the length of the shadow cast by

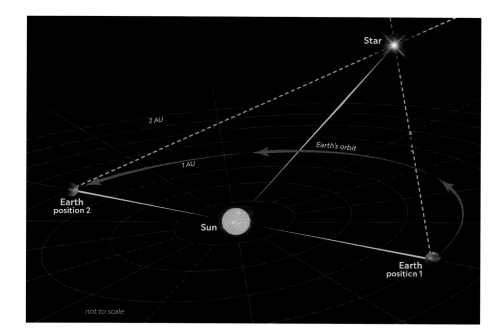

a pole of known height in Alexandria on the same day. From these two measurements he concluded, again using simple geometry, that the distance between Alexandria and Syene was 1/50 of the circumference of Earth.

We don't know how Eratosthenes determined this distance, but he gave a number for the radius of Earth: 250,000 stades. The stade (a length from which we derive the word "stadium") was a unit of length in the ancient world. Unfortunately, there were several different stades in common use at that time. Using the most generous interpretation of his data, we find that Eratosthenes got a value of 46,671 kilometers (29,000 mi) for Earth's circumference (compared to the actual value of about 40,030 kilometers/25,000 mi). Not bad for a guy whose only instrument was a stick pushed into the ground!

Measuring distance by parallax: After measuring the angles of the two blue lines, simple geometry reveals the distance, which is the length of the line from the sun to the star.

The famous well near Aswan in Egypt shows light from the sun reaching the bottom. Data from Eratosthenes's study of this phenomenon allowed him to measure the diameter of Earth.

Using this baseline, Greek astronomer Hipparchus was later able to estimate the distance to the moon to be about 63 times Earth's radius—not too far off the actual distance of about 60 times Earth's radius. However, with the naked-eye instruments of the day, and given the best baseline available to him, there is no way that Hipparchus could have measured the distance to the sun, much less the distance to the stars.

This doesn't mean, however, that the stars were neglected in Greek cosmology. The Greeks created their picture of the universe based on two assumptions—assumptions that they felt were so obvious that they never questioned them. The first assumption, which they shared with most early astronom-

THE PROBLEM OF THE STADE

It should come as no surprise that there is some ambiguity in the definition of the stade. For instance, in modern times we have two different lengths we call "miles": the statute mile at 5,280 feet and the nautical mile at 6,076 feet. In the same way, there were several stades in use in Alexandrian times. The most commonly used were:

- **OLYMPIC STADE**
  Equivalent to 176.4 meters (579 ft)

- **ITALIAN STADE**
  Equivalent to 184.8 meters (606 ft)

- **BABYLONIAN STADE**
  Equivalent to 195.1 meters (640 ft)

- **EGYPTIAN STADE**
  Equivalent to 209.2 meters (681 ft)

For reference, a football field is 91.4 meters (300 ft) long, so roughly speaking, a stade is about two football fields in length.

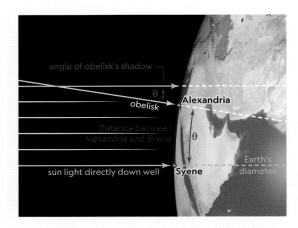

By measuring a shadow in Alexandria (top of the red line) when sunlight reached the bottom of a well in Syene, Eratosthenes was able to estimate Earth's radius.

ical civilizations, was that Earth was the unmoving center of the universe. It's not hard to see the justification for this dictum. After all, our senses tell us that the solid ground under us is totally stationary. What more evidence do you need?

Actually, since you started reading this chapter the rotation of Earth has carried you hundreds of miles from where you started, the revolution of Earth around the sun has carried you thousands of miles around our planet's orbit, the rotation of the Milky Way has carried you even farther than that, and who knows how far you've moved due to the expansion of the universe. Because they didn't know about any of this, the Greeks were able to put together relatively simple models of the universe with Earth at the center. Ignorance can be bliss, after all.

PERFECT SPHERES

Their second unquestioned assumption is a little harder to justify on the basis of experience, because it was more about philosophy than astronomy. Basically, the Greeks felt that although Earth was by its nature imperfect, the heavens, the home of the gods, were perfect and unchanging. Since to their minds the most perfect geometrical shapes were the circle and the sphere, it followed that the planets and stars must move in circular orbits. In the original models of the universe, such as the one proposed by Plato's student Eudoxus of Cnidus, the moon, the sun, and the planets were carried along on solid crystal spheres that rotated around Earth, and the sphere of the stars was simply the outermost of all those spheres. Thus, in Greek cosmology, the stars defined the outer limits of the universe. Later astronomers, particularly Claudius Ptolemy in Alexandria around A.D. 150, refined this simple model to include spheres rolling within spheres to account for various details of their observations. His book, the *Almagest,* came to Europe after having been translated into Arabic, and it remained the accepted picture of the heavens until the 15th century.

Interestingly enough, the phenomenon of parallax (or, more precisely, the lack of parallax) was used during the Renaissance to argue against the heliocentric universe proposed by the great Polish astronomer Nicolaus Copernicus. The argument went like this: If Earth is really moving, then nearby

The winter sky as seen from Germany. The constellation Orion (The Hunter) is in the center, with the line of three bright stars forming his belt.

A radio receiver forms part of the Deep Space Network. Instruments like this located all around the world allow scientists to stay in touch with spacecraft, even those around distant planets.

stars should appear to move against the background of stars farther out, much as a nearby fence post moves against the background when you drive past it in a car. Since no such parallax was seen, the conclusion was that Earth wasn't moving. However, scientists like Galileo pointed out that the lack of measurable parallax could also mean that the stars were very far away.

He could not have known how right he was! So long as astronomers were confined to baselines smaller than the diameter of Earth and to naked-eye observations, the fundamental question about the size of the universe could never be answered. To improve on this situation, astronomers had to do two things. First, they had to get instruments that could measure angles more accurately, and then they had to find a longer baseline.

THE ASTRONOMICAL UNIT

The obvious candidate for a longer baseline is the radius of Earth's orbit. You could, for example, take measurements of a distant object in July and January and play the parallax game. But this simple suggestion leads to a fundamental problem. To measure Earth's orbit, and hence the distance to the stars, we need the distance from Earth to the sun, a measurement now called the astronomical unit, or AU. This can't be determined with Renaissance-era instruments and only Earth's diameter as the baseline. We had to wait for the development of good telescopes. In the end, it wasn't until 1672 that scientists measured the AU.

In that year, the French Academy of Sciences sent an expedition to Cayenne in French Guiana to measure the position of the planet Mars at the same time that measurements of that planet were being made in Paris. The idea was that if we know the distance between two planets (such as Earth and Mars), Kepler's laws of planetary motion allow us to calculate the distance from the sun to both of these planets. This expedition was equipped with telescopes capable of measuring angles to within several seconds of arc (an arc second is 1/3600 of a degree). In the end, the scientists came within about 10 percent of the current accepted value of the AU. For the record, today this sort of measurement would be made by bouncing radio signals off other planets and measuring the time of flight when the radio waves return.

The Cayenne expedition produced another important (and totally unexpected) scientific result. They found that the day at Cayenne, as measured by pendulum clocks synchronized in Paris, was two minutes too short. This was the first proof that Earth is not a sphere, but instead bulges a little at the Equator. The added distance from the center of Earth at Cayenne changed the period of the pendulum being used as a clock and hence the length of the day that the clock measured.

This period of expanding our baseline outward from the radius of Earth to the AU, with its need for sending astronomers to far-off locales, produced some interesting stories. My favorite involves an

A copper engraving from the 1500s shows the Earth-centered spheres of Eudoxus. On the right is the French astronomer and mathematician Oronce Finé.

The standard candle method of determining distances depends on knowing how much light a distant object is emitting and comparing that to the amount of light received.

■ For more on Cygnus, see page 224.

The relative positions of objects in the solar system, measured in astronomical units (AU). Note that the scale is not uniform, so outer objects are much farther away from the sun than they appear to be here.

expedition sent by the English Royal Society to St. Helena in 1761. It was headed by the British astronomer Nevil Maskelyne. Upon his return, he submitted an expense statement that included 141 guineas for liquor. This was a considerable sum, since the average laborer earned about 18 pounds (a guinea was worth just slightly more than a pound) per year at the time.

Now that, my friends, is what I call an expedition!

STELLAR DISTANCES

Unfortunately, the expansion of the baseline in and of itself still wasn't enough to get us to the stars. These distances are so great that astronomers had to wait until the early 19th century for telescopes to be improved to the point where a stellar parallax measurement could be made. This period also saw what may be called the first "space race," in which

different countries entered a kind of competition to show that they had the best science and would therefore harvest the prestige gained from being the ones to determine the size of the universe.

The winner of this race was a man whose name will be recognized by all astronomers and physicists: Friedrich Bessel. Bessel is best known today for the development of the mathematical function that bears his name. Working at the observatory in Königsberg (now Kaliningrad) with a telescope capable of measuring angles as small as .02 seconds of arc, he compiled measurements of the position of the star 61 Cygni (the 61st brightest star in the constellation Cygnus, the Swan). In 1838, he published the results of 18 months of observations, reporting a distance to this star of 657,000 AU, or 10.3 light-years. To Bessel's credit, this number is within experimental error of the currently accepted value. Within a year of Bessel's publication, Friedrich Georg Wilhelm von Struve in Russia (measuring the star Vega) and Thomas Henderson in South Africa (measuring Alpha Centauri) reported similar findings.

With Bessel's measurement, we began to get a sense of the immense size of the universe we inhabit. The sheer expansion of our mental picture that followed the determination of the distance to nearby stars was incredible. After all, the distance to even the nearest star is more than 200,000 times the distance between Earth and the sun. If Earth

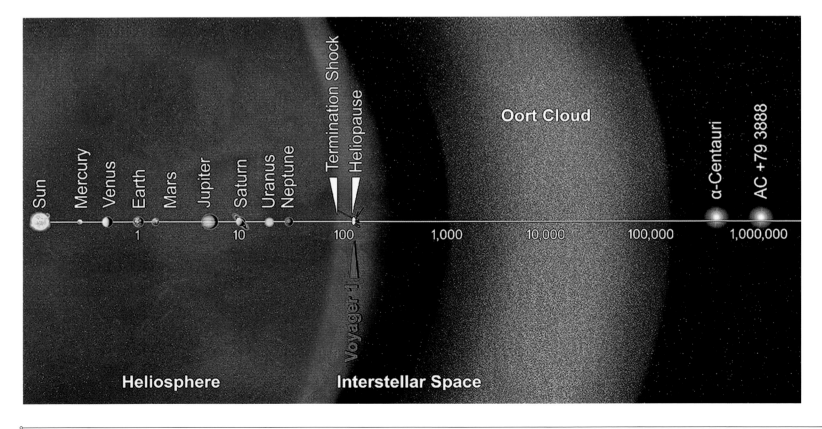

were a few city blocks from the sun, for example, the nearest star would be a continent away. Clearly, our own neighborhood is pretty small stuff compared to the vastness of space.

STANDARD CANDLES

If we want to go from the distance to nearby stars to the distance to stars on the other side of the Milky Way galaxy, however, we're going to have to add another rung to our distance ladder. Parallax just won't get us that far. The new rung was supplied by astronomer Henrietta Swan Leavitt, working at the Harvard College Observatory in the late 19th century. She studied a class of stars called Cepheid variables, so named because the first star of this type was seen in the constellation Cepheus (a Northern Hemisphere constellation representing an ancient king). Over a period of weeks or months these stars were observed to become brighter and dimmer in a regular cycle. Using the best astronomical distance ladder available to her, Leavitt determined the distance to some nearby Cepheid variables. Knowing the distance to the star and the amount of

light from the star that reached earthly telescopes, Leavitt was then able to measure the total amount of light each star emitted—a quantity called the star's luminosity. She found that the longer the star took to go through its brightening-dimming curve, the more energy it was emitting. In astronomical jargon, a Cepheid variable is called a "standard candle" because it has a known brightness. By watching

The distance from Earth to a star was first measured at the Königsberg Observatory in Prussia (now Kaliningrad, Russia).

■ For more on Cepheus, see page 214.

HIPPARCOS AND THE DISTANCE LADDER

In 1989, the European Space Agency launched a satellite called Hipparcos, a name that pays tribute to the Greek astronomer Hipparchus. For some reason, satellite builders have fallen into the habit of finding complex ways to give their craft recognizable acronyms. The name of the Hipparcos satellite, for example, comes from HIgh Precision PARallax COllecting Satellite. During its brief lifetime (the satellite stopped transmitting in 1993), Hipparcos performed the first accurate parallax measurements of nearby Cepheid variables. It could do this because it was capable of measuring angles to a thousandth of an arc second (i.e., one thousandth of 1/3600 of a degree). Thus the satellite succeeded in bringing the first two rungs of the astronomical distance ladder into alignment

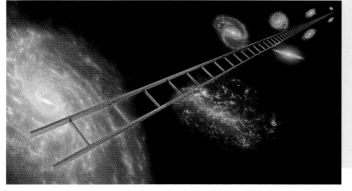

An artist's conception depicts the cosmic distance ladder that allows us to determine the size of the universe.

Taken with the Hubble Space Telescope, this scene is part of the sharpest and most detailed image of the Andromeda galaxy ever made.

For more on the Andromeda galaxy, see page 196.

(Opposite) Star-forming clouds glow in red and blue in nearby Barnard's galaxy.

it and timing its cycle we can determine how much light it is emitting.

Here's how the new rung worked: You would find a Cepheid variable and watch it go through its cycle. You would then use Leavitt's results to find the luminosity of that star and then, knowing the total amount of light we receive from the star, determine how far away it is.

THE EXPANDING UNIVERSE

In the 1920s, the American astronomer Edwin Hubble, working at what was then the world's largest telescope on Mount Wilson in California, used the Cepheid variable technique to establish two fundamental facts about the universe. First, he measured the distance to several nebulae, structures that looked like hazy patches of light in the sky when seen by earlier telescopes. One such object was what we now call the Andromeda galaxy—one of the few objects outside the Milky Way visible to the naked eye. Because he had access to a powerful telescope, Hubble was able to see individual Cepheid variables in these nebulae and show that the distance to various nebulae was measured in millions of light-years. They were much too far away to be part of the Milky Way, whose diameter was known to be about 100,000 light-years. Thus, Hubble's measurements showed that these hazy patches were in fact gigantic stellar structures, galaxies like our Milky Way, and that we live in an enormous universe made up of galaxies.

But that wasn't all. It had been known for some time that the light received from nebulae was shifted to the red (long-wavelength) end of the spectrum compared to light from the same atom emitted in the laboratory. This so-called redshift—a

consequence of the Doppler effect—was evidence that the nebulae were moving away from us. When Hubble correlated his distance measurements with the redshift data, he discovered that the farther away a galaxy was, the faster it was receding from us. In equation form, the Hubble law says that

$$v = H_0 d$$

where v is the velocity at which the galaxy is moving away, d is its distance from us, and H_0 is a number known as the Hubble constant.

In other words, Hubble discovered that the universe is expanding. Imagine running the film of the expansion backward in your mind, and you realize that Hubble's result implies that the universe began at a specific time in the past. At the moment, the best estimate of the age of the universe is about 13.8 billion years.

Because of the work of Hubble and his successors, we now believe that our universe began as a titanic explosion and has been expanding and cooling ever since. This is known as the "big bang"

picture, where "big bang" refers both to the initial event and the subsequent expansion. Having said that, we have to note that the big bang is not like the explosion of an artillery shell or fireworks, with fragments flying outward through space. Instead it is an expansion of space itself, with galaxies carried along as passengers.

Here's a common way of visualizing the Hubble expansion: Imagine that you are cooking a loaf of raisin bread, and that you are standing on one of the raisins as the dough rises. What would you see?

Since your raisin isn't moving with respect to the neighboring dough, as far as you are concerned you are stationary. Look at a neighboring raisin, however, and you will see it moving, because the dough between you and it is expanding. Look at a raisin twice as far away and it will be moving twice as fast, because there's twice as much dough between it and you. Substitute raisins for galaxies and intergalactic space for the dough and you have a picture of the Hubble expansion.

SPEARMEN AND SPEARWOMEN

The fourth-century B.C. Greek philosopher Archytas of Tarentum was thinking about the question of whether or not the universe had an edge—a boundary—and produced an image we can find useful. If there is an edge, he argued, you could send a spearman to that edge and have him throw his spear outward. By definition, it would have to land outside of the boundary you had placed for the edge of the universe. Therefore, he argued, the universe had to be infinite.

You can think of each of the people we've discussed—Eratosthenes, Hipparchus, Copernicus, Bessel, Leavitt, Hubble—as examples of Archytas's spearmen and spearwomen. Each stepped up to the edge of the unknown and expanded our universe beyond imagining. The story of parallax measurements, then, is really the story of humanity's understanding of the universe in which we live and of the relative insignificance of Earth and the sun in that universe. Each successive measurement that allowed us to increase the baseline or measure angles better pushed the limits of our universe farther away. And in this grand chain of accomplishment, that day in Germany in 1838 when Friedrich Bessel first measured the distance to 61 Cygni stands out as a crucial milestone, opening the universe to our imagination.

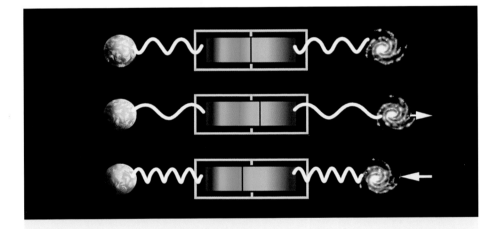

THE DOPPLER EFFECT

In the early 1840s, Austrian physicist Christian Doppler examined the behavior of waves emitted by a moving source, proposing a mechanism that is found in everything from medical ultrasounds to the analysis that led to the big bang picture. Let's look at sound waves as a familiar example. If the source of a sound is stationary—a speaker on a table, for example—then what strikes our ear will be a series of regularly spaced high pressure crests in the air. The closer together the crests are spaced, the higher pitched the sound.

If the source is moving toward you, however, the distance between the crests that reach your ear will be shorter and you will hear an even higher pitched sound. If the source is moving away from you, the distance between crests increases and the sound you hear will be lower in pitch.

This phenomenon will occur for any wave, including light. In the case of light, if the source is moving away from you the wavelengths will shift toward the red (long-wavelength) end of the spectrum, producing the redshift that Hubble used to determine the speed at which distant galaxies are moving away from us.

The big bang started 13.8 billion years ago, as shown in the lower right, and progressed through various stages until our familiar universe of galaxies and stars (upper left) evolved.

WHAT ARE THEY?

The nature of the distant stars is revealed through the story of starlight.

Now that we understand how we can convert the two-dimensional sky we see into a three-dimensional display, we can move on to our second great question. What, exactly, are those lights in the sky? Regardless of how far away they are, the simple fact that we can see them tells us something important.

Stars, like everything else in the universe, are born and die.

Although this sounds profound, the very fact that you can see a star at night proves it must be true. Look at it this way: The star is emitting the light you see, and the emission process requires energy. On the other hand, the star is a finite object, which means that it must have a finite amount of energy. Sooner or later, then, the energy required for the emission of light will exceed what is left in the star. The star's energy will give out and the star will have to stop shining. In other words, it will die.

Of course, that dying can (and usually will) take a long time. In what follows we will look at stars that live "only" a few tens of millions of years and others that live for billions. Fast or slow, however, every star will die eventually.

So the simple act of looking at a star at night teaches us something fundamental about the universe. This is only the beginning, however. We can learn a lot more just by thinking about what happens when we look at a star. For example, the fact that the light has to come through several miles of air to get to you on the ground tells you that Earth's atmosphere is transparent to light. You may have come to the same conclusion yourself from other observations. The fact that you can see these words, for example, means that light can travel unimpeded from the page or screen to your eye. Alternatively, you may have had the experience of seeing the lights of a distant city from an airplane window at night, in which case the light will have traveled miles through the air to get to you.

WAVELENGTHS

We know the atmosphere is made (mostly) of molecules of nitrogen and oxygen, so the transparency we see must be a property of the way light interacts with matter at the atomic level. In fact, Albert Einstein explained the properties of light at this level in 1905, when he showed that certain phenomena could be understood if light was considered to be a packet of electrical and magnetic waves now called a photon. (Most people are surprised to learn that it was this work, and not the theory of relativity, that earned Einstein his Nobel Prize in 1922.) Because of its composition, light is often referred to as an electromagnetic wave.

Waves are ordinarily characterized by the distance

(Opposite) A stellar breeding ground in the Tarantula Nebula is located 170,000 light-years away.

The lights of northern Europe at night, seen from the International Space Station

between crests, a distance called the wavelength. In the case of visible light, the difference between colors corresponds to differences in wavelength. Red light, for example, has a wavelength about 8,000 times the diameter of a typical atom, while the more energetic blue light has a wavelength of about 4,000 atomic diameters (shorter wavelengths pack more energy). While visible light is very important to humans, it covers only a small fraction of all the possible wavelengths that could exist. By the late 19th century, our theories were telling us that there had to be other electromagnetic waves—waves with different wavelengths from those visible to the human eye.

In 1888, the German physicist Heinrich Hertz produced one of these anticipated waves in his laboratory, and the use of Hertz's waves for communication was demonstrated soon after by the Italian scientist Guglielmo Marconi. These waves had very large wavelengths and soon acquired the name radio. It is these waves that your radio and cell phone detect and turn into sound. Radio waves are built in exactly the same way as light waves, and the two differ only in the size of their wavelength. If you

The electromagnetic spectrum ranges from radio waves to gamma rays. Only a small percentage of these waves gets through Earth's atmosphere to the ground.

take a photon of visible light, in other words, and stretch it out so that the wavelength is measured in feet or miles instead of atomic diameters, you'd have a radio wave.

THE ELECTROMAGNETIC SPECTRUM

In the years that followed Hertz's discovery, an entire family of electromagnetic waves was discovered. This wasn't the result of a planned search—many of the discoveries were accidental. The French physicist Henri Becquerel, for example, discovered what we now call x-rays by accidentally exposing a photographic plate to a radioactive sample.

The full range of these newly discovered waves is called the electromagnetic spectrum. As is the case with radio waves, all these other electromagnetic waves are built like visible light but have different wavelengths. In addition, all these waves travel through a vacuum at the same speed—a speed customarily referred to as the speed of light and denoted by the letter c. (This is the same c, incidentally, that appears in Einstein's famous equation $E = mc^2$.)

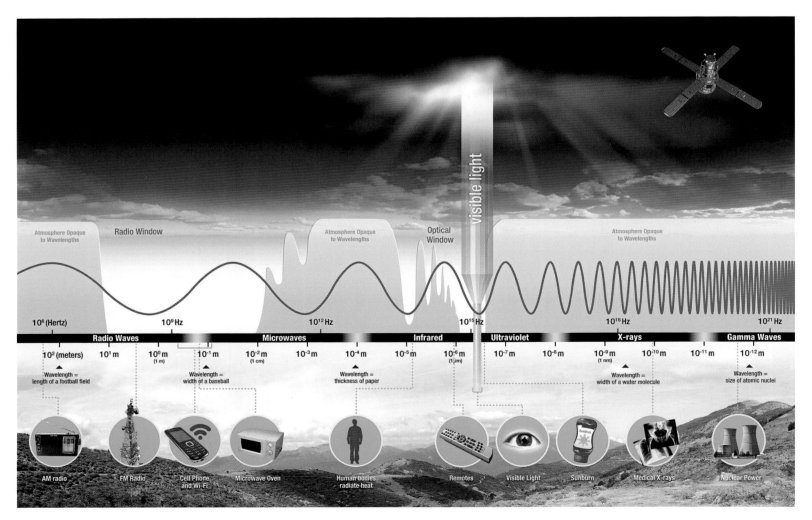

There is, however, one important difference between some waves and others in the electromagnetic spectrum. With the exception of visible light and radio waves, the atmosphere is opaque to all of them. If a star emits one of these other waves, in other words, the photon will travel all the way to Earth through the vacuum of space only to be absorbed by Earth's atmosphere before it gets to our telescopes. This is why we build astronomical observatories and put them in orbit above the absorbing atmosphere. It also explains why optical and radio astronomy were the first areas of the science to be developed.

You might be wondering why visible light is so important to us when, in fact, it constitutes such a small part of the electromagnetic spectrum. There are two facts to take into account in answering this question. First, as we have already seen, the atmosphere is transparent to these wavelengths, so detecting them is an efficient way for any organism to be aware of its surroundings. Second, it turns out that an object like the sun, with a surface temperature of about 5500°C (10,000°F), radiates most of its energy at these wavelengths. In other words, on our planet the most abundant source of light comes to us through a transparent atmosphere. Is it any wonder, then, that the evolution of life on Earth produced organs capable of detecting visible light? In fact, evolutionary biologists estimate that eyes sensitive to visible light have evolved independently no fewer than 40 times in various creatures over the course of Earth's history.

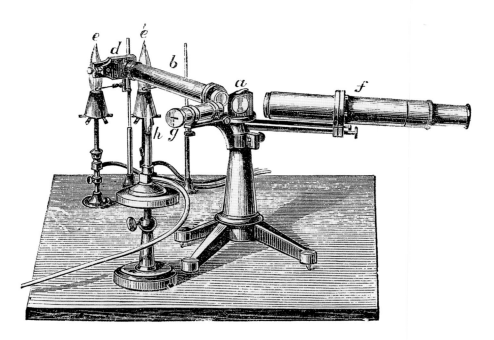

A simple spectrograph of the type used by early astronomers. Such instruments enabled scientists to determine the chemical content of stars.

Having said this, we must point out that this sort of correlation between source and atmospheric transparency need not be found on other planets. Extraterrestrials that evolve in different environments might just as well "see" radio waves or x-rays as visible light.

ATOMIC FINGERPRINTS

Throughout the early development of modern science, the consensus was that the stars would always be distant and mysterious. The 19th-century French philosopher Auguste Comte, for example, included learning the chemical composition of stars in his list of problems that would never be solved. His argument was simple and, for the time, reasonable. To learn the chemical composition of any object at that time, you had to bring that object into a laboratory and subject it to various chemical tests. Obviously, no one was going to bring a star into a terrestrial laboratory, so these tests could never be run. Consequently, the star's composition would never be known.

There is, however, a fatal flaw in this argument. It is certainly true that no one was ever going to bring a bucket of starstuff into a laboratory, then or now. Something else, however, comes to us from the stars, and that is the light they emit. Is it possible that starlight itself could allow us to get around Comte's argument?

There is actually a lot of information in the light we get from a star—information we can dig out if we have the proper tools. In 1859, two German scientists at Heidelberg, each well known for other

■ For more on the sun's surface, see page 57.

Stars come in many sizes: The letters underneath the stars (below) correspond to the standard classification scheme, dependent on temperature and spectra. In the illustration on the bottom, several very large stars are depicted together to express their relative size. From smallest to largest: Rigel (blue), Antares, Betelgeuse, and VY Canis Majoris.

discoveries, discovered how to do a chemical analysis of stars. Gustav Kirchhoff and Robert Bunsen were studying the light emitted by heated chemical elements with an instrument that would today be called a spectroscope. Modern spectroscopes can cost hundreds of thousands of dollars, but Kirchhoff and Bunsen cobbled theirs together from a couple of surplus field telescopes, a prism, and, believe it or not, a cigar box.

With this rudimentary apparatus, Kirchhoff and Bunsen discovered two fundamental facts about light emitted by a heated material. First, each chemical element gives off a series of well-defined colors (i.e., wavelengths) called a spectrum. In essence, the spectrum is a kind of atomic fingerprint that indicates the presence of a specific element in the material. Second, it was seen that if an atom emit-

ted a particular wavelength of light when heated, it would absorb that same wavelength as well.

The spectrum, then, becomes a way of learning about the chemical composition of a heated material. The atomic fingerprint—the colors of the spectrum—is imprinted on the light being emitted, and it stays there until that light is absorbed. It doesn't make any difference if the light crosses a room or a galaxy: Wherever it goes it carries the information about its source with it. By studying spectra, using a technique called spectroscopy, we can learn the chemical composition of the stars we see in the sky.

This is important because earlier in the 19th century, as we have seen, Auguste Comte had predicted that we would never be able to determine the chemical composition of the planets or the stars because doing so would require testing a sample in a laboratory, something that seemed impossible at the time. What Kirchhoff and Bunsen showed was that this assumption simply wasn't true because the information could be carried to us by light.

So now we know two important things about stars: We know how far away each one is, and we know what each is made of. At this point we are in a position familiar to scientists. We have an impressive collection of objects out there. Some stars are

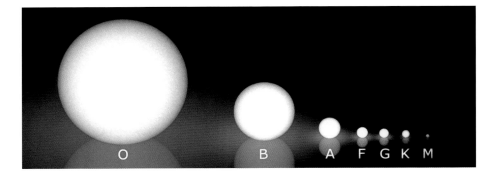

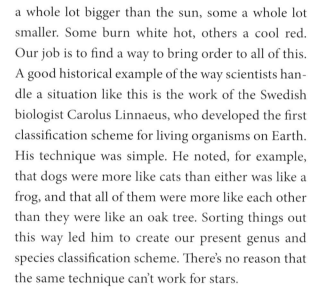

Astronomer Annie Jump Cannon (near left) developed the current stellar classification system. (Far left) The spectral signatures of variable stars, photographed in the 1890s

a whole lot bigger than the sun, some a whole lot smaller. Some burn white hot, others a cool red. Our job is to find a way to bring order to all of this. A good historical example of the way scientists handle a situation like this is the work of the Swedish biologist Carolus Linnaeus, who developed the first classification scheme for living organisms on Earth. His technique was simple. He noted, for example, that dogs were more like cats than either was like a frog, and that all of them were more like each other than they were like an oak tree. Sorting things out this way led him to create our present genus and species classification scheme. There's no reason that the same technique can't work for stars.

CLASSIFYING THE STARS

In 1886, astronomers at Harvard University began the long and laborious task of classifying the stars. They developed a new technique in which they recorded the spectra of many stars on a single photographic plate. Using this database, astronomer Annie Jump Cannon, with astonishing single-mindedness, analyzed no fewer than 350,000 spectra of individual stars. She would look, for example, at how many electrons were stripped from atoms in a star's atmosphere as a way of categorizing the star. This is a marker that we now recognize as a way of measuring temperature. The hottest stars, which we now know have surface temperatures about 50,000°C (90,000°F), were called O-type stars. For

Cannon, these were stars whose spectra indicated the presence of highly ionized atoms of substances such as helium, nitrogen, and silicon. Next came stars classified as B, A, F, G, K, and M, from hottest to coolest, a scheme that has caused generations of university lecturers to teach their students the old-fashioned mnemonic "Oh, Be A Fine Girl/ Guy—Kiss Me." Our sun is a G-type star.

You might think that a strange sequence of letters like this arose from a well-thought-out scheme based on sound physical principles. Unfortunately, scientists are rarely so rational when they assign names to phenomena. When Cannon first took over the Harvard catalog, she found a straightforward alphabetical scheme in place—stellar classes were labeled A, B, C, and so on. As her work progressed, she came to realize that these classes just didn't work. Hundreds of thousands of stars later, she found herself with the modern scheme, without ever really having planned to get there. I often wonder if she ever wanted to redo the whole system.

As it happens, some important stellar properties are not taken into account in a scheme that

Annie Jump Cannon was the first woman to be named an officer in the American Astronomical Society.

In this typical stellar spectrum, dark lines represent wavelengths that are absorbed in the star's atmosphere.

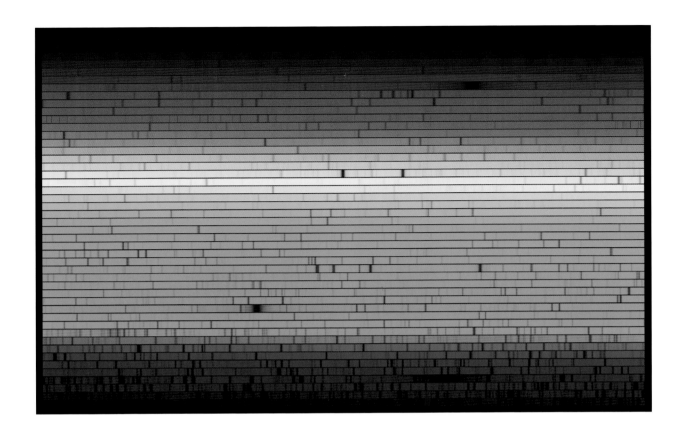

only looks at surface temperature, and this fact was taken seriously by the Danish chemical engineer Ejnar Hertzsprung. He was interested in the chemistry of photographic films, which led him to look at the discussions of Cannon's collection of photographs at Harvard. As we mentioned above, the apparent brightness of a star can depend on its distance from Earth—stars that are bright and far away may look the same as stars that are dim and close. Hertzsprung noticed that some red stars (i.e., stars with a low surface temperature) exhibited parallax and others did not. He realized that the former were relatively close and the latter were far away. He also noted that some of the nearby red stars were relatively bright.

He also realized that for a star with a low surface

temperature to appear bright, it would have to be giving out a lot of light. In astronomers' jargon, it would have to have a high luminosity. In such a star, each square foot of surface area could emit only a relatively small amount of light, so to get the high luminosity it would need to have a lot of surface area. It would have to be what is now called a red giant star.

What Hertzsprung established, then, was that there were actually two different kinds of red stars—normal and giant—and that mixing them together was like mixing apples and oranges. This was an important discovery, and it cleared the way for our current understanding of the evolution of stars. By one of those strange quirks of history, however, it also led to a common situation that occurs when two scientists, working independently, make the same discovery.

Unfortunately, when Hertzsprung decided to publish his work in 1905 and 1907, he picked a journal that, to put it politely, was unlikely to be widely known among astronomers—it was called *Zeitschrift für Wissenschaftliche Photographie* (*Journal of Scientific Photography*). Consequently, Hertzsprung's early work was largely unknown in the scientific community. Years later, the British astronomer Sir Arthur Eddington told him, "One of the sins of your youth was to publish important results in inaccessible places."

■ For more on red giants, see page 67.

This diagram shows how light from a star is passed through a spectrograph by a series of lenses and mirrors.

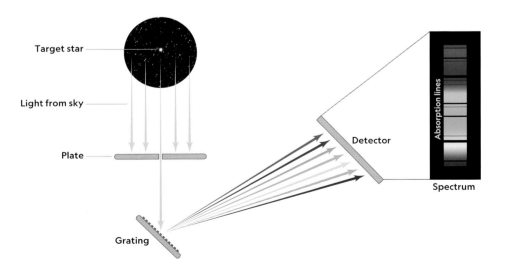

An 1890 photograph shows Harvard "computers" at work.

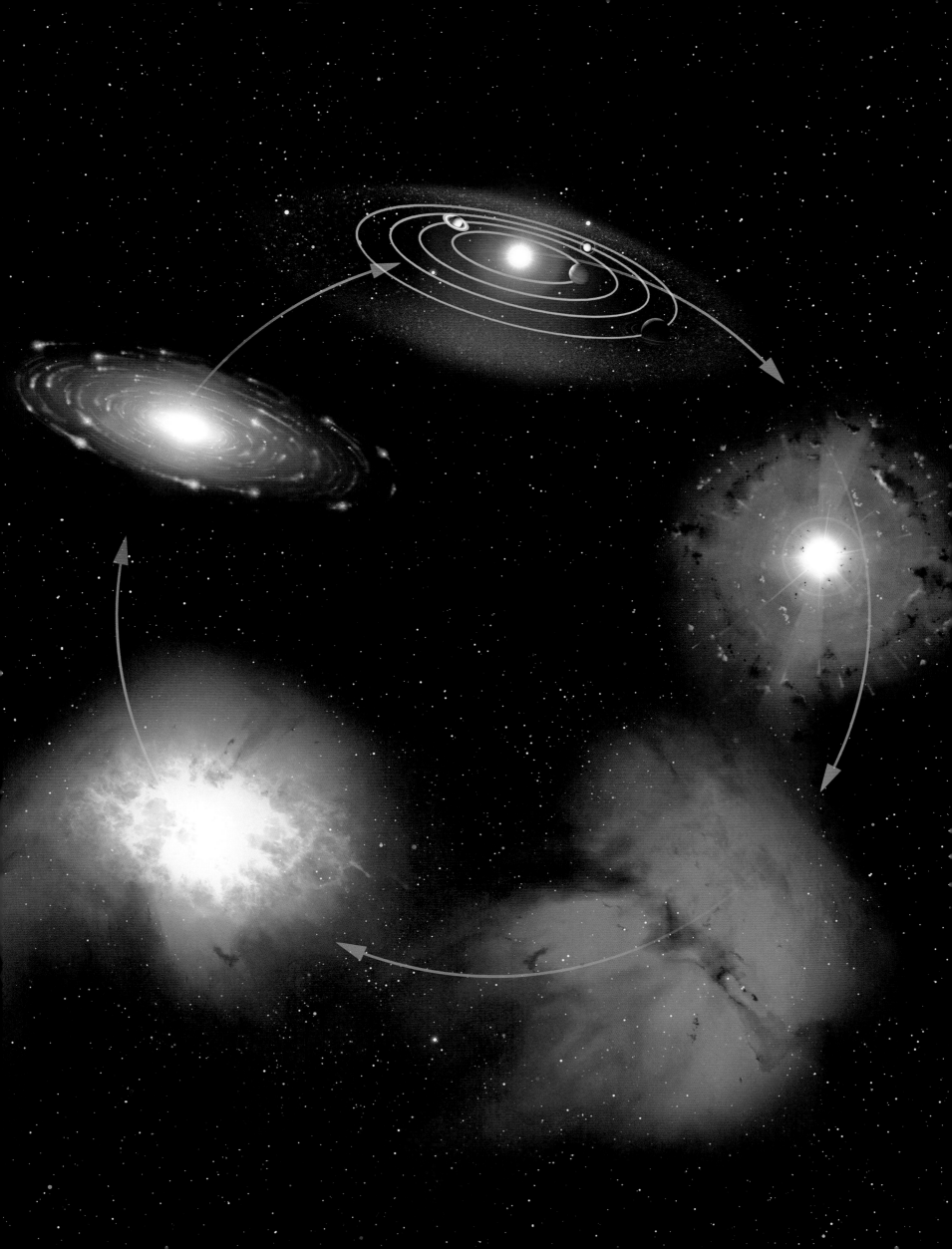

THE STELLAR LIFE CYCLE

As stars are born, grow, and die over time, their mass is their destiny.

While Hertzsprung was working in Europe, the American astronomer Henry Norris Russell was starting to look at the same kinds of problems at Princeton University. He was not motivated, as Hertzsprung was, by trying to straighten out the stellar classification scheme. Instead, he was trying to understand the life cycle of stars: how they are born and how they die. According to Russell's notes, he had understood that there were two types of red stars—normal and giant—by 1909 and, like Hertzsprung, realized that for the vast majority of stars, the higher the surface temperature, the more energy that was being radiated (i.e., the higher the luminosity). In 1913, he presented what is now called the Hertzsprung-Russell diagram (usually abbreviated H-R diagram) at meetings in London and Atlanta.

On an H-R diagram, the vertical axis on the graph is the luminosity of a given star, and the horizontal axis is the star's surface temperature. For some reason, surface temperatures in the H-R diagram are customarily plotted so that they get lower as you move to the right.

The usefulness of the H-R diagram should now become apparent. Every star is represented by a single point on the graph: for example, the sun, with a surface temperature of around 5500°C (10,000°F) and an energy output of about 4 x 10²⁶ watts (that's a 4 followed by 26 zeroes). Every other star is represented in the same way. Furthermore, as we shall see below, as stars change and evolve over time the point representing them moves around on the diagram, so that the life cycle of each star can be represented by a path on the diagram.

Looking at the diagram, we see that most stars fall into one of three categories. The band running from the upper left (high temperature and luminosity) to lower right (low temperature and luminosity) is called the main sequence. Most stars (including the sun) are on the main sequence. Note that the hotter a main sequence star is, the more energy it is pouring into space.

In addition, there are two other clusters in the diagram. The one in the upper right contains stars that are cool and bright. These are the red giants that both Hertzsprung and Russell discovered independently of each other. Another cluster in the lower left is made up of stars that are hot and dim. These are called white dwarves, and they weren't added to the diagram until well after the H-R diagram was first published.

With the H-R diagram, then, Hertzsprung and

The Hertzsprung-Russell diagram depicts the main classification groups of stars, arranged by luminosity and temperature

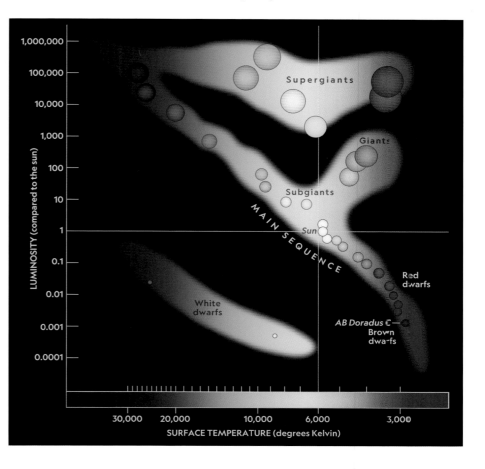

(Opposite) A stellar system evolves from nebula (at bottom right) to supernova.

Russell completed the task of organizing the stars. We now have a way of characterizing every one of those bright lights in the sky.

GREAT MINDS

These discoveries are interesting, of course, but there's something about this story that is even more fascinating. That is the fact that two men, working on different continents and approaching the problem of classification in different ways, came up with the same solution to the problem. This isn't an unprecedented situation in the history of science. Isaac Newton and Gottfried Leibniz developed calculus independently of each other in the 17th century, for example. Maybe the H-R diagram was just "in the air" in the early 20th century. After all, astronomers had only recently learned how to measure distances to stars, and data on luminosities and temperatures were pouring in. Sooner or later somebody was bound to put the two together.

But how could two prominent astronomers, each working at a mainstream research center, not know what was going on with the other? Hertzsprung's choice of journal in which to publish may have had something to do with it. In addition, both communication and travel were a lot more difficult in the early 20th century than they are now. It took the

Pierre-Simon, Marquis de Laplace

better part of a week to cross the Atlantic, and nothing like the internet existed.

Given that Russell introduced his diagram at major scholarly meetings, it is not surprising to learn that the work was referred to as the Russell diagram until 1933, when Hertzsprung's publications were finally brought to light. Had they been working today, the two scientists would surely have run into each other at a conference or, at the very least, been aware of each other's work via the standard academic grapevine. You can imagine a couple of online conversations once they discovered their mutual interests, most likely followed by a joint publication.

HOW STARS ARE BORN

In any case, now that we have a way of categorizing stars, and now that we understand that stars, like everything else, must be born and die, we can start to dig a little deeper into our fundamental "What are they?" question. In particular, we can ask exactly how stars are born and how they die. As we hinted at the beginning of this chapter, gravity is involved in both of these processes.

Let's start with the birth of stars. The basic idea behind our theory of stellar birth was proposed by a number of scientists in the 18th century. The most modern version of the theory was published in 1796 by the French scientist and mathematician Pierre-Simon Laplace. One of the most important scientists who ever lived, Laplace was a confidant of Napoleon, who enjoyed being surrounded by scholars.

MULTIPLE STAR SYSTEMS

When a cloud of interstellar dust starts to collapse to form a star, any spin it has will be magnified during the contraction. There are basically two ways the nascent star can deal with this increased spin. One way was followed by our own solar system. As we have seen, when the sun formed, a so-called protoplanetary disk was spun out. From the point of view of an astronomer, the rotation of this disk allowed the sun to form without being torn apart by its own rotation. Another way the system could have evolved would be for the contracting cloud to split into two or more stars—stars that revolve around each other in different patterns, depending upon their masses. Roughly speaking, about one-half of the stars you see in the sky are actually multiple star systems.

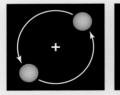
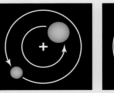
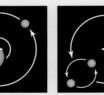

Identical Masses
If two binary stars have equal masses, the center of gravity is directly between the two.

Different Masses
If two binary stars have different masses, the star with the higher mass will be closer to the center of gravity.

Large Differences
If the two stars have greatly different masses, the center of gravity might be within the more massive star.

Quadruple System
In a system with two pairs of binary stars, the center of gravity will be between the two pairs.

To understand this theory, known as the nebular hypothesis, we have to start with a discussion of what a nebula is. The word means "cloud" in Latin, and it is applied to cloudlike structures in the night sky that are typically seen with binoculars or telescopes. (A few, such as the Orion Nebula, can be seen with the naked eye.) Nebulae are basically clouds of dust and small particles, mixed with primordial hydrogen from the big bang. They are found throughout the Milky Way.

If you think about the structure of such a dust cloud, you realize that there must be some regions of the cloud that contain more mass than others—in effect, the cloud will be lumpy. These regions of high mass will exert a greater gravitational force on their surroundings than other regions will, and hence will pull in more material and become more massive. This, in turn, will increase the gravitational

force they exert and pull in even more material. The effect of the gravitational force acting in the nebula, then, is to break it up into a relatively small number of concentrations of mass. This is how stars are born.

But the key point about gravity is that it never quits. As the material in the nebula is pulled inward, two things happen. First, as the matter gets more and more compressed, the collisions between its constituent atoms and molecules get more energetic, more violent: The material gets hotter. Second (and less important for the nascent star at the nebula's center) any small net rotation that was present in the cloud will become magnified. Like an ice skater who spins faster when she pulls in her arms, the rotation rate of the contracting cloud will increase. Eventually, a small fraction of the material in the cloud will get spun out into a disk, much like clay

Stellar nursery Gum 29, about 20,000 light-years away. Most of the stars in this cloud are only a few million years old.

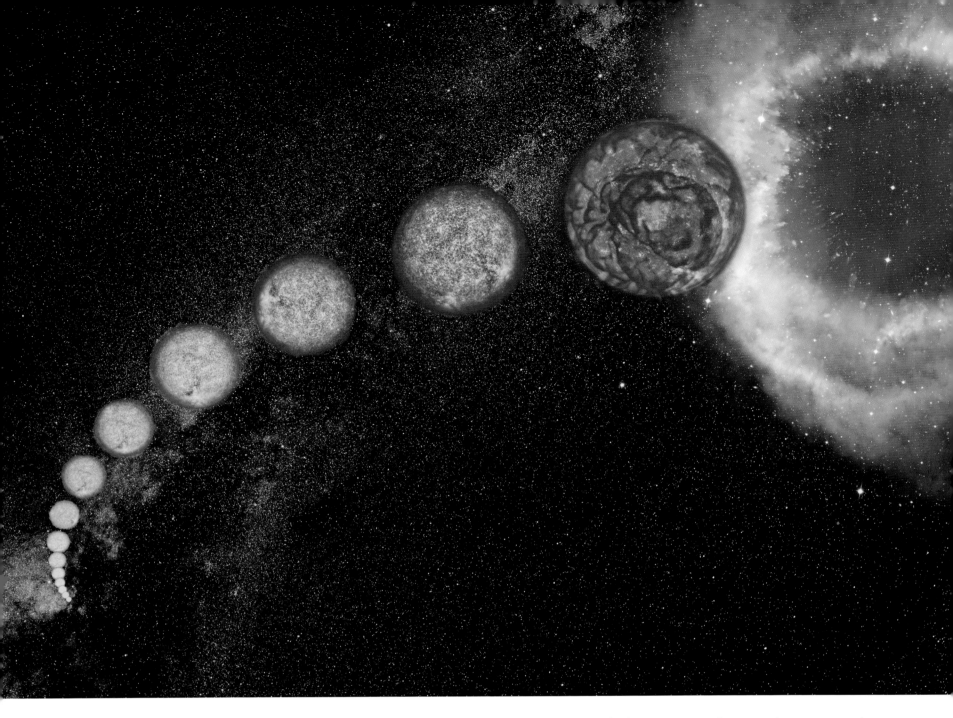

At the end of their lives, stars like our sun move from the main sequence through the red giant phase.

on a potter's wheel. This disk is where planets will eventually form.

The main body of the cloud just keeps contracting, however. Eventually the collisions between atoms will become so violent that the electrons are torn loose and the material becomes a plasma, a form of matter in which positive and negative electrical charges wander around independently of each other.

FUSION

Gravity doesn't stop there. The inward compression continues until the temperature at the center of the body passes 14,000,000°C (more than 25,000,000°F). At this temperature something totally new starts to happen. The loose protons that used to be the nuclei of hydrogen atoms start to undergo a series of interactions that we call nuclear fusion.

Normally, if two protons get near each other, they will experience a repulsion that drives them apart. At very high temperatures, however, the protons can be moving fast enough to overcome this repulsion and actually come into contact with each other. When this happens, the subatomic particles undergo a series of rapid reactions in which smaller nuclei are combined to make a larger one: nuclear fusion.

The net effect of all of these nuclear shenanigans is to convert four protons into a helium nucleus and a spray of miscellaneous particles carrying energy. Note that a helium nucleus—the end of the chain— has less mass than the four protons that started it. The difference in mass can be converted, according to Einstein's famous $E = mc^2$, into energy, and this energy starts flowing out from the star's core. In doing so, it creates an outward pressure that can counter the inward pull of gravity, and once it reaches the star's outer surface it moves out into space. An outside observer would see the object start to shine.

A star has been born.

HOW STARS STAY ALIVE

So long as the star has enough hydrogen to feed the nuclear fusion reaction, the relentless inward pull of gravity can be kept at bay. In fact, one way of thinking about the life of a star from this point forward is to see it as the deployment of one strategy after another to counteract gravity.

A word of caution: Astronomers usually refer to the fusion reaction that produces helium as hydrogen "burning." This kind of burning, which involves the nuclei of atoms, is different from burning a piece of wood. When wood burns, the reaction involves electrons in the wood's atoms—we call this "chemical burning." Nuclear burning involves atomic nuclei, and it releases a lot more energy than chemical burning does.

On an H-R diagram, the path by which a star arrives at the stage of hydrogen burning is represented by a line that starts in the lower right (cool and dim nebula) and moves upward and to the left until it reaches the main sequence. In fact, now that we understand why a star starts to shine, we can define the main sequence as the place where stars that are burning hydrogen are found. Stars spend most of their lifetimes on the main sequence.

You might expect that large stars, because they have more hydrogen to burn, would stay longer on the main sequence than smaller stars. Just the opposite is true—a moderate-size star like the sun can burn its stored hydrogen for more than 10 billion years, while much larger stars might burn through their hydrogen in some tens of millions of years. The reason is simple: The force of gravity is greater in large stars, and hence more hydrogen is required to counterbalance it.

It's important to realize that only the inner core of a star is hot enough to burn hydrogen—hydrogen away from the core doesn't undergo fusion. Inside the core, however, hydrogen is consumed prodigiously. The sun, for example, burns about 544 million metric tons (600 million tons) of the stuff per second. (Not to worry, though—it still has over five billion years' worth of fuel left.)

HOW STARS DIE

Eventually, however, every star uses up all of the hydrogen in the core, and the core is filled with helium: the "ash" of the nuclear fire. And as those nuclear fires die down, gravity takes over again and

THE LONG TRAIN JOURNEY

By the end of the 19th century, astronomers were starting to think seriously about where stars get their energy. Mechanisms involving fuel, such as coal, or gravity—material falling into the star or a slow contraction—simply didn't work. For a while, the problem of a star's energy seemed insoluble.

Then, in March 1938, the Carnegie Institution of Washington convened a small conference attended by a German émigré physicist named Hans Bethe. Years later, in his Nobel Prize lecture, Bethe described the meeting:

"At this conference the astrophysicists told us physicists what they knew about the internal constitution of stars. This was quite a lot, and all of their results were derived without knowledge of a specific source of energy. All that was necessary was that most of the energy was produced near the center of the star."

According to physics folklore, Bethe listened to what the astronomers had to say and then, on the train back to his home base at Cornell, produced the first fully worked out theory of how nuclear fusion could produce the energy that makes stars shine. In 1967, this work won Bethe the Nobel Prize.

the long-averted collapse resumes. This raises the temperature of the unburned hydrogen around the stellar core, and it starts to generate energy. This energy streams outward and pushes on the outer envelope of the star. The envelope swells up and, because the same amount of energy is radiating out through a larger surface, the outer surface of the star cools down and turns red. Our main sequence star has become a red giant, the kind found in the upper right hand corner of the H-R diagram.

While in this corner of the diagram, the star goes

The final stage in the life of stars like the sun is shown in Kohoutek 4-55, about 4,600 light-years away. As the star loses its atmosphere, it reveals the burning core, which shines as a white dwarf.

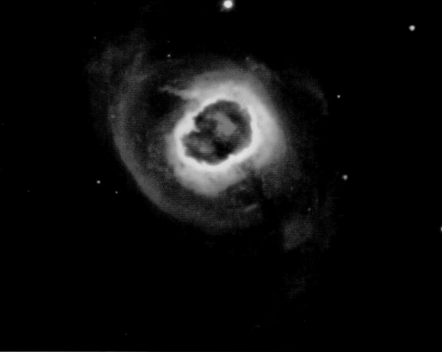

A series of images taken by the European Southern Observatory's Very Large Telescope shows (left to right) the constellation Orion and two increasingly close-up views of its supergiant star Betelgeuse.

through a number of complex changes, but we will look at only two of them. As the temperature of the core goes up, it eventually reaches the point where another nuclear fire can be ignited. Three helium nuclei come together to form the nucleus of a carbon atom (six protons, six neutrons). This process is completed in a matter of minutes and, because it is muffled by the star's interior, probably wouldn't even be visible from outside the star. It's called the helium flash.

Toward the end of its tenure as a red giant, our star will blow about half of its mass out into interstellar space in what is called a planetary nebula. At this point, there isn't enough mass left in the star to trigger further nuclear reactions. Once again, gravity takes over and the star's final collapse begins.

YOU ARE STARSTUFF

The universe's simplest elements were made during a brief period early in the big bang. The Hubble expansion quickly separated subatomic particles, however, so the process of making more complex nuclei was quenched. The universe entered its life made mostly of hydrogen and helium.

You can think of the stars, then, as the factories in which all the rest of the chemical elements are made. Massive stars, with their short lifetimes, blow their contents into the interstellar medium as they die. The heavy elements they contain can then be incorporated into new stellar systems and their planets. The calcium in your bones, the iron in your blood, and the oxygen in your DNA were all created by some long gone supernova. Some scientists estimate, in fact, that the material in our own solar system represents the output of up to three different supernovae.

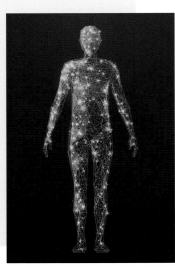

The star has to deploy its last strategy. The electrons that were torn from their atoms when the collapse started now assert their authority. It turns out that electrons can't be crammed together beyond a certain limit—they need elbow room. (This is a result of what is called the "exclusion principle.") When our star has shrunk down to about the size of Earth, the electrons are packed as closely together as they can be. Gravity can't compress them anymore and the collapse stops forever. The star is now a cooling ember in space—a white dwarf. Its life story ends up in the lower left-hand corner of the H-R diagram.

SUPERNOVAE

This main sequence–red giant–white dwarf progression will be followed by every star that is up to about eight times the mass of our sun. More massive stars play out a different endgame. Like their lightweight colleagues, they go through hydrogen burning and put in time on the main sequence.

They are big enough that when the collapse at the end of hydrogen burning starts, the temperature in the core gets high enough that helium can be burned to create carbon, just as it does in smaller stars. In more massive stars, though, the nuclear reactions keep going, with the ashes of each nuclear fire serving as the fuel for the next. From carbon they move on up through oxygen and silicon, and keep going until they get to iron. The star starts to look like the layers of an onion, with a different nuclear reaction going on at each level.

An artist's interpretation of recent images of the supergiant star Betelgeuse includes its huge plume (larger than our solar system) and the gigantic bubble on the stellar surface.

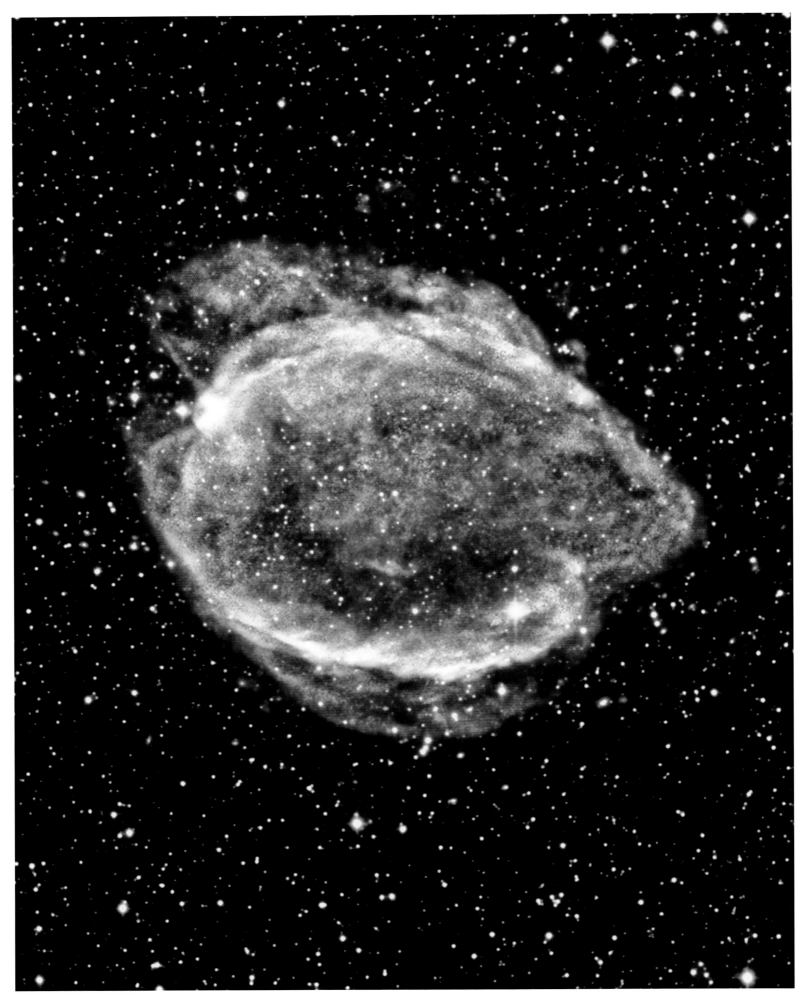

The debris field left behind by supernova G299

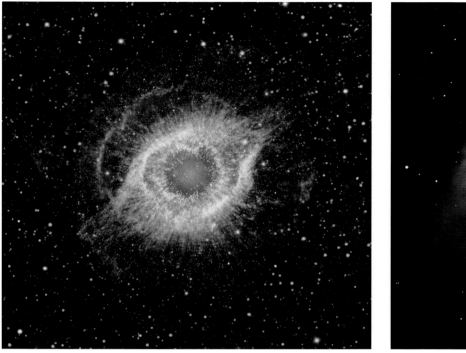

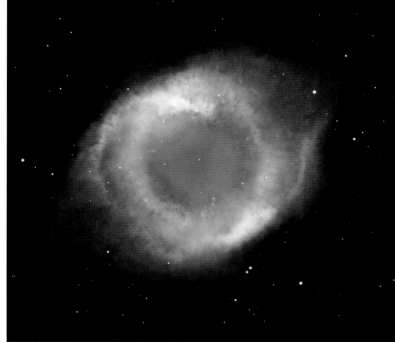

The fusion stops at iron, however. Iron is the ultimate ash—you just can't get energy from it. Consequently, the core begins to fill up with iron nuclei. When the temperature gets high enough, electrons start to be forced into the iron nuclei, changing its protons into neutrons. At this point, all hell breaks loose.

The core, now consisting entirely of neutrons, goes into a precipitous collapse. The outer envelope, the rug pulled out from under it, starts to collapse as well. When the core collapse stops it sends out shock waves, and those shock waves literally tear the star apart. The outer envelope, with its complement of heavy elements, is blown into space in a supernova. The shock waves also supply the energy needed to create nuclei all the way up to uranium.

In the meantime, the star's collapsing core of neutrons gets smaller and smaller. It turns out that, like electrons, neutrons need elbow room. When the diameter of the core gets down to about 20 kilometers (12 mi), the neutrons can't be squeezed any further and the collapse stops. We are left with an object known as a neutron star.

The collapse has two consequences. First, the neutron star will be spinning rapidly (think of the ice skater). Second, the collapsing core will drag the star's magnetic field with it, concentrating that field in the process. A rapidly spinning star with a high magnetic field is called a pulsar.

For stars more than 15 times as massive as the sun, the gravitational force is strong enough to overcome the outward pressure of the neutrons and the system collapses into a black hole: the ultimate triumph of gravity.

TYPE Ia BLASTS

The kind of stellar explosions we've just described are called Type II supernovae. Another kind of supernova, called a Type Ia, is created in an entirely different way.

Picture a double star system in which one of the stars has gone through its life cycle and is now a white dwarf. Picture further that the white dwarf is pulling material off its partner. When that material has accumulated to the point that the mass of the white

Two views of the Helix Nebula reveal different regions of gas thrown off by a dying star.

An artist's conception of a black hole surrounded by an accretion disk and emitting a jet of hot plasma

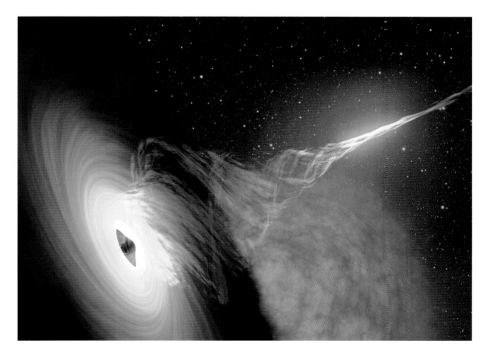

(Opposite) A view of 10,000 galaxies as seen by the Hubble Space Telescope. The smallest and dimmest of them may be the oldest and most distant galaxies ever detected, formed when the universe was only 800 million years old.

dwarf exceeds the mass that can be supported by the star's electrons, the star will explode like a huge thermonuclear bomb. This critical mass is called the Chandrasekhar limit, after the 20th-century Indian-American astrophysicist Subrahmanyan Chandrasekhar. When a Type Ia supernova goes off, it can, for a period of weeks or months, outshine the entire galaxy in which it is located. It is estimated that one will occur every century or so in a typical galaxy. This doesn't sound like many events, but you have to remember that there are an awful lot of galaxies in the universe. Scientists estimate that there is a supernova like this going off somewhere within sight of Earth every few seconds—hundreds since you started reading this chapter, in fact.

HOW DOES THE UNIVERSE END?

Because a Type Ia supernova can be used as a standard candle, it gives astronomers a way to measure distances even to remote galaxies. By observing Type Ia supernovae in different locations, astronomers are able to approach a fundamental question about the Hubble expansion. The question: Will it ever stop? Does the universe have an end, as it has a beginning?

Look at it this way: A galaxy that is being carried along by the Hubble expansion will feel the gravitational attraction of other galaxies pulling it back. If there is enough mass to exert a big enough gravitational force, the expansion will eventually stop and reverse itself. We will live in what astronomers call a closed universe. If, on the other hand, the

THE ASTRONOMICAL DISTANCE LADDER REVISITED

Each rung in the astronomical distance ladder gives out at a certain point. Parallax, for example, fails when we can no longer measure the difference between lines of sight at the ends of the baseline, and the Cepheid variable technique will fail when galaxies are too far away for us to be able to distinguish individual stars (a distance of about 56 million light-years at the moment). Fortunately, the Type Ia supernova provides us with the next rung.

By looking at these types of supernovae in nearby galaxies, astronomers were able to establish a regular pattern. The longer a supernova lasted, the brighter it was. This means that astronomers can watch a Type Ia supernova go off, record the amount of time of the brightening, and from this deduce how much light the event is actually producing. Like the Cepheid variable before it, the Type Ia supernova can be used as a standard candle. In the late 1990s, when the results of such supernova observations began to be published, a new rung was added to the distance ladder, taking it out to about nine billion light-years.

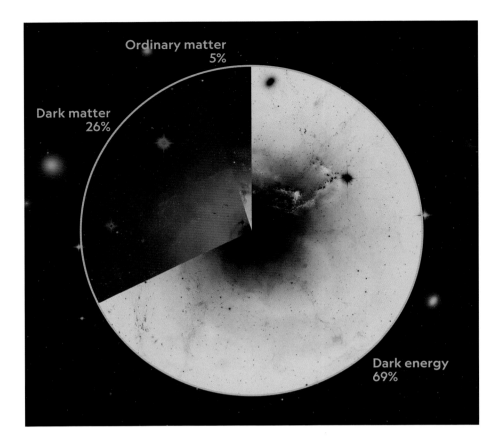

Ordinary matter
5%

Dark matter
26%

Dark energy
69%

A pie chart shows the relative abundance of ordinary matter, dark matter, and dark energy in the universe.

■ For more on dark energy, see page 370.

rate was actually slower in the past than it is now. The only way to explain this astonishing result is to say that there's something else out there, something capable of supplying a force that can overcome gravity's inward pull, something causing the galaxies to be repelled from each other. That something has been labeled dark energy. We don't know what it is or how it works, but we know it's out there.

SUMMING UP

The authors of this book hope that once you have read it, you will, in fact, go out and look at the night sky in all its majesty and grandeur. We hope you will see the stars not as simple points of light, but as giant factories creating the elements from which more planets—and perhaps life—will someday form. But we have to warn you. For all its glory, what you see in the sky is perhaps 5 percent of what's actually out there.

We've already seen this in our discussion of the electromagnetic spectrum. Our eyes detect only visible light, but the universe is sending us messages in every other electromagnetic form, from radio waves to gamma particles. They constitute one part of the unseen universe.

In 1970, another big part of the unseen universe came to light when astronomer Vera Rubin at the Carnegie Institution of Washington discovered that the visible parts of galaxies were encased in a large sphere of matter. This matter was capable of exerting a gravitational force but did not interact with the electromagnetic spectrum. Since then, we have found that this type of material, called dark matter, makes up about a quarter of the mass of the universe.

The dark energy described above is yet another piece of the unseen universe. It acts as a kind of antigravity, pushing galaxies apart and accelerating the Hubble expansion. We don't know what it is, but we know it's there because we can see what it does. Despite the similarity in their names, dark matter and dark energy are not related to each other. Dark matter pulls stuff together, while dark energy pushes it apart.

So that's the universe we live in: 5 percent familiar matter, 26 percent dark matter, and 69 percent dark energy. How exciting that 95 percent of the universe is still unknown and is out there waiting to be explained and explored!

gravitational force isn't big enough to reverse the expansion, the expansion will slow down but never end. We will be in what astronomers call an open universe. (The boundary between an open and closed universe is called "flat.")

We can determine what kind of universe we live in by measuring how fast the expansion is slowing down—a quantity astronomers call the deceleration parameter. All we need is a way to find out how quickly distant galaxies are slowing down, and this is where Type Ia supernovae come in, because they provide us with a way of measuring the distance to those decelerating galaxies.

Then comes the big surprise. When the measurements are made, it turns out that the expansion isn't slowing down a lot (as you would need for a closed universe), and it isn't slowing down a little (as you would need for an open universe). In fact, it isn't slowing down at all—*it is speeding up!*

Galaxies farther from us seem to be moving faster than those closer in. This means that the expansion

The discovery of **dark energy** showed just how much we still have to learn about our universe.

In the collision between two galaxy clusters, known jointly as the Bullet Cluster, ordinary matter is shown in pink, dark matter in blue.

Combining photographs taken with several different telescopes, scientists created an image of x-rays streaming off the surface of the sun.

CHAPTER TWO

SUN
MOON
EARTH

LOOKING OUTWARD

When we look at the sky, two celestial objects command our attention—the sun and the moon. Although quite different from each other, they dominate the earliest folklore associated with the appearance of the sky. Together with Earth, the platform from which all observations were once made, they form a kind of heavenly triumvirate that we need to understand before we can go on to a more detailed description of astronomical objects. Let's start our exploration, then, with the sun, the acknowledged supporter of life on our planet, and arguably the most important object in our sky. We can begin with a simple statement.

The sun is a star.

This statement, so obvious to us, actually represents millennia of thought and research. The sun is so essential to life on Earth that it's not surprising it occupies a special place in our thoughts. There is scarcely a society anywhere that doesn't have a branch of mythology devoted to explaining the sun's daily movement across the sky and other solar phenomena.

In Peru and northern Chile, for example, the sun god descends into an east-flowing river at sunset and emerges refreshed the next morning. The Aztec sun god Tonatiuh, on the other hand, was thought to make an epic journey in which he would die every night, join a coterie of warriors slain in battle and women who died in childbirth, and then be reborn in the morning.

On a less bloody, but equally catastrophic, note, the Greeks believed that the sun god Helios drove the sun across the sky each day in a golden chariot. One day he yielded to the pleading of his son Phaethon and let the boy drive the chariot. Phaethon quickly lost control and the chariot first went high in the sky, burning out the arc of the Milky Way, and then descended too low, scorching Africa and creating the Sahara desert. All of which, as one author put it, probably made Helios the first parent who regretted giving a teenager the keys to the family vehicle.

It wasn't until the 16th century that the Polish astronomer Nicolaus Copernicus produced the first detailed model of the solar system in which the sun was stationary and the planets (including Earth) moved. People finally realized that the apparent daily motion of the sun across the sky was actually due to the rotation of Earth.

Aztec sun god Tonatiuh

THE SUN

When we learn about the sun, we are learning about all the stars we see.

Given that the sun is one star among hundreds of billions in the Milky Way galaxy, we can learn a lot about the universe by studying the sun up close. As we pointed out in chapter 1, the sun gets its energy from the nuclear fusion process in which hydrogen is "burned" to produce helium. This process takes place in the sun's innermost core, a region that extends out to about 25 percent of the sun's radius. In the core, the pressure is high enough that the density of material is about 150 times that of water at Earth's surface and the temperature is close to 16 million K (almost 28 million degrees F).

The energy generated by nuclear reactions in the core moves outward in the form of x-rays and gamma rays. These photons collide with particles in the sun's radiative zone, causing it to re-emit high-energy photons. The radiative zone extends out to about 70 percent of the sun's radius. It takes energy about 170,000 years to get through that zone, and at its outer edge the sun's density has dropped to about 20 percent of the density of water, with a temperature of about 2 million K (3.6 million degrees F). At this point, the energy enters the last major energy-transfer layer of the sun—the convective zone. In this region, which extends almost to the solar surface, the material in the sun essentially boils, bringing the energy created by fusion to the sun's surface.

THE SOLAR SURFACE

This gets us to the outer layer of the sun—the part we can actually see, often called the solar atmosphere. Looking into the sun is a little like looking down into a murky pond. Eventually the material in the water (or, in the case of the sun, the solar

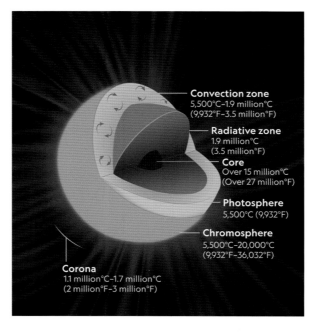

A cutaway view peels back the structure of the sun, from core to corona.

Convection zone
5,500°C–1.9 million°C
(9,932°F–3.5 million°F)

Radiative zone
1.9 million°C
(3.5 million°F)

Core
Over 15 million°C
(Over 27 million°F)

Photosphere
5,500°C (9,932°F)

Chromosphere
5,500°C–20,000°C
(9,932°F–36,032°F)

Corona
1.1 million°C–1.7 million°C
(2 million°F–3 million°F)

plasma) deflects and absorbs light trying to get out, preventing you from seeing anything farther down. The lowest of the sun's outer layers is called the photosphere. It's several hundred miles deep, and its bottom marks the lowest layer with visible light that we can see. The photosphere is a complicated kind of place, with heat and magnetic fields rising up from the sun's interior. When these magnetic fields break through the surface of the photosphere, they produce sunspots. Although (as we shall see in a moment) sunspots appear dark, they are actually quite hot, and look darker only because they are surrounded by much hotter material.

The next layer in the solar atmosphere is called the chromosphere ("sphere of color"), because super-heated hydrogen moving out from the sun causes

TEMPERATURE SCALES

When talking about objects like the sun, temperatures are generally reported in what are called degrees Kelvin, or simply kelvins, written K. The unit in the Kelvin scale is the same as a degree Celsius, but zero K is absolute zero (-273°C). Thus, for example, 0°C (the freezing point of water) is 273K.

(Opposite) The sun sets over Africa's Serengeti Plain.

it to glow a dull red. It is very faint and normally seen only during eclipses. The final layer of the solar atmosphere, also seen only during an eclipse, is the corona. The corona (Latin for "crown") is created by hot gases streaming away from the sun: gases that eventually slow down and become the solar wind. The solar wind extends the sun's atmosphere out beyond the orbit of Pluto. In a sense, then, we can say that all the planets are inside the sun.

The solar wind—the stream of particles that the sun throws out into space—produces one of the few phenomena associated with the sun's magnetic field that are visible to the naked eye. When particles in the solar wind encounter Earth's magnetic field, most of them are diverted away from the planet, much as river water flows around a rock. A few, however, are able to continue inward, and Earth's magnetic field guides them toward the North and South Poles. As they enter the atmosphere, they collide with atoms of oxygen and nitrogen, and these atoms convert the energy of those collisions into visible light: the spectacular displays we call the **northern and southern lights** (aurora borealis and aurora australis).

Given its complex structure, it's no wonder that the sun has played a major role in our understanding of the universe. When Galileo Galilei first turned a telescope to the sky, for example, one of the first things he looked at was the sun. To his surprise, he saw sunspots (or, more probably, a sunspot group). The accepted idea at the time was that objects in the heavens were supposed to be pure, beautiful, and unchanging. Yet here on the sun were dark blemishes that moved over time. These observations of the sun, along with Galileo's observations of the moon discussed later, were part of the beginning of the end for the old astronomical ideas of the Greeks.

SUNSPOTS

Consider the following two statements:

The sun is a ball of plasma.

The sun has a magnetic field.

A plasma, you might remember from chapter 1, is a material in which some or all of the electrons have been stripped from their parent atoms. Consequently, it is a material in which positive and negative charges can move around independently of each other. When a plasma interacts with a magnetic field, the field and the particles lock together, so that if the plasma is moved the field moves along with it, and if the field is moved the plasma does the same.

This means that in a region such as the solar surface, where processes like convection are moving hot plasma around, we expect to see the effects of the sun's magnetism. As the plasma moves, it drags the magnetic field lines with it, stretching and twisting them like rubber bands. When a rubber band snaps, energy is released, and the same thing hap-

■ For more on the northern and southern lights, see page 315.

(Below left) A striking view of a sunspot, the first taken with special equipment on the Inouye Solar Telescope, located in Hawaii. (Below right) Sunspots as sketched by Galileo in 1612

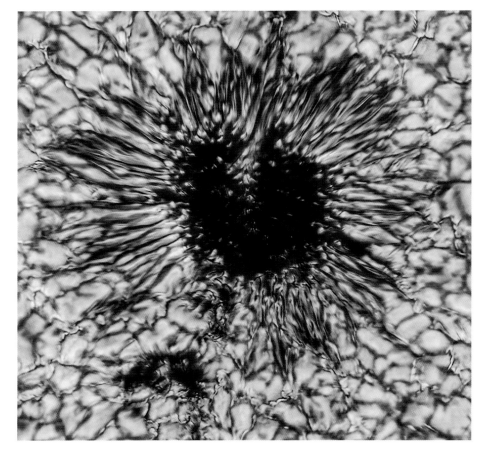

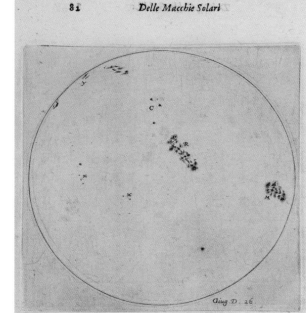

pens to magnetic field lines at the solar surface. This gives rise to some spectacular structures, such as solar prominences, and effects such as coronal mass ejections, which we'll discuss in a moment.

We know now that the sun rotates, but it rotates in a strange way. It takes about 35 days for material near the north and south poles of the sun to make a complete circuit, but only 25 for material at the equator to do the same. If Earth rotated this way, every day Miami would move a little farther east than Boston. Since the magnetic field is locked in to the plasma in the sun's body, this differential rotation strains the magnetic field lines (think of the rubber band again). Over time, the fields begin to deform and produce sunspots.

At a sunspot, the magnetic field gets intense—as much as 1,000 times stronger than the general solar field. This prevents hot material from the solar interior from coming to the surface. The region of the intense field, then, is slightly cooler than its surroundings. Because the surroundings are so hot, we see this region as dark. In fact, the typical sunspot has a temperature near 4000K (about 6700°F), and appears dark only because the surrounding material is at a temperature of about 5800K (about 10,000°F).

Another interesting peculiarity is associated with the solar magnetic field. As the sun's differential rotation distorts the magnetic field lines, the number of sunspots we see increases. Eventually, however, the system just can't keep going. The solar magnetic field reverses itself, and the process starts again. This progression, from start to finish, takes 11 years.

When we say the magnetic field reverses itself, what we mean is that the north magnetic pole becomes the south magnetic pole and vice versa. It's as if every 11 years the magnetic pole near Greenland switched from being north to being south and the magnetic pole near Antarctica did the opposite. Once the reversal is complete, the slow buildup of sunspots begins again until, 11 years later, another reversal starts things off again. This process is called the solar cycle. In fact, we can talk about two solar cycles: one 11 years long associated with the appearance of sunspots, and the other 22 years long, in which both sunspot activity and the orientation of the solar magnetic field return to their starting points.

THE DISCOVERY OF THE SOLAR CYCLE

Heinrich Schwabe started his professional life as a pharmacist with an avid interest in astronomy. He decided to participate in what was, at the time, a common astronomical investigation—the search for another planet orbiting between Earth and the sun. This required that he monitor the sun, looking for the regular transit of a dark spot across its surface. The problem, however, was that there were other dark spots on the sun—sunspots. To deal with this problem, Schwabe kept careful records of the appearance of sunspots, eventually selling his pharmacy and becoming a full-time astronomer. By 1843, he had accumulated data over two solar cycles and published his results. It wasn't until 1851, however, that the astronomical community started to pay attention to his results and the solar cycle became part of the growing body of knowledge about the sun. "If all this be not premature," wrote astronomer John Herschel to the physicist Michael Faraday, "we stand on the verge of a vast cosmical discovery such as nothing hitherto imagined can compare with." Then, as now, however, no one understood why the cycles exist.

THE MAUNDER MINIMUM

Nothing illustrates the problems we yet have to solve about the workings of the sun more than the so-called Maunder Minimum. In one of those coincidences that could only happen in real life, it happens that during the reign of the French king Louis XIV, the "Sun King," there was a prolonged disappearance of sunspots.

By the late 19th century, telescopic studies of the sun were well advanced, and a man named Edward Walter Maunder was in charge of the solar studies program at the Greenwich Observatory in England. (This is the place where the Eastern and Western Hemispheres officially meet, and clocks read Greenwich Mean Time, or GMT.) Maunder had access to all the old notebooks and publications produced by scholars at the observatory, and a close study revealed that between 1645 and 1715 there

The number of **solar flares** each year waxes and wanes according to the solar cycle.

were almost no recorded sightings of sunspots. He published his discovery in 1894, and this period has since come to be called the Maunder Minimum.

The reaction of scientists to his findings seems very strange to us today—his studies were basically ignored until the 1970s. The main argument against Maunder was that the absence of sunspot sightings need not have been the result of an actual absence of sunspots. In the 18th century, scientists thought of sunspots as clouds in the solar atmosphere and might simply have ignored them. There was another, deeper reason for ignoring Maunder's work, however. Scientists were just getting used to the idea of the sun as a mighty engine generating a huge energy output. The new science of thermodynamics explained many features of that engine, and the sunspot cycle would have been seen as analogous to the turning of a wheel in a locomotive. The last thing you would have wanted to hear was that this engine just shut itself off every once in a while,

An image of one area of the sun, as seen in detail by the Inouye Solar Telescope, shows the cells of superheated gas that cover its surface—each about the size of Texas.

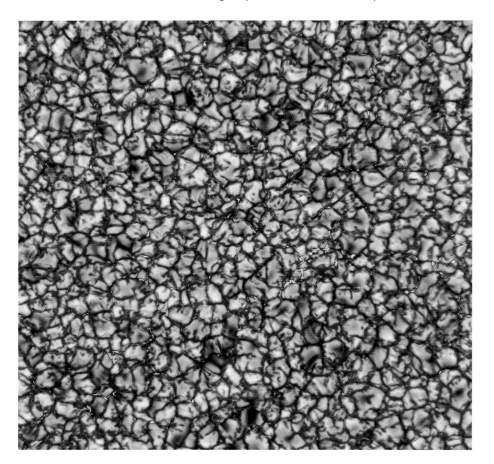

only to start up again on its own. Better to question your predecessors than to have to deal with that problem!

On the other hand, several pieces of evidence supported Maunder's claim. For example, in 1671, the astronomer Giovanni Cassini wrote the following after observing a sunspot: "It is now about twenty years since astronomers have seen any considerable spots in the sun, though before that time . . . they have from time to time observed them." This comment was considered so important that it was reproduced in the *Transactions* of the (British) Royal Academy, arguably the most prestigious scientific publication of its time. Couple this to Cassini's standing in the astronomical community—he discovered the so-called Cassini division in the rings of Saturn—and you have a pretty strong case for the reality of the Maunder Minimum. This is important because astronomers still do not have a clear picture of the way the sun works. It tells us that when we do have such a theory, it will need to be capable of explaining phenomena such as the Maunder Minimum.

SOLAR FLARES AND CORONAL MASS EJECTIONS

Today we monitor sunspot activity carefully, not because we want to change our ideas about the universe, but because events on the sun can have an enormous effect on our lives here on Earth. The surface of the sun is a roiling, dynamic system. As chunks of plasma move around, they carry parts of the solar magnetic field with them. As we outlined above, you can think of what happens to the magnetic fields as similar to what happens when a rubber band is stretched and twisted. Eventually the rubber band will snap, releasing energy when it does. In the same way, the twisting, churning magnetic fields at the sun's surface can release bursts of energy.

One such burst is called a solar flare. Think of a solar flare as a ball with an intense concentration of electromagnetic waves. Changes in the sun's magnetic field produce the flare, which looks like a brighter spot on the sun's surface. If the radiation reaches Earth, this "space weather" can interfere with things like radio transmissions and other communication systems, but that is pretty much the extent of the damage it can cause. A simple

THE SUN

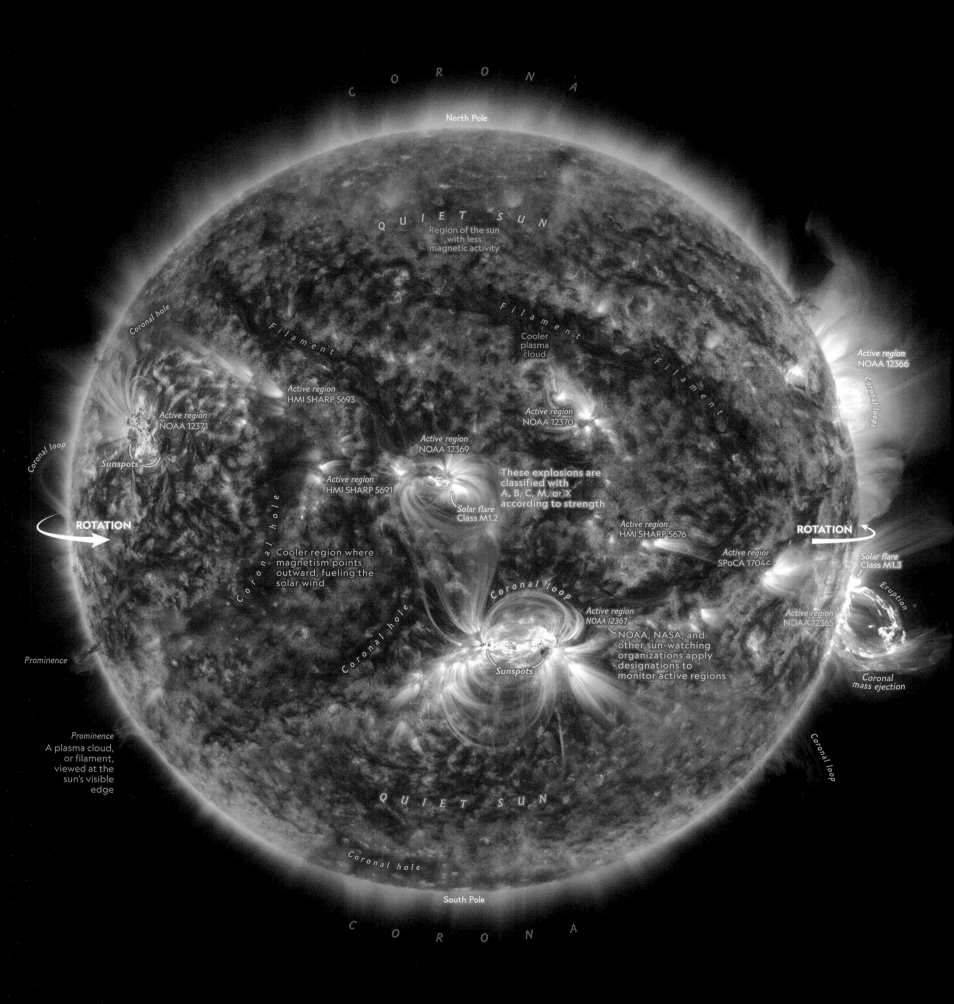

CORONA

North Pole

QUIET SUN
Region of the sun
with less
magnetic activity

Filament

Cooler
plasma
cloud

Coronal hole

Filament

Active region
NOAA 12366

Active region
HMI SHARP 5693

Coronal loops

Active region
NOAA 12371

Active region
NOAA 12370

Coronal loop

Active region
NOAA 12369

Sunspots

Active region
HMI SHARP 5691

These explosions are
classified with
A, B, C, M, or X
according to strength

ROTATION

Solar flare
Class M1.2

Active region
HMI SHARP 5676

ROTATION

Coronal hole

Cooler region where
magnetism points
outward, fueling the
solar wind

Active region
SPoCA 17044

Solar flare
Class M1.3

Coronal hole

Coronal loop

Eruption

Active region
NOAA 12367

Active region
NOAA 12365

Coronal loop

Sunspots

NOAA, NASA, and
other sun-watching
organizations apply
designations to
monitor active regions

Prominence

Coronal
mass ejection

Prominence
A plasma cloud,
or filament,
viewed at the
sun's visible
edge

Coronal loop

QUIET SUN

Coronal hole

South Pole

CORONA

solar flare isn't the worst thing that the sun can throw at us.

Another sort of bundle emitted from the sun's busy surface, a coronal mass ejection, or CME, is a lot more dangerous. As the name implies, during a CME the sun emits a huge cloud of particles. The cloud, which can have a mass of a billion tons or more, consists of a plasma in which negatively charged electrons and positively charged nuclei can move independently of each other. Usually (but not always) the appearance of a CME is preceded by a solar flare. CMEs from the sun are emitted in random directions, since they are associated with the snapping and twisting of the solar magnetic field. Consequently, it is unlikely (but not impossible) that Earth would be in a CME's path.

If, however, a CME is directed toward Earth, the

A coronal mass ejection from the sun in February 2015 passed by without hitting Earth.

sequence of events would look like this: Eight minutes after the eruption on the sun, a burst of gamma rays and x-rays would arrive. An hour or so later, protons and electrons accelerated by the sun's magnetic field would reach Earth; finally, a day or so later, the main plasma cloud would show up.

We know that this sequence of events can occur because in 1859 the British astronomer Richard Carrington described the collision of a CME with Earth. This was before the invention of electrical generators and the construction of power grids, so the consequences of the CME collision were mild. The main effect was on the telegraph system (sometimes referred to as the "Victorian Internet"). Operators were shocked by sparks leaping from their telegraph keys and setting fire to nearby papers. Auroras were seen as far south as Cuba, and peo-

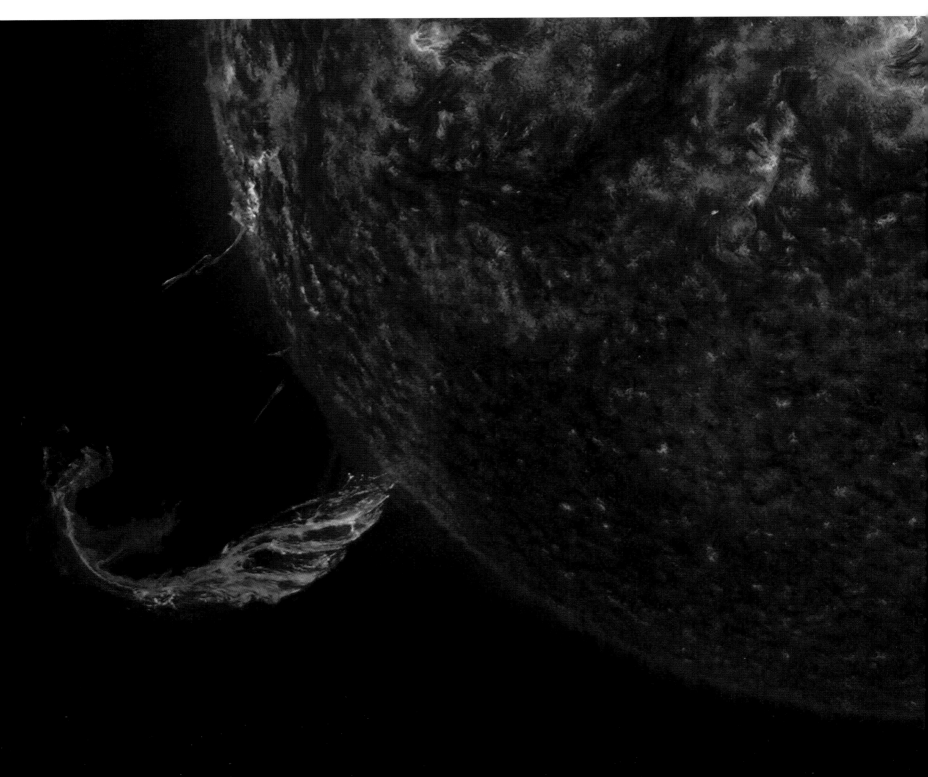

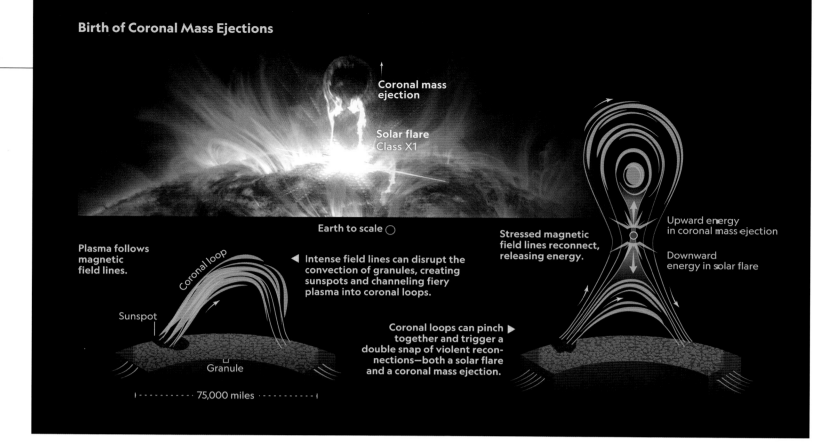

Birth of Coronal Mass Ejections

Coronal mass ejection

Solar flare
Class X1

Earth to scale ○

Plasma follows magnetic field lines.

Coronal loop

Sunspot

Granule

◀ Intense field lines can disrupt the convection of granules, creating sunspots and channeling fiery plasma into coronal loops.

Coronal loops can pinch ▶ together and trigger a double snap of violent reconnections—both a solar flare and a coronal mass ejection.

Stressed magnetic field lines reconnect, releasing energy.

Upward energy in coronal mass ejection

Downward energy in solar flare

⊢- - - - - - - - - 75,000 miles - - - - - - - - - ⊣

ple in Boston could read newspapers at midnight by their light. But the consequences of a Carrington event today would be nothing short of disastrous.

Today, the initial blasts of gamma rays and x-rays would probably fry many of the satellites in orbit around the planet and cause serious health issues for astronauts on the International Space Station. However, the effect of the plasma cloud slamming into Earth would be a catastrophe. The blast of charged particles would produce induced currents (see sidebar this page) in every piece of metal, including electrical wires, on the side of the planet where the blast hit.

The first system to be affected would be the power grid, because all those long, exposed wires would be a perfect target for a CME. Anything plugged into a wall socket—computers, TVs, refrigerators—would be fried as massive induced currents followed power lines into homes. More important, though, would be the effects on a part of the grid that we seldom think about: the huge transformers that convert the power brought in by the grid into the (much lower voltage) current that is distributed around a city. A transformer is basically a device that has a lot of copper wire wrapped into coils. The induced currents would melt those wires, the transformers would stop functioning, and cities would go dark.

This wouldn't be the kind of temporary power outage with which we're all familiar. Engineers estimate that it might take years to replace the damaged transformers. In the interim, there would be no electricity. As one scientist pointed out, people wouldn't

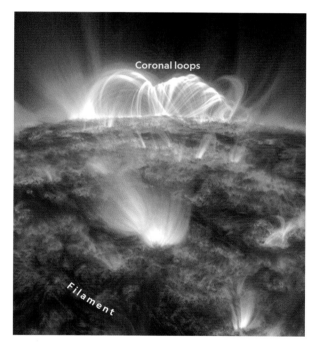

Coronal loops

Filament

(Above) A diagram illustrates how the sun's magnetic field helps to produce the eruptions and filaments seen on the solar surface. (Left) Coronal loops like these are only one example of the structures found on the sun's roiling surface.

even be able to flush their toilets, since most urban water systems depend on electric motors to pump water around.

Just for the record: A Carrington-level CME went through Earth's orbit on July 23, 2012, a few days after Earth passed through that spot. Had that CME happened a few days earlier, we would still

INDUCED CURRENTS

The basic laws of electricity and magnetism tell us that electrical currents can produce magnetic fields and that changing magnetic fields can produce electrical currents. If charged particles go by a wire, a magnetic field will be created near the wire, since the particles constitute an electric current. This changing magnetic field will induce an electric current in the wire. This is called an induced current.

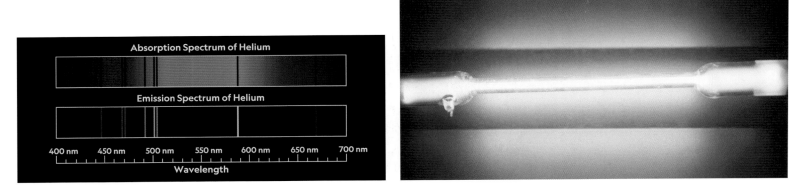

(Above left) A diagram shows the absorption and emission spectra of helium, one of the most common elements found in stars like the sun. (Above right) By studying emission spectra of gases in tubes like this one, scientists began to understand the chemical composition of many extraterrestrial objects.

■ For more on spectroscopy, see page 34.

be digging our way out of the consequences. Who knows—I might even be writing this text on an old manual typewriter.

THE STORY OF HELIUM

There have been many mysteries posed by studies of the sun. Perhaps none has had such an important philosophical impact as the great helium controversy of the late 19th century. By that time, the techniques of spectroscopy we talked about in chapter 1 were becoming standard tools for astronomers. And, of course, if you want to know how stars are made, what better object to study than our own sun, the nearest star?

Astronomers and chemists set about compiling information about the spectra of all the elements they thought might be in the sun. They also gathered information on the spectra of ions: atoms from which electrons have been stripped in energetic collisions. Thus they had ready information about the spectral fingerprint of ions such as iron IV—iron atoms from which three electrons had been removed. (The numbering system counts iron I as the normal atom.) Armed with this information, astronomers could then assign a known source to each line in the complex spectra they received from the sun.

In 1868, an amateur astronomer by the name of Joseph Norman Lockyer noticed something strange. Tucked into the solar spectrum between two bright yellow lines associated with sodium was a third line that did not seem to come from any atom that could

be made in the laboratory. Lockyer argued that this line came from an as yet unknown element that was present in the sun, but for which no evidence could be found on Earth. He named the element helium, from *helios*, the Greek word for the sun.

You have to know something about Lockyer to understand the impact this discovery had on the scientific world. In the first place, referring to him as an "amateur astronomer" is a little misleading. The term is used here just to indicate that he did not hold an official position. He was, however, well known in the British astronomical community. Just before his discovery of the helium line, for example, he had detected and named the sun's chromosphere by means of spectroscopy.

The second thing to know about Lockyer is that his greatest contribution to science was not the discovery of helium, but his founding and editing of the scientific journal *Nature*, still one of the world's most prestigious scientific journals—perhaps the most prestigious such journal. The journal premiered in 1869, and Lockyer was its editor for 50 years thereafter.

In any case, Lockyer's discovery of helium created a serious problem for the scientific community. One of scientists' deepest beliefs is that there is nothing unique about Earth or our solar system. This is usually referred to as the "principle of mediocrity" or the "Copernican principle." What it says is that if we measure a certain property of nature here and now in our laboratories, the result we get would be the same as if we had made the measurement anywhere in the universe. There are, in other words, no special conditions that apply only to Earth.

If you believe this principle, then any chemical element found anywhere in the universe should also be found on Earth. The natural question, then, was this: What was Lockyer's element doing on the sun if it wasn't here on Earth? The existence of

Helium on Earth seeps up to the surface from **radioactive elements** deep in the planet's crust.

helium seemed to threaten one of the foundation stones of science. Some scientists suggested that there was some as yet unknown flaw in the whole process of spectroscopy. Others suggested that the high pressures and temperatures in the sun distorted some normal spectra to produce Lockyer's lines. A few (including Lockyer) argued that helium would eventually be found on Earth. And there the controversy stayed for 30 years.

Then, in 1895, the British chemist William Ramsay, who had collaborated in the discovery of argon in Earth's atmosphere, began examining the gases given off by minerals containing uranium. After removing known gases like oxygen and nitrogen from his sample, he ran a spark through the remaining gas and examined the spectrum of emitted light. Sure enough, there were the helium lines that Lockyer had seen in the sun 30 years before. Earth is indeed just like the rest of the universe!

PROBING THE SUN'S INTERIOR: THE SOLAR NEUTRINO PROBLEM

It's all very well to discuss theories about how the sun works, but the fact remains that we can't see more than a few hundred miles into the sun's interior to verify these theories. The so-called solar neutrino problem arose in the late 20th century as a result of trying to "see" deeper into the sun.

You will remember that when we introduced the notion of nuclear fusion as the source of a star's energy in chapter 1, we said that the fusion of four protons would produce helium "and a spray of miscellaneous particles." As it happens, one of those miscellaneous particles gives us a way of learning about the sun's interior.

The particle is called a neutrino ("little neutral one"). It was named by the Italian-American physicist Enrico Fermi in 1933, although its existence wasn't verified until 1956. The neutrino has no electric charge, almost no mass, and travels at close to the speed of light. It interacts very weakly with matter, which makes it extremely difficult to detect. For example, as you read this, billions of neutrinos are passing through your body, as they have ever since you were born. Perhaps one or two times during your life, one of these neutrinos will interact with one of your atoms.

The production of neutrinos in the fusion reaction thus presents us with a classical good news/ bad news situation. On the one hand, once the

A British team led by Norman Lockyer (white beard) prepares to observe a solar eclipse in 1896. It was on an expedition like this in 1868 that Lockyer first saw evidence of helium.

Scientists first measured neutrinos from the sun in this giant tank, located in the Homestake gold mine in South Dakota.

neutrinos are produced in the sun's core they will have no problem passing through all the matter in the sun and reaching Earth. This means we can "see" all the way to the sun's core. This, in turn, means that by finding neutrinos that are emitted in fusion reactions, we can check our theories about the sun's energy source. On the other hand, once the neutrinos are here they are very difficult to detect because they interact so weakly with the matter in a detector.

In the early 1960s, the American chemist Raymond Davis, Jr., began to think about ways to detect neutrinos created in the sun's core. Because neutrino interactions would be rare in any detector, his first problem was to find a site that wouldn't flood his instruments with extraneous signals. Interactions such as those produced by cosmic rays could easily hide the events he was looking for. His solution was elegant. The Homestake Gold Mine in the Black Hills of South Dakota had been operating for many years. A large chamber in the mine was 1,478 meters (4,850 ft) underground—a perfect spot for his experiment, since the thick layer of rock would shield any apparatus he built from cosmic rays. Only neutrinos would be able to get through.

With the cooperation of the mining company, Davis constructed a 379,000-liter (100,000-gal) tank and filled it with perchloroethylene (ordinary dry-cleaning fluid). He then waited while the flood of neutrinos interacted with the chlorine atoms in the fluid, converting a few of those atoms into

argon. Every few months he would analyze the fluid and count the argon atoms, thereby discovering how many times the solar neutrinos had interacted with his target. Typically, he would see a few dozen argon atoms every few months.

At the same time, theorists were working out the details of how many neutrinos were being produced in the sun, and that's when the trouble began. Davis (and eventually other people who built neutrino detectors around the world) were only getting about a third of the number of neutrino interactions the theory predicted. This became known as the "solar neutrino problem."

In a situation like this there are three possible explanations: Either our ideas about the sun are wrong, our ideas about nuclear physics are wrong, or our ideas about neutrinos are wrong. After almost a decade of uncertainty, advances in the physics of elementary particles allowed us to settle on the third alternative. It turns out that there are actually three types of neutrinos: the electron, muon, and tau neutrinos. These neutrinos can transform into each other, especially if given enough time. On their eight-minute journey from the sun, then, the original electron neutrinos created in fusion reactions began changing into the other two. The supply of electron neutrinos would dwindle as the other two kinds appeared. As it happened, Davis's apparatus

Tracks of elementary particles recorded in a bubble chamber

detected only electron neutrinos. By the time those original solar neutrinos reached the Black Hills, the number of detectable neutrinos had been reduced to a third of its original value. Davis received the Nobel Prize for his work in 2002.

As was the case with Galileo and his telescope, studying the sun by monitoring its neutrino output taught us something fundamental about the world we live on.

THE SOLAR FUTURE

Despite the fact that the sun consumes about 544 million metric tons (600 million tons) of hydrogen per second and has been doing so for the last four and half billion years, it will be able to keep burning hydrogen for another 5.5 billion years. As in all other matters, the sun is an average star as far as its lifetime is concerned. As with many other stars, the amount of energy it emits has been increasing slowly over time. Today, for example, the sun is about 30 percent brighter than it was in the beginning. It will increase by a comparable amount in the future, but the increase will be slow enough that life-forms on the planet will have ample time to adjust.

The same is not true of the next step in the sun's life—the transition to its red giant phase. As the sun swells up, it will swallow the innermost planets, Mercury and Venus. It will also spew a large fraction of its mass out into space in an augmented solar wind. Current theories suggest that Earth itself will not be incorporated into the sun, but it will be turned into a planet that no one living today would recognize.

Someone on Earth when the sun is in its red giant phase would see a huge red ball rising above the horizon, eventually filling half the sky. The planet's surface temperature would be so high that rocks would melt, producing a red-hot lava ocean. On this ocean "icebergs" made of metal would float, perhaps passing "islands" made of materials with a high melting point. When the sun completes its final collapse into a white dwarf, a suddenly cooled Earth will pass into oblivion, circling a dying cinder in space.

An artist s conception reveals the fate of Earth's surface during the sun's red giant phase.

THE MOON

The moon shines only by reflected light and has no energy source of its own.

Like the sun, the moon is a part of the sky with which everyone is familiar. No matter how much visual pollution impairs your view, you can always see the moon if you look for it. This means that almost every culture has folklore associated with the moon. Unlike the sun, which looks the same every day, the moon goes through phases, completing an entire cycle in 29.53 days (this is called a "synodic month"). Thus, much of the folklore associated with the moon deals with change—the moon is linked to the constant alternation of birth and death, creation and destruction. There are old farmers' tales, for example, that identify the waxing phases of the moon with fertility and advise planting as the moon approaches full.

For the Romans, the moon goddess Luna drove the moon across the sky each night in a silver chariot. The correspondence between the lunar month and the menstrual cycle caused many cultures to associate the moon with women and motherhood. In Persia, for example, the moon was Mangha, the moon goddess. But the moon could also be feared: The Maori called the moon "man eater." Similarly, the mottled appearance of the moon was interpreted in different ways. We are used to the "man in the moon," but many Asian societies saw a rabbit instead.

One of the most interesting aspects of lunar folklore has to do with the assignment of gender to the sun and moon. To us, the association of masculinity with the sun and femininity with the moon may seem obvious, but there are many cultures that assign the moon a masculine identity. Our own man in the moon probably comes from Germanic tribes who, among many others, saw the moon as masculine. Others who made this choice include

(Opposite) A supermoon rises over Sakhalin Island in eastern Russia.

the Norse, Indigenous Australians, Mongolians, Armenians, and many more. In fact, there seems to be a link between geography and the assigning of gender roles to the sun and moon. Desert people, for whom the sun was a constant source of danger, tended to picture the moon as a gentle female god-

A stone carving of the Aztec moon goddess Coyolxauhqui

dess who could shelter them from the harsh sun. Northern people, on the other hand, saw the cold stillness of the moon as a danger and welcomed the warming feminine sun.

WHERE DID IT COME FROM?

If we're going to understand what we're seeing when we look at the moon, we need to understand how it came to be what it is. The quest for an answer to this question involved a long and arduous search by astronomers.

Here's the problem: The moon and Earth are obviously close to each other, yet they are very different. The moon, for example, has an average density of about 3.34 grams/cubic centimeter. (For reference, the density of liquid water is 1 g/cc.) The moon's density, therefore, is approximately the same as the densities of rocks like granite and basalt, which make up the outer layers of Earth. Our own planet, on the other hand, has an average density of 5.5 g/cc, a fact that reflects the great amount of iron (density 7.9 g/cc) in its core. The question is obvious—how could two objects so different in density have formed so close to each other?

There are basically two ways to solve this problem, both of which were explored extensively during the mid-20th century. One approach is to assume that the two bodies formed in different places, and that Earth captured the moon at a later date. It turns out that the moon's size makes such a capture very unlikely. The other approach is to assume that the early Earth possessed some special property that influenced the formation of the moon. For example, in the 1950s it was quite common for textbooks to claim that the Pacific Ocean was a kind of "birth scar," marking the spot from which the moon had been torn.

In chapter 1 we introduced the nebular hypothesis, which says that the solar system formed from an interstellar cloud collapsing under the influence of gravity. In this picture, the small, rocky inner planets formed after the solar wind generated by the nascent sun had blown volatile materials out past the orbit of Mars, leaving only non-volatile material (think sand grains) to form planets such as Earth. As time went on, these small grains began to stick together, forming mountain-size objects known as planetesimals, which in turn began to come together to form Mars-size protoplanets. Computer models suggest that there could have been more than a dozen of these protoplanets whizzing around the solar system, one of which would eventually become Earth.

In this phase of its life, the proto-Earth was going through its orbit, sweeping up material that had not yet been incorporated into other bodies. Had you been on Earth's surface at that time, you would have seen a constant rain of meteorites. Each impact brought more material and, more importantly from our point of view, more energy that was converted into heat. Earth heated, perhaps melting all the way through. Like a salad dressing left on a shelf too long, the material in the protoplanet began to separate, with the heaviest substances sinking to the center and the lightest floating to the top. It turns out that this process, known as differentiation, plays a crucial role in the formation of the moon.

By the time Earth had cooled enough for a solid crust to form, it had acquired its current geology, with a core of iron and nickel at the center, a mantle about 2,900 kilometers (1,800 mi) thick above it, and a thin crust forming the outer surface. It was at this point that a critical event occurred. One of those Mars-size protoplanets crashed into Earth, struck it a glancing blow, and shattered into pieces. This collision threw a huge amount of material from Earth's mantle into near-Earth orbit. It was from this orbital material, as well as the material in the Mars-size object, that the moon formed.

Notice that, although the moon formed near Earth, it formed from terrestrial materials from which the heaviest elements had been removed by differentiation. Thus, the problem of explaining the

■ For more on solar system formation, see page 97.

PHASES OF THE MOON

Like the planets, the moon shines by reflected light. This means that only the side of the moon facing the sun will shine in the night sky. When the moon is between Earth and the sun, the illuminated face will face away from us and we will have a night with no visible moon (known as the new moon). When the moon is on the other side of its orbit, away from the sun, observers on Earth will see the entire illuminated face and we will have a full moon. During the two weeks between these two situations, we will see more and more of the illuminated face each night. This is called the "waxing" phase. After the full moon we will see less and less of the illuminated face each night until the moon disappears. This is called the "waning" phase.

different densities of these two heavenly objects is solved. This "giant impact hypothesis" is now generally accepted by scientists and is often referred to as the "Big Splash." The colliding object has been named Theia, after the Titan in Greek mythology who was the mother of Selene, the goddess of the moon. The collision is thought to have occurred between 20 million and 100 million years after the formation of the planets started—a mere blink of an eye in astronomical time.

Since it was formed, the moon has engaged in a complicated gravitational *pas de deux* with Earth, a dance whose main effects are to increase the distance between the two bodies and to slow the rotation of Earth. At the time of the dinosaurs, the day was about 23 hours long, for example, and even today the length of the day is increasing at the rate of a few milliseconds per century.

LUNAR GEOLOGY

The moon is not a perfectly smooth crystal sphere, as ancient philosophers thought. A cursory look at a full moon will convince you that there has to be variation in the lunar surface, since some parts appear dark and others appear light. We can understand this variation if we think a little about the early history of the moon. Like the early Earth, the moon must have been heated by the processes that led to its formation. Whether it had a magma ocean on its surface or only partially melted areas remains a subject of debate among scientists, but in either case the cooling process would have produced a body that had molten magma underneath a thick upper crust. It was at this stage that the variations in the lunar surface began to be created.

The impact of large meteorites or volcanic

Continued on page 74

An illustration portrays the dramatic collision that formed the moon, early in Earth's history.

EARTH'S MOON

NEAR SIDE

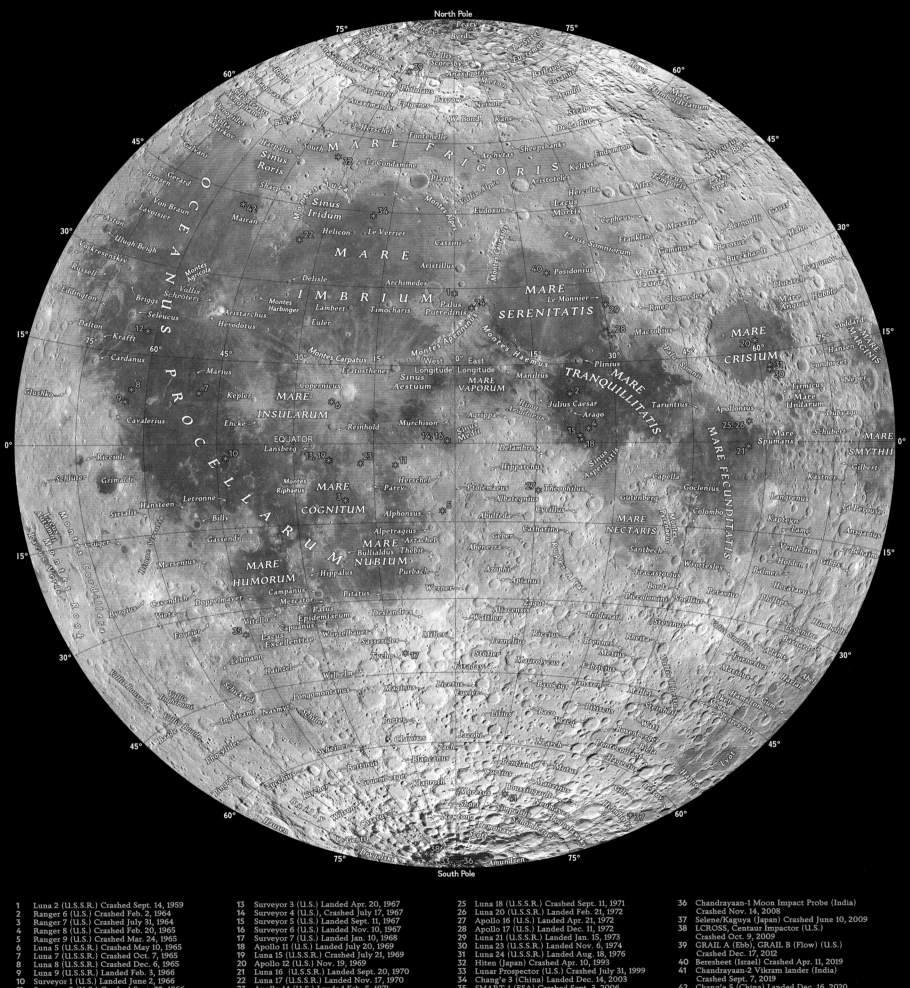

North Pole

South Pole

1 Luna 2 (U.S.S.R.) Crashed Sept. 14, 1959
2 Ranger 6 (U.S.) Crashed Feb. 2, 1964
3 Ranger 7 (U.S.) Crashed July 31, 1964
4 Ranger 8 (U.S.) Crashed Feb. 20, 1965
5 Ranger 9 (U.S.) Crashed Mar. 24, 1965
6 Luna 5 (U.S.S.R.) Crashed May 10, 1965
7 Luna 7 (U.S.S.R.) Crashed Oct. 7, 1965
8 Luna 8 (U.S.S.R.) Crashed Dec. 6, 1965
9 Luna 9 (U.S.S.R.) Landed Feb. 3, 1966
10 Surveyor 1 (U.S.) Landed June 2, 1966
11 Surveyor 2 (U.S.) Crashed Sept. 22, 1966
12 Luna 13 (U.S.S.R.) Landed Dec. 24, 1966

13 Surveyor 3 (U.S.) Landed Apr. 20, 1967
14 Surveyor 4 (U.S.), Crashed July 17, 1967
15 Surveyor 5 (U.S.) Landed Sept. 11, 1967
16 Surveyor 6 (U.S.) Landed Nov. 10, 1967
17 Surveyor 7 (U.S.) Landed Jan. 10, 1968
18 Apollo 11 (U.S.) Landed July 20, 1969
19 Surveyor 3 (U.S.S.R.) Crashed July 21, 1969
20 Apollo 12 (U.S.) Nov. 19, 1969
21 Luna 16 (U.S.S.R.) Landed Sept. 20, 1970
22 Luna 17 (U.S.S.R.) Landed Nov. 17, 1970
23 Apollo 14 (U.S.) Landed Feb. 5, 1971
24 Apollo 15 (U.S.) Landed July 30, 1971

25 Luna 18 (U.S.S.R.) Crashed Sept. 11, 1971
26 Luna 20 (U.S.S.R.) Landed Feb. 21, 1972
27 Apollo 16 (U.S.) Landed Apr. 21, 1972
28 Apollo 17 (U.S.) Landed Dec. 11, 1972
29 Luna 21 (U.S.S.R.) Landed Jan. 15, 1973
30 Luna 23 (U.S.S.R.) Landed Nov. 6, 1974
31 Luna 24 (U.S.S.R.) Landed Aug. 18, 1976
32 Hiten (Japan) Crashed Apr. 10, 1993
33 Lunar Prospector (U.S.) Crashed July 31, 1999
34 Chang'e 3 (China) Landed Dec. 14, 2003
35 SMART-1 (ESA) Crashed Sept. 3, 2006

36 Chandrayaan-1 Moon Impact Probe (India)
    Crashed Nov. 14, 2008
37 Selene/Kaguya (Japan) Crashed June 10, 2009
38 LCROSS, Centaur Impactor (U.S.)
    Crashed Oct. 9, 2009
39 GRAIL A (Ebb), GRAIL B (Flow) (U.S.)
    Crashed Dec. 17, 2012
40 Beresheet (Israel) Crashed Apr. 11, 2019
41 Chandrayaan-2 Vikram lander (India)
    Crashed Sept. 7, 2019
42 Chang'e 5 (China) Landed Dec. 16, 2020

FAR SIDE

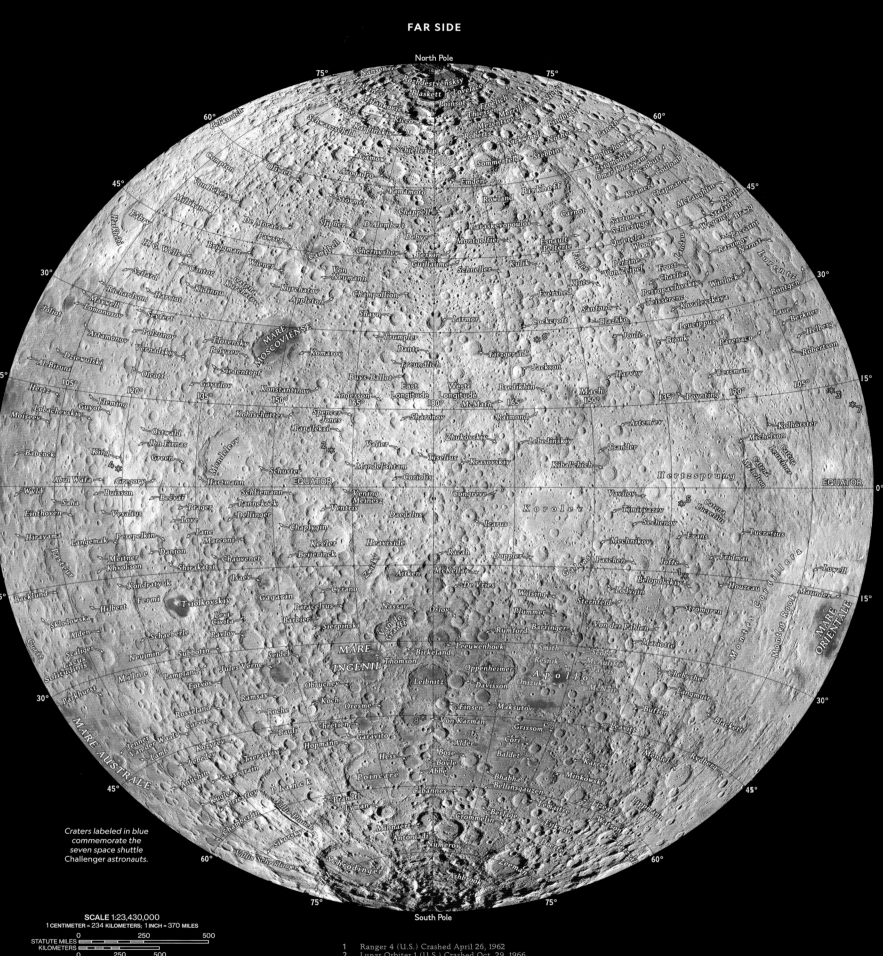

Craters labeled in blue commemorate the seven space shuttle Challenger astronauts.

SCALE 1:23,430,000
1 CENTIMETER = 234 KILOMETERS; 1 INCH = 370 MILES

STATUTE MILES
KILOMETERS

0   250   500

* Spacecraft landing or impact site

1   Ranger 4 (U.S.) Crashed April 26, 1962
2   Lunar Orbiter 1 (U.S.) Crashed Oct. 29, 1966
3   Lunar Orbiter 3 (U.S.) Crashed Oct. 9, 1967
4   Lunar Orbiter 2 (U.S.) Crashed Oct. 11, 1967
5   Lunar Orbiter 5 (U.S.) Crashed Jan. 31, 1968
6   Okina, Selena Orbiter (Japan) Crashed Feb. 12, 2009
7   LADEE (U.S.) Crashed April 18, 2014
8   Chang'e 4 (China) Landed Jan. 3, 2019

(Above) Scientist-astronaut Harrison Schmitt stands on the lunar surface during 1972's Apollo 17 mission. Note the lunar roving vehicle to the right of the boulder. (Right) Astronaut Charles Duke (Apollo 16) examines the boulder nicknamed Outhouse Rock on the moon. The white objects hanging from his hand are sample bags.

Continued from page 71

forces within the moon itself would have created openings in the crust, and the magma beneath the crust flowed out. These outflows could have been quite large and would have covered the lighter-colored surface rocks with darker magma. When the magma cooled, it created the dark splotches we see on the lunar surface. The ancients referred to these as *maria*, Latin for "seas," and we still use that name today, even though we know that the dark areas are not oceans. The paler areas are known as the lunar highlands, and they are heavily covered with craters created by ancient impacts.

The moon, being a very small world, quickly lost whatever atmosphere it might have had originally, and whatever internal heat it possessed at the beginning quickly radiated out into space. The moon, then, like the planet Mercury, is an airless, dead world.

This fact explains a great deal about our neighboring world. For one thing, it tells us that there are no processes on the moon that correspond to the weathering we see on our own planet. Once an impact crater forms on the lunar surface, it stays there. Thus, the entire record of meteorite collisions with the moon remains there to be read. In addition, without a heated interior, the moon has no magnetic field—a fact that will become important when we discuss lunar colonization.

The absence of any atmosphere or magnetic field means that particles from the solar wind and other kinds of cosmic rays, as well as all sorts of micrometeorites, have been bombarding the lunar surface since the moon's formation. Small meteorites routinely burn up when they enter Earth's atmosphere—we call them "shooting stars." On the moon, however, they produce a constant churning of the surface. This so-called space weathering, together with the occasional large impact, has produced a covering of small chunks of rocks and dust that varies between 5 meters (16 ft) in thickness in the young maria to 15 meters (49 ft) in the old highlands. Furthermore, the heat associated with impacts tends to make material in the layer of loose rock known as regolith stick together—think of it as a layer of sugar that has been too long in damp air.

TIDES

One effect the moon has on our planet can be detected even when the moon is not visible: the ocean tides. In most coastal locations on Earth, we see a periodic rise in sea level twice each day. This is a result of the gravitational force that the moon exerts on Earth's oceans.

Tides are complicated, so let's start with a simple example in which neither Earth nor the moon is moving (we'll work our way up to full complexity as we go). As shown at right, the moon raises what is known as a tidal bulge on Earth's oceans. On the side of Earth that faces the moon, the moon pulls the water away from Earth, and on the opposite side

it pulls Earth away from the water. We say that the tidal bulge has two lobes.

Now let's allow Earth to rotate but keep the moon stationary. An observer on Earth is carried around once a day. She sees the water rise and fall twice each day—once as she is carried through each lobe.

Now let's add the final complication. The tidal bulge should be highest at a spot right underneath the moon. This spot travels around Earth once every 24 hours. As it happens, Earth's oceans are relatively shallow (they have an average depth of about 3.7 kilometers/2.3 miles or so). The friction between the water and the ocean floor prevents that water from moving fast enough to follow the spot directly

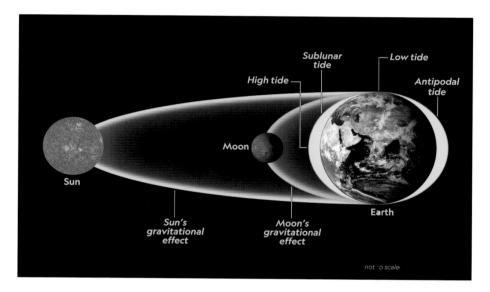

A diagram illustrates how the moon raises tides in Earth's oceans.

An artist (center) paints during low tide in St. Ives harbor in Cornwall, England.

In some places on Earth, **tides** vary by up to 12 meters (40 ft) between low and high tides.

under the moon, so the ocean tides are highest when the moon is on the horizon, rather than when it is directly overhead. This is called an inverted tide, as opposed to a direct tide, in which the high tide would occur when the moon is directly overhead. As mentioned, ocean tides on Earth are the result of a complex interplay of forces. You would be amazed at how many standard astronomy textbooks get this little bit of astrophysics wrong!

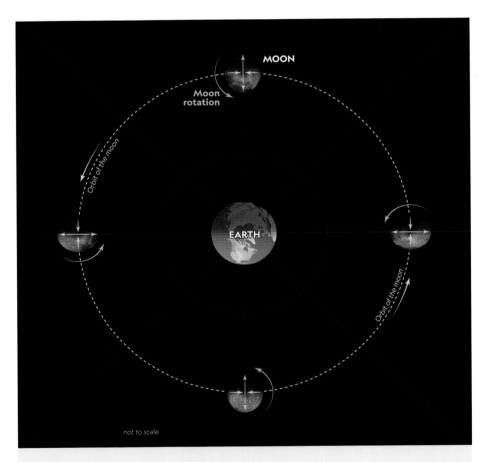

not to scale

EARTH'S GRIP ON THE MOON

Just as the gravitational forces exerted by the moon raise land tides on Earth, gravitational forces exerted by Earth raise tidal bulges on the moon. Consider two situations. First, assume that the moon rotates faster than would be required to keep its same face toward Earth. In this case, at the end of a lunar day the tidal bulge would have gone past the line joining the centers of Earth and the moon. In this case, Earth's gravity would pull back on the moon and slow its rotation. A similar argument shows that Earth's gravity would speed up the moon's rotation if that rotation were too slow. Only if the moon kept one face toward Earth would there be no speeding up or slowing down of the rotation due to tidal forces.

Now that we are discovering planets circling other stars, we know that there are exoplanets out there with oceans 80 to 160 kilometers (50 to 100 mi) deep. In these exoplanets, the forces of friction are overcome by the motion of the water, and tides are direct.

Although they aren't as noticeable as ocean tides, there are tides in the solid earth as well. Both the moon and the sun produce tidal deformations in the solid earth—deformations that have the same two-lobed structure we see in the oceans. These are called Earth tides or land tides, and they involve actual motion of Earth's crust. Unlike ocean tides, however, Earth tides are not normally noticeable. They involve lifting Earth's surface by less than a meter over an area of thousands of square kilometers—an operation equivalent to lifting up land about the size of Texas. In this situation, all the land around a normal building moves up and down together, so land tides have no effect on buildings or other structures. At a place like the Large Hadron Collider in Geneva, Switzerland, however, scientists need to know the circumference of their circular machine, a device almost eight kilometers (5 mi) in diameter, to within less than a millimeter. In situations involving large-scale scientific instruments like this one, Earth tides become important and have to be dealt with.

LOCKED IN

We know that the moon always keeps the same face toward Earth. This phenomenon means that the time it takes for the moon to rotate around its axis is the same as the time it takes to circle Earth in its orbit. In other words, the moon's "day" is exactly as long as its "year." It turns out that this is not a strange coincidence, but a result of tidal forces as well (see sidebar this page). In the jargon of astronomers, we say that the moon is "tidally locked," or "despun." In fact, many exoplanets exhibit the same behavior with respect to their star, especially if they are close to a small star. On such a world, the sunward side would be very hot and the spaceward side would be very cold. The question of whether life could develop on such a planet is an interesting one.

As far as the moon is concerned, however, tidal locking produces a situation in which, in most places on the moon, the sun is above the horizon for two Earth weeks and below the horizon for the next

Astronaut Alan Bean prepares to step onto the lunar surface during the Apollo 12 mission in 1969.

two Earth weeks. This produces enormous swings in temperature—at these spots the temperature can go from 127°C (261°F) at high noon to -173°C (-279°F) at night. For reference, 127°C is above the temperature most ovens need to slow-cook beef, and the coldest temperature ever recorded on Earth (at Vostok Station in Antarctica) is -89°C (-128°F). Only a few small areas near the lunar poles escape this sort of temperature swing, because they stay in sunlight all the time.

EXPLORING THE MOON

As the nearest celestial body to Earth, the moon has always had a special attraction for those who would like to pay a visit. It has also been a place where people believed we would find living inhabitants. As early as 1638, the English clergyman John Wilkins wrote a book titled *The Discovery of a World in the Moone,* in which he suggested that there might be inhabitants on the moon but admitted that "of what kinde they are is uncertaine." In 1865, the celebrated author Jules Verne began his trilogy of moon travel stories with *From the Earth to the Moon,* a novel in which members of the fictional Baltimore Gun Club build a huge cannon to send a spaceship to the moon. (Spoiler alert: They don't make it.) And then, in 1901, H. G. Wells published his novel *The First Men in the Moon,* in which his protagonists meet a race of "Selenites" on a moon that has an atmosphere of breathable air. The notion of the moon as a feasible target for visitors seems to have been around for a long time.

The first stage in the assembly of a launch vehicle for the Artemis I mission is put into place in the NASA Vehicle Assembly Building at Florida's Kennedy Space Center.

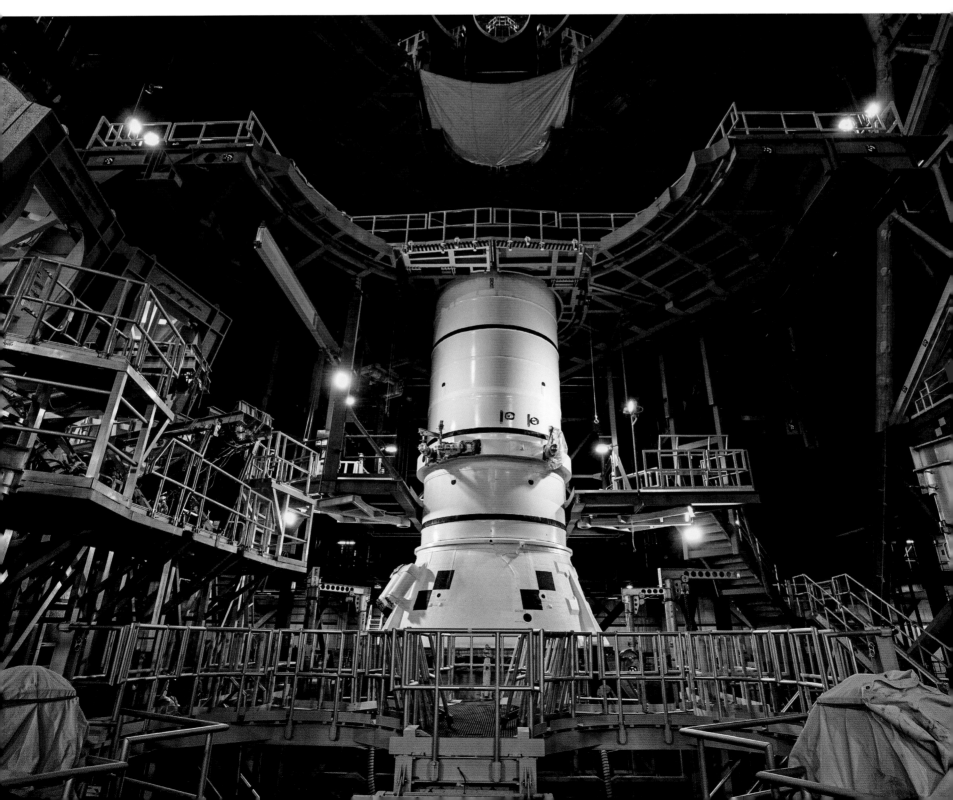

A model of the Soviet lunar rover Lurokhod 1 on display in Moscow. The original was the first remote-controlled rover placed on the moon, having landed there in 1970.

The shift from the moon as a source of science fiction to the moon as a reasonable target for a technological society can be set pretty precisely. The launch of the Soviet satellite Sputnik 1 in 1957 jolted the scientific self-esteem of the United States. It seemed that the Soviet Union had achieved mastery of a technical field and thus scored a major victory in the Cold War. The consensus among historians is that the United States got into what is now known as the space race because of the Soviet challenge, not because the country suddenly became interested in space.

In any case, in 1962, President John F. Kennedy made his famous speech in which he promised that America would land an astronaut on the moon by 1970. Virtually overnight the spigots opened and money poured into space programs. Seven years and $25.4 billion later ($206 billion in current dollars), astronaut Neil Armstrong set foot on the moon. In the following years there were five more landings of humans on the moon, and then . . .

And then . . . maybe because the "race" was over, maybe because people just got bored with NASA's success, maybe because people's attention moved elsewhere, the crewed missions to the moon stopped. Since 1972, the moon has been allowed to return to its historical role as Earth's close (but unvisited) companion. As robotic space probes explored the rest of the solar system, landing on Mars and visiting Pluto, the moon stayed in the sky as a kind of celestial afterthought.

Lately, however, this situation has been changing. Many countries besides the United States have started to display an interest in exploring the moon. In fact, the 21st century has seen an enormous growth in international interest not only in exploring the moon, but also in establishing a human

"Selenites" line up in an illustration from the H. G. Wells novel *The First Men in the Moon,* published in 1901.

(Right) China's Yutu-2 lunar rover, as seen from the Chang'e 4 spacecraft, which landed on the far side of the moon in 2018

■ For more on advances in space exploration, see pages 405–413.

The proposed Gateway space station, an international project, would provide a stopping place for astronauts visiting the moon.

presence there. A summary of some of the most important advances made by various countries is given below:

Soviet Union

In 1959, the Soviet Union's spacecraft Luna 2 was the first human-made object to reach the surface of the moon. In that same year the Soviet spacecraft Luna 3 sent back the first images of the far side of the moon. In 1976, Luna 24 brought the last lunar samples carried to Earth until the launch of Chang'e 5 (see below).

United States

Following its successful Apollo program, NASA launched a number of lunar missions, including the Lunar Prospector (1998) and Lunar Reconnaissance Orbiter (2009). NASA's Project Artemis is built around the goal of landing the first woman, and the next man, on the moon by 2025 at a cost of $30 billion. It seems unlikely that this ambitious date will be met, but the project is symbolic of a change in attitude about exploring the moon.

China

China's moon probes are called Chang'e (pronounced roughly as "Chahng-uh"), after the goddess of the moon in Chinese folklore. In 2019, Chang'e 4 made the first successful landing on the

far side of the moon. The mission deployed a rover and communicated to Earth through a relay satellite. In 2020, the Chang'e 5 mission sent a lander to the moon's surface, where it drilled into the ground and used a scoop arm to collect a 1.7-kilogram (3.7-lb) rock sample. The sample was placed in an ascent vehicle, which then rendezvoused with the mission orbiter. There the sample was added to a capsule that returned it to Earth. This was the first return of a lunar sample since 1976.

The long-term goal of the Chinese space program is to establish a crewed base at the moon's south pole sometime in the 2030s.

India

In 2008, the Indian Space Research Organisation sent the Chandrayaan-1 orbiter to the moon. The craft released an impactor near the moon's south pole and detected water ice in the debris of the impact. The orbiter mapped the mineral abundances on the lunar surface until it stopped communicating in 2009. Unfortunately, an attempt in 2019 to put a lander on the moon during the Chandrayaan-2 mission failed, although the Chandrayaan-2 orbiter is still functioning, producing a more detailed mineralogical map of the lunar surface.

Japan

In 2007, Japan launched the SELENE lunar orbiter, which created a topographical map of the moon's surface, as well as capturing stunning images of Earthrise over the lunar surface.

One design for a lunar base includes solar cells for power and sealed buildings covered in lunar regolith.

Russia

After a long hiatus in moon landings, Russia plans to launch an orbiter and lander called Luna 25 to the moon in 2022 and aims to place an astronaut on the lunar surface by 2030. Russia's general strategy is to concentrate its space program on the moon and leave Mars largely to NASA.

HUMAN SETTLEMENTS ON THE MOON

It seems so obvious. The moon is Earth's nearest celestial body—we can see it every night. Even our creaky 1970s spaceships could get there in a few days. Humans have already been there. Isn't it clear that when the human race finally decides to leave Earth, the moon will become our next stomping ground?

Maybe, but then again, maybe not. If we're going to talk about human settlement, there is one fact about the moon that we have to understand:

The moon is a really dangerous place.

We can make this point by considering the dangers in order. Let's start with gravity. Because the moon is so small, every person on its surface will weigh only about 16.6 percent of what he or she weighs on Earth. A 90-kilogram (198-lb) man, for example, will weigh in at 15 kilograms (33 lbs).

This isn't quite as radical a departure from the terrestrial environment as astronauts experience on the International Space Station, but it is likely to have serious health effects nonetheless. The problem is that the bones in the human body are programmed to retain only enough material to maintain their function in the environment in which they find themselves. In a low-gravity environment, they will lose inevitably lose mass and weaken. Stay on the moon long enough, in other words, and you will probably never be able to return to Earth—your

A building block produced by 3D printing from simulated lunar dust

bones would simply shatter on your return.

If physicians don't find a way of dealing with this problem, human visits to the moon will probably be restricted to a few years at the most. And just to emphasize this point, remember that a lunar colony would lead, eventually, to families. What effect would the low-gravity environment have on pregnant women and growing children?

To be blunt, we have no idea.

So much for gravity—let's move on to radiation. Because Earth has a magnetic field and an atmosphere, most of the particles in the solar wind and cosmic rays don't reach us on the surface. The moon has neither of these amenities, so anyone on the lunar surface would be subject to a constant bombardment of high-energy particles, which can cause radiation poisoning and cancer. This means that "Selenites" would have to spend most of their time shielded from radiation. They could, for example, cover their dwellings with a layer of the lunar regolith about two meters (6.6 ft) thick. Alternatively, they could move underground, most likely in open underground spaces known as lava tubes: leftover structures from the early geological history of the moon. We know of many such structures—Japan's SELENE spacecraft, for example, found one that was a full 50 kilometers (31 mi) long and 75 meters (246 ft) wide, a little less than twice the width of a football field.

LUNAR RESOURCES

On the other hand, assuming that problems like low gravity and high radiation can be dealt with, the moon has many resources that colonists could use. We know that there is ample water in the shaded areas of craters near the poles, for example, and the so-called peaks of eternal sunlight that stay in the light in polar areas could provide an abundant supply of solar energy. Engineers have suggested that

Astronauts on the moon's surface are exposed to **200 times as much radiation** per hour as they would receive on Earth.

the moon's low gravity might allow it to serve as a supply station for building structures in space—structures whose materials would otherwise have to be hauled into orbit from inside Earth's deep gravitational well. Launching building material into space would be a lot easier to do from the moon than it would be from Earth.

In addition, if we ever learn how to harness nuclear fusion, the moon turns out to be an excellent source of a version of the helium atom known as helium-3. This atom, which may be very useful in fusion reactors, has been raining down on the moon as part of the solar wind for billions of years. At the same time, Earth's magnetic field has deflected these atoms, causing them to miss us and flow outward into space. Whether settlers mine helium-3 or just plain old water and aluminum, the moon seems to have resources that will be valuable to future spacefarers. There could well be, in other words, economic reasons to colonize the moon and make use of its resources.

Given the kind of problems we have outlined, we can guess that the first Selenites will probably be like the scientists who travel to research stations in

THE LUNAR VILLAGE

Just to show how seriously many scientists are taking the possibility of a human presence on the moon, let's talk about a recent project taken on by the European Space Agency and the architectural firm of Skidmore, Owings, & Merril . (In case you don't recognize the name, SOM has designed some of the world's most impressive buildings, including the Burj Khalifa in Dubai, the world's tallest bui ding.) The task they set themselves was to design buildings that could be used to house a permanent lunar settlement. Their solution was a series of four-story inflatable habitats that would house living quarters, laboratories, and hydroponic farms for future inhabitants.

Antarctica today. They aren't planning to stay there, and they are willing to put up with inconveniences to be where they can do their job. Most likely there will be only a few at a time at first—the inflatable habitats discussed in the sidebar above are designed to hold only four people, for example. After that we can expect installations like South Pole Station and, eventually, tourist hotels and possibly even real colonies.

This means that the next time you see the moon in the sky, you may or may not be seeing a future home of humanity.

SKYLIST

MANY MOONS

Over the centuries, many names have been applied to the moon's appearance at different times. Some of the best known of these are:

■ **BLUE MOON**
There are 29.53 days in a lunar month, but 30 or 31 days in a calendar month, so it is possible to have two full moons in a single calendar month. The second full moon in a calendar month is called a blue moon. (An alternate definition for a so-called seasonal blue moon describes the third full moon in a season of four full moons.) Blue moons occur every two or three years, which is why the phrase "once in a blue moon" is used to describe unlikely events.

■ **HARVEST MOON**
This is the full moon closest to the autumnal equinox. It normally occurs in September (and occasionally in October), which is harvest time in the northern temperate regions. Before headlights on equipment, farmers used the extra light to harvest after sunset.

■ **WOLF MOON**
The wolf moon is the first full moon in January. The name apparently comes from the fact that farmers would hear wolves howling in midwinter.

Full moon in eclipse behind Chile's Paranal Observatory

■ **BLOOD MOON**
During a lunar eclipse, the moon acquires a reddish cast. This is a blood moon.

■ **SUPERMOON**
At times the moon's orbit brings it as close to Earth as it ever gets—we call this a perigee. At perigee, the moon appears larger in the sky than usual. The first new or full moon that meets this condition in a year is called a supermoon.

THE VIEW FROM EARTH

Our planet is the spinning, orbiting platform from which we view the heavens.

Today's understanding suggests that the solar system formed from the collapse of an interstellar cloud, with Earth and the moon created by a series of collisions in the collapsing disk. In the end, it was these events that created the world we live on today.

Think for a minute, though, about what it means to look at the sky from the surface of Earth. You are standing on a spinning platform that is moving around the sun in an elliptical orbit, and the planets you see are also moving in their own elliptical orbits. There is some pretty complicated geometry here—is it any wonder that it took people so long to figure it out? In what follows, we will look at some basic facts about Earth and see how they influence what we observe when we start stargazing.

EARTH IS ROTATING

The first thing that every mythological system tries to explain about the sky is the fact that the sun and moon appear to move from east to west on a daily basis. Whether we thought they were drawn by a chariot, carried by dragons, or moving on crystal spheres, celestial objects moved and we, standing on an unchanging Earth, watched them go by. Once Nicolaus Copernicus taught us that it was the sun and not Earth that was stationary, we were able to dispense with the chariots and dragons. Today we understand that the daily progression of celestial objects is a consequence of our own motion. Just as roadside objects appear to move from in front of you to in back of you as you drive by, objects in the sky appear to move because the ground we stand on is part of a rotating planet.

(Opposite) The European Southern Observatory's Very Large Telescope in Chile. The laser beam being shot into the sky is used to correct for changes in the motion of the atmosphere—part of a system called "adaptive optics."

Historically, the length of time it takes for Earth to complete one revolution—the day—and the time it takes Earth to complete its orbit—the year—have been our basic time standards. It turns out, however, that natural effects like the shifting of ground during an earthquake or the action of the tides can affect Earth's rotation. The tides, for example, increase the length of the day by a few milliseconds every century. Unfortunately for traditional Earth-

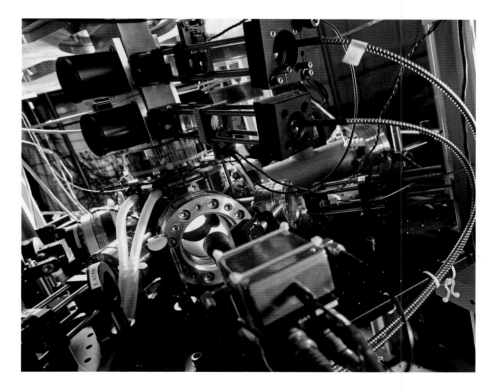

The motion of electrons in cesium atoms in a cesium clock (above) is used to define the second.

based time standards, by the start of the 20th century clocks could be built with enough accuracy to detect these sorts of changes. To everyone's consternation, our "time constants" turned out not to be as constant as we had thought. Consequently, in 1967, the second was formally defined in terms of the motion of an electron in a cesium atom—the atomic clock—and the motion of Earth was dethroned as our main timekeeper.

Telling time from the rotation of Earth leads to another problem. We have referred loosely to the

A composite, time-lapse photo shows the path of the sun across the sky during midsummer (top), autumn (middle), and midwinter (bottom).

When the sun returns to the same position above the ground, that is a solar day. When the stars return to the same position, that is a sidereal day: slightly shorter than a solar day.

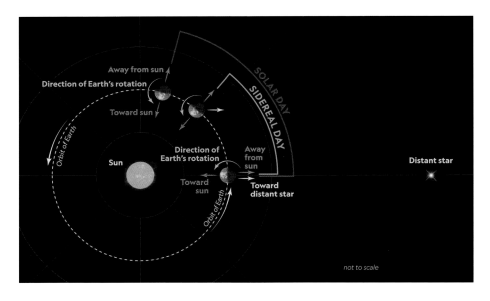

time it takes Earth to complete one rotation, but how do we determine that? We could, for example, start our clock ticking when a particular star is overhead and see how long it takes for that star to be overhead again. This would be what is called a sidereal day, which is 23 hours and 56 minutes long. If, on the other hand, you had started your clock when the sun was in a particular position in the sky, the fact that during the day Earth moves in its orbit means that you would have to let your clock run for four minutes more for the sun to be back where it

was at the start of your timing period. This is called a solar day, and it is 24 hours long.

There are, in other words, two definitions of "day," and humanity, by and large, has chosen the longer one.

EARTH'S AXIS IS TILTED

In the earliest phases of the beginning of the solar system, material that would later form the planets was spun out into a disk around the newborn sun. This means that after the eight planets formed, their orbits were in a single plane. This plane is called the ecliptic. (Technically, the ecliptic is defined to be the path that the sun appears to take across the sky over the course of a year as seen by an observer on Earth.)

If you think about the rough-and-tumble process of planetary formation, you will note that the rotation of any planet will most likely be determined by the last big collisions that take place during its formation. The speed of rotation and the direction of the axis of rotation of any planet, in other words, are basically the result of random processes. The tilt of Earth's axis with respect to the ecliptic is one possible outcome, but so is the rotation of Uranus, which lies on its side with its axis of rotation pointing toward the sun once each Uranian year. Venus, on the other hand, rotates "backward"—clockwise when viewed from above as opposed to the counterclockwise rotation of everything else. This sort of variation is perfectly normal.

The tilt of Earth's axis gives rise to all sorts of effects that the stargazer can see. For example, when the Northern Hemisphere is tilted away from the sun, an observer in that hemisphere will have to look farther down (i.e., farther south) to see the sun. The sun, in other words, will appear to be low in the southern sky. On the other hand, when the Northern Hemisphere is tilted toward the sun, the sun will appear to be higher in the sky. Thus, over the course of a year the sun in its path through the sky seems to move up (during the Northern Hemisphere summer) and down (during the Northern Hemisphere winter). Because the time that the sun is above the horizon changes while this is going on, the length of daylight varies as this cycle progresses. This apparent motion of the sun, of course, is a consequence of Earth's tilt and rotation—the sun doesn't really move at all as far as the planets are concerned.

In any case, over the course of a year an observer

TRACKING THE SUN

You can set up a simple set of observations to track the path of the sun in the sky as Earth moves around in its orbit. First, find a marker whose height will not change over a period of time: the top of a building, for example, or a flagpole in a large open square. Make sure you can see the place where the marker's shadow hits the ground at a specific time of day—high noon is a good choice, but any time will do Then every month or so go by and see where the shadow touches the ground at that specific time. You will see the shadow shorten as we move into summer and the sun climbs higher in the sky and then lengthen as we move into winter.

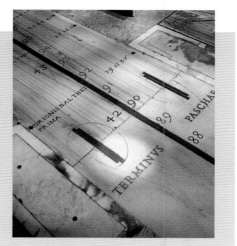

In 1702, Pope Clement XI inaugurated this meridian line, now in the church of Santa Maria degli Angeli in Rome. The sunlight from a small aperture in the wall falls on the meridian and indicates the date.

will see the sun's daily track in the sky move upward, stop and reverse itself, move downward, stop and reverse itself again, move upward and so on. The point at which the sun reverses itself is called a solstice (from "sun standing"). Solstices traditionally define the beginning of summer (when the

Aymara people in Bolivia celebrate a sunrise festival to mark their new year on June 21, 2013.

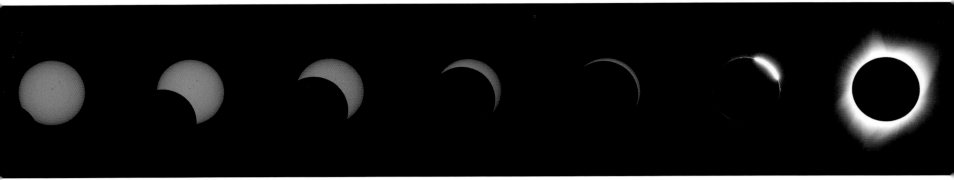

sun is high in the sky) and winter (when it is low). In between there will come a time when daylight and nighttime each last for 12 hours. This is called an equinox ("equal night"). Many folk traditions revolve around the summer equinox, or Midsummer's Eve.

One consequence of the tilt of Earth's axis is that parts of our planet experience seasons. When the Northern Hemisphere is tilted away from the sun, the energy emitted by one square foot of the sun's surface will fall onto more than a square foot of land or water on Earth's surface in that half of Earth. In this situation, the Northern Hemisphere receives less energy from the sun and therefore will be cooler. It is the Northern Hemisphere winter. Similarly, when the Northern Hemisphere is tilted toward the sun, more of the sun's energy falls on a given area of Earth's surface there, making it summer.

Another way of thinking about the apparent motion of the sun in the sky is to think about a point on Earth directly under the sun at high noon. This spot will move north and south over the course of a year. At the summer solstice, the point under the sun is as far north as the spot will ever go. This spot marks what we call the Tropic of Cancer. Similarly, the farthest south the point will move is called the Tropic of Capricorn.

THE PRECESSION OF THE EQUINOXES

Because it is spinning, Earth acts like a giant gyroscope. So does a child's top. We can, in fact, learn a great deal about Earth by considering the child's toy. Just as the axis of rotation of a top will move around in a lazy circle, so will the axis of rotation of Earth. This motion is called precession.

It takes about 26,000 years for Earth's axis to describe a full circle. Right now, for example, the axis points toward Polaris, the North Star. In 12,000 years, it will point toward the star Vega, and in

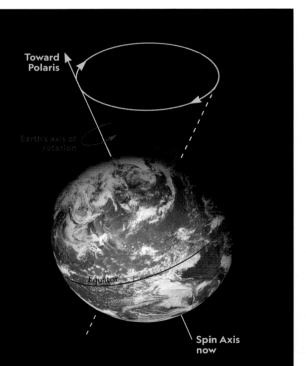

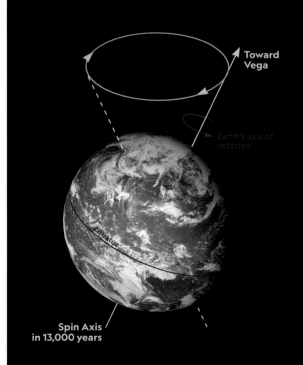

Because Earth's axis moves in a slow circle over time, it will eventually point to a new North Star, Vega.

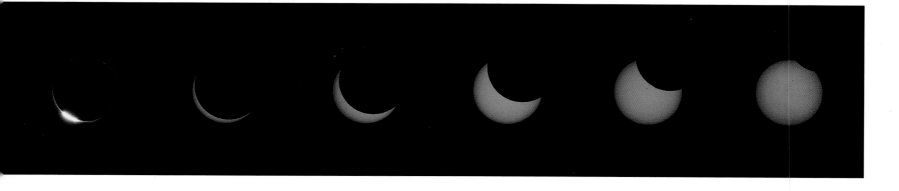

24,000 years, it will point toward Polaris again. This motion can be detected by observations that span centuries, and it was well known to ancient Greek (and possibly Babylonian) astronomers.

ECLIPSES

There is probably no more awe-inspiring sight in the sky than a solar eclipse. The familiar sun is replaced by a dark circle with a flaming edge and the world seems to stand still. Eclipse events arise because of a special arrangement of Earth, the sun, and the moon. Basically, a solar eclipse happens when the moon comes between the sun and Earth, blocking the light from the sun. In effect, Earth moves into the moon's shadow.

It's important to realize that the moon always casts a shadow, which means that there is always a solar eclipse going on somewhere. The types of eclipse we're interested in, however, are the ones when the shadow falls on Earth. If the plane of the moon's orbit was the same as the plane of Earth's, we would have a solar eclipse every new moon. However, the moon's orbit is tilted about five degrees with respect to the ecliptic, which means that for

an eclipse to be visible from the surface of Earth the moon must not only be between Earth and the sun but must be in the ecliptic plane as well. If the moon happens to be a little above or below the ecliptic, its shadow will miss Earth. Because of these constraints, there are usually two (and never more than five) solar eclipses visible from Earth's surface each year.

Depending on the relative distances from the moon to Earth and the sun, an eclipse may be total (i.e., the moon may block out the entire solar disk), partial (the moon blocks out only a part of the solar disk, basically taking a chunk out of it), annular (the moon is too far away to cover the entire disk, so a ring of solar light remains visible), and, more rarely, hybrid (an eclipse starts annular and progresses to total).

There is another kind of eclipse—one somewhat less dramatic than a solar eclipse. It happens when Earth comes between the sun and the moon. Since the moon shines by reflected sunlight, when the moon enters Earth's shadow, the portion of the lunar disk in Earth's shadow will stop shining. Such an event is called a lunar eclipse. Note that unlike

Phases of a total solar eclipse, as seen above Madras, Oregon, in 2017

For more on eclipses, see pages 347–348.

THE SOLAR ECLIPSE

The first thing to know about observing a solar eclipse is that you should *never look directly at the sun*. Doing so can cause serious and permanent damage to your eyes. With a suitable filter, however, you will see the following events:

- **First contact:** A curved chunk begins to be taken out of the solar disk.
- **Diamond ring:** Just before totality, you may see a bright light at the edge of the disk. This is caused by light being channeled through valleys on the moon.
- **Totality:** Light from the sun's disk is completely blocked and the solar corona is visible. During totality, take a moment to glance at the horizon. Its reddish color is caused by the fact that light has to travel a long way through the atmosphere to your eye—the same effect that causes sunsets to be red.

Children from an elementary school in Avon, Colorado, observe the 2017 eclipse.

An image of the Tarantula Nebula assembled with data from orbiting optical, infrared, and x-ray telescopes

a solar eclipse, which is typically visible to only a small number of people, a lunar eclipse will be visible to all observers on the part of the planet facing away from the sun.

ORBITING OBSERVATORIES

No matter how refined their instruments may be, astronomers using telescopes on Earth's surface face many obstacles. For one thing, all radiation from the heavens except for visible light and radio waves is absorbed by the atmosphere before it can get to the ground. For another, as the light from a star passes through the atmosphere, it is diffracted and bent by moving pockets of air at different temperatures. This produces the familiar twinkling appearance of stars and sets limits on the accuracy of astronomical measurements. Thus, there is a lot to be gained by putting our observatories in space, above the atmosphere.

The easiest place to put an observatory in space is in Earth orbit. The Hubble Space Telescope—arguably the most productive astronomical instrument ever built—orbits about 547 kilometers (340 mi) above Earth's surface. Being this close to Earth, the telescope can be upgraded and repaired by astronauts. These missions have allowed the Hubble to keep functioning for decades beyond what might have been expected when it was launched in 1990.

There are, however, other places to park satellite observatories. It turns out that there are five places in the Earth-moon system where the gravitational forces balance in such a way that an object placed at one of these points will follow the motion of Earth and the moon. These spots are called Lagrange points, after the Italian-French physicist and mathematician Joseph-Louis Lagrange. In the jargon of physicists, these are points of stability—if something is placed at one of these points, it will stay put. Note that there are five Lagrange points for the Earth-moon system, as shown at right, and five more for the Earth-sun system. That gives us lots of places to park satellites out there!

Actually, the points labeled L1, L2, and L3 are places where the equilibrium is unstable—that is, if an object in one of these places is moved slightly it will not return to equilibrium. Think of a marble balanced on top of a watermelon. The position of a satellite at any of these points has to be adjusted every 23 days or so to keep it from drifting away. Nevertheless, the advantages gained from locating observatories at L1 and L2 are enough to make them prime locations for observatories. In these locations, for example, telescopes can deploy a solar

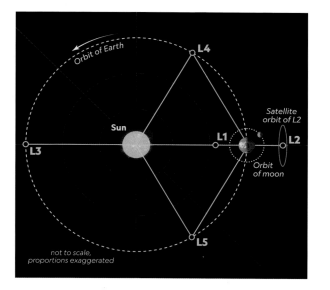

Lagrange points are locations where the gravitational forces of Earth and the sun allow objects to be parked. In the diagram, they are shown as points L1–L5. The James Webb Telescope orbits at the point known as L2.

shield to block out light from the sun, Earth, and the moon. The James Webb Space Telescope—the successor to the Hubble—is located at L2 in the Earth-sun system, for example. For all the advantages of this location, however, the telescope is a million miles from Earth. This is far too distant for it to be visited by astronauts should something need to be repaired. NASA engineers have to get everything right the first time!

The Hubble Space Telescope in orbit, in a photo taken by an astronaut on the 2009 servicing mission

An Atlas V rocket carrying the
Perseverance rover begins its trip to
Mars from Cape Canaveral Air Force
Station, Florida, on July 30, 2020.

CHAPTER THREE

THE SOLAR SYSTEM

OUR PLANETARY FAMILY

One particularly dark night, more than 2,000 years ago, the moon crept into Earth's shadow—and Greek astronomer Aristarchus of Samos was watching. Using the time it took for the moon to pass through this zone of darkness, Aristarchus calculated the relative sizes and distances of the moon, Earth, and sun. The records of these remarkable observations are among the few bits of information about Aristarchus that have survived through the millennia. But some scholars believe that the study of these cosmic ratios led him to an even more groundbreaking revelation: Earth was not stationary, but rather orbited the smoldering sun.

Though the idea might seem elementary today, it took centuries before it caught on among other scholars. Polish astronomer Nicolaus Copernicus revived this sun-centric view—known as a heliocentric system—just before his death in 1543 when he published a book that declared "the sun is the center of the universe." This concept lies at the heart of our solar system's story, from its birth some 4.6 billion years ago to its evolution into the cosmic setup we know today.

While many other systems of planets have been found orbiting distant stars, ours is the only one with the name "solar." This moniker comes from the name of the life-giving sun at its center, around which the eight planets orbit mostly in one plane—Mercury, Venus, Earth, Mars, Jupiter, Saturn, Uranus, and Neptune. Tiny worlds—including asteroids, meteoroids, comets, moons, and dwarf planets—dot the space around the bigger planets. (Pluto is now a dwarf planet, losing its official planetary status in 2006.)

Hundreds of robotic spacecraft have explored the riches of the solar system beyond our home world. These missions have opened scientists' eyes to a variety of alien landscapes once unimaginable: rocky hellscapes hot enough to melt lead, volcanoes that tower many miles into the sky, glaciers of exotic non-water ice, winds that whip into Earth-size hurricanes, and much more.

Let's take a journey back to a tumultuous time in our solar system's history, some 4.6 billion years ago, when our sun was just starting to form. Then we can zip through the system as we know it today to visit its many wild worlds—from the hellish landscape of Mercury to the icy objects of the Oort cloud.

The Milky Way galaxy forms a hazy band of light across clear nighttime skies above Yellowwood Lake in Indiana.

ORIGINS OF THE SOLAR SYSTEM

A swirling cloud of gas and dust birthed our sun, as well as the sundry worlds of the solar system.

The story of our solar system began nearly 4.6 billion years ago, when a swirling cloud of dust and mostly hydrogen gas collapsed under its own gravity. The event may have been propelled by the arrival of shock waves from a nearby supernova, which sent ripples out into space.

Over millions of years, the gas and dust raced inward as a result of an increasingly strong gravitational tug, forming a core that grew ever hotter and denser. Eventually, the pressure in the core became so high that it sparked reactions known as nuclear fusion. During this process, atoms of hydrogen combined into helium, releasing a gargantuan amount of energy and the first light that illuminated our cosmic neighborhood. Our sun was born.

The sun would come to contain more than 99.8 percent of the solar system's mass. As the remaining cloud shrank, it also began to spin faster and faster, like a high diver tucking into a tight ball. Bits of the remaining matter in this spinning nebula began to merge, forming a collection of growing clumps that flattened out into a ring. These early processes laid the foundation for the structure of our solar system today, setting planets yet to be formed on their nearly circular orbits in the same direction and plane around our central star.

ROCKY WORLDS AND GAS GIANTS

Scientists are still attempting to pin down the specifics of how each world in our solar system came to be. But it's thought that the dusty clumps within our system's protoplanetary disk combined, which turned once microscopic tidbits into planetesimals, or the building blocks of planets, just a few miles wide. The formation of our planets within this disk

is the reason why they all orbit in nearly the same plane today.

Scientists got a close-up look at one of these ancient building blocks in January 2019, when NASA's New Horizons spacecraft flew by the snowman-shaped Arrokoth (2014 MU69). This space rock orbits at the edge of our solar system beyond Neptune and appears to be two orbs joined

■ For more on Arrokoth, see page 365.

In the early solar system, a nebula surrounded the sun (top left). The gas and dust condensed into a disk (center) that eventually formed planetesimals (bottom).

at the waist. By studying its geology and structure, scientists crafted computer simulations of its formation, which revealed that Arrokoth grew not from a high-speed collision but instead from a leisurely dance between the two spheres. They circled one another, growing ever closer, until eventually merging into one 35-kilometer-long (22-mi) space rock. The find hints that other planetesimals formed from similar choreography.

As the planetesimals grew larger, so did their gravity, drawing in more material. Some of these rocky clumps eventually grew to planets—but not all.

(Opposite) An artist's conception of the newborn sun's earliest days, when it sat within a swirling disk of gas and dust

The asteroid belt, which lies between the orbits of Mars and Jupiter, is littered with rubble from failed planets. Other rocky balls became moons or meteoroids, or even comets, if they were particularly icy.

Our solar system has two types of planets: small, rocky worlds and gaseous giants. The rocky worlds orbit closest to the sun, in a range that was likely too hot for volatile molecules such as water and methane to linger during planetary formation. The growing bodies in this zone were built of metals, such as iron and nickel, as well as silicates that eventually became the "terrestrial" planets—Mercury, Venus, Earth, and Mars.

As the rocky planets formed, they grew quite hot, which drove the process known as differentiation. While parts of differentiation remain a mystery, the result is a separation of a planet into layers, with heavier metals collecting at the planet's core and the lighter magnesium-rich silicate rocks enveloping the center in a thick layer known as the mantle. Other silicates rich in elements such as sodium or potassium make up the outermost layer, known as the crust.

Farther out in the solar system, where temperatures were cool enough for volatile molecules to linger, are the gas and ice giants. They likely began growing early in the protoplanetary disk, accumulating gases and water ice in remote regions far from the sun's hot winds. But not all planets necessarily stayed where they first formed. Some recent models suggest that Jupiter took shape beyond the current orbits of Neptune and Pluto, then migrated inward to its location today.

The solar system sprawls much farther than its last official planet, Neptune. An array of rocky oddballs orbits in the Kuiper belt, which includes the famous dwarf planet Pluto. Beyond the Kuiper belt lies the icy collection of objects known as the Oort cloud, which forms a shell around our solar system.

This solar system illustration shows the relative planet sizes, and their orbits in astronomical units (AU) at a condensed scale.

Our solar system orbits the Milky Way's center **every 230 to 250 million years.**

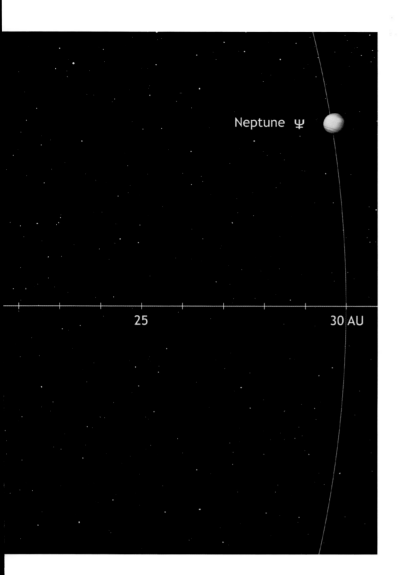

Neptune ♆

25   30 AU

ters (864,300 mi) across. If Earth were a blueberry, the sun would be wider than most refrigerators are tall. Even the largest planet in our solar system, Jupiter, is 10 times smaller than the sun.

INVESTIGATING THE PLANETS

For thousands of years, humans have cast their eyes to the sky, untangling the wonders of our cosmic neighborhood. But humanity entered a new era of exploration in 1957 when the Soviet Union launched Sputnik 1, the first of many artificial satellites to orbit Earth. Many space firsts have followed.

■ For more on early space exploration, see pages 78–80.

The young dog Laika became the first animal to orbit Earth in 1957, though the journey ended in tragedy soon after launch when a failure led to her overheating. Russian cosmonaut Yuri Gagarin became the first human to travel to space in April 1961. American astronaut Neil Armstrong was first to set foot on the moon in July 1969. The mission also brought back the first samples ever collected from another body in our solar system—a whopping 21.6 kg (47.6 lbs) of rocks, dust, and soil that is still studied today. Scientists have also sent spacecraft to explore every planet in our solar system, as well as asteroids, moons, comets, and dwarf planets.

These missions taught scientists around the world immense amounts about our cosmic neighborhood and beyond. Next up, we'll look at the ever sharpening portraits of each member of our cosmic family, starting with our solar system's innermost planet, Mercury.

We've yet to directly observe this region, but scientists have predicted its presence through mathematical models—and they believe we've already met some emissaries of the Oort cloud in the form of passing comets.

The Oort cloud stretches as far as our sun's gravitational reach likely extends, capturing bodies for trillions of miles. Yet some define the edge of our solar system as slightly closer to home, at the border of the heliosphere. This protective bubble is driven by solar winds and encompasses the sun and all the planets, with its closest side hovering some 120 AU (1.8 billion km/1.1 billion mi) from the sun.

No matter how its edge is defined, the vastness of our solar system is difficult to comprehend. The distance between the sun and Earth is 150 million kilometers (93 million mi), which is just a smidge shorter than 3,740 times around our planet's Equator. Neptune orbits at about 30 times this distance from the sun. The differences in the relative sizes of solar system bodies are also quite great. The sun is by far the biggest at an average of 1,391,000 kilome-

During the Late Heavy Bombardment, about 3.8 to 4 billion years ago, a destructive shower of space rocks rained down on Earth.

SOLAR SYSTEM IN ACTION

The astronomer Nicolaus Copernicus theorized in 1543 that Earth orbited the sun, and scientists have been expanding our knowledge of the solar system ever since. Digital technology has provided a significant boost in recent decades, helping us identify previously unknown, heliocentric, and highly dynamic objects that shed new light on how our solar system evolved—and how it can be orderly and chaotic at the same time.

PLANET SIZE, SPACING, AND ALIGNMENT ARE NOTIONAL.

Lurking danger
Objects crossing Earth's orbit are potentially hazardous. Experts are studying how to use spacecraft to divert them before they impact Earth.

ROCKY Vesta

CARBON-RICH Ceres Psyche

Banding together
Clusters of asteroids with similar orbital properties, known as asteroid families, are evidence of ancient collisions. They orbit in bands with others of the same composition.

Sun

Mercury

Venus

Earth
1 AU
One astronomical unit (AU) equals the distance between Earth and the sun.

Mars

Minding the gap
Astronomers once thought a planet might be found here. It's home to a ring of minor planets, called asteroids.

Jupiter's gravity creates spaces in the belt known as Kirkwood gaps

ASTEROID BELT

MAIN ASTEROID BELT

HILDA ASTEROIDS

ROCKY RING
MAIN ASTEROID BELT

More than a million bits of space rubble—and counting—orbit the sun in the space between Mars and Jupiter known as the main asteroid belt. These mostly rocky and metal leftovers hold clues to the formation of the planets some 4.6 billion years ago.

FROST LINE

Jupiter
5.2 AU

PLANET FORMATION
Swirling material collides, creating the beginnings of planets, called planetesimals. Gravity draws these objects together; the greater the mass, the greater the force of attraction.

Fragments from catastrophic impacts are held together by gravity as collections of rubble.

Basalt
Stone
Metal

More collisions meant small bodies broke off from parent planetesimals.

The biggest of all
Jupiter, a massive ball of hydrogen and helium, exists almost like a mini solar system. It has four inner planet-size moons and swarms of outer moons.

Planetesimals likely began as tiny clumps of dust that gradually [grew]; experts theorize [they] may have been "born big" when gravity forced clumps together.

In some planetesimals minerals separated out, forming distinct layers as gravity continued to shape them. Icy layers formed around cores found beyond the frost line.

Metallic cores formed if heated by radioactive decay or collisions.

These planetary embryos continued to grow into planets if debris was available. If not, they remained as large asteroids, dwarf planets, or moons.

Saturn
9.5 AU

Moons, oceans, and rings
Captured by gravity, multiple moons and rings of dust and rock orbit around each of the outer planets. New observations of moons and dwarf planets indicate many have oceans under ice.

COLLISION AND CREATION
Our star, the sun, was formed by the collapse of an interstellar cloud. In the wake of its creation a volatile and rotating disk of dust and gas was left behind—the beginnings of our solar system.

Once the planets took shape, the chaos was not over. Interactions with shifting planetesimals pushed Jupiter and Saturn around. A possible fifth gas giant may have been ejected.

formation: 4.6 billion years ago

Rocky material

Possible fifth gas giant

Neptune

Uranus

Icy material

migration: 4.1 to 3.7 billion years ago

Scatter

Planet migration

Ejected

Gravitational interaction with planetesimals

Uranus

Neptune and Uranus may have swapped positions

present: 3.7 billion years ago to today

SUN

M V E M

AU

5 AU

15 AU

20 AU

URANUS

TERRESTRIAL PLANETS
Craters, canyons, and volcanoes mark these rocky, windswept planets. Earth's water may have come from asteroid collisions.

JUPITER
Twice as massive as all the other planets combined, its dark rings come from meteoroids striking its moons.

Tipped on its side from a suspected collision, Uranus rotates from east to west (as does Venus).

INNER PLANETS
ROCKY CRUSTS, SOLID CORES, NO RINGS

OUTER PLANETS
ICE AND GAS WITH SOLID CORES, MANY MOONS AND RINGS

DIAGRAMS ARE TO SCALE.

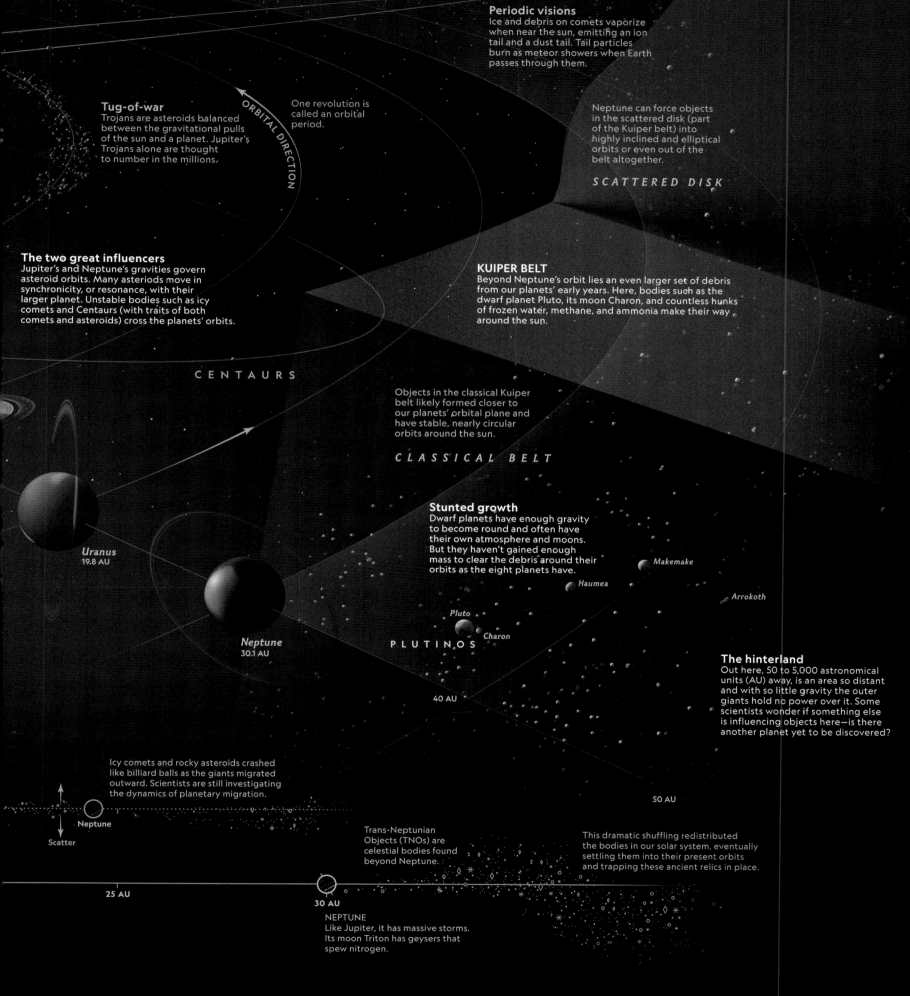

Periodic visions
Ice and debris on comets vaporize when near the sun, emitting an ion tail and a dust tail. Tail particles burn as meteor showers when Earth passes through them.

ORBITAL DIRECTION

One revolution is called an orbital period.

Tug-of-war
Trojans are asteroids balanced between the gravitational pulls of the sun and a planet. Jupiter's Trojans alone are thought to number in the millions.

Neptune can force objects in the scattered disk (part of the Kuiper belt) into highly inclined and elliptical orbits or even out of the belt altogether.

SCATTERED DISK

The two great influencers
Jupiter's and Neptune's gravities govern asteroid orbits. Many asteriods move in synchronicity, or resonance, with their larger planet. Unstable bodies such as icy comets and Centaurs (with traits of both comets and asteroids) cross the planets' orbits.

KUIPER BELT
Beyond Neptune's orbit lies an even larger set of debris from our planets' early years. Here, bodies such as the dwarf planet Pluto, its moon Charon, and countless hunks of frozen water, methane, and ammonia make their way around the sun.

CENTAURS

Objects in the classical Kuiper belt likely formed closer to our planets' orbital plane and have stable, nearly circular orbits around the sun.

CLASSICAL BELT

Stunted growth
Dwarf planets have enough gravity to become round and often have their own atmosphere and moons. But they haven't gained enough mass to clear the debris around their orbits as the eight planets have.

Makemake

Haumea

Arrokoth

Uranus
19.8 AU

Pluto

Charon

Neptune
30.1 AU

PLUTINOS

The hinterland
Out here, 50 to 5,000 astronomical units (AU) away, is an area so distant and with so little gravity the outer giants hold no power over it. Some scientists wonder if something else is influencing objects here—is there another planet yet to be discovered?

40 AU

50 AU

Icy comets and rocky asteroids crashed like billiard balls as the giants migrated outward. Scientists are still investigating the dynamics of planetary migration.

Neptune

Scatter

Trans-Neptunian Objects (TNOs) are celestial bodies found beyond Neptune.

This dramatic shuffling redistributed the bodies in our solar system, eventually settling them into their present orbits and trapping these ancient relics in place.

25 AU

30 AU

NEPTUNE
Like Jupiter, it has massive storms. Its moon Triton has geysers that spew nitrogen.

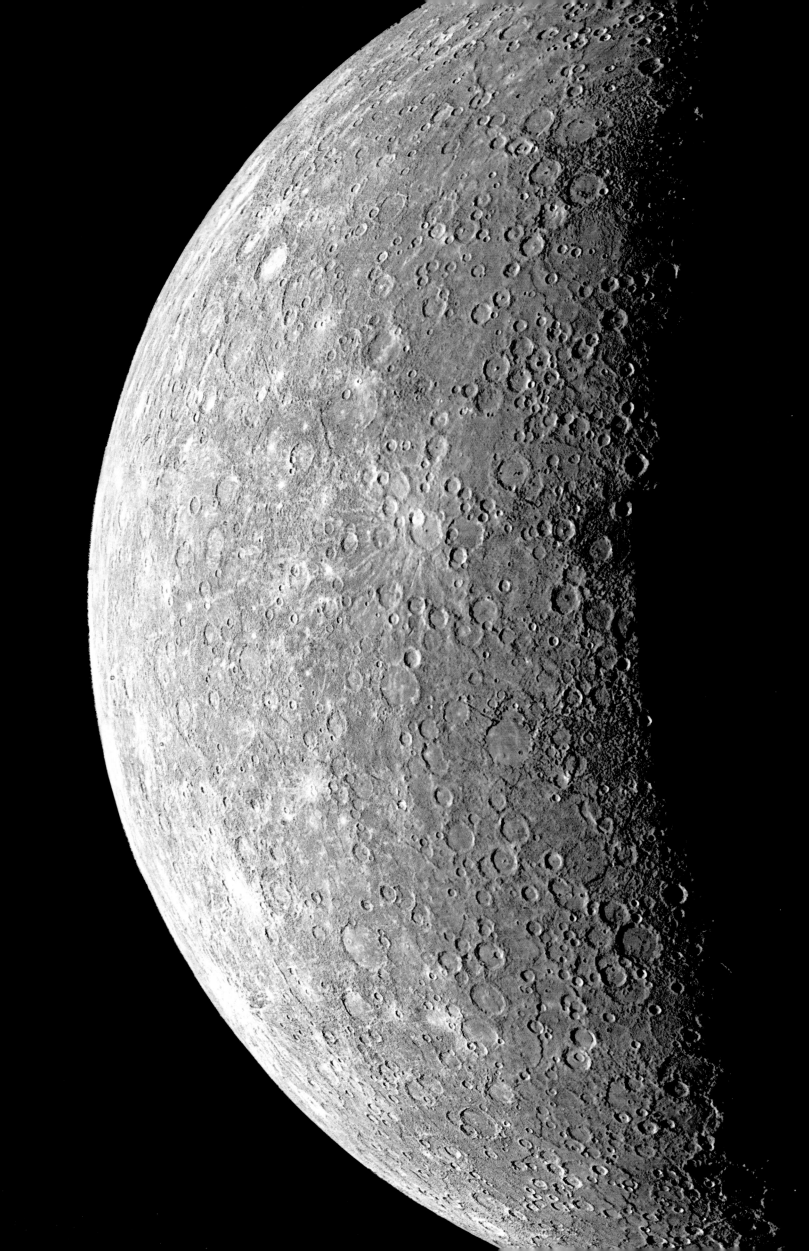

MERCURY

While the solar system's innermost planet may be the smallest, it offers up big intrigue.

I f you could watch daybreak from the surface of Mercury—the closest planet to the sun—you'd be in for an unusual treat: a double sunrise. From many places on the tiny orb's surface, the sun rises over the horizon, then appears to reverse course in a short-lived sunset, before finally rising again. Another double feature can happen at sunset: The sun sets, rises, then sets again.

The spectacle is the result of Mercury's zippy orbital speed and slow rotation, combined with its egg-shaped path around the sun. The planet rotates once every 59 Earth days, but it orbits the sun faster than any other world in our solar system, completing one trip every 88 Earth days. This speed probably earned Mercury its name, taken from the wing-footed messenger to the Roman gods. When the planet moves through the point in its orbit closest to the sun, known as the perihelion, the speed of its orbit exceeds that of its rotation and in its sky the sun moves in retrograde—an illusion that makes the sun appear as if it's traveling backward.

Standing on the surface of Mercury, however, is not possible, at least with today's technology. The planet's close position to the sun means temperatures can rise to a scorching 430°C (800°F) at the equator during the day. The planet also lacks the thick blanket of an atmosphere, which traps heat and distributes it across the surface of some planets. This means that nighttime temperatures on Mercury drop to around -180°C (-290°F).

OBSERVING MERCURY

Its nearness to the sun also makes Mercury notoriously difficult to study from Earth. While the planet is one of six that can be spotted in the sky with the naked eye, it only graces Earth's darkened horizon

(Opposite) Mercury's cratered surface was captured by NASA's Mariner 10 in 1974, the first spacecraft sent to study the planet.

Average distance from sun: 57.9 million kilometers (36 million mi)

Mass: 0.055 x Earth's

Length of day (one rotation): 58.7 Earth days

Length of year: 88 Earth days

Average temperature: 167°C (333°F)

Average radius: 2,440 kilometers (1,520 mi)

Number of moons: 0

Planetary ring system: No

for a short time, setting soon after sunset and rising just before sunrise. Daytime viewing is possible but is challenging because of the risk that intense solar rays could fry sensitive equipment.

Yet the tiny planet has likely been seen by earthbound viewers for thousands of years, from the time our ancient ancestors first looked to the skies. Mercury is represented in many ancient writings, with Babylonian and Assyrian cuneiform tablets among the oldest.

Continued on page 106

Artwork depicts a typically intense sunrise on Mercury, with Earth and Venus appearing as bright stars to the upper right.

MERCURY

Impact craters speckle Mercury's surface, evidence of the planet's lack of a thick, protective atmosphere. Without water and tectonics, Mercury also lacks forces to erase the scars of past collisions.

CALORIS PLANITIA

The map above highlights Caloris Planitia, a sprawling plain within the planet's largest impact crater.
Created by an ancient collision, the crater spans about 1,550 kilometers (963 mi).

SCALE 1:32,832,000
1 CENTIMETER = 328 KILOMETERS; 1 INCH = 518 MILES

STATUTE MILES
KILOMETERS

A 2013 image of Mercury's Kuiper crater highlights its strikingly different terrains: smooth zones from impact melt and rugged central peaks.

Continued from page 103

Just as the sun occasionally appears to reverse motion in Mercury's sky, three to four times each year Mercury seems to move backward in our own sky, trekking westward rather than in its usual eastward direction. This retrograde motion phenomenon plays a prominent role in astrology and has been blamed by astrologers for a host of problems, from computer crashes to terrible traffic. But there is no scientific basis for such claims. Instead, the about-face is an optical illusion that stems from the differing speeds of Earth and Mercury as they circle around the sun. Mercury races around the sun faster than our home planet, so

it will eventually catch up to and pass Earth, causing the orb to appear as if it is briefly moving backward across Earth's skies.

EXPLORING MERCURY

The first detailed observations of Mercury started with a bright streak in the Floridian night sky at 12:45 a.m. local time on November 3, 1973, when the Mariner 10 spacecraft was launched into space. The craft flew for three months en route to Venus, where it used that planet's gravity to slingshot over to Mercury.

While the mission wasn't without its technical difficulties, the craft eventually returned more than 7,000 images of not only Mercury but also Venus, Earth, and the moon. Mariner 10 revealed the complexity of Mercury's twisted and cratered surface. Among the many striking features discovered was Caloris Basin, a 1,550-kilometer (963-mi) crater surrounded by mountainous ridges, created by the impact of an ancient space rock.

After this momentous first, however, Mercury fell to the space exploration wayside. The next venture to explore this tiny world didn't launch until 2004, with the MESSENGER (MErcury Surface, Space Environment, GEochemistry and Ranging) spacecraft. This robotic voyager was the first to enter Mercury's orbit, studying the planet for more than four years. Among its many accomplishments was helping to unravel the planet's geologic history and confirming the presence of water ice at its poles.

MERCURY'S GEOLOGY

Much like our lunar neighbor, Mercury is riddled with pockmarks from meteoroids, asteroids, and comets colliding with its surface. The force of each impact expelled bursts of pulverized rock that show up as bright rays radiating from the crater.

Caloris Basin is the largest impact crater yet found on Mercury. It was probably created when an asteroid at least 100 kilometers (62 mi) wide careened into the planet's surface. The force of the impact can even be seen on the opposite side of the planet in a jumble of unusual hills and fractures known as the "weird terrain." Scientists think that this strange layout could have come from the impact's shock waves,

SEE FOR YOURSELF

INSCRIBED IN STONE

People have gazed upon Mercury for thousands of years, with the earliest skywatchers likely mistaking the planet's appearance in the morning and evening skies as the glow from two different stars. But experts believe that by at least 1500 B.C. Babylonian and Assyrian astronomers realized that these winks of light came from the same orb, which we now know as Mercury. The MUL.APIN tablets—one of which is shown here—contain some of the earliest written descriptions of Mercury's movements. Like ancient encyclopedias, the tablets reflect knowledge accumulated over many years, which was then copied for populations across Babylonia and Assyria. The oldest surviving copies of these tablets date to roughly 700 B.C., but when and where the first one was made remains uncertain.

The first clay tablet of the series MUL.APIN, compendiums of ancient knowledge about the heavens

which ricocheted through the tiny world and eventually focused on one spot with enough energy to buckle the ground. Alternatively, the terrain could come from the collision of ejecta from the impact that traveled around the planet, meeting on the opposite side.

Like Earth, Mercury is thought to have a solid metallic inner core surrounded by a molten outer core. But Mercury's core makes up nearly 85 percent of the planet's volume. The churning molten metal may generate the planet's magnetic field, but the result is oddly lopsided: The field is three times stronger in the north than in the south.

More information about Mercury is soon to come. BepiColombo—a joint effort between the European Space Agency and the Japan Aerospace Exploration Agency—launched in 2018. Two spacecraft make up the mission: the Mercury Planetary Orbiter (MPO) and the Mercury Magnetospheric Orbiter (Mio). The pair got their first glimpse of Mercury up close in October 2021, but they won't enter orbit around the planet until the end of 2025. The many years of travel required for BepiColombo to enter orbit is due to the crafts' great speed. If they went straight to Mercury, the BepiColombo spacecraft would sail right by. Instead, they shed speed by swinging around Venus and Mercury until the planet's gravity can capture them. BepiColombo will brave temperatures up to 350°C (662°F) for years, offering fresh insights into this tiny world.

An artist's impression of MESSENGER, which traveled six and a half years from Earth to become the first spacecraft to orbit Mercury

VENUS

The second planet from the sun may once have been
similar to Earth, but today it's an inferno.

When you gaze up at the stars on a dark evening or early morning, chances are you'll catch a glimpse of the piercing glow from the second planet from the sun, Venus. Named after the Roman goddess of love and beauty, this alien world is the brightest object in the night sky, apart from the moon, thanks to its highly reflective, pale yellow shroud of sulfuric acid clouds.

A STIFLING ATMOSPHERE

Underneath that cloudy blanket lies a hellishly hot world. The planet's hefty atmosphere is largely made up of the heat-trapping gas carbon dioxide. Temperatures at Venus's surface are a sizzling average of 464°C (867°F), hot enough to melt a hunk of lead—making Venus the hottest planet in our solar system despite not being the closest to the sun. Heat on Venus persists day and night and throughout each year. The planet tilts only about three degrees on its axis, which means fall and winter offer no reprieve from the punishing heat. The dense atmosphere also leads to surface pressures about 90 times that of Earth's, which would make visitors feel as though they were swimming half a mile deep in the ocean.

Hurricane-force winds whip the top levels of the atmosphere around the planet at speeds of 360 kilometers an hour (224 mph), sending the clouds whizzing around the planet every four Earth days, a process known as super-rotation. Yet things move a bit slower at Venus's surface. Not only do the winds die down, but the planet itself rotates at stunningly slow speeds, taking 243 Earth days for each rotation. This means a day on Venus is longer than a Venu-

Average distance from sun: 108 million kilometers (67.2 million mi)

Mass: 0.815 x Earth's

Length of day: 243 Earth days (retrograde rotation)

Length of year: 225 Earth days

Average temperature: 464°C (867°F)

Average radius: 6,050 kilometers (3,760 mi)

Number of moons: 0

Planetary ring system: No

sian year, which takes 225 Earth days to complete.

To add to the list of oddities about this second planet out from the sun, it rotates backward (retrograde) compared to Earth and other planets. Scientists aren't sure how the planet started on its reverse spin. Perhaps an ancient collision flipped it upside down, or its own internal friction and other forces caused it to slow and then reverse course.

Continued on page 112

Hinode, a joint JAXA and NASA mission, snapped this image of Venus's transit in 2012 as the planet passed between Earth and the sun.

(Opposite) An image from the Magellan spacecraft shows layers of volcanic activity and a trio of pristine-looking craters in a region dubbed the "Crater Farm."

VENUS

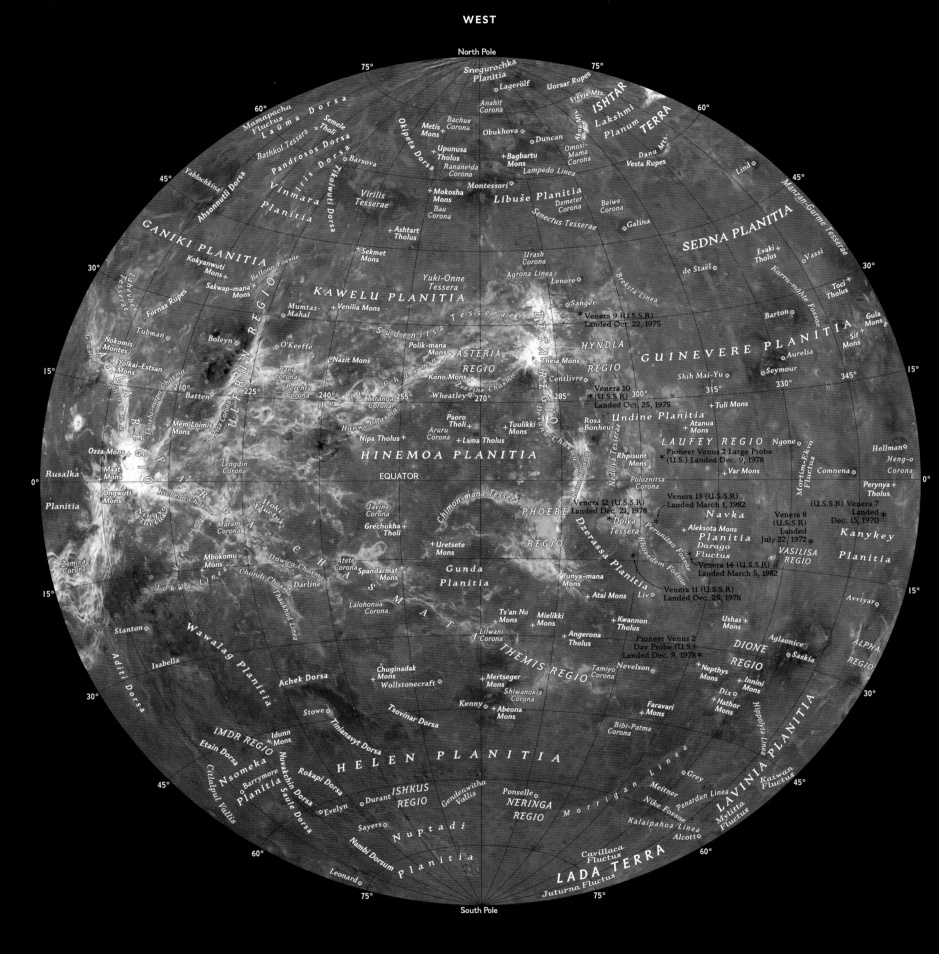

Many of Venus's surface features are sculpted by volcanism, such as the winding bright zones
dubbed Atla Regio and Beta Regio in the map above.

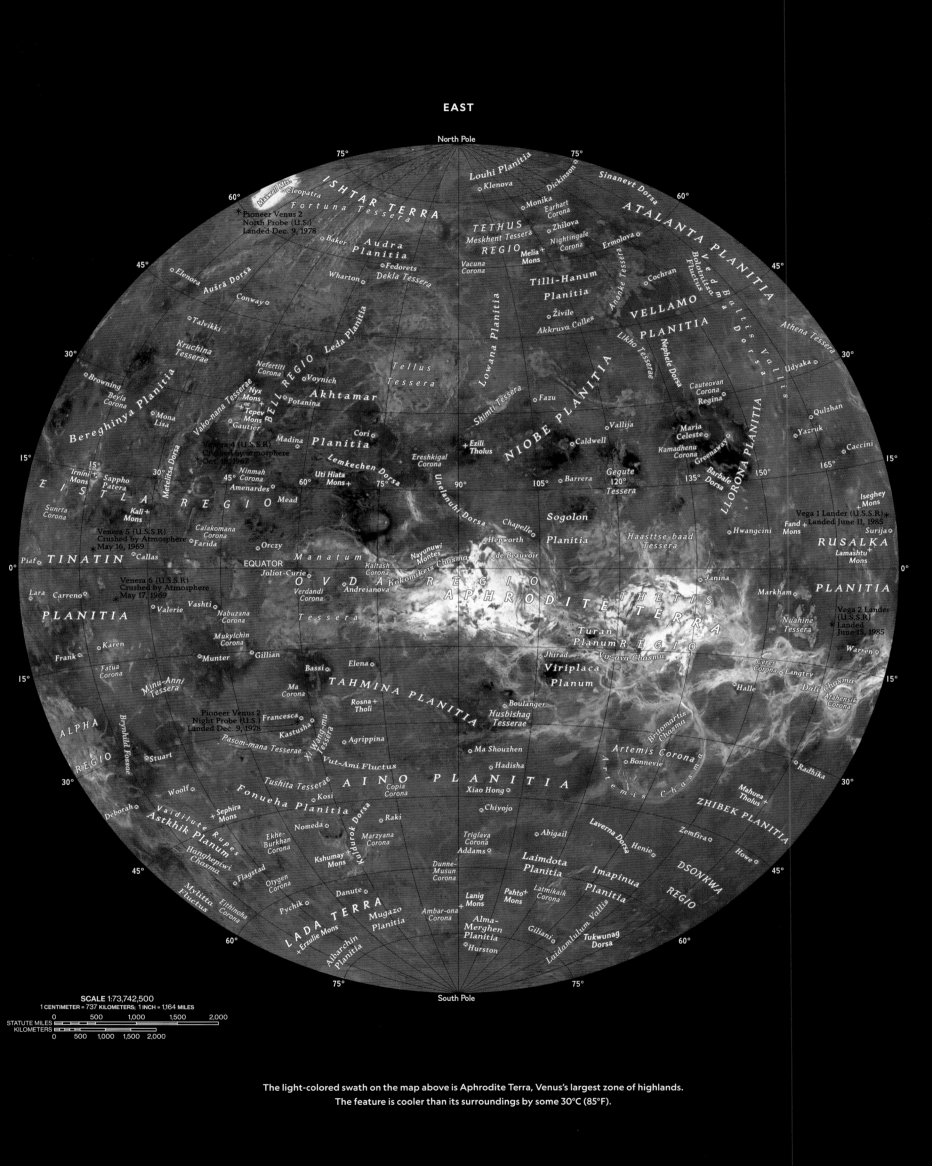

The light-colored swath on the map above is Aphrodite Terra, Venus's largest zone of highlands.
The feature is cooler than its surroundings by some 30°C (85°F).

A stargazer peers up at a hazy Venus while standing between pillars of volcanic rock at the Giant's Causeway in Northern Ireland.

Venera 13 landed on Venus's surface in March 1982 and survived the harsh conditions for more than two hours, snapping the picture below (with part of the spacecraft visible in the foreground).

EARLY VENTURES TO THE INFERNO

When space agencies set their sights on exploring the cosmos, Venus was an early target. The year 1962 marked a momentous first for interplanetary exploration, when NASA's Mariner 2 whizzed by Venus, scanning the planet for 42 minutes and beaming back data on its atmosphere and surface.

The Venera probes of the Soviet Union, which flew from the 1960s to the 1980s, tried to get an even closer look, but the intense Venusian atmosphere crushed many of these craft as they headed toward the surface. Venera 3 earned the title of first spacecraft to touch down on another planet, but its communication systems failed in the weeks before its landing, so it returned no information on its surroundings. Venera 4 was the first to take in-place measurements of the atmosphere of a planet other than Earth, telling of skies flush with carbon dioxide.

In 1970, Venera 7 set mechanical foot on Venusian rock. Though the landing was rough, the probe beamed back a weak stream of data that included measurements of the surface temperature and pressure—the first time a craft transmitted data from the surface of any alien planet.

DWINDLING VENUS VISITORS

In the following decades, we've sent other spacecraft to spy on Venus, including NASA's Magellan, which reached Venusian orbit on August 10, 1990, and collected data for more than four years. The results provided a nearly complete global map of the planet's surface via radar. These images revealed the diversity of this volcanic world, from rugged mountaintops to smooth plains.

Trips to the planet next door have grown scarce in recent years. Magellan was NASA's last mission to the sizzling world, and just one craft is now orbiting Venus: Akatsuki from Japan's Aerospace Exploration Agency (JAXA). In 2010, the spacecraft zipped right by Venus, missing the point at which it was supposed to swing around the planet. But JAXA recovered the mission, and Akatsuki entered orbit in December 2015. Among its findings are hints of what may be Venusian lightning.

This pressure cooker of a planet is not an easy place for a robotic emissary to survive, and it is inhospitable to life as we know it. Scientists have suggested that if life were to exist on Venus today, it might take the form of high-flying microbes that would waft through a cloudy zone some 48 kilometers (30 mi) high.

We may get a closer look at this world again soon. In 2021, not just one but three missions were approved to return to Venus: NASA's DAVINCI+ and VERITAS, as well as the ESA's EnVision. All three missions have one focus: to untangle Venus's past complexities of climate.

PEEKING BENEATH THE CLOUDS

Venus is often called Earth's twin, thanks to its similar size, makeup, and distance to the sun. Its surface is largely composed of rocks with a similar chemistry and density to volcanic rocks known as basalts on Earth. For many years, Venus may have once hosted an Earthlike climate, with oceans and rivers on its surface.

Yet billions of years ago, climate chaos turned Venus into the inferno we see today, which sent any fluids fleeing into space. What led to this dramatic change has long remained a mystery—and cracking the case could give clues to our own planet's warmer future as we continue to pump greenhouse gases into the sky.

To search for clues, the three future missions will study Venus in a slew of different ways, from analyzing its atmospheric composition, to mapping the geology of its surface, to sussing out the layers below.

A REPAVED PLANET

Without water or global plate tectonics, which have shaped the features on Earth, Venus's surface is strikingly flat. While it has varied terrains—from high plateaus and volcanoes to canyons and impact craters—the total elevation difference is low. Only a few of Venus's features stand taller or dive deeper than a couple kilometers (1.2 mi).

Many craters also speckle Venus's surface, but they're conspicuously few compared to other bod-

Venus has the most volcanoes of any planet in our solar system, with more than **1,600 major fiery peaks.**

ies in our solar system, which hints at a youthful surface—but the precise age remains uncertain. Scientists once thought the planet underwent a global resurfacing from a surge of volcanic activity roughly 500 million years ago. But more recent studies suggest the volcanism may have continued until 150 million years ago.

Some volcanoes may still be active on Venus. The European Space Agency's Venus Express has spotted some hints of current eruptions, including bright spots that might be fresh lava flow.

Scientists still have much to learn about this planetary neighbor. Its bright glint in the skies is a reminder of how far humans have come in interplanetary exploration—and how much further we have to go.

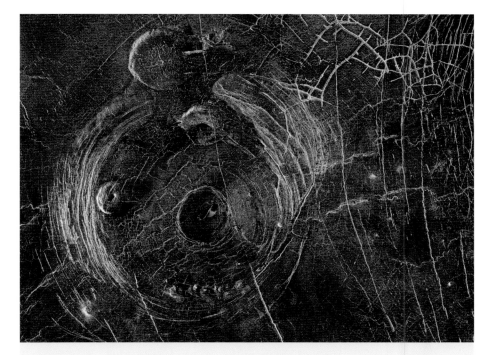

VENUS'S WEIRD VOLCANOES

While many of Venus's volcanoes tower in peaks like those on Earth, some take on strange forms, such as the Fotla Corona above. Coronae are concentric rings of wrinkles and fractures perhaps born from giant plumes of hot rising mantle that later drain and deflate the surface. Some of these features sit among weblike networks of cracks, earning them the informal name of arachnoids. Still other odd structures known as ticks may come from the surface pulling apart, an action that forms domes with wispy lava flows breaching the edges in eerily bug-like shapes.

EARTH

From its expansive oceans to its lush lands, our home world is unlike any other in the system.

We all live on the third planet from the sun, Earth—and by solar system standards it's a remarkably weird world.

While we might be used to rivers flowing aboveground or waves sloshing in the sea, no other planet in our solar system boasts the same set of features at its surface today. Water covers nearly 70 percent of our planet, collecting in vast oceans that are around 3.7 kilometers (2.3 mi) deep on average. They reach their greatest depth in the Pacific Ocean's Mariana Trench, which drops nearly 11,000 meters (6.8 mi) below sea level.

Our world has also been shaped by its restless surface, which is cracked into a series of tectonic plates that constantly jostle for position. Their collisions have sculpted the planet, raising mountains as well as opening deep basins and valleys. The grinding of the plates past one another sends the surface bucking in earthquakes and drives volcanoes that resurface the landscape as they spew molten rock. While other planets have tectonic activity, no other known world boasts a similar global system of tectonic plates. (Some evidence hints at plate tectonics on Jupiter's frigid moon Europa, but with ice rather than rock.)

The uniqueness of our planet gives rise to something else we've not yet found anywhere else in the universe: life. From ancient primordial stews, plants and animals winked into being and evolved over billions of years into the diverse array that grows, gallops, swims, and flies around the planet today.

GETTING TO KNOW OUR PLANET

Earth is just slightly larger than Venus, making it the largest of the four rocky planets that compose

Average distance from sun: 150 million kilometers (93 million mi)

Length of day: 23.9 hours

Length of year: 365.26 days

Maximum temperature: 56.7°C (134°F)

Minimum temperature: -89.2°C (-128.6°F)

Average temperature: 15°C (59°F)

Average radius: 6,371 kilometers (3,959 mi)

Number of moons: One

Planetary ring system: No

the inner solar system. It orbits the sun at an average of about 150 million kilometers (93 million mi) away, just far enough to allow for liquid water at its surface.

In years past, explorers risked their lives venturing to Earth's highest peaks and coldest realms. Today we study the planet in high-tech fashion as well. A fleet of satellites peers down at our home world from space, helping scientists better understand shifts in climate, pollution, and the planet's

Continued on page 118

Continued on page 118

The collision of Earth's restless tectonic plates can raise mountain ranges, such as the stunning Himalayan peaks shown here.

(Opposite) The sun beats down on the Caribbean Sea, shown in this image taken by astronaut Terry Virts in January 2015 from the International Space Station.

EARTH

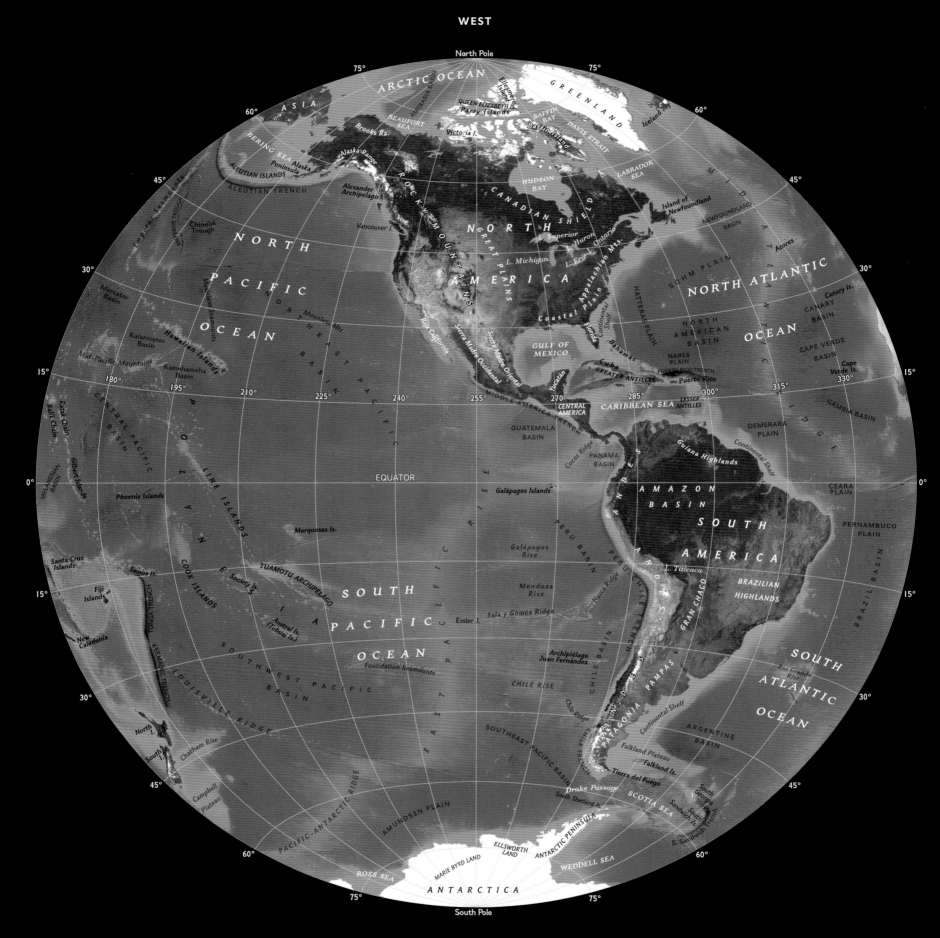

North Pole

75° · 75°

60° · 60°

ARCTIC OCEAN

GREENLAND

Iceland

ASIA

IMBRASIA BASIN

Ellesmere Island

QUEEN ELIZABETH IS.
Parry Islands

BAFFIN
BAY

Baffin Island

Davis Strait

Brooks Ra.
Alaska

Bering Sea

Victoria I.

45° · 45°

Peninsula
ALEUTIAN ISLANDS

Alaska Range

Alexander
Archipelago

HUDSON
BAY

CANADIAN SHIELD

LABRADOR
SEA

Island of
Newfoundland

NEWFOUNDLAND

ALEUTIAN TRENCH

Emperor Seamounts

Chinook
Trough

Vancouver I.

NORTH

NORTH

L. Superior

Azores

30° · 30°

NORTH

PACIFIC

Moonless Mts.

ROCKY MOUNTAINS

GREAT PLAINS

AMERICA

L. Michigan

L. Huron

L. Erie

L. Ontario

Appalachian Mts.

SOHM PLAIN

NORTH ATLANTIC

Mercator
Basin

Mid-Gems Seamounts

OCEAN

NORTHEAST

Sierra Madre Occidental

Coastal Plain

Continental
Shelf

HATTERAS PLAIN

NORTH
AMERICAN
BASIN

OCEAN

CANARY
BASIN

Canary Is.

Kalaniopuu
Basin

Hawaiian Islands

Baja California

Sierra Madre Oriental

GULF OF
MEXICO

Florida

CAPE VERDE
BASIN

Mid-Pacific Mountains

Kamehameha
Basin

Bahamas

Cuba

GREATER ANTILLES

NARES
PLAIN

PUERTO RICO TRENCH

Puerto Rico

Cape
Verde Is.

15° · 15°

180° 195° 210° 225° 240° 255° 270° 285° 300° 315° 330°

Rafak Chain

CENTRAL PACIFIC
BASIN

Gilbert Islands

MIDDLE AMERICA TRENCH

CENTRAL
AMERICA

CARIBBEAN SEA

LESSER
ANTILLES

GAMBIA BASIN

MELANESIAN
BASIN

NORTH PACIFIC RISE

GUATEMALA
BASIN

Cocos Ridge

PANAMA
BASIN

DEMERARA
PLAIN

Continental Shelf

0° EQUATOR

Phoenix Islands

LINE ISLANDS

Galápagos Islands

Guiana Highlands

CEARA
PLAIN

0°

POLYNESIA

Marquesas Is.

Galápagos
Rise

AMAZON
BASIN

SOUTH

PERNAMBUCO
PLAIN

Santa Cruz
Islands

TUAMOTU ARCHIPELAGO

Society Is.

SOUTH

PERU BASIN

Nazca Ridge

AMERICA

15° · 15°

Fiji
Islands

Samoa Is.

COOK ISLANDS

Mendoza
Rise

L. Titicaca

BRAZILIAN
HIGHLANDS

New
Caledonia

Austral Is.
(Tubuai Is.)

PACIFIC

Sala y Gómez Ridge

Easter I.

PERU-CHILE TRENCH

ANDES

BRAZIL BASIN

TONGA TRENCH

OCEAN

Foundation Seamounts

CHILE RISE

Archipiélago
Juan Fernández

GRAN CHACO

PAMPAS

SOUTH

Rio Grande
Rise

ATLANTIC

30° · 30°

KERMADEC TRENCH

SOUTHWEST PACIFIC
BASIN

CHILE BASIN

Chile Ridge

Continental Shelf

PATAGONIA

ARGENTINE
BASIN

OCEAN

LOUISVILLE RIDGE

EAST PACIFIC RISE

SOUTHEAST PACIFIC BASIN

CHILE TRENCH

Falkland Plateau

Falkland Is.

North I.

Tierra del Fuego

South I.

Chatham Rise

Drake Passage

SCOTIA SEA

South
Georgia

45° · 45°

Campbell
Plateau

South Shetland Is.

S. Sandwich Is.

S. Sandwich Trench

PACIFIC-ANTARCTIC RIDGE

AMUNDSEN PLAIN

ANTARCTIC PENINSULA

60° · 60°

ROSS SEA

ELLSWORTH
LAND

MARIE BYRD LAND

WEDDELL SEA

75° · 75°

ANTARCTICA

South Pole

Earth's Western Hemisphere highlights our planet's vast oceans. Water covers more
than 70 percent of the planet's surface.

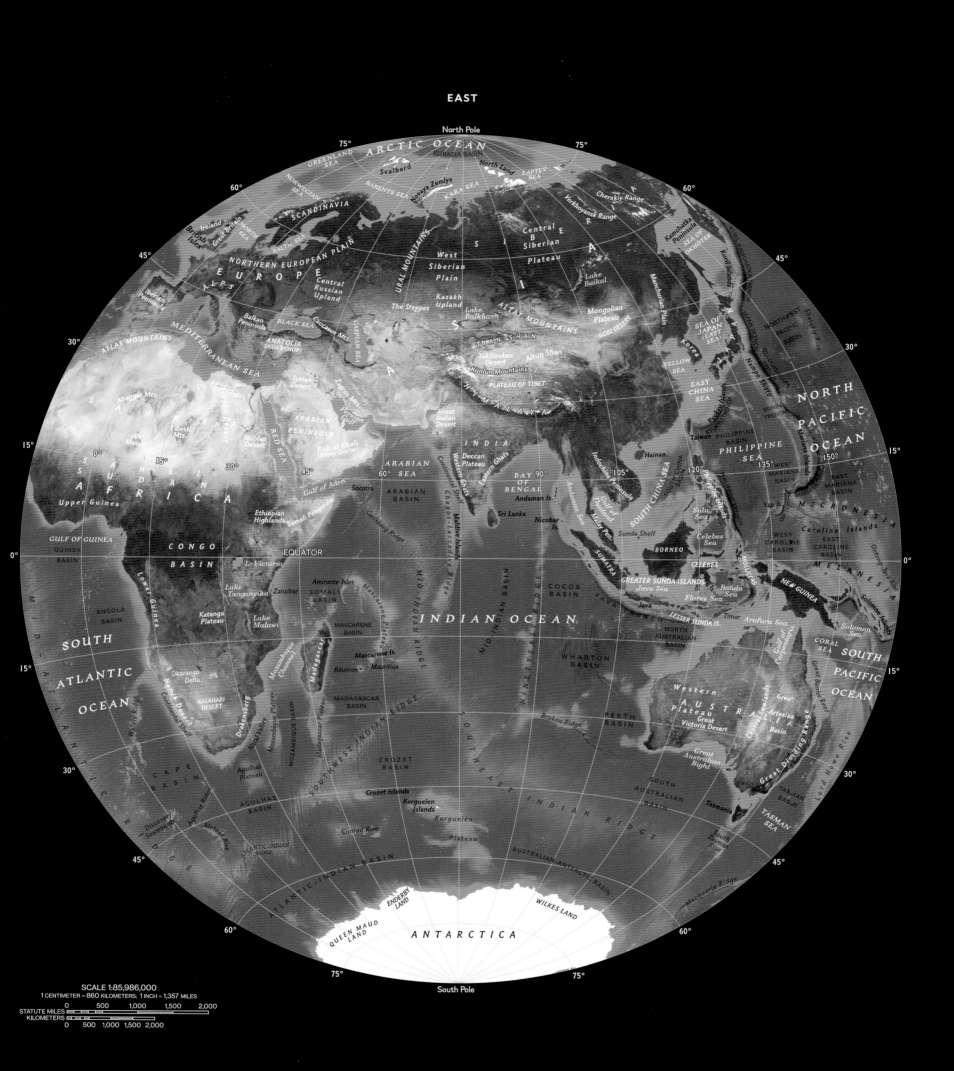

This view of Earth contains our planet's highest point (Mount Everest in Nepal and Tibet) and its lowest (Challenger Deep in the North Pacific Ocean).

SCALE 1:85,986,000
1 CENTIMETER = 860 KILOMETERS; 1 INCH = 1,357 MILES

0   500   1,000   1,500   2,000
STATUTE MILES
KILOMETERS
0   500   1,000   1,500   2,000

Continued from page 115
changing surface. We can also see beneath our planet's rocky terrain by harnessing the rumbles of earthquakes near and far. These seismic waves change speed and direction as they move through rocks of different compositions, densities, and temperatures. Scientists can monitor these changes, using the waves like a planetary ultrasound to image Earth's innards.

EARTH'S MOON

Many insights about our home world have come from studying our moon. As we learned in chapter 2, our lunar companion formed while the solar system was still taking shape, when a Mars-size body collided with proto-Earth. Unlike Earth's surface, which is constantly eroding from wind and water, as well as being recycled by the churn of plate tectonics, the moon's surface has had no large-scale resurfacing. This allows our satellite to act as our solar system's historian, cataloging billions of years of cosmic events.

■ For more on the moon's formation, see pages 70–71.

EARTH'S ATMOSPHERE

A gaseous blanket envelops our planet, protecting life at the surface. The atmosphere helps control our planet's temperature and shields us from debris such as space rocks, as well as UV radiation from the sun.

The atmospheric layer in which we all live is the troposphere, which extends from the surface to roughly 10 to 15 kilometers (6 to 9 mi) high, depending on season and latitude. This contains the air we breathe, which is made up of 78 percent nitrogen, 21 percent oxygen, and one percent argon, water vapor, and carbon dioxide. Commercial planes fly just above this layer in the stratosphere, a dry zone of the atmosphere that reaches 50 kilometers (31 mi) above the ground.

Next is the mesosphere, reaching up to about 85 kilometers (53 mi), where most meteors burn up as they plunge toward the surface. Then comes the thermosphere, which extends some 600 kilometers (373 mi) high and filters out much of the harmful space radiation. The final layer is the thin air of the exosphere, reaching around 10,000 kilometers (6,200 mi) above Earth's surface, though exactly where this layer ends remains uncertain.

GEOLOGIC GOBSTOPPER

Centuries of exploration and study have revealed much about our home world. We've learned that Earth is layered like an everlasting gobstopper candy, though some of its layers are a bit gooey. At its center lies a solid ball of iron and nickel heated to 5400°C (9800°F). Enveloping the solid center is a liquid outer core. While it's similar in composition to the inner core, the shallower depth of the outer core lowers the pressure on the metals so much they turn molten. The swirling and churning outer core charges the planet's geodynamo. This geologic engine powers the protective magnetic bubble known as the magnetosphere, which shields our home world from radiation and prevents solar winds from stripping our atmosphere.

Next up is the mantle, a 2,900-kilometer-thick (1,800-mi) layer of silicate rock. While this layer is largely solid, it's also viscous like taffy, slowly mixing over time. A rigid outer layer, known as the crust, caps the planet. A combination of the crust and the upper mantle make up the tectonic plates, or lithosphere.

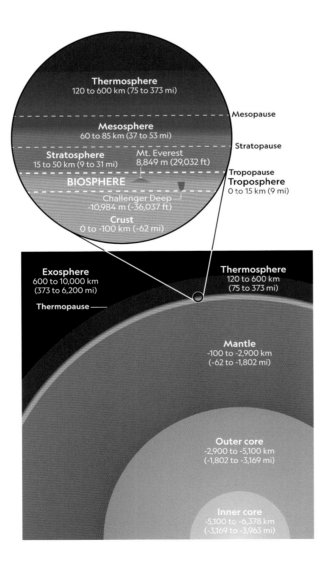

A cutaway image shows Earth's layers inside and out. Rocky layers form the planet's interior structure, and gaseous layers make up the atmosphere.

THE TECTONIC BALANCING ACT

Our global system of shifting plates is vital and unique to our planet, sculpting Earth's surface as we know it today. The movements of tectonic plates stoke volcanic blasts as one slab of rock dives beneath another. While volcanoes can be destructive, the material they release is also a life-giving force, dredging up nutrients in rocks that break down into fertile soil.

Plate tectonics are also thought to act as a kind of planetary thermostat over millions of years. Over geologic timescales, volcanic eruptions can heat the planet by spewing carbon dioxide into the skies. Yet tectonics can also draw this greenhouse gas back down through weathering, in which carbon dioxide dissolved in rain reacts with rock, and then subduction traps the carbon deep underground. Some scientists believe that this tectonic balancing act has been a major player in Earth's lasting habitability.

However, human actions are now overriding Earth's planetary thermostat with the relentless release of greenhouse gases into the skies. Reducing our emissions will be vital to protecting our little blue marble for generations to come.

Our home world is bursting with a diversity of life, from elephants to microbes. But are we alone in the universe? Generations of scientists have attempted to find out.

EARTH'S SHIFTING SURFACE

It was once thought that our churning mantle drove the movement of tectonic plates like a conveyor belt. But scientists have discovered the system is much more complicated, with other pushes and pulls at the surface also influencing the speed and direction of plate movement. The plates move apart at the mid-ocean ridge, where the mantle wells up to form new land in towering underwater mountain ranges. But the rocks at this ridge

Spreading

Subduction

are still relatively hot, which means the rigid layer making up the lithosphere is thin. As the plate cools and thickens, gravity forces it downhill, away from the mid-ocean ridge. The plates are also tugged along at the end of their life, as they plunge back into the mantle in a subduction zone where dense oceanic plates plunge beneath more buoyant continental plates. The descending slab of rock is much cooler and denser than the surrounding mantle, which causes it to sink and pull at the plate above

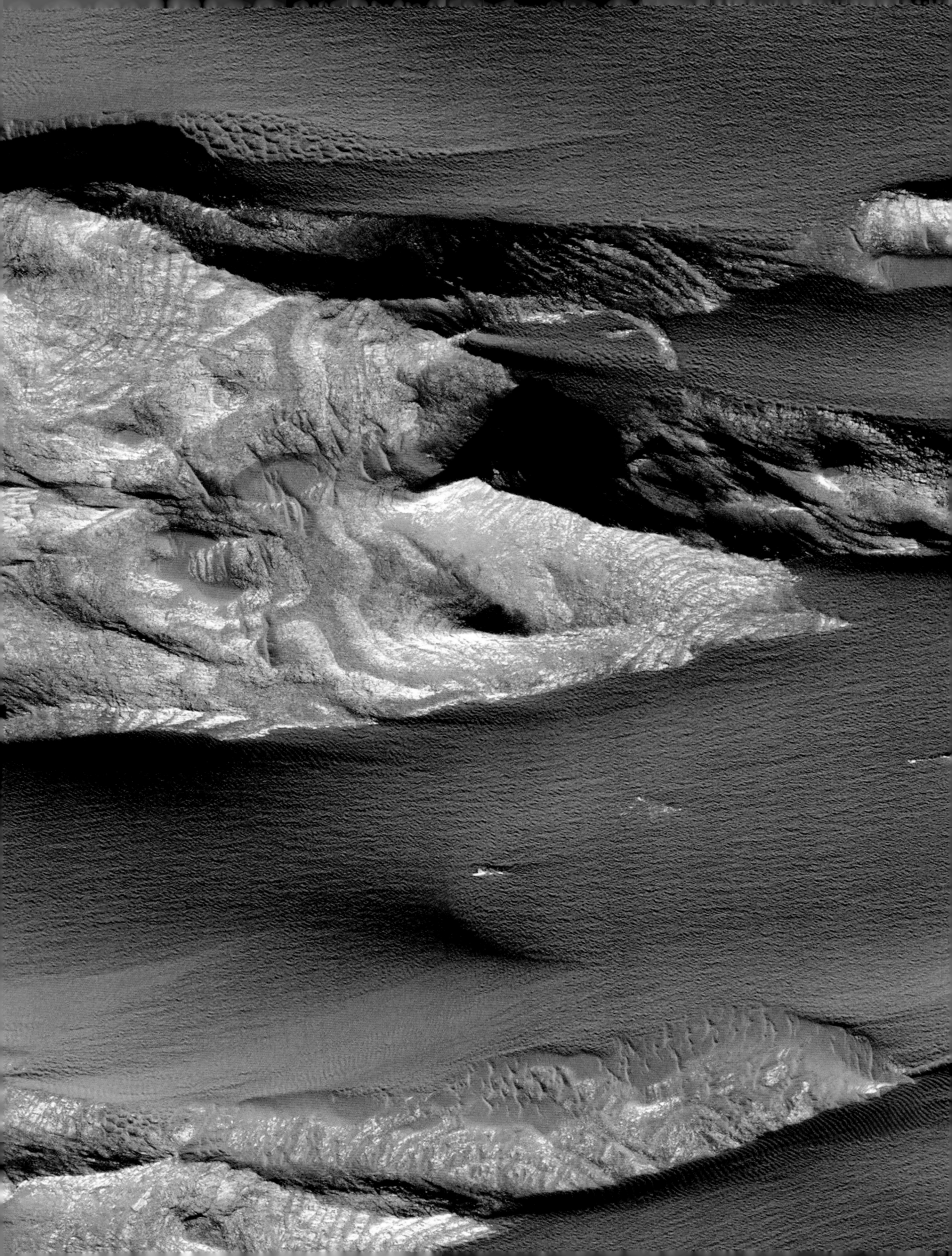

MARS

Once thought to be a frigid, dead orb, Mars serves up new mysteries at every turn.

Among the specks of light in the night sky, you might spot one with a steady glow tinged with red. That's the fourth planet from the sun, Mars.

The planet gets its rosy hue from the iron found in many rocks at its surface, which rusts like an old wheelbarrow left out in the elements. Gusts of Martian wind kick the reddish dust into the atmosphere. The fearsome, fiery coloring led many past civilizations to name the world after their gods of war or conflict, with the moniker "Mars" taken from the Roman deity.

However, Mars is much more than a ball of red dust. Rocks not only red, but brown, tan, and even slightly green are sculpted into vast valleys, canyons, hills, and peaks. Bright white glaciers cap Mars's poles, while wispy blue clouds of water ice float across its skies.

The planet has captured the imagination of generations of scientists and the wider public, and it's a prime target in our hunt for life on other worlds. While we know little green people don't race around its surface, as science fiction once suggested, the question of microbial life remains open. Did life once thrive—and does it still, somehow—on this rusty world?

A DYNAMIC WORLD

Mars was once thought to be dead, but the planet is far from quiescent. Wind whips across Mars as dust devils scour crisscrossing tracks in the Martian dirt. Occasionally a windstorm turns global, enveloping the planet in a thick dusty blanket. In June 2018, one such storm killed NASA's Opportunity rover, which had roamed the red planet for a stunning 14

(Opposite) Mars's landscapes are eerily similar to many on Earth—a feature apparent in the windblown sand of Ganges Chasma.

Average distance from sun: 228 million kilometers (142 million mi)

Mass: 0.107 x Earth's

Length of day: 24.6 hours

Length of year: 687 Earth days

Average temperature: -65°C (-85°F)

Average radius: 3,390 kilometers (2,110 mi)

Number of moons: Two

Planetary ring system: No

years, by blotting out the sunlight used to charge its electronics.

The planet's thin atmosphere, made up mostly of carbon dioxide, inexplicably changes composition with the Martian seasons. Some variation is expected as the planet heads into winter, when some carbon dioxide gas turns into ice at the poles. Yet oxygen and methane defy the expected flow, mysteriously increasing during the spring and summer.

The planet also has a strikingly strong geologic

Continued on page 124

NASA's Ingenuity helicopter made history in April 2021 when it took to Martian skies, becoming the first powered, controlled flight on another planet.

MARS

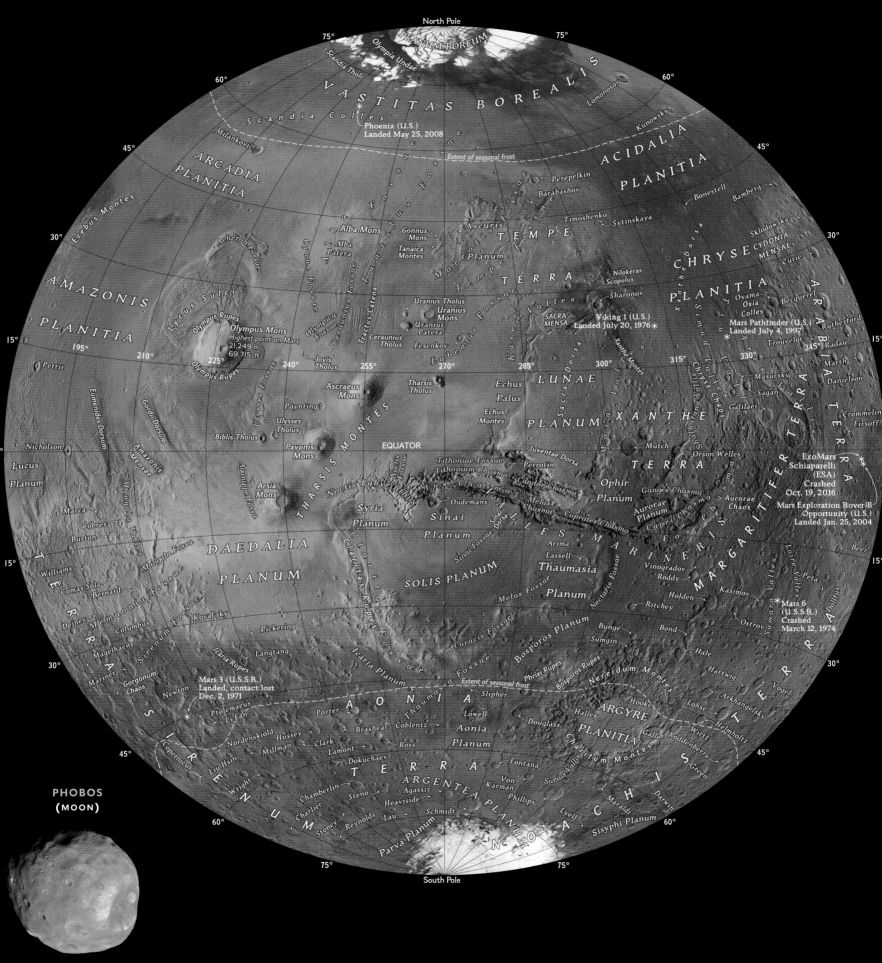

North Pole

PLANUM BOREUM

VASTITAS BOREALIS

Olympia Undae

Scandia Tholi

Scandia Colles

Milankovic

Phoenix (U.S.)
Landed May 25, 2008

Extent of seasonal frost

Kunowsky

ACIDALIA
PLANITIA

ARCADIA
PLANITIA

Perepelkin

Barabashov

Bonestell

Bamberg

Erebus Montes

Alba Mons

Gonnus
Mons

Ascuris
Planum

TEMPE

Timoshenko

Sytinskaya

Sklodowska

CHRYSE

CYDONIA
MENSAE

Alba
Patera

Tanaica
Montes

Curie

Acheron Fossae

AMAZONIS

Lycus Sulci

Uranius Tholus

TERRA

Nilokeras
Scopulus

PLANITIA

Oyama

Oxia
Colles

Becquerel

Rutherford

Uranius
Mons

Sharonov

Trouvelot

Radau

PLANITIA

Olympus Rupes

Olympus Mons
Highest point on Mars,
21,249 m
69,715 ft

Uranius
Patera

Ceraunius
Tholus

Fesenkov

SACRA
MENSA

Viking 1 (U.S.)
Landed July 20, 1976

Mars Pathfinder (U.S.)
Landed July 4, 1997

Marth

Danielson

Pettit

Olympus Rupes

Jovis
Tholus

Ascraeus
Mons

Tharsis
Tholus

Echus
Palus

LUNAE

Masursky

Sagan

Galilaei

Crommelin
Firsoff

Nicholson

Poynting

PLANUM

Echus
Montes

Mutch

Orson Welles

ExoMars
Schiaparelli
(ESA)
Crashed
Oct. 19, 2016

Lucus
Planum

Ulysses
Tholus

Pavonis
Mons

EQUATOR

Tithoniae Fossae

Juventae Dorsa

Perrotin

TERRA

Ophir
Planum

Aurorae
Chaos

Mars Exploration Rover-B
Opportunity (U.S.)
Landed Jan. 25, 2004

Biblis Tholus

Arsia
Mons

Syria
Planum

Noctis Labyrinthus

Oudemans

Candor Chasma

Ganges Chasma

Capri Chasma

Aurorae
Planum

Marca

DAEDALIA

Sinai
Planum

Coprates Chasma

Eos
Chasma

Cobres

Burton

PLANUM

SOLIS PLANUM

Arima

Lassell

Thaumasia

Vinogradov

Roddy

Holden

Beer

Williams

Comas Solà

Bernard

Dejnev

Melas Fossae

Planum

Nectaris Fossae

Ritchey

Kasimov

Mars 6
(U.S.S.R.)
Crashed
March 12, 1974

Columbus

Magelhaens

Pickering

Coracis Fossae

Bosporos Planum

Bunge

Sumgin

Bond

Hale

Hartwig

Ostrov

Icaria Rupes

Langtang

Icaria Planum

Phrixi Rupes

Bosporos Rupes

Nereidum Montes

Lohse

Arkhangelsky

Mariner

Gorgonum
Chaos

Newton

Extent of seasonal frost

Slipher

Hooke

ARGYRE

Gallo

Roddenberry

Wirtz

Helmholtz

Mars 3 (U.S.S.R.)
Landed, contact lost
Dec. 2, 1971

Ptolemaeus

Li Fan

Lowell

Halley

Douglass

PLANITIA

Green

Porter

Brashear

Coblentz

Aonia

Fontana

Charitum Montes

Nordenskiöld

Hussey

Clark

Ross

Planum

Von
Kármán

Surius Vallis

Maraldi

Darwin

Liu Hsin

Millman

Lamont

Dokuchaev

Agassiz

Phillips

Vogel

Wright

Chamberlin

Steno

ARGENTEA PLANUM

Lyell

Sisyphi Planum

Charlier

Heaviside

Schmidt

Parva Planum

Stoney

Reynolds

Lau

South Pole

PHOBOS
(MOON)

Mars's western hemisphere hosts a vast system of canyons known as Valles Marineris
and the volcanic peaks of Tharsis Montes and Olympus Mons.

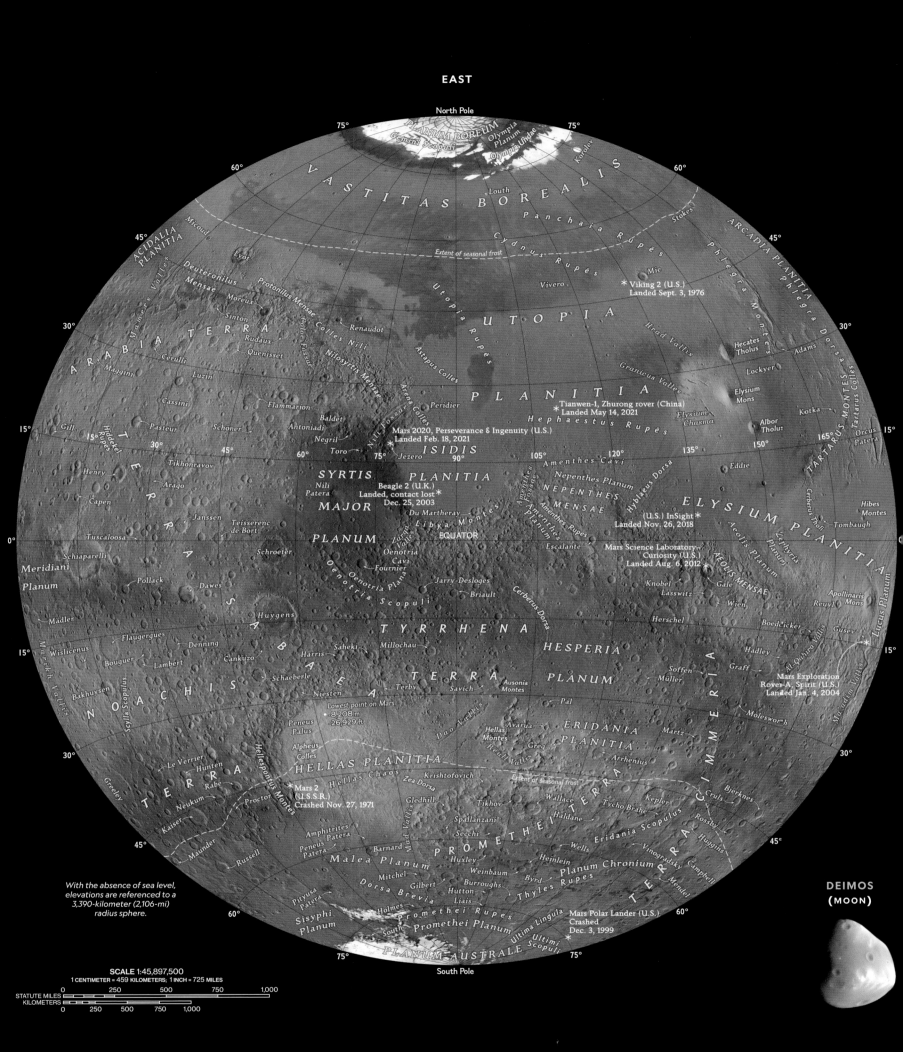

North Pole

PLANUM BOREUM
Gemini Scopuli
Olympia Planum
Olympia Undae
Korolev

V A S T I T A S   B O R E A L I S

ACIDALIA
PLANITIA

Micoud
Lyot

Panchaia

Louth

Cydnus   Rupes
Stokes

ARCADIA PLANITIA
Phlegra   Montes

Extent of seasonal frost

Utopia Rupes

Vivero
Mie
★ Viking 2 (U.S.)
Landed Sept. 3, 1976

Mamers Valles
Deuteronilus
Mensae
Moreux

Protonilus Mensae Colles Nili
Renaudot

U T O P I A

Hrad Vallis

Hecates
Tholus
Adams

Phlegra Dorsa

Sinton
Rudaux

Nili Fossae

Astapus Colles

P L A N I T I A
Lockyer

A R A B I A   T E R R A
Maggini
Cerulli
Luzin
Queniisset

Flammarion

Arena Colles
Peridier

Elysium
Mons

Elysium
Chasma

Elysium
Albor
Tholus
Kotka

Cassini

Baldet
Antoniadi

Nilosyrtis Mensae

Tianwen-1, Zhurong rover (China)
Landed May 14, 2021

Hephaestus Rupes

TARTARUS MONTES
Tartarus Colles

Pasteur
Schöner
Negril
Toro

Mars 2020, Perseverance & Ingenuity (U.S.)
Landed Feb. 18, 2021

Amenthes Cavi
Eddie
Orcus
Patera

Hibes
Montes

Henry
Tikhonravov
Arago

Nili
Patera

S Y R T I S   P L A N I T I A
Jezero

I S I D I S
PLANITIA

Nepenthes Planum

Amenthes
Fossae

N E P E N T H E S
M E N S A E

Hyblaeus Dorsa

Cerberus Tholi

Zephyria
Planum

E L Y S I U M   P L A N I T I A
Tombaugh

Capen
Janssen
Teisserenc
de Bort

M A J O R

Beagle 2 (U.K.)
Landed, contact lost
Dec. 25, 2003 ★

Du Martheray

Amenthes Rupes
Amenthes
Planum

(U.S.) InSight ★
Landed Nov. 26, 2018

Aeolis Planum

Gale

Apollinaris
Mons

Tuscaloosa
Schiaparelli

Schroeter

P L A N U M

Libya Montes

Targa Valles

Oenotria
Cavi
Fournier

Escalante

Mars Science Laboratory—
Curiosity (U.S.)
Landed Aug. 6, 2012 ★

Knobel
Lasswitz
Wien

Reuyl

Meridiani
Planum
Pollack
Dawes

Oenotria Plana
Oenotria Scopuli

Jarry-Desloges
Briault

Cerberus Dorsa

Herschel

Boeddicker
Gusev

Lucus Planum

Mädler
Flaugergues
Denning

Huygens

T Y R R H E N A

H E S P E R I A

Hadley
Graff

Mars Exploration
Rover-A, Spirit (U.S.)
Landed Jan. 4, 2004

Wislicenus
Bouguer
Lambert
Cankuzo

Harris
Saheki
Millochau

T E R R A

P L A N U M

Soffen
Müller

Al-Qahira Valles

Ma'adim Valles

Bakhuysen
Schaeberle
Niesten
Terby
Savich

Ausonia
Montes
Pal

Molesworth
Martz

Scylla Scopulus
Peneus
Palus

• 8,208 m
26,929 ft

Dao Vallis

Hellas
Montes

Avarua
Greg

E R I D A N I A
P L A N I T I A

Arrhenius

Bjerknes

Le Verrier
Hunten
Rabe

Alpheus
Colles

Lowest point on Mars

H E L L A S   P L A N I T I A
Hellas Chaos

Reull Vallis

Cruls

Neukum
Kaiser

Mars 2
(U.S.S.R.)
Crashed Nov. 27, 1971 ★

Zea Dorsa
Krishtofovich

Tikhov
Wallace

Tycho Brahe
Kepler
Rossby

Huggins

Amphitrites Patera
Peneus
Patera

Gledhill

Spallanzani
Secchi

Haldane

Matunder
Russell

Barnard
Mad Vallis
Huxley
Weinbaum
Burroughs

Heinlein
Byrd
Wells
Planum Chronium

Vinogradsky
Campbell

Mendel

Pityusa
Patera
Mitchel
Gilbert
Hutton
Liais

Thyles Rupes

Sisyphi
Planum
Holmes

Promethei Rupes
Promethei Planum

Ultima Lingula
Ultimi
Scopuli

Mars Polar Lander (U.S.)
Crashed
Dec. 3, 1999 ★

South

P L A N U M   A U S T R A L E

South Pole

*With the absence of sea level,
elevations are referenced to a
3,390-kilometer (2,106-mi)
radius sphere.*

**DEIMOS
(MOON)**

SCALE 1:45,897,500
1 CENTIMETER = 459 KILOMETERS; 1 INCH = 725 MILES

STATUTE MILES 0   250   500   750   1,000
KILOMETERS 0   250   500   750   1,000

Mars's eastern hemisphere contains the dark volcanic terrain of Syrtis Major Planum and
many vast plains within large impact craters, such as Utopia Planitia.

A composite image from the Curiosity rover reveals sedimentary rocks on Mars, color adjusted to show how they would look in daylight on Earth.

Continued from page 121

pulse, as measured by NASA's InSight lander. The first marsquake it detected rumbled from the Martian interior on April 6, 2019, and since then the robotic geologist has detected hundreds more. Exactly what is causing all these marsquakes is uncertain, since Mars lacks the global system of plate tectonics that drives temblors here on Earth. While not all the quakes seem to come from the same source, researchers have traced two of the quakes back to their origins: a series of deep gashes known as Cerberus Fossae, which is the first active fault zone found on Mars.

Other aspects of the seismic data have also left scientists scratching their heads: Beneath its surface, Mars is mysteriously humming in a quiet but steady drone.

MARS'S WATERY PAST

Mars's surface is scribbled over with features that point to a past rich with water, including branching river valleys, fan-shaped deltas, and floodplains. And where there is water, scientists often see the possibility of life.

In its infancy, Mars's atmosphere was probably dense and thick, trapping enough heat to prevent liquid water from freezing as well as providing enough air pressure to stop the liquid from turning

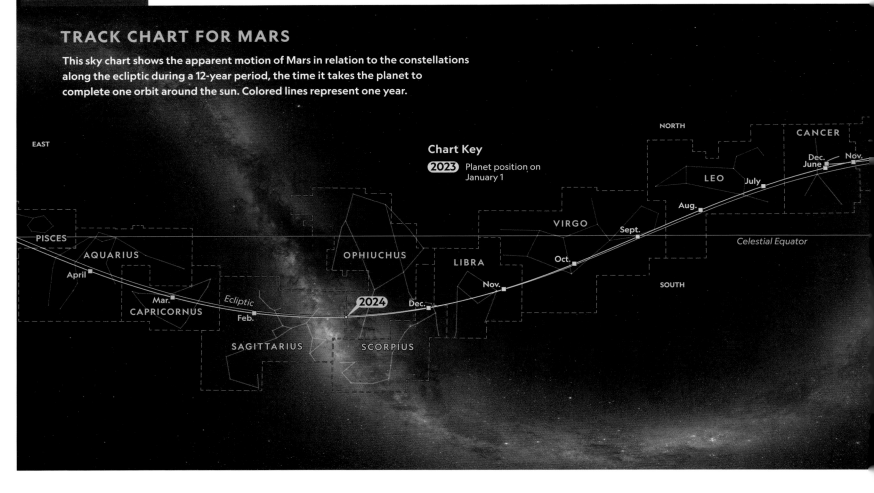

SEE FOR YOURSELF

TRACK CHART FOR MARS

This sky chart shows the apparent motion of Mars in relation to the constellations along the ecliptic during a 12-year period, the time it takes the planet to complete one orbit around the sun. Colored lines represent one year.

Chart Key
2023 Planet position on January 1

THE MARTIAN CANALS

Much of the mythos of intelligent life on Mars can trace its roots to a map drawn in 1877 by Italian astronomer Giovanni Schiaparelli. Like others of the time, he saw and sketched out what he believed were continents and seas in the light and dark splotches on the Martian surface, as seen through his telescope, but he also included a series of starkly linear features he dubbed *canali* in Italian. The term translates in English to either channel or canal, but the latter term was the one that took off in the media. While Schiaparelli didn't intend to imply that Martians built the linear features, the translation of the term as canals sparked fanciful tales of intelligent life and alien constructions that spread like wildfire.

Schiaparelli drew this map of Mars (top), highlighting the diverse terrain. Modern analyses brought the planet into sharper focus (bottom).

Wealthy Bostonian Percival Lowell helped whip the Mars mania into an outright frenzy. In 1894, he built his own observatory in Arizona to study the planet. He mapped out a vast network of these supposed "canals," which he believed Martians used to draw water from the polar ice caps to the equator. In an address to the Boston Scientific Society, he described the features as likely "the result of the work of some sort of intelligent beings."

Eventually, scientists realized that the canals were mere optical illusions, created by human brains reaching for familiar shapes in disconnected splotches. Modern Mars is a frigid and desiccated world, lacking any liquid water at its surface. But canals or not, interest in the red planet hasn't waned.

---

straight to gas. But at some point, this lush environment turned into the frigid desert we see today.

The reason behind the drastic turn remains uncertain, but scientists think it may be related to the mysterious loss of Mars's magnetosphere as early as four billion years ago. As on Earth today, Mars's magnetic field could have prevented solar winds from whisking away its atmosphere. Without a magnetosphere, the planet's atmosphere thinned, and along with it went the liquid water. Today, Martian water exists as ice underground, as well as ice mixed with frozen carbon dioxide in polar glaciers. Some research hints it may even pool in salty subglacial lakes.

THE MOON DUO

Mars's two moons, Phobos and Deimos, are among the smallest in the solar system. Both are irregularly shaped, thanks to gravities too weak to draw them into spheres, with Phobos 27 kilometers (17 mi) across and Deimos a mere 15 kilometers (9 mi) at their widest. Their tiny sizes have led to a long-held debate about their origins. Perhaps they're asteroids captured in Mars's gravity, or else chunks of Mars that broke away during an impact. A JAXA mission slated to launch in 2024 will search for clues to the duo's unknown past.

Similar to our own lunar companion, both moons always present the same face to their host planet. But the pair encircle Mars at vastly different distances. Phobos orbits closer than any other known moon encircles its planet—and it's spiraling ever closer. Each century, Phobos inches roughly 1.8 meters (6 ft) closer to Mars, which means that in the next 50 million years it could smash into the dusty planet. Alternatively, Phobos might shred to pieces as it draws closer, scattering to form a ring around Mars.

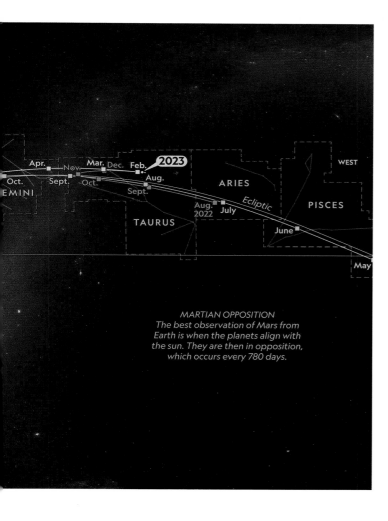

MARTIAN OPPOSITION
The best observation of Mars from Earth is when the planets align with the sun. They are then in opposition, which occurs every 780 days.

EXPLORING THE RED PLANET

Dozens of missions have brought Mars and its pair of tiny moons into ever sharper focus. After many unsuccessful early flights, Earthlings got their first look at the red planet up close in 1965 as Mariner 4 made its nearest pass by Mars, snapping pictures along the way.

In 1971, Mariner 9 became the first spacecraft to orbit the red planet, mapping 85 percent of its surface. Among the landscape's many breathtaking features was Olympus Mons, a volcano larger than any other in our solar system: 21.9 kilometers (13.6 mi) tall and more than 600 kilometers (373 mi) wide. Because of its expanse, the volcano's overall slope is fairly gentle, and its size would be difficult to take in from the surface. If you stood at the mountain's base and gazed at the peak, its summit would be hidden behind the horizon.

Mariner 9 also spotted a deep system of chasms, dubbed Valles Marineris, which stretches more than 4,000 kilometers (2,500 mi)—roughly the distance between Los Angeles and New York City. At its deepest, Valles Marineris plunges 7 to 10 kilometers (4 to 6 mi) into the planet's surface.

The many missions that followed gave us planetary views that once seemed unimaginable. Orbiters, landers, and six rovers have explored vast stretches of Mars, making it the best studied place in our solar system, aside from Earth. Millions of miles away, one of the newest robotic emissaries, Perseverance (or "Percy" for short) is now collecting data with a slew of the highest-tech instruments ever sent to another world. The rover has also begun scooping up samples of Mars rocks and soils for a future mission to grab and return to our home world. One of Percy's primary missions: to hunt for signs of ancient life.

These missions have revealed that Mars has—or once had—many of the ingredients for life as we know it, including water and complex carbon-containing molecules, as well as sulfur, nitrogen, oxygen, hydrogen, and phosphorus.

Yet for every new discovery, Mars serves up another mystery. Someday humans may explore these mysteries in person.

Researchers are developing the next generation of planetary exploration space suits to help future explorers venturing to Mars and beyond.

ROVERS ON MARS

■ SOJOURNER landed 1997
Part of the Mars Pathfinder mission, Sojourner was the first rover to set wheels on another planet. It was the tiniest of the Martian rovers, about the size of a carry-on suitcase, and it spent 83 days exploring the red planet, snapping pictures and analyzing its chemistry.

■ SPIRIT AND OPPORTUNITY landed 2004
This pair of rovers landed on opposite sides of Mars, carrying an identical set of instruments. Together they beamed back hundreds of thousands of images, as well as chemical and mineralogic data, which revealed Mars's watery past.

■ CURIOSITY landed 2012
This car-size rover is equipped with 10 scientific instruments, including a laser to vaporize small bits of rock, to study an ancient lakebed nestled within Gale crater. Among its many discoveries were organic molecules (carbon-rich building blocks of life), as well as many of the ingredients necessary for life as we know it (carbon, sulfur, nitrogen, oxygen, and phosphorus).

This "selfie" of Curiosity combines multiple images taken in April and May 2014.

■ PERSEVERANCE landed 2021
"Percy" is exploring Jezero crater—an ancient impact basin that was likely once flooded with water—and beyond for signs of ancient life. Powered by nuclear energy, the rover is not only photographing and analyzing the surface; it is also collecting samples for a future mission to return to Earth.

■ ZHURONG landed 2021
In a major milestone for the Chinese space program, the Zhurong rover joined the growing fleet of rolling robots on Mars. The rover began transmitting data in the summer of 2021. It will explore and study the rocks of Utopia Planitia, a vast plain that might have once hosted an ancient sea.

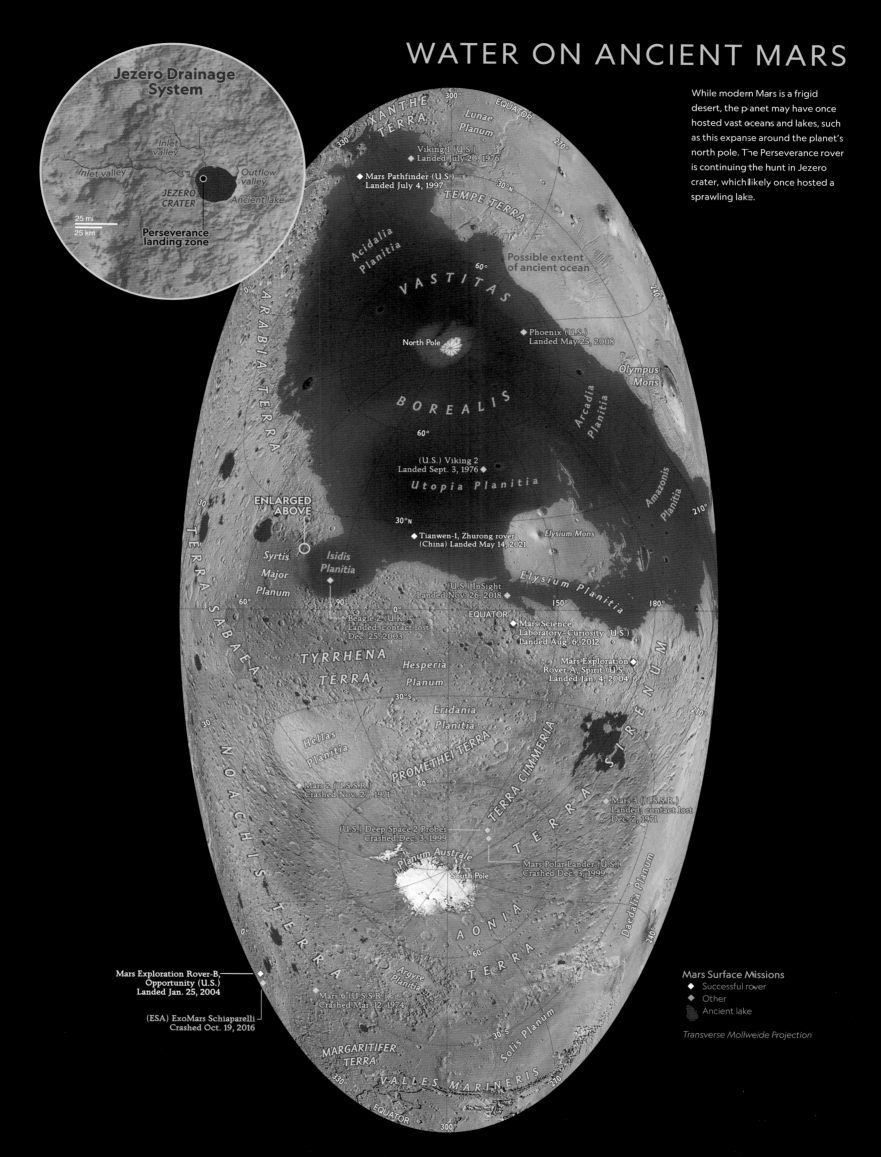

WATER ON ANCIENT MARS

While modern Mars is a frigid desert, the planet may have once hosted vast oceans and lakes, such as this expanse around the planet's north pole. The Perseverance rover is continuing the hunt in Jezero crater, which likely once hosted a sprawling lake.

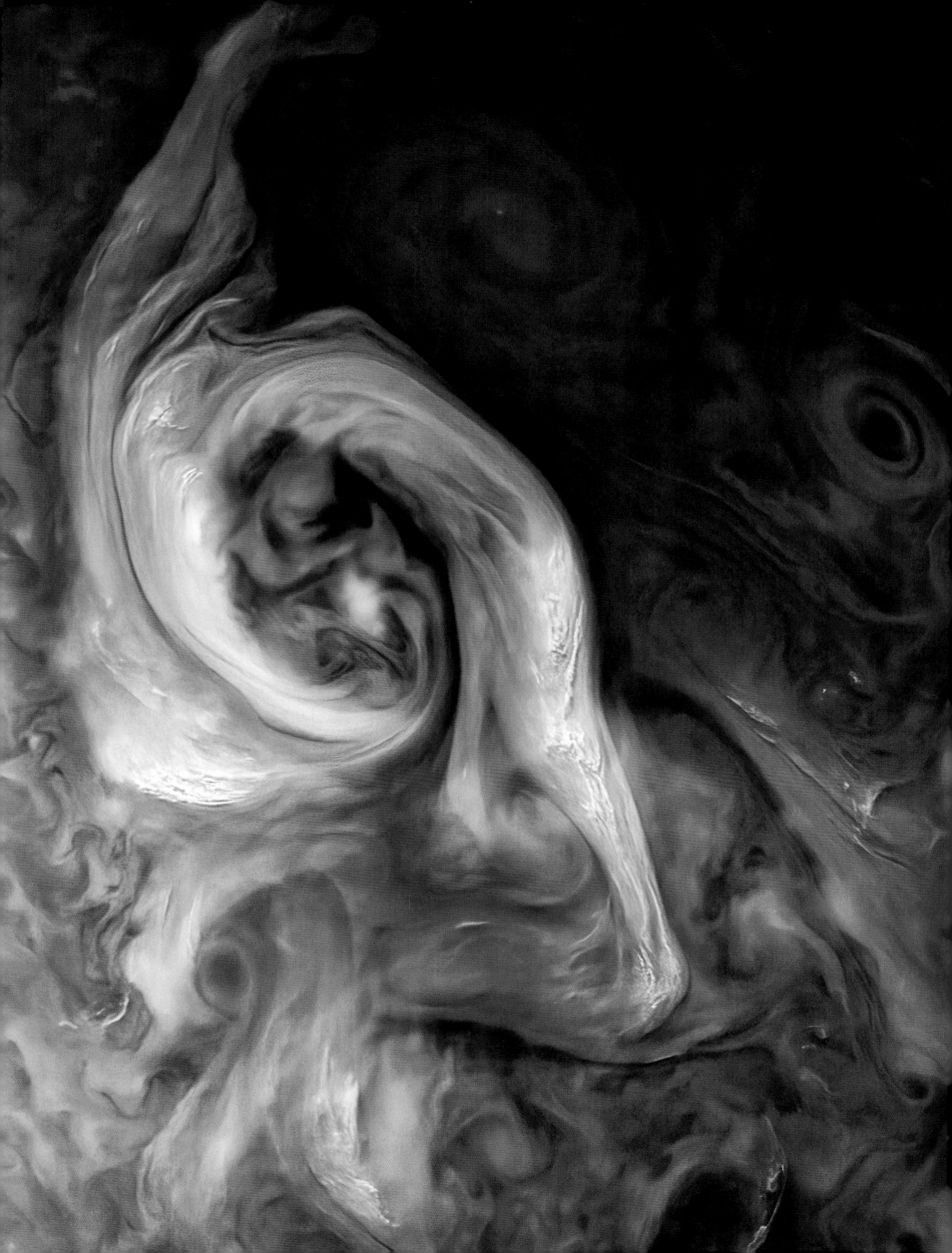

JUPITER

Colorful bands swirl through Jupiter's atmosphere, revealing the stormy secrets of this gas giant.

Jupiter is a king among planets. This giant ball of swirling gas is the largest planet in our solar system and could encompass more than 1,300 Earths. Its mass is more than twice that of the other planets combined. Accordingly, the supersize world was named after the most powerful Roman deity, Jupiter, the ruler of the classical gods.

The name for the fifth world from the sun may be even more fitting than the ancients knew. Like the Roman god, who controlled the weather, the planet is known for intense storms, which on Jupiter are driven by heat from the planet's interior. Cloudy streaks of red, peach, and gray mark winds that whip around the world at hundreds of miles an hour—faster than any found on Earth.

The gas giant, the first of the four outer planets, may also act as gatekeeper for the four rocky inner worlds. Jupiter's large size and strong gravity may help it consume or deflect many comets and asteroids, protecting Earth and our planetary neighbors—though it might also occasionally fling these space rocks our way.

PULLING BACK THE CLOUDY VEIL

Jupiter's atmosphere is largely hydrogen and helium, but the topmost clouds seen from afar are likely made up of ammonia ice. The streaky vibrant bands, perhaps colored by sulfur and phosphorus gases, reveal both eastward and westward jets that ring the planet. These jet streams may extend as deep as 3,000 kilometers (1,864 mi) beneath the cloud tops.

The most prominent storm is a crimson megahurricane known as the Great Red Spot, which turns counterclockwise at faster than 640 kilome-

(Opposite) As it swooped over Jupiter's cloud tops, NASA's Juno spacecraft snapped this image of a swirling storm, later color enhanced.

Average distance from sun: 778 million kilometers (484 million mi)

Mass: 318 x Earth's

Length of day: 9.93 hours

Length of year: 4,332 Earth days

Average temperature: -110°C (-166°F)

Average radius: 69,900 kilometers (43,400 mi)

Number of moons: 79

Planetary ring system: Yes

ters an hour (400 mph). Scientists estimate that the spot extends at least 200 kilometers (1,200 mi) into the planet's atmosphere and may be as deep as 500 kilometers (310 mi). Its billowing clouds encompass an area wider than our home planet and may have been in motion as early as the 1600s. Over time, however, the spot has shrunk, and scientists aren't sure why.

JUPITER'S STRANGE CENTER

The Great Red Spot and other long-lasting storms on the planet are driven by the planet's internal heat, likely left over from its formation. Their endurance comes, in part, from the planet's lack of solid land, which would weaken the storms—as seen when hurricanes make landfall on Earth.

Scientists are still working out precisely what lies at the planet's center. Beneath the vibrant clouds, the pressure increases with depth, eventually soaring so high that the gaseous atoms of hydrogen are forced together to form liquid in a planetwide ocean. At even greater depths, the intense pressure forces the electrons away from the hydrogen atoms, turning the liquid metallic. Exactly what lies at

Jupiter in ultraviolet light, as seen by the Hubble Space Telescope in 2017

JUPITER

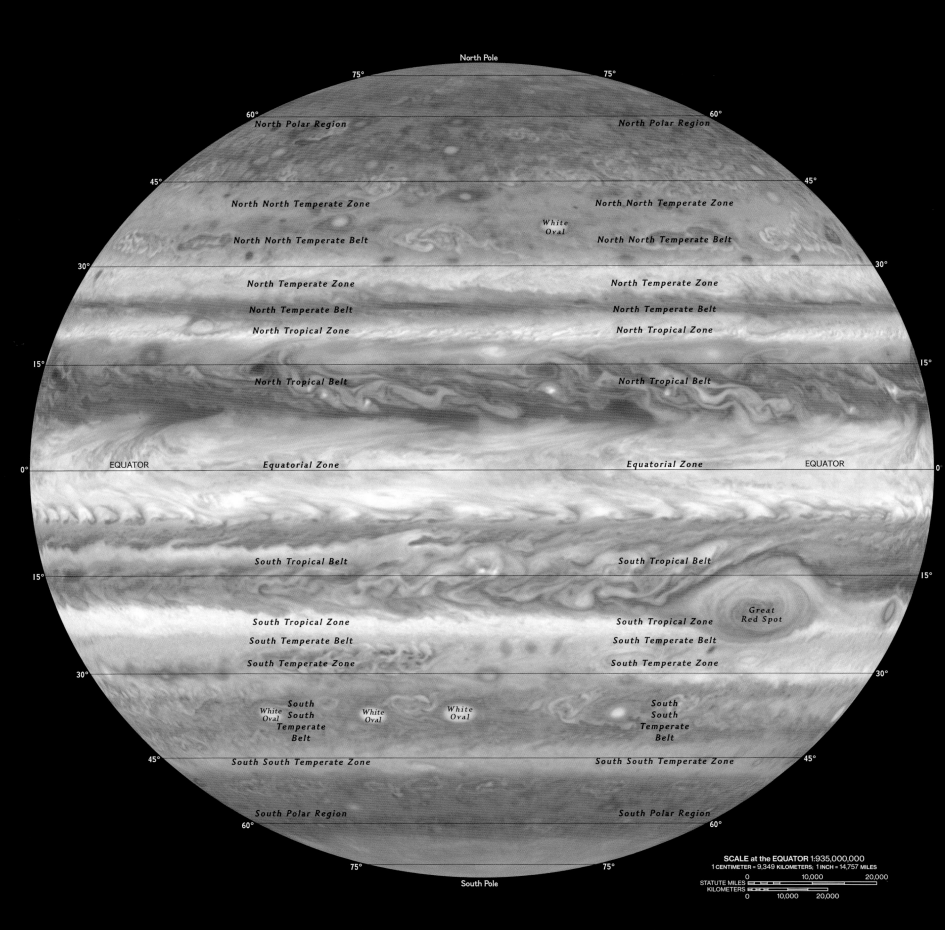

North Pole

75°   75°

60°   North Polar Region   North Polar Region   60°

45°   45°

North North Temperate Zone   North North Temperate Zone

White Oval

North North Temperate Belt   North North Temperate Belt

30°   30°

North Temperate Zone   North Temperate Zone

North Temperate Belt   North Temperate Belt

North Tropical Zone   North Tropical Zone

15°   15°

North Tropical Belt   North Tropical Belt

EQUATOR   Equatorial Zone   Equatorial Zone   EQUATOR   0°

South Tropical Belt   South Tropical Belt

15°   15°

South Tropical Zone   South Tropical Zone   *Great Red Spot*

South Temperate Belt   South Temperate Belt

South Temperate Zone   South Temperate Zone

30°   30°

White Oval   South South Temperate Belt   *White Oval*   *White Oval*   South South Temperate Belt

45°   South South Temperate Zone   South South Temperate Zone   45°

South Polar Region   South Polar Region

60°   60°

75°   75°

South Pole

SCALE at the EQUATOR 1:935,000,000
1 CENTIMETER = 9,349 KILOMETERS; 1 INCH = 14,757 MILES

0   10,000   20,000
STATUTE MILES
KILOMETERS
0   10,000   20,000

Rotating stormy bands encircle Jupiter, driven by jet streams that race both east and west around the gaseous world. Differences in cloud chemistry create a variety of hues.

Jupiter's core remains a mystery, but many believe it is a collection of remnants from the solar system's early formation.

Data from NASA's Juno spacecraft revealed that while that may be the case, the bits are not combined in one dense mass, but in a diffuse cloud of heavy elements spread across half of the planet's radius. One possible explanation for this is an early collision between a baby Jupiter and a planetary embryo, which shattered the young planet's once compact core and distributed heavy elements into the surrounding blanket of gas.

THE JOVIAN MOONS

Jupiter's many moons—79 and counting—sit at the extremes, from fiery to freezing, tiny to tremendous. Most are small and rocky, with the smallest less than a kilometer (.06 mi) across. These small moons are lumpy and irregular; some might be asteroids captured by Jupiter's strong gravitational pull.

The four largest satellites, known as the Galilean moons (see the sidebar this page), are Io, Europa, Ganymede, and Callisto, and each one is a remarkably different world. Ganymede is the largest moon in the solar system. It even has its own magnetic field, complete with glowing auroras that dance in step with shifts in Jupiter's magnetic field nearby. What's more, study of this auroral sway revealed strong evidence that a vast, subsurface, salty ocean hides beneath Ganymede's thick icy crust.

Jupiter's second largest moon, Callisto, is roughly the size of Mercury. Speckled with pockmarks from colliding space rocks, this moon boasts the most heavily cratered surface in our solar system. Like Ganymede, Callisto might harbor an ocean beneath its surface, but the details remain hazy. Both Ganymede and Callisto's icy surfaces could be quite thick, concealing their salty secrets beneath thick icy crusts.

Io is the most volcanically active body in our solar system. Hundreds of volcanoes eject gases and molten rock onto Io's surface, some so powerful they can be seen through the telescopes of earthbound viewers. The abundant volcanism is powered by varying tugs from Jupiter's intense gravity and that of Io's two neighboring moons, Europa and Ganymede. These tidal pulls stretch and release Io's surface, generating so much subterranean heat that large sections of rock become molten. Perhaps Io

even hosts a vast magmatic ocean under its rocky surface. Future missions to this dynamic world will be necessary to fill in the details.

Europa is among the most promising places to look for life beyond Earth. Beneath the moon's icy surface may lurk a saltwater ocean estimated to be 60 to 150 kilometers (40 to 100 mi) deep that could harbor twice the water of all Earth's oceans combined. This water may even vent at the moon's surface, jetting an estimated 160 kilometers (100 mi) high in thin plumes. Returning spacecraft may have the chance to taste this ocean spray to test it for clues to alien life. Scientists have also postulated that an icy form of plate tectonics could have reworked swaths of Europa's surface, similar to the rocky plate tectonics that sculpt Earth's crust.

An artist's concept of the view from the icy surface of Jupiter's moon Europa

■ For more on Europa, see pages 400–401.

Continued on page 134

THE STARS THAT WEREN'T

On January 7, 1610, Galileo Galilei spotted what he initially thought were three stars clustering near the planet Jupiter. As he studied the skies over several nights, he realized that these stars weren't moving in step with others in the sky, but rather tagging along with Jupiter while continually shifting their positions. He also found a fourth, seemingly erratic, pinprick of light. By the following week he realized these stars were not stars at all but moons, the planet's four largest satellites. The realization was a key find in dispelling the lingering notions of the time that Earth sat at the center of the solar system. Galileo published the finds in a 1610 book entitled *Sidereus Nuncius,* which translates as *The Starry Messenger.*

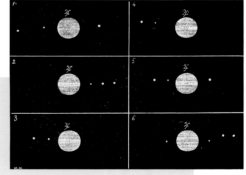

Jupiter's four moons as seen in Galileo Galilei's Sidereus Nuncius *(1610)*

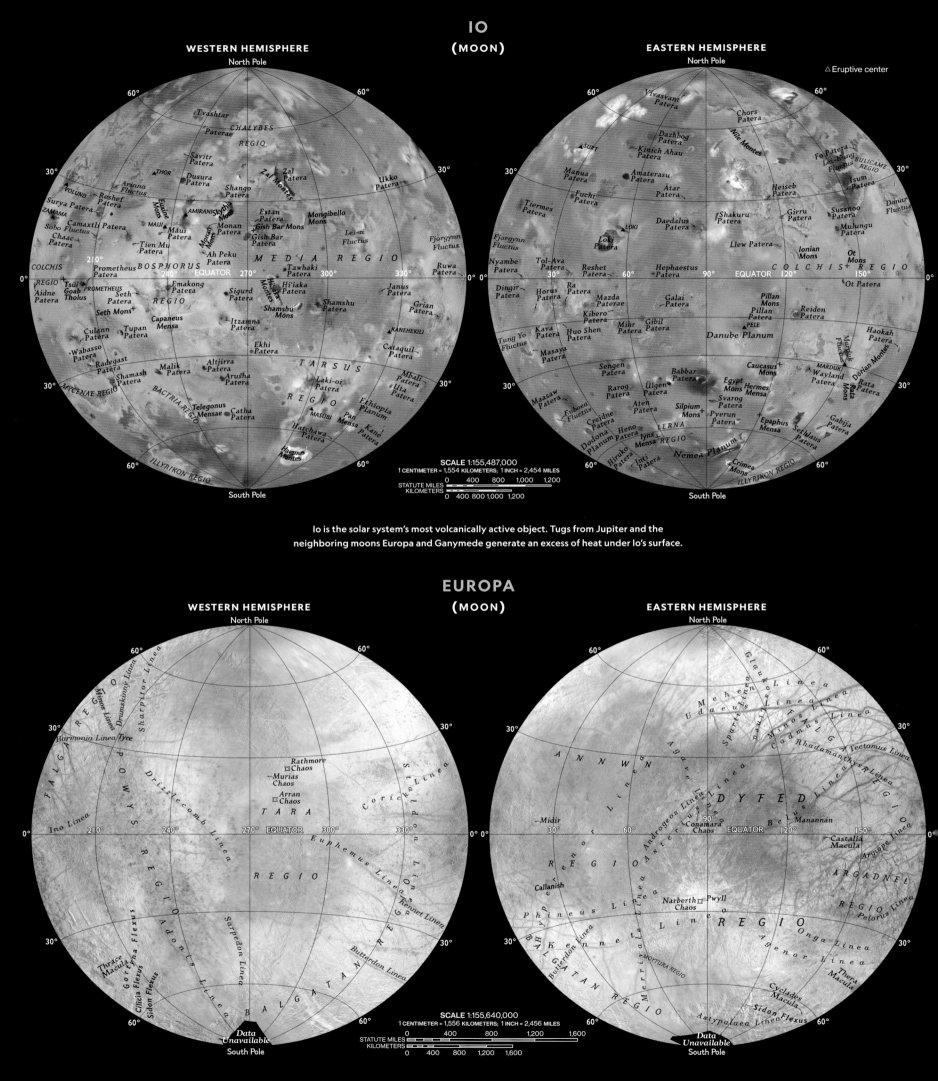

IO
(MOON)

WESTERN HEMISPHERE
North Pole

EASTERN HEMISPHERE
North Pole

△ Eruptive center

SCALE 1:155,487,000
1 CENTIMETER = 1,554 KILOMETERS; 1 INCH = 2,454 MILES
STATUTE MILES 0 400 800 1,000 1,200
KILOMETERS 0 400 800 1,000 1,200

South Pole

Io is the solar system's most volcanically active object. Tugs from Jupiter and the neighboring moons Europa and Ganymede generate an excess of heat under Io's surface.

EUROPA
(MOON)

WESTERN HEMISPHERE
North Pole

EASTERN HEMISPHERE
North Pole

SCALE 1:155,640,000
1 CENTIMETER = 1,556 KILOMETERS; 1 INCH = 2,456 MILES
STATUTE MILES 0 400 800 1,200 1,600
KILOMETERS 0 400 800 1,200 1,600

Data Unavailable
South Pole

Europa's surface is mostly composed of water ice. Reddish lines mark cracks and ridges that crisscross the moon's crust, perhaps colored by salts rising from a proposed subsurface ocean.

GANYMEDE
(MOON)

WESTERN HEMISPHERE
North Pole

EASTERN HEMISPHERE
North Pole

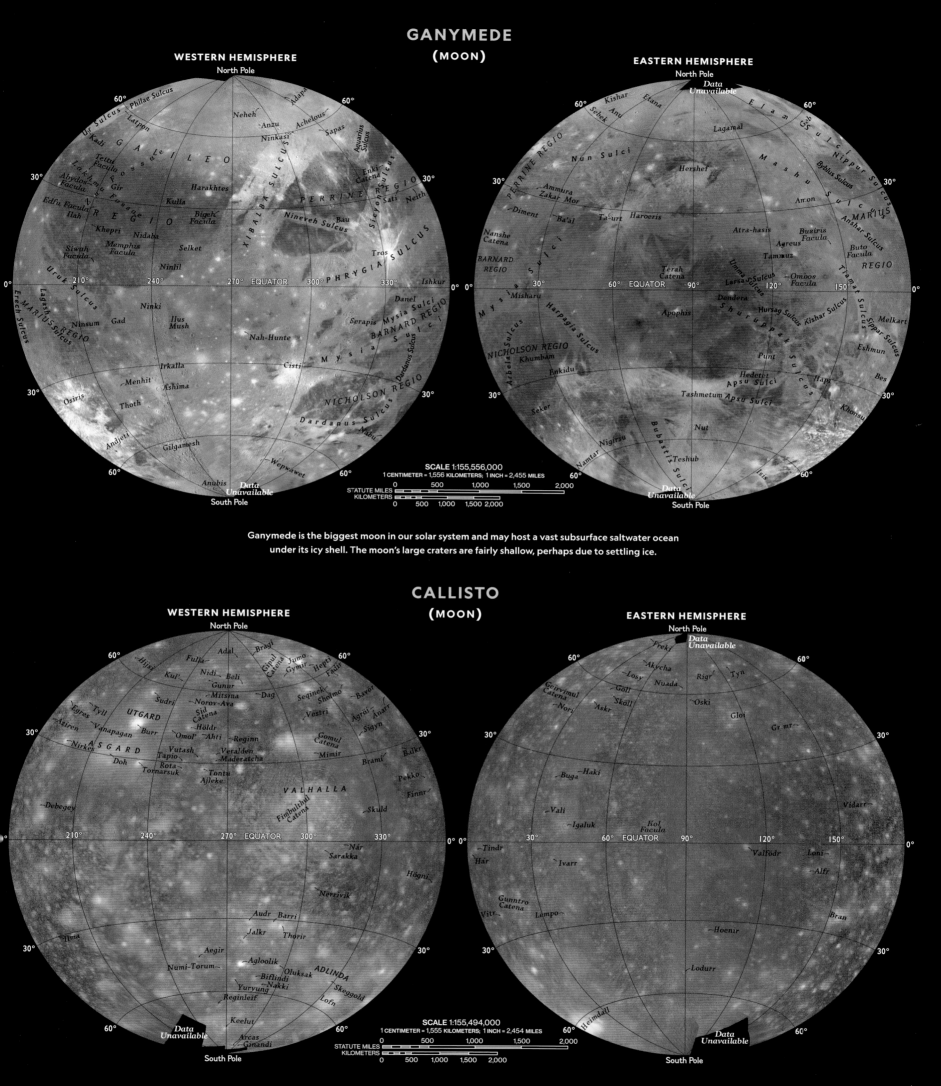

Ganymede is the biggest moon in our solar system and may host a vast subsurface saltwater ocean under its icy shell. The moon's large craters are fairly shallow, perhaps due to settling ice.

CALLISTO
(MOON)

WESTERN HEMISPHERE
North Pole

EASTERN HEMISPHERE
North Pole

Callisto, Jupiter's second largest moon, is a world of rock and ice. The abundance of craters hints that the moon largely lacks geologic activity, which would have erased some impact traces over time.

THE PLANETARY SLINGSHOT

The Voyager missions capitalized on a rare cosmic alignment. Once every 176 years, the position of the solar system's outer four planets allows a passing spacecraft to use each planet's gravity like a slingshot, pinging from world to world without consuming any additional fuel. And the late 1970s presented this once-in-a-lifetime chance. The Voyager launch vehicles would propel the craft to Jupiter, where they could skim some of the orb's energy to soar toward Saturn. Voyager 2 continued on to visit all four gas giants. Both Voyager craft are now whizzing through interstellar space, with Voyager 1 traveling farther than any human-made object has before.

Continued from page 131

The four innermost moons—Metis, Adrastea, Amalthea, and Thebe—as well as the four largest moons all circle the planet in the same direction as Jupiter's rotation. But a collection of Jupiter's outer moons orbits in retrograde, which means that they revolve around the planet in the opposite direction. Such an oddball orbit hints that these bodies may have been captured by Jupiter's intense gravity. The red-hued, retrograde satellites of the Carme group, for example, may have started as one asteroid smashed to pieces by a collision before or after Jupiter snagged them.

Jupiter may still be hiding even more moons.

Though it is not officially confirmed, an amateur astronomer discovered yet another potential satellite while looking at images from 2003.

HAZY RINGS

When the topic of ringed worlds arises, Saturn is usually what comes to most people's minds. But all our solar system's giants—Jupiter, Saturn, Uranus, and Neptune—host diverse systems of rings. Unlike Saturn's bold hoops, which owe their shine to their icy composition, Jupiter's comparatively subtle rings are made of dust.

These tiny particles may be kicked up by small space rocks colliding with the planet's four innermost moons, after which Jupiter's gravity captures the haze in its orbit. The dust makes the rings dark, and thus difficult to see against the inky black backdrop of space. The rings weren't discovered until 1979, when Voyager 1 snapped a shot of the planet backlit by the sun.

The dust clusters into four main rings: the halo, the main ring, and the two outer gossamer rings, the farthest of which peters out in a haze known as the Thebe extension. Small rings or arcs may still continue to form. In 2006, NASA's New Horizons

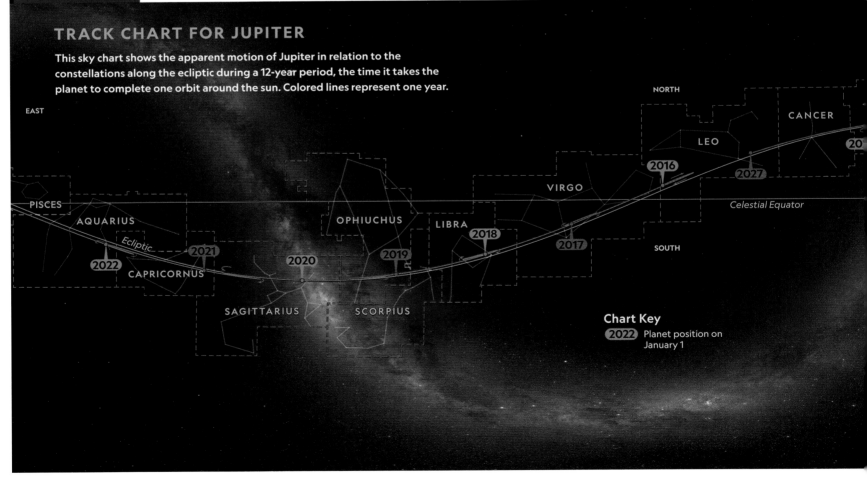

SEE FOR YOURSELF

TRACK CHART FOR JUPITER

This sky chart shows the apparent motion of Jupiter in relation to the constellations along the ecliptic during a 12-year period, the time it takes the planet to complete one orbit around the sun. Colored lines represent one year.

EAST

NORTH

CANCER

LEO

2016

2027

VIRGO

20

PISCES

Celestial Equator

AQUARIUS

OPHIUCHUS

LIBRA

2018

2017

SOUTH

Ecliptic

2021

2019

2022

2020

CAPRICORNUS

SAGITTARIUS

SCORPIUS

Chart Key

2022 Planet position on January 1

spacecraft spotted a cloud near the moon Himalia, which may have come from a collision—though the details remain fuzzy.

EXPLORING THE WINDY WORLD

Scientists got their first look at Jupiter up close during NASA's Pioneer 10 mission, which launched the first spacecraft to escape the inner solar system and visit any of the outer planets. Powered entirely by nuclear energy, Pioneer 10 zipped within 130,000 kilometers (81,000 mi) of Jupiter on its closest approach on December 4, 1973, snapping hundreds of pictures of the windy world.

Since then, eight more spacecraft have studied Jupiter up close. Voyager 1 and 2 passed by Jupiter in 1979 as well as Saturn in the following few years. Between the two spacecraft, the Voyager mission explored all the giant outer planets in addition to 48 of their moons. Among their many finds, the missions discovered Jupiter's faint rings of dust, as well as lightning crackling in the Jovian clouds.

Juno, the most recent mission to Jupiter, entered orbit around the planet in summer 2016 and has revealed a series of stunning details about its atmosphere, structure, and magnetosphere. The craft

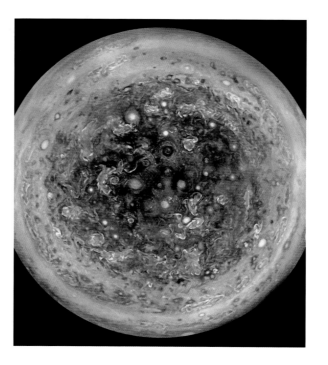

NASA's Juno spacecraft snapped this image of cyclones—some as large as 1,000 kilometers (600 mi) across—swirling on Jupiter's south pole.

gave scientists their first clear glimpse of the swirling blues of the Jovian north pole, where it spotted an octagonal cluster of megastorms. The craft also revealed a similar hexagonal storm cluster near the south pole.

Juno also measured Jupiter's magnetic field, revealing that it's roughly 10 times stronger than Earth's. This magnetic field contributes to the complex—and not yet fully understood—process that generates the glowing auroras that dance at the planet's poles. Jupiter's swift rotation, the fastest in the solar system, likely creates electrical currents that charge the magnetic field.

As Juno continues to collect data, the next generation of explorers is readying for launch. Planned for the mid-2020s, NASA's Europa Clipper and ESA's JUICE (JUpiter ICy moons Explorer) will head to the giant world to get a close look at some of Jupiter's impressive collection of moons.

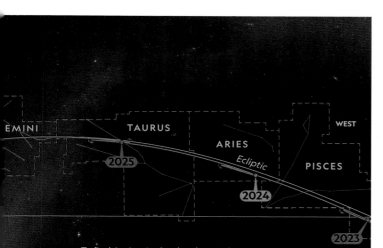

To find Jupiter in the sky, choose the current year, estimate the month along that colored line, and identify the constellation in which it will appear. Refer to the seasonal sky chart for your hemisphere to locate that constellation. If local sky conditions are favorable, Jupiter will be be bright enough to spot with the naked eye.

With steadily held binoculars, you should be able to see Jupiter as a bright, creamy-white disk, as well as its four largest moons—Io, Europa, Ganymede, and Callisto. Because the innermost moon, Io, orbits in less than two days, you can actually witness its position relative to Jupiter change in just an hour or so, while its neighboring moons change position night to night.

Watching Jupiter through time, you may notice that it appears at some times to reverse course and move backward. This is an illusion caused by its slower relative speed around the sun.

COMET COLLISION

In July 1994, scientists watched with anticipation as 21 fragments of comet Shoemaker-Levy 9 slammed into Jupiter with the force of hundreds of millions of atomic bombs. Plumes of material burst from Jupiter's lower atmosphere, generating dark, cloudy rings that slowly dispersed over time. The event was a wake-up call that such collisions can still happen in our solar system.

SATURN

The sixth planet from the sun boasts the solar system's brightest rings and strongest winds.

Encircled by rings in creamy hues, Saturn is perhaps the most recognizable planet in our solar system. While it's not the only world with rings, the sixth planet from the sun lays claim to the biggest and brightest; they're so prominent that earthbound stargazers can see them using even small telescopes. This Saturnian skirt is made up mostly of ice chunks in different sizes, from tiny specks to house-size lumps.

Saturn itself is a gaseous ball of hydrogen and helium and is second only to Jupiter in its size. Without its rings, the planet is about the same width as nine Earths lined up side by side.

Because of Saturn's abundant hydrogen, which is the lightest element, and lower gravity than Jupiter, the planet is the least dense of any in the solar system. It's often said that Saturn could float on water—that is, if you could find a big enough pool and the gaseous planet was somehow made rigid.

Recent data from Cassini revealed that Saturn's core is surprisingly slushy and large, making up 60 percent of the planet's width. It's likely composed of ice and rocks, as well as hydrogen and helium compressed to a liquid under the intense pressures deep in the planet.

JET STREAMS

Saturn also sports jet streams that speed around the world with winds as fast as 500 meters (1,600 ft) per second—almost 4.5 times faster than the strongest winds measured on Earth. The dominant winds swirl around the planet to the east, in the same direction as the body rotates. On rare occasions, observers spot large white storm clouds of ammo-

Average distance from sun: 1.43 billion kilometers (886 million mi)

Mass: 95.2 x Earth's

Length of day: 10.7 hours

Length of year: 29 Earth years

Average temperature: -140°C (-220°F)

Average radius: 58,200 kilometers (36,200 mi)

Number of moons: 82

Planetary ring system: Yes

nia ice billowing up through a tawny layer of cloud-top particles.

One of Saturn's many remarkable features is the hexagonal jet stream that speeds around its north pole. The feature is some 30,000 kilometers (20,000 mi) across and encompasses a rotating mega-hurricane at its center with scattered clusters of smaller gyres. Scientists have not yet found any other storm like it in our solar system, and the precise reason for its unexpected shape remains unknown. Computer simulations hint it might

Cassini was intentionally plunged into Saturn's atmosphere in 2017, orienting its antenna toward Earth to continue sending data as it descended, as depicted in this illustration.

(Opposite) NASA's Cassini spacecraft took this vibrant image of Saturn while passing through the planet's shadow. The sun lit up the planet's rings from behind, creating a stunningly detailed view of the delicate structures.

Saturn's tiny moon Daphnis kicks up waves within the planet's rings, shown in an enhanced-color image mosaic. A thin strand of ring trails in the moon's wake.

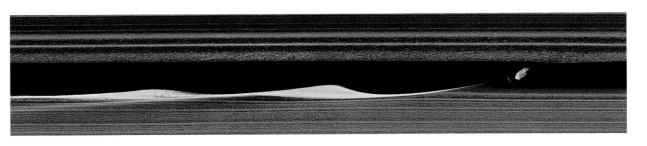

Saturn boasts the solar system's most impressive system of rings, which are made up of ice and rock chunks of vastly different sizes.

come from turbulence whose roots extend deep below the planet's amaretto clouds.

EXPLORING THE RINGED GIANT

On September 15, 2017, scientists gathered with tears and cheers to bid a tiny spacecraft farewell. Some 1.5 billion kilometers (932 million mi) away, the Cassini orbiter was taking a deadly—but planned—plunge into Saturn's atmosphere. It had circled the planet for 13 years, snapping more than 450,000 images and opening scientists' minds to the remarkable diversity of worlds that exist in our solar system.

Scientists got their first look at Saturn up close in 1979, when NASA's Pioneer 11 spacecraft flew by the planet, thanks to a rare alignment of the outer worlds. More glimpses swiftly followed as Voyager 1 and 2 rocketed past in 1980 and 1981. These spacecraft gave scientists stunning views of the planet's storms, discovered new moons, and provided clues to the planet's magnetic field and atmospheric makeup.

Cassini launched in October 1997, entering orbit around Saturn in the summer of 2004. The craft carried the European Huygens lander that touched down on Saturn's moon Titan in 2005. Meanwhile, Cassini spent more than a decade collecting data on Saturn and its menagerie of moons, including sampling liquid spewed from geysers from one moon's surface. Once Cassini's fuel began running low, scientists orchestrated 22 thrilling orbits through the gap between Saturn and its rings, before ending with the robotic emissary's plunge into the planet's atmosphere. The suicide mission prevented the craft from possibly contaminating Saturn's moons, which just might harbor life, and allowed the craft to collect a remarkable set of data—right up to the moment it disintegrated in a hot flash.

THE PLANETARY HOOPS

Saturn appears to have seven rings, but each is made up of thousands of ringlets that vary in size.

Continued on page 142

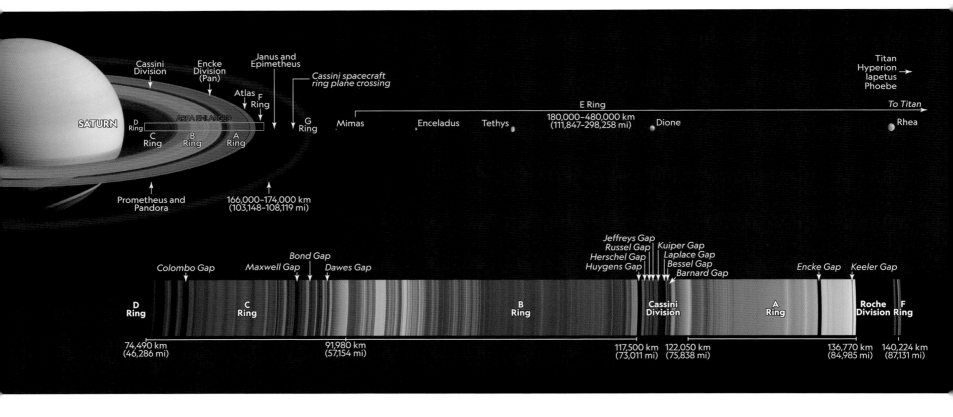

SATURN

North Pole

75°   75°

60°   North Polar Region   North Polar Region   60°

45°   North North Temperate Zone   North North Temperate Zone   45°

North North Temperate Belt   North North Temperate Belt

30°   North Temperate Zone   North Temperate Zone   30°

North Temperate Belt   North Temperate Belt

North Tropical Zone   North Tropical Zone

North Tropical Belt   North Tropical Belt

15°   15°

EQUATOR   Equatorial   Equatorial   EQUATOR   0°

Zone   Zone

15°   South Tropical Belt   South Tropical Belt   15°

South Tropical Zone   South Tropical Zone

South Temperate Belt   South Temperate Belt

South Temperate Zone   South Temperate Zone

30°   South South Temperate Belt   South South Temperate Belt   30°

South South Temperate Zone   South South Temperate Zone

45°   45°

South Polar Region   South Polar Region

60°   60°

75°   75°

South Pole

Orthographic Projection
SCALE at the EQUATOR 1:788,260,000
1 CENTIMETER = 7,883 KILOMETERS; 1 INCH = 12,422 MILES

0   10,000   20,000
STATUTE MILES
KILOMETERS
0   10,000   20,000

Saturn's surface is faintly striped in gold and yellow bands of clouds from storms and jet streams. Brewing storms occasionally cause white spots or streaks to appear.

MIMAS
(MOON)

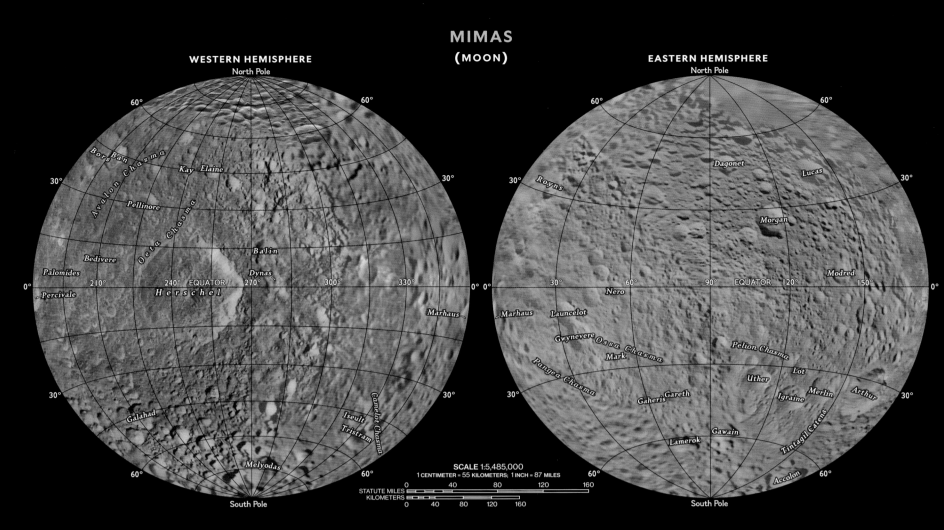

WESTERN HEMISPHERE
North Pole

Bors Ban
Avalon Chasma
Kay
Elaine
Pellinore
Octa Chasma
Balin
Bedivere
Palomides
Percivale
Dynas
Herschel
Marhaus
Galahad
Camelot Chasma
Iseult
Tristram
Melyodas
South Pole

60° 60° 30° 30° 0° 0° 30° 30° 60° 60°
210° 240° EQUATOR 270° 300° 330°

EASTERN HEMISPHERE
North Pole

Dagonet
Lucas
Royns
Morgan
Modred
Nero
Marhaus
Marhaus
Launcelot
Gwynevere Ossa Chasma
Mark
Pelion Chasma
Pangea Chasma
Lot
Uther
Gaheris Gareth
Igraine
Merlin
Arthur
Tintagil Catena
Lamerok
Gawain
Accolon
South Pole

60° 60° 30° 30° 0° 0° 30° 30° 60° 60°
30° 60° 90° EQUATOR 120° 150°

SCALE 1:5,485,000
1 CENTIMETER = 55 KILOMETERS; 1 INCH = 87 MILES
STATUTE MILES   0   40   80   120   160
KILOMETERS   0   40   80   120   160

Saturn's tiny moon Mimas is almost entirely water ice and features an eye-catching impact crater that extends across a third of its diameter. The impact may even have sent shock waves that fractured the moon's opposite side.

ENCELADUS
(MOON)

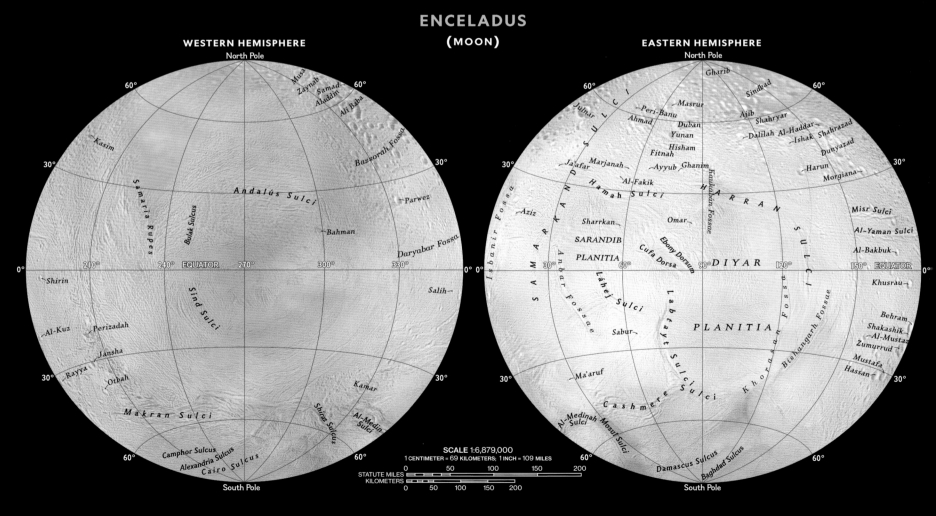

WESTERN HEMISPHERE
North Pole

Musa
Zaynab
Samad
Aladdin
Ali Baba
Kasim
Bassorah Fossa
Samaria Rupes
Andalús Sulci
Bulak Sulcus
Parwez
Bahman
Daryabar Fossa
Shirin
Salih
Al-Kuz
Perizadah
Sind Sulcus
Jansha
Rayya
Otbah
Kamar
Makran Sulci
Shiraz Sulcus
Al-Medin Sulci
Camphor Sulcus
Alexandria Sulcus
Cairo Sulcus
South Pole

60° 60° 30° 30° 0° 0° 30° 30° 60° 60°
210° 240° EQUATOR 270° 300° 330°

EASTERN HEMISPHERE
North Pole

Gharib
Sindbad
SAMARKAND SULCI
Juhar
Peri-Banu
Masrur
Ajib
Shahryar
Ahmad
Duban
Dalilah Al-Haddar
Ishak Shahrazad
Yunan
Hisham
Dunyazad
Fitnah
Ayyub Ghanim
Harun
Ja'far
Marjanah
Morgiana
Al-Fakik
Hamah Sulci
Fankalah Fossae
Misr Sulci
Aziz
Al-Yaman Sulci
Sharrkan
Omar
Al-Bakbuk
SARANDIB PLANITIA
Ebony Dorsum
DIYAR
Anbar Fossae
Cufa Dorsa
Labtayt Sulci
Khusrau
Lahej Sulci
PLANITIA
Behram
Sabur
Shakashik
Al-Mustaz
Zumurrud
Ma'aruf
Mustafa
Hassan
Al-Medinah Mosul Sulci
Cashmere Sulci
Khorasan Fossa
Bishangarh Fossae
Sulci
Damascus Sulcus
Baghdad Sulcus
South Pole

60° 60° 30° 30° 0° 0° 30° 30° 60° 60°
30° 60° 90° DIYAR 120° 150° EQUATOR

SCALE 1:6,879,000
1 CENTIMETER = 69 KILOMETERS; 1 INCH = 109 MILES
STATUTE MILES   0   50   100   150   200
KILOMETERS   0   50   100   150   200

Enceladus is a small, icy world that harbors a subsurface ocean. A series of deep crevasses near its south pole, known as "tiger stripes," spurt water vapor, gases, and organic molecules into the sky.

TETHYS
(MOON)

WESTERN HEMISPHERE
North Pole

EASTERN HEMISPHERE
North Pole

South Pole

South Pole

SCALE 1:14,433,000
1 CENTIMETER = 144 KILOMETERS; 1 INCH = 228 MILES

STATUTE MILES 0 100 200 300 400
KILOMETERS 0 100 200 300 400

Tethys is mostly water ice and hosts two primary features: Odysseus crater and Ithaca Chasma valley. The crater-forming impact may have cracked open the valley, or it could have formed as the orb's internal water froze and expanded.

IAPETUS
(MOON)

WESTERN HEMISPHERE
North Pole

EASTERN HEMISPHERE
North Pole

South Pole

South Pole

SCALE 1:20,201,000
1 CENTIMETER = 202 KILOMETERS; 1 INCH = 319 MILES

STATUTE MILES 0 200 400 600
KILOMETERS 0 200 400 600

Iapetus is made up of a mixture of ice and rock and has intriguingly dark and light sides. The stark coloring may partly come from the moon's slow rotation, which allows the dark side to absorb excess heat, driving away light-colored ice.

THE COSMIC TEACUP

When Galileo first spotted Saturn's rings in July of 1610, he mistook them for a pair of moons or planets. After further observation, he concluded the planet had handles, like a cosmic teacup. It took four more decades of observation before another astronomer, Christiaan Huygens, finally realized that Saturn wasn't part of a planetary tea set at all, but rather was decorated with a ring. The low power of early telescopes didn't resolve the rings clearly, making Galileo's early mix-up understandable. Want to see Saturn's rings yourself? While it's tough to get a perfect view, the rings can still show up on small telescopes with just 25 times magnification and an aperture of at least 50 mm (2 in).

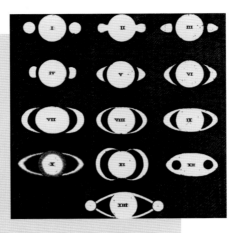

Evolving conceptions of Saturn and its rings through the 17th century: The first is by Galileo Galilei, the last, as shown, is by Christiaan Huygens.

Continued from page 138

For all we've learned about these striking features, scientists still don't agree on exactly how or even when they formed. Data from Cassini's daring dives between Saturn and its rings hinted that the rings are a recent addition, forming less than 100 million years ago, when dinosaurs still roamed Earth. Perhaps they are captured comets and asteroids that were shredded in Saturn's crushing gravity, or even moons that crashed into other bodies. Other scientists believe they are as old as the planet itself. The debate over their age and origin is far from settled.

Some of Saturn's moons, known as "shepherd moons," still actively shape the rings today through their gravitational influence. The tiny orbs Prometheus and Pandora flank the planet's outermost ring, which keeps it strikingly slender. Some moons also orbit within the rings, their gravity clearing small gaps in the bands. Daphnis, for example, cuts a narrow band through Saturn's large "A" ring, sculpting ripples along one edge as it zooms by.

MANY MARVELOUS MOONS

With 82 orbiting worlds, Saturn currently boasts the most moons in our solar system. The largest are round bodies like our own planet's companion. But Saturn's smaller satellites take on all kinds of shapes: potato-shaped Pandora, ravioli-shaped Pan and Atlas, and more.

Each moon is a world apart. Craters speckle the surface of the oddball Hyperion, making it look like a sponge. As it orbits, Hyperion tumbles in an unpredictable pattern. Another satellite, Phoebe,

TRACK CHART FOR SATURN

This sky chart shows the apparent motion of Saturn in relation to the constellations along the ecliptic during a 12-year period, the time it takes the planet to complete one orbit around the sun. Colored lines represent one year.

EAST

NORTH

CANCER

Chart Key
2022 Planet position on January 1

LEO

2038

2036

2035

2037

VIRGO

2040

Celestial Equator

PISCES

2041

2039

AQUARIUS

OPHIUCHUS

LIBRA

2042

2026

Ecliptic

2022

2021

2043

SOUTH

2025

2020

2019

2018

2017

2016

2044

2024

2023

2045

CAPRICORNUS

SAGITTARIUS

SCORPIUS

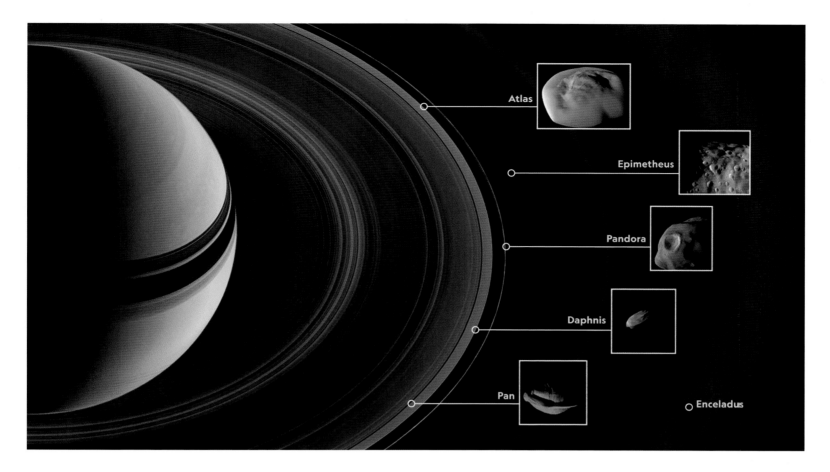

The following moons are labeled in the image: Atlas, Epimetheus, Pandora, Daphnis, Pan, Enceladus.

orbits in the opposite direction from most other moons—a hint that it may have been captured by Saturn sometime after the planet's formation. Some scientists believe the contrary moon was once a "centaur," a primordial object from the Kuiper belt that migrated toward the inner solar system.

Saturn's largest moon, Titan, is the only satellite in the solar system known to have its own dense atmosphere: a hefty brew made mostly of nitrogen and methane. Titan also has clouds, rain, rivers, lakes, and seas—but they're all made of hydrocarbons such as methane and ethane. Underlying its fluid features is a crust of water ice that may cap a subsurface ocean of water that's slightly salty. The moon's watery resources make this world a prime candidate in the search for life.

Enceladus is also on the list of potentially life-friendly worlds. The moon is the whitest object in the solar system—the result of a water-ice layer that coats this world. In 2005, Cassini spotted icy water spurting from the moon's surface supplying material to one of Saturn's nearby rings. Most of the spray falls back down as bright white snow.

NASA's Cassini mission flew through this misty spray to discover a surprising mix of components, including not just water but also silica, salts, nitrogen-containing molecules, and even carbon compounds. Further study of this world hints at the presence of hydrothermal vents, which on Earth support a marvelous menagerie of life-forms. Taken together, the discoveries suggest the moon hosts the ingredients necessary for life as we know it.

Images of Saturn's moons taken by NASA's Cassini spacecraft: Pan, Daphnis, Pandora, Epimetheus, and Atlas. Enceladus bathes Epimetheus in an icy spray (not pictured).

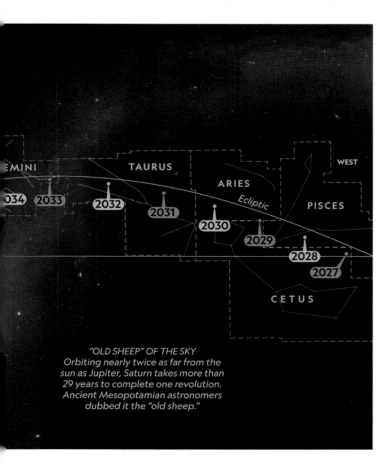

"OLD SHEEP" OF THE SKY
Orbiting nearly twice as far from the sun as Jupiter, Saturn takes more than 29 years to complete one revolution. Ancient Mesopotamian astronomers dubbed it the "old sheep."

URANUS

Once thought to be a bland and boring world, the seventh planet from the sun is full of surprises.

U ranus is about four times wider than our home world, but as it orbits at a stunning 2.9 billion kilometers (1.8 billion mi) from the sun, this giant planet is barely visible to the naked eye. British astronomer William Herschel first spotted its faint glow in his telescope in March 1781, but he initially mistook it for a comet. It took two more years of observations from other astronomers for scientists to accept that the tiny glimmer was actually a planet.

Herschel attempted to name the newfound world Georgium Sidus ("George's star"), after King George III. But the scientific community settled on the name Uranus, after the Greek god of the sky, who was the son and then husband of the goddess of Earth, Gaea.

Both Uranus and Neptune are known as ice giants, thanks to interiors that differ from their gas giant neighbors. While all these giants have atmospheres containing hydrogen and helium, Uranus and Neptune also have ice-forming vapors of water, methane, and ammonia. Yet deep under their cloud tops, temperatures soar far too hot for ice to form. Pressure also increases and eventually forces the vapors to shift to a dense, scalding fluid that envelops a small rocky core. Uranus's core could be hotter than 5000°C (9000°F).

The composition of this pair of planets has posed a bit of a mystery. As the planetary embryos took shape in the early days of our solar system, the far reaches of the protoplanetary disk—where Neptune and Uranus are today—would not have had the right mix of elements to form these giant worlds. One possible explanation is that the planetary duo didn't form in place but instead took shape closer

Average distance from sun: 2.87 billion kilometers (1.78 billion mi)

Mass: 14.5 x Earth's

Length of day: 17 hours

Length of year: 84 Earth years

Average temperature: -195°C (-320°F)

Average radius: 25,400 kilometers (15,800 mi)

Number of moons: 27

Planetary ring system: Yes

to the sun before sailing to the farthest edges of our cosmic neighborhood. Yet not all models agree, and a closer look at these distant worlds is necessary to unwind the planets' past.

THE (NOT SO) PLAIN PLANET

Compared to Saturn's sweeping rings or Jupiter's stormy marbled surface, Uranus's uniform azure

A diagram shows Uranus's rings (white), magnetic field lines (narrow blue lines), and axis of rotation (red). The magnetic field is offset from the planet's center and tilted relative to the planet's rotation, creating a corkscrew-shaped magnetotail as it spins.

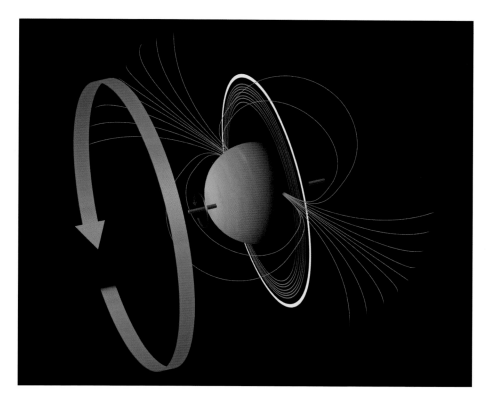

(Opposite) An illustration shows Uranus and its rings, displaying the off-kilter rotation that puts it nearly on its side.

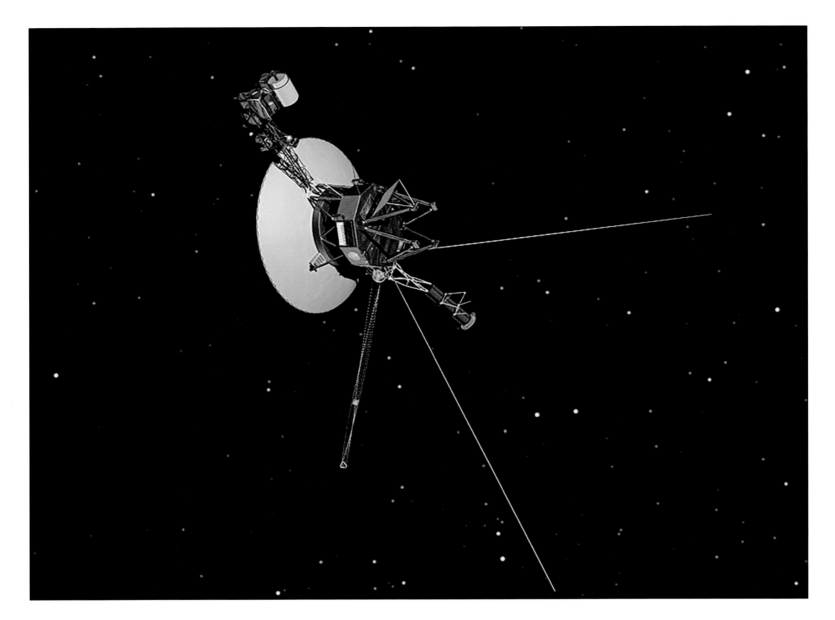

An illustration of NASA's Voyager spacecraft shows its trek through space, stars glinting in the distance.

hues may seem downright bland. But the turquoise world has many oddities to offer, including the fact that it isn't always so plainly colored.

First there's Uranus's strange sideways spin. The phenomenon might be the result of an ancient collision with an Earth-size body, which toppled the planet onto its side. The result is an axis that lies nearly parallel to the plane in which the planet transits the sun. Similar to Venus, Uranus also has a retrograde spin, the opposite of other worlds.

Uranus's magnetosphere also reflects some of the planet's wild ways. On other worlds in the solar system, the magnetic field aligns with the planet's axis of rotation, but Uranus's magnetosphere sits off kilter by some 60 degrees and is offset from the planet's center. This topsy-turvy alignment means that the planet's magnetotail—the trail of magnetic field lines blown back by the solar winds—twists into a lengthy corkscrew.

Voyager 2 is the only spacecraft ever to visit the seventh planet from the sun; it flew by in January 1986 while on its grand tour of the outer planets. The spacecraft snapped thousands of images and recorded other data, whizzing past as close as 81,500 kilometers (50,600 mi) above the planet.

These images depict a quiescent world with few cloudy features. The planet partly owes its uniform looks to methane in its upper atmosphere, where it forms a haze that swaths the planet and obscures the potentially storm-forming lower atmosphere from view. The haze comes from two different sources, including methane turning to particulates from reactions sparked by UV rays, as well as from the gas freezing in the chilly temperatures at the outer reaches of Uranus's atmosphere.

Yet there's more to this uniform world than initially met the eye. In the years that followed Voyager's flyby, scientists spotted cream-colored clouds and dark vortices whipping up in the planet's atmosphere.

URANUS

North Pole

75°   75°

60°   60°

45°   45°

30°   30°

15°   15°

0°   EQUATOR   EQUATOR   0°

15°   15°

30°   30°

45°   45°

60°   60°

75°   75°

South Pole

SCALE at the EQUATOR 1:352,160,000
1 CENTIMETER = 3,521 KILOMETERS; 1 INCH = 5,558 MILES

0   2,000   4,000   6,000   8,000
STATUTE MILES
KILOMETERS
0   2,000 4,000 6,000 8,000

Similar to its gaseous neighbors, Uranus has abundant hydrogen and helium, but the
planet is also composed of ice-forming vapors of water, methane, and ammonia.

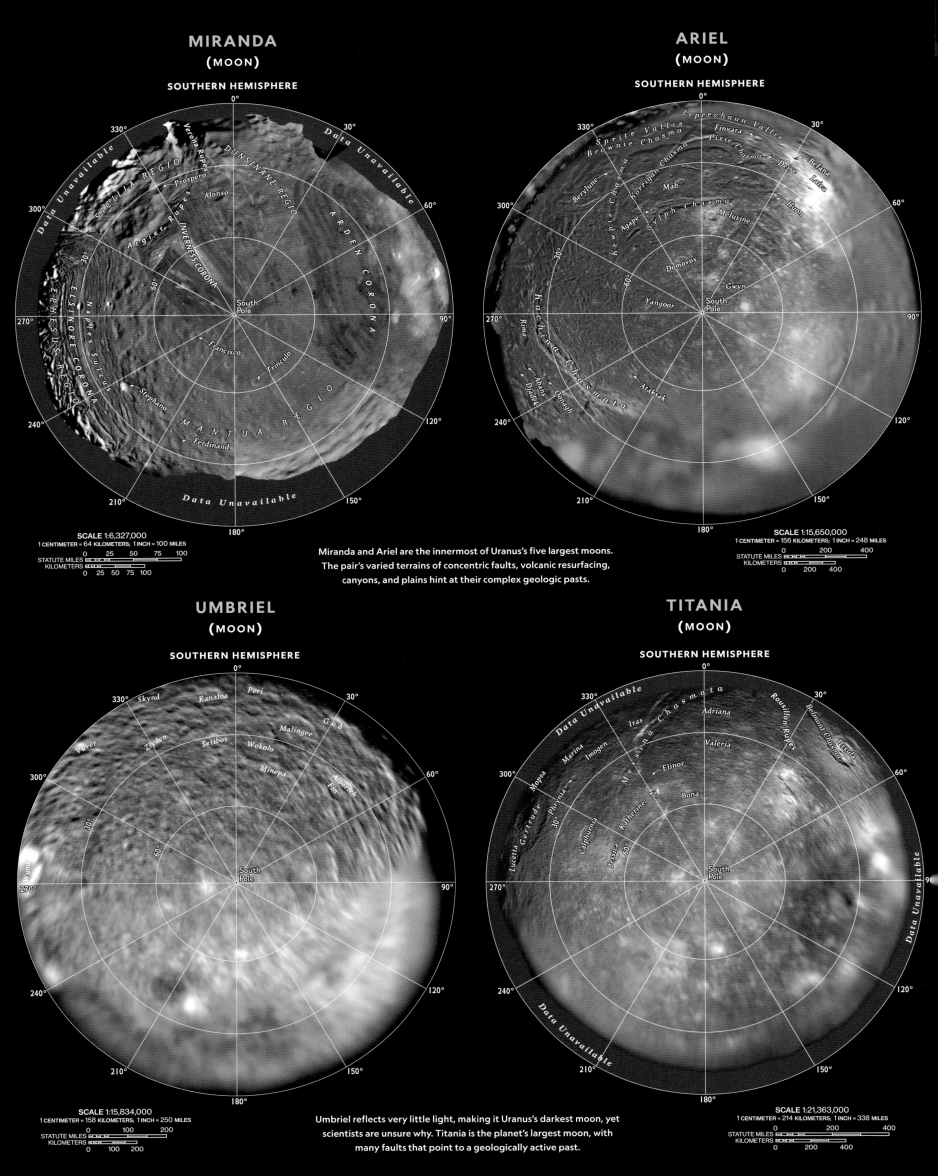

MIRANDA
(MOON)

SOUTHERN HEMISPHERE

0°

330°

Vecoria Rupes

Data Unavailable

30°

SICILIA REGIO

Prospero

DUNSINANE REGIO

Data Unavailable

60°

300°

Argier Rupes

Alonso

INVERNESS CORONA

ARDEN CORONA

Data Unavailable

ELSINORE CORONA

Naples Sulcus

EPHESUS REGIO

270°

South Pole

90°

Francisco

240°

Trinculo

120°

Stephano

MANTUA REGIO

Ferdinand

210°

Data Unavailable

150°

180°

SCALE 1:6,327,000
1 CENTIMETER = 64 KILOMETERS; 1 INCH = 100 MILES

0 25 50 75 100

STATUTE MILES
KILOMETERS

0 25 50 75 100

ARIEL
(MOON)

SOUTHERN HEMISPHERE

0°

330°

Sprite Vallis

Brownie Chasma

Leprechaun Vallis

Finvara

Pixie Chasma

Deive

Befana

30°

Berylune

Kewpie Chasma

Korrigan Chasma

Mab

Huon

Laica

60°

300°

Agape

Sylph Chasma

Melusine

Kachina Chasmata

Domovoy

Gwyn

270°

Yangoor

South Pole

90°

Rima

Albans

Djadek

Oonagh

Ataksak

240°

120°

210°

150°

180°

SCALE 1:15,650,000
1 CENTIMETER = 156 KILOMETERS; 1 INCH = 248 MILES

0 200 400

STATUTE MILES
KILOMETERS

0 200 400

Miranda and Ariel are the innermost of Uranus's five largest moons.
The pair's varied terrains of concentric faults, volcanic resurfacing,
canyons, and plains hint at their complex geologic pasts.

UMBRIEL
(MOON)

SOUTHERN HEMISPHERE

0°

330°

Skynd

Kanaloa

Peri

30°

Gob

Zlyden

Setibos

Malingee

Wokolo

Vuver

60°

300°

Minepa

Alberich

Fin

270°

South Pole

90°

Wunda

240°

120°

210°

150°

180°

SCALE 1:15,834,000
1 CENTIMETER = 158 KILOMETERS; 1 INCH = 250 MILES

0 100 200

STATUTE MILES
KILOMETERS

0 100 200

TITANIA
(MOON)

SOUTHERN HEMISPHERE

0°

330°

Data Unavailable

Iras

Adriana

Rousillon Rupes

Belmont Chasma

30°

Marina

Imogen

Messina Chasmata

Valeria

Ursula

Mopsa

Elinor

60°

300°

Lucetta

Gertrude

Phrynia

Bona

Calphurnia

Jessica

Katherine

270°

South Pole

Data Unavailable

90°

240°

120°

210°

Data Unavailable

150°

180°

SCALE 1:21,363,000
1 CENTIMETER = 214 KILOMETERS; 1 INCH = 338 MILES

0 200 400

STATUTE MILES
KILOMETERS

0 200 400

Umbriel reflects very little light, making it Uranus's darkest moon, yet
scientists are unsure why. Titania is the planet's largest moon, with
many faults that point to a geologically active past.

EXTREME SEASONS

Uranus's occasional spate of storms may, in part, come from the planet's sideways spin. During Uranus's summer and winter, each of which lasts 21 Earth years, the sun directly roasts one pole while the opposite side languishes in darkness. But as Uranus orbits to its equinox the sun illuminates the planet's equator where it produces days and nights of nearly equal length.

As the planet approached its equinox in 2007, the weather seemed to worsen as sunlight bathed once dark zones. But Uranus surprisingly hit a stormy peak in 2014, when scientists expected rough weather to wane. Unusually bright clouds popped up in the planet's upper atmosphere, and storms swirled across the northern hemisphere. Scientists are still working out why.

POETIC MOONS AND RINGS

Uranus has more rings than scientists initially thought, with the current count at 13. Orbiting among the rings are a total of 27 moons, many of which are tiny—some estimated to be a mere 18 kilometers (11 mi) across. The silvery moons that orbit other planets typically get their names from ancient mythology, but Uranus's moons have a bit more poetic flair. Most are named after Shakespearean characters, with a couple from the writings of Alexander Pope.

Uranus's largest moons are Titania and Oberon, the latter of which is heavily pockmarked. They were the first of Uranus's satellites discovered and were identified by William Herschel in 1787. William Lassell, the astronomer who discovered Neptune's first moon, found Uranus's next two, Ariel and Umbriel.

A century later, Gerard Kuiper discovered Miranda, one of the planet's strangest moons. A massive canyon up to 12 times as deep as the Grand Canyon on Earth splits Miranda's surface. But no more moons were found until NASA's Voyager 2 sailed by. In 1986, that spacecraft found an additional 10 moons—the inner satellites Juliet, Puck, Cordelia, Ophelia, Bianca, Desdemona, Portia, Rosalind, Cressida, and Belinda— all of which are less than about 154 kilometers (96 mi) across. (An additional moon, Perdita, was also in Voyager's

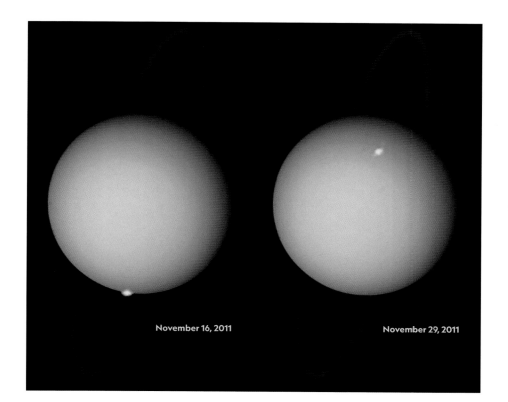

November 16, 2011   November 29, 2011

These composite images are among the first clear pictures showing Uranus's aurorae (shown as white spots), as captured by the Hubble Space Telescope.

images but it wasn't identified until 1999.) Cordelia and Ophelia are shepherd moons, keeping one of Uranus's thin rings in check. Beyond lies a swarm of other tiny inner moons that's so crowded it's a surprise to scientists they don't collide. It's possible this cluster shepherds some of Uranus's other rings.

Some of Uranus's moons seem to be composed of half water and half rock, but exactly what many of them are remains unknown. The planet might even be hiding two more moons around its narrow dusty rings—one of many mysteries this distant world still holds.

CAROLINE AND WILLIAM HERSCHEL

Siblings Caroline and William Herschel—born in 1750 and 1738, respectively, in Hanover, Germany—were trailblazers in charting the night skies. Both trained as musicians, following in their father's footsteps, but they also had a keen interest in astronomy. William began building telescopes that refined the optics of Isaac Newton's designs. In 1781, after both siblings had moved to England, William Herschel discovered the planet Uranus, initially believing it was a comet. The following year, King George III appointed him his private astronomer. Caroline began work as William's assistant and she made many notable discoveries, including—by her estimation—14 nebulae in less than a year and a half. The two astronomers filled in many blanks. William cataloged thousands of nebulae and star clusters, while Caroline published a list of corrections to past stellar observations and added hundreds of stars.

An illustration of William Herschel's 12-meter (40-ft) telescope, constructed between 1785 and 1789

NEPTUNE

The solar system's eighth planet orbits so far away that sunlight barely reaches it.

If you have good eyesight and dark clear skies, you can see almost every planet in our solar system with the naked eye—except for Neptune. Even Uranus is visible without a telescope when it makes its closest pass to Earth. But at some 4.5 billion kilometers (2.8 billion mi) from the sun, Neptune is the most distant planet in our solar system and is a dark and frigid world. Temperatures near its cloud tops are hundreds of degrees below zero, and even at Neptunian high noon the skies are lit only with a dim glow.

Using his telescope, Galileo identified Neptune as a star in the early 1600s. Scientists didn't realize it was a planet until more than two centuries later. As astronomers tracked the newly discovered Uranus in its orbit around the sun, they noticed it didn't quite follow the path mathematics had predicted. The English astronomer John Couch Adams and the French mathematician Urbain Jean Joseph Le Verrier independently proposed that the deviation came from a yet undiscovered planet that lay even farther away from the sun.

In the fall of 1846, Le Verrier sent Johann Gottfried Galle at the Berlin Observatory a missive with the calculations for his planetary prediction. The night Galle received the letter, he turned a telescopic eye to the sky and quickly spotted Neptune, almost precisely where Le Verrier thought it would be. The world was named Neptune, per Le Verrier's suggestion, after the Roman god of the sea.

THE BLANKET OF BLUE

Despite its mythological name, Neptune's blue color doesn't come from global seas, but from methane gas in its atmosphere. Its thick atmospheric blan-

Average distance from sun: 4.50 billion kilometers (2.80 billion mi)

Mass: 17.1 x Earth's

Length of day: 16 hours

Length of year: 165 Earth years

Average temperature: -200°C (-330°F)

Average radius: 24,600 kilometers (15,300 mi)

Number of moons: 14

Planetary ring system: Yes

ket also contains hydrogen and helium, similar to its neighbor Uranus. The fact that it is a slightly brighter blue compared to Uranus's blue-green hue, however, hints at an additional but unknown atmospheric ingredient.

Winds whip around the planet at the fastest speeds in the solar system—more than 2,000 kilometers an hour (1,200 mph)—a supersonic rate on Earth. Solar heating drives Earth's winds, but the sun's rays barely reach this distant world. How can the planet's winds blow stronger than Jupiter's and

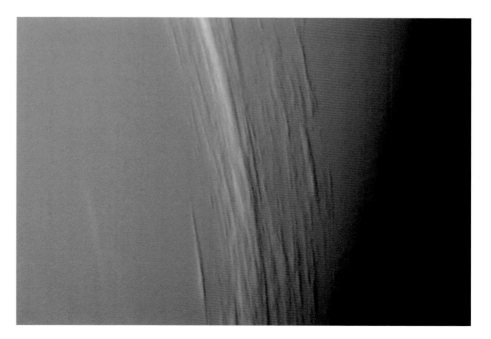

This high-resolut on image taken by Vcyager 2 shows prominent cloudy streaks in Neptune's atmosphere near the region where day turns to night.

(Opposite) Neptune's tiny moon Hippocamp, which has an average width of 34 kilometers (21 mi), looms large in this illustration. Scientists identified the little world in images taken by Hubble between 2004 and 2016.

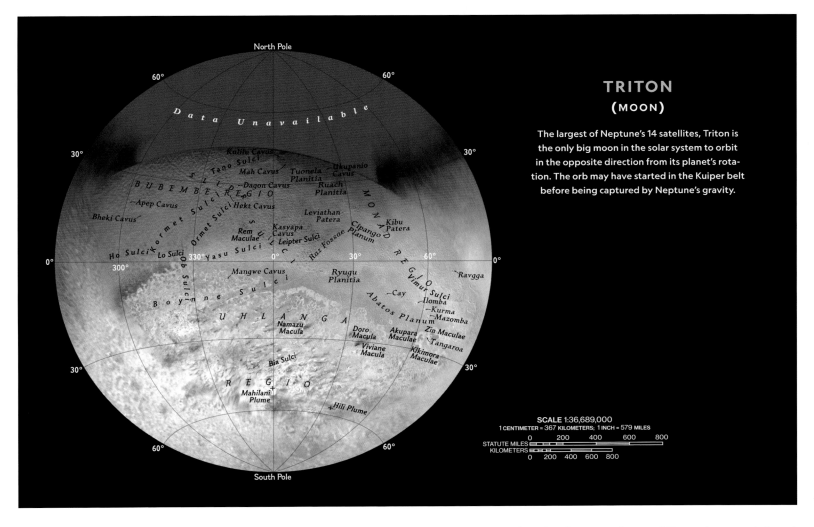

TRITON
(MOON)

The largest of Neptune's 14 satellites, Triton is the only big moon in the solar system to orbit in the opposite direction from its planet's rotation. The orb may have started in the Kuiper belt before being captured by Neptune's gravity.

SCALE 1:36,689,000
1 CENTIMETER = 367 KILOMETERS; 1 INCH = 579 MILES

almost five times stronger than Earth's? The driving force behind this churn may instead lie in the temperature and composition beneath the clouds—but scientists have yet to explore this windy world. As with Uranus, the only craft to visit Neptune was Voyager 2, which sped by around 4,900 kilometers (3,000 mi) above the planet's north pole in 1989, 12 years after it launched from Earth.

Neptune's magnetic field is about 27 times stronger than Earth's, but it sits crooked at about 47 degrees from the planet's axis of rotation.

MOONS AND RINGS

A mere 17 days after Neptune's discovery, beer brewer and amateur astronomer William Lassell (who also discovered two of Uranus's moons) found its largest moon, Triton, using a telescope he made himself.

The more scientists learn about Triton, the weirder it seems. The moon is remarkably similar to the dwarf planet Pluto; it's only slightly larger in size and has similar density and a surface that is rich in nitrogen, methane, and carbon monoxide. This similarity may mean that Triton originated from the Kuiper belt along with Pluto but was later captured by Neptune's gravity—an idea supported by Triton's retrograde orbit, opposite to Neptune's spin.

When Voyager 2 soared by Triton, it captured evidence of active ice volcanoes, features that spew geysers of watery slush into the skies. Triton is just one of the planet's 14 known moons, the last of which, Hippocamp, was discovered in 2013—and there may still be more to find.

Voyager also confirmed the presence of at least five thin rings around Neptune, the outermost of which contains four dusty clumps called arcs—two of which have faded since 1989. Exactly what keeps the remaining pair intact, rather than spreading evenly into a cosmic hoop, is uncertain.

The oval shape of Pluto's orbit means that for 20 years out of every 248-year-long trip around the sun, Pluto is **closer to the star than Neptune.**

NEPTUNE

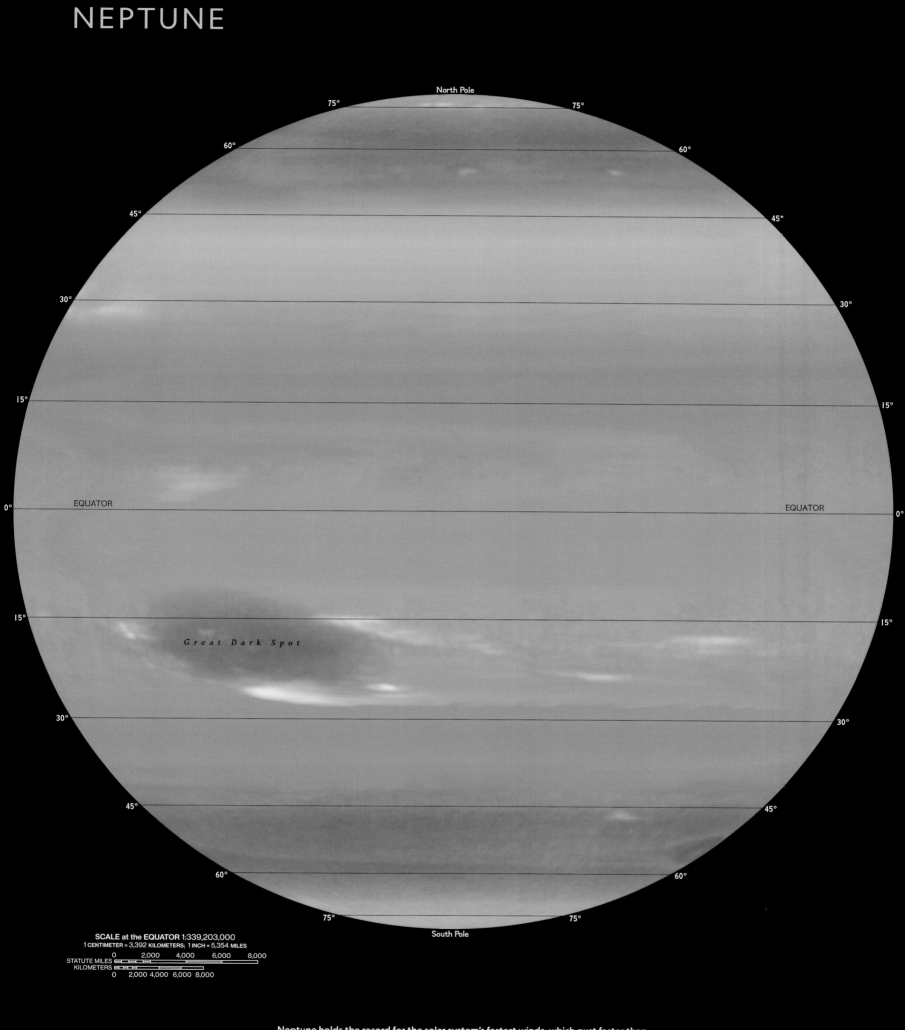

North Pole

75°   75°

60°   60°

45°   45°

30°   30°

15°   15°

0° EQUATOR   EQUATOR 0°

15° *Great Dark Spot*   15°

30°   30°

45°   45°

60°   60°

75°   75°

South Pole

SCALE at the EQUATOR 1:339,203,000
1 CENTIMETER = 3,392 KILOMETERS; 1 INCH = 5,354 MILES
  0 2,000 4,000 6,000 8,000
STATUTE MILES
KILOMETERS
  0 2,000 4,000 6,000 8,000

Neptune holds the record for the solar system's fastest winds, which gust faster than
sound travels on Earth. Neptune's storms show up as dark spots, but unlike those on
Jupiter, they may be born and die within several years.

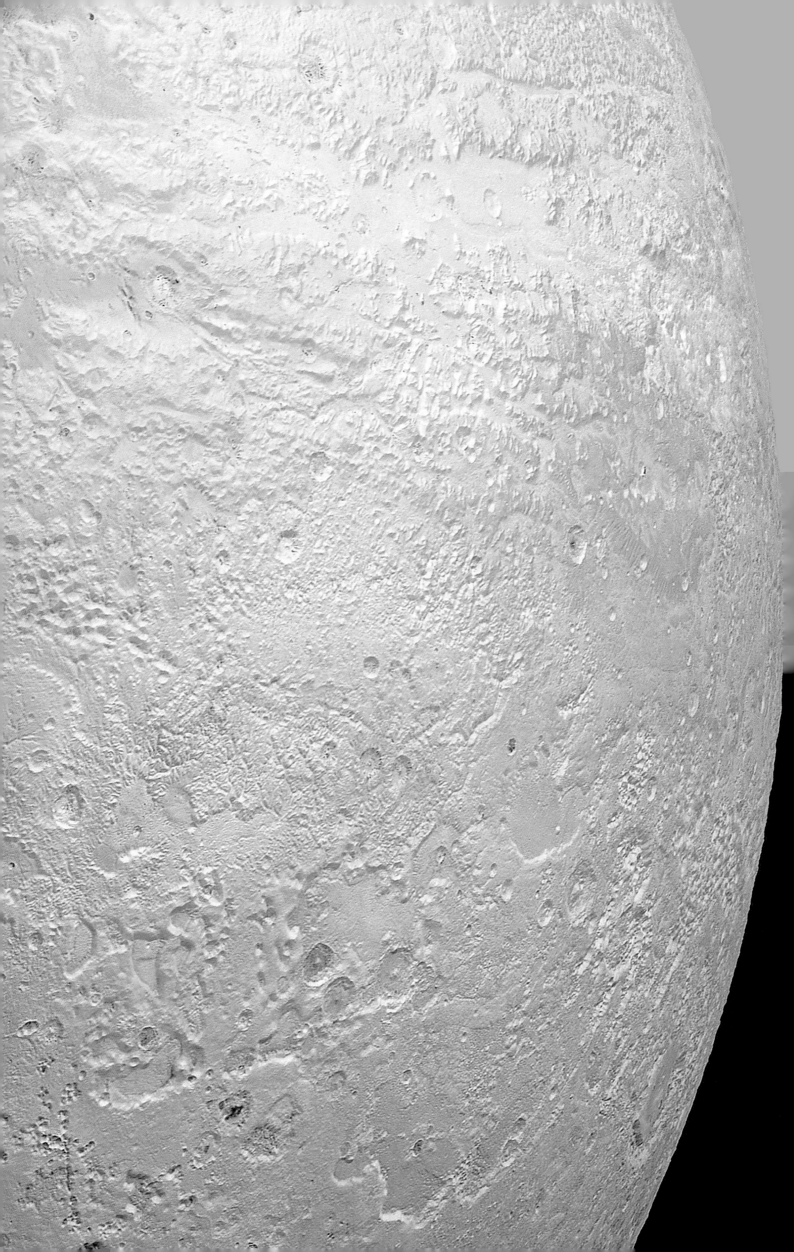

DWARF PLANETS

These tiny worlds blur the planetary lines and offer an oversize dose of intrigue.

I n 2006, our solar system lost a planet. After 76 years of holding the title of ninth planet in our system, Pluto was downgraded to dwarf planet.

The announcement came on the heels of the 2005 discovery of Eris, a Pluto-size body found in the Kuiper belt outside Neptune's orbit. Eris was initially hailed as our solar system's 10th planet. Yet its discovery kicked up a cloud of questions about what truly constitutes a planet. A growing number of smaller bodies in the zone beyond Neptune, including the planet-like Makemake discovered just a few months after Eris, began blurring the lines between planets and all other space rocks—and researchers couldn't agree on a cutoff.

The following year, a vigorous debate over the definition of a planet kicked off at a meeting of the International Astronomical Union (IAU), the keeper of names for all things celestial. The resulting vote moved not only Pluto but also Eris into the new category of "dwarf planet."

The new rules require an object to meet three conditions to be considered a planet. First, it must orbit the sun. Second, it must have enough mass that its gravity pulls it into a nearly round shape. And finally, it must be large enough that its gravity clears away other objects in its orbital path around our star. This final characteristic is the one Pluto doesn't achieve. As of 2021, the IAU recognizes five dwarf planets: Pluto, Eris, Ceres, Makemake, and Haumea. But many more are suspected to exist.

PLUTO: A TINY WORLD WITH BIG SURPRISES

On February 18, 1930, former farm boy Clyde Tombaugh spotted a moving black spot in a set of photographic plates at the Lowell Observatory in Flagstaff, Arizona. He had found what would come to be known as Pluto.

Orbiting an average of 5.9 billion kilometers (3.7 billion mi) from the sun, the dwarf planet largely remained a mystery to scientists until the arrival of the New Horizons spacecraft to it in 2015. The mission captured images of diverse features in an actively changing landscape, ranging from deep valleys to towering crags of water ice.

The rainbow of hues in this false-color image from NASA's Dawn mission reveals materials flowing inside and outside a crater on the asteroid Vesta.

With temperatures hovering around -225°C (-375°F), the dwarf planet's surface is made up of a variety of ices—nitrogen, carbon monoxide, methane, water, and ammonia. Splotches of red color the world, but scientists still aren't sure what causes the color.

One of New Horizons' earliest glimpses of Pluto revealed a bright heart-shaped zone, later named Tombaugh Regio. The western lobe, known as Sputnik Planitia, is an ancient impact crater now filled with a glacier roughly four kilometers (2.5 mi) thick, some 1,000 kilometers (620 mi) across—the largest glacier yet found in our solar system.

(Opposite) NASA's New Horizons spacecraft took this strikingly detailed image of the complex geology in Pluto's north polar region.

The **icy filling** of Pluto's heart-shaped feature grew so massive it may have caused the orb to tip over.

■ For more on asteroids,
see pages 159–161.

Beneath Pluto's frozen surface may lie a liquid water ocean, which computer models suggest could have arisen early in the planet's history. This liquid might drive some of Pluto's modern tectonic activity, splitting open deep fractures as the subsurface liquid freezes and expands.

Pluto is surrounded by five moons—Charon, Styx, Nix, Kerberos, and Hydra—which are likely rubble from an ancient collision similar to the one that created our own moon. Four are small and asteroid-like. Pluto's largest moon, Charon, is almost half Pluto's size, which is why the pair is sometimes called a double dwarf planet system. Charon has its own wild landscape of valleys and mountains, and it may have also been shaped by the activity of a subsurface ocean. The moon also sports a reddish polar cap composed of methane, possibly fed by Pluto's atmosphere.

CERES: ASTEROID BELT RULER

Orbiting between Mars and Jupiter is a swath of rubble known as the asteroid belt. We'll return to this zone in the next section, but let's start by zeroing in on its queen: Ceres. This round, rocky body was initially called an asteroid, but it is quite large compared to its rocky neighbors, making up roughly a third of the asteroid belt's total mass. So, when scientists devised a dwarf planet category in 2006, they included Ceres in those ranks. It's the only dwarf planet yet found in the inner solar system.

While large in comparison to other asteroid-belt dwellers, Ceres is only some 950 kilometers (590 mi) across and is about 14 times less massive than Pluto. Like the rocky planets, it has a layered internal structure, but is much less dense. This low density has led scientists to suggest that the dwarf planet hosts a subsurface layer rich in water ice.

Among Ceres's most intriguing features are glimmering spots seen by NASA's Dawn spacecraft in 2015, the brightest of which sits within the ancient Occator impact crater. Scientists believe they are made of reflective salts left over from brines that seeped up from watery pockets deep below the surface—and may continue to do so to this day.

An illustration of the five officially recognized dwarf planets (shown here with their moons): from left to right, Pluto, Eris, Makemake, Ceres, and Haumea

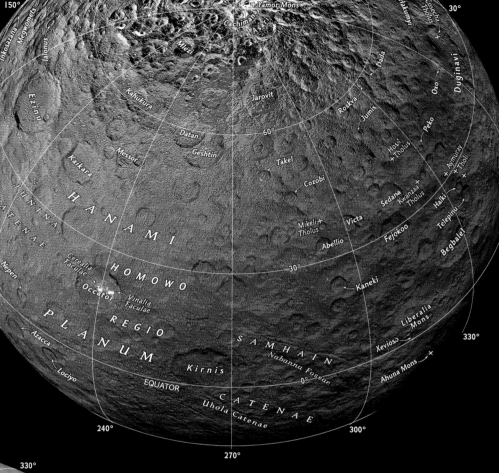

CERES

Dwarf planet Ceres is by far the largest object in the asteroid belt, lurking between the orbits of Mars and Jupiter. Bright spots, which are likely the result of salts, are one of Ceres's most curious features.

90°
120°
Xochipilli
Omonga
Ernutet
60°
Kiriamma
Jaja
Shennong
150°
Fukurokuju
Color Data Unavailable
Ghanan
Libra
30°
Dikhan
Cosecha
Dada
Hakumyi
Asari
Cosecha Tholus
North Pole
Yamor Mons
Cachimana
Mlezi
Oxo
Duginavi
Tinkoasatana
Jalonus
Kahukura
Jarovit
Peko
Megwomets
Ezinu
Datan
Geshtin
Roskva
Hosi Tholus
Takel
Jumis
180°
Naw ish
Kaikara
Messor
Cozobi
Sedana
Kwanzaa Tholus
Aymuray
Tholi
0°
Henebo
Mikeli Tholus
Victa
Fejokoo
Halki
Telepinu
JUNINA CATENAE
Cerealia Facula
Abellio
Begbalel
HANAMI
Nepen
Occator
Vinalia Faculae
Kaneki
HOMOWO
Liberalia Mons
210°
Azacca
REGIO
SAMHAIN
Xevioso
330°
PLANUM
Nabanna Fossae
Lociyo
Kirnis
CATENAE
Ahuna Mons
EQUATOR
240°
Uhola Catenae
300°
270°

SCALE at the EQUATOR 1:6,480,000
1 CENTIMETER = 64.8 KILOMETERS; 1 INCH = 102.2 MILES

STATUTE MILES 0   50   100
KILOMETERS 0   50   100

0°
30°
330°
60°
North Pole
300°
LOWELL REGIO
VEGA TERRA
Djanggawul Fossae
VOYAGER
VENERA
TERRA
TERRA
HAYABUSA TERRA
270°

PLUTO

Pluto has a remarkable range of curious features—rugged water-ice mountains, dunes of methane ice, churning nitrogen glaciers, and perhaps even frosty eruptions of cryovolcanoes. Pluto and Ceres were both named dwarf planets in 2006.

90°
Khare
Pirri Rupes
Burney Crater
Hunahpu Vallis
60°
Morando Fossa
Alcyonia Lacus
Al-Idrisi Montes
Meng-p'o Fossae
TARTARUS DORSA
Virgil Fossae
Sleipnir Fossa
Elliot Crater
Kiladze
30°
120°
SPUTNIK PLANITIA
Pigafetta Montes
Hekla Cavus
Hillary Montes
240°
EQUATOR
TOMBAUGH REGIO
0°
Wright Mons
Tenzing Montes
Adlivun Cavus
150°
210°
180°

SCALE at the EQUATOR 1:15,634,000
1 CENTIMETER = 156.3 KILOMETERS; 1 INCH = 246.8 MILES

STATUTE MILES 0   200   400
KILOMETERS 0   200   400

SMALL WORLDS

Remnants of our solar system's formation linger in a slew of tiny orbiting worlds.

I n a scattered ring beyond Neptune sprawls a broad swath of icy rubble, known as the Kuiper belt, which once made up the outer edge of the disk that birthed our sun and planets. Exactly what all this cosmic flotsam is remains uncertain, so the bits are simply called Kuiper belt objects, or KBOs.

Combined, the material has a mass less than 2 percent that of Earth, according to a recent estimate. If Neptune hadn't formed, the KBOs might have glommed together into their own complete planet, but Neptune's gravity acts like a cosmic blender, stirring up the rubble at our solar system's outer edge and preventing it from coalescing.

Kuiper belt objects spread across a wide zone from 30 to 50 AU from the sun. This main section of the belt—where most of the KBOs dwell—is about 20 AU thick, but its full extent may sprawl to nearly 1,000 AU.

The KBOs seem to be made of rock, ice, and dust in a range of different shapes and sizes. On the larger end of the size range for KBOs are the dwarf planets. All known dwarf planets except for Ceres orbit in the Kuiper belt—Pluto, Makemake, Eris, and Haumea. The region also holds some suspected dwarf planets, including 2015 RR245, first spotted in February 2016 in images taken the prior year.

Thousands, or maybe hundreds of thousands, of icy bodies circle in this zone, occasionally zinging toward the inner solar system. These frosty chunks are known as short-period comets, which means they take less than 200 years to make a single trip around the sun.

Despite these large numbers, scientists have studied only a tiny fraction of these objects. Some researchers believe there were once many more

(Opposite) The Kuiper belt object Arrokoth, shown in an illustration, has given scientists insights into the solar system's earliest days.

bodies in the Kuiper belt, objects now lost thanks to the four giant planets shifting their orbital paths in the early days of the solar system.

THE ASTEROID BELT

While many of the Kuiper belt objects are rocky and some are quite asteroid-like, most of them contain larger amounts of frosty volatile compounds— water, methane, and ammonia—than the airless space rocks classified as asteroids. The vast majority of these rocky worlds orbit in what is called the asteroid belt, which forms a ring between the orbits of Mars and Jupiter and divides the inner and outer solar system.

Not all asteroids are the same, as they are composed of varying amounts of clays, rocks, and metal—and some of these ingredients may have undergone little change since our solar system's earliest days. The most common type of asteroid is

The OSIRIS-REx spacecraft captured this mosaic image of asteroid Eennu from just 24 kilometers (15 mi) away.

Comet 67P/Churyumov-Gerasimenko (left) is an icy world from the Kuiper belt that orbits the sun once every 6.5 years. Eros (right) is the first near-Earth asteroid ever discovered.

carbonaceous, made up of a mixture of clay and silicate rocks. Others can be metallic, made mostly of iron and nickel.

The asteroid Psyche is one such metallic world. This potato-shaped body is about 225 kilometers (140 mi) across, a little more than the distance from Los Angeles to San Diego. What process would create such a world? Some scientists suggest the asteroid was once the core of an early planet, a doomed world that got caught up in collisions that crushed its outer rocky layers. NASA's Psyche mission will soon be on its way to study this curious body.

While there may be more than a million asteroids in the main belt, their mass combined is less than that of our planet's moon. Like KBOs, these bodies are probably primordial remnants of our early solar system and likely the building blocks of

planets. Studying these bodies could help unravel the secrets of how our solar system formed.

The asteroid belt's residents can sometimes be bumped from their orbits after encounters with Jupiter's intense gravity or other passing objects. These interactions send them soaring in all directions, and potentially toward our home world. If they make it through Earth's atmosphere as a meteorite, these space rocks can have deadly consequences. Scientists believe that an asteroid belt exile was responsible for one of the most catastrophic days in our planet's history, when, some 66 million years ago, a space rock slammed into Earth. Known as the Chicxulub impact, this event set around 75 percent of all plant and animal species on a path toward extinction, ending the reign of the non-avian dinosaurs.

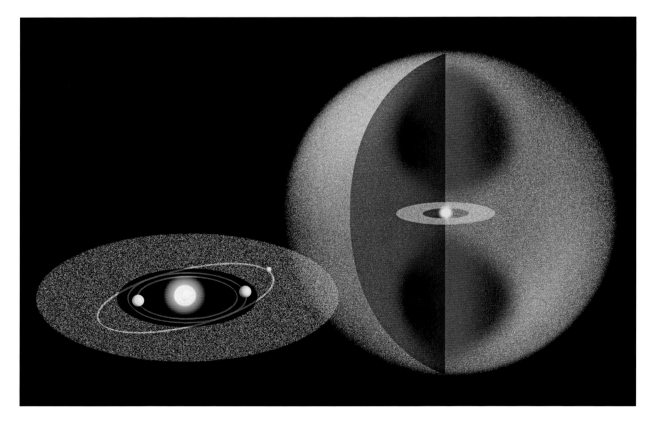

This illustration depicts the Kuiper belt (lower left) and the Oort cloud (right), which encompass our solar system and may extend to 100,000 AU or more.

With these dangers in mind, scientists continuously monitor the night skies for potentially hazardous asteroids, and in 2021 NASA launched the Double Asteroid Redirection Test (DART), the first test of a craft designed to redirect future earthbound asteroids.

THE OORT CLOUD

At the edges of our solar system, beyond the known planets and the icy bodies of the Kuiper belt, beyond even the point at which the sun's magnetic field ends, lies the Oort cloud. Astronomers believe this is a region of trillions of icy worlds at the limits of our sun's gravitational reach, but no one has yet directly observed them. Combined, the Oort cloud bodies may make up a mass many times that of Earth.

These bodies probably form a spherical layer, a kind of rocky bubble enveloping our solar system, starting at some 2,000 to 5,000 times the distance between Earth and the sun and extending out to 100,000 AU or even more. Our farthest robotic space traveler—the Voyager 1 spacecraft—will take another 300 years or more even to reach the Oort cloud's inner edge.

Dutch astronomer Jan Oort predicted the cloud's existence in the 1950s based on observations of so-called long-period comets. These icy bodies take at least 200 years, and as long as tens of millions of years, to complete one trip around the sun. They can follow unusual orbits that don't necessarily lie in the same path as the planets.

The objects in the Oort cloud are so far away that they're only weakly influenced by our sun's gravity. They can be readily disturbed by other forces, such as passing stars, which may send them hurtling inward toward the solar system's planets. Others may whizz away into the great beyond, where even more cosmic oddities wait to be explored.

An illustration depicts 'Oumuamua, the solar system's first known interstellar visitor. Its name is Hawaiian for "a messenger from afar arriving first."

The **Oort cloud** is so distant that sunlight takes nearly a month to reach its inner edge.

COSMIC JOURNEYS

Explorations of the solar system have advanced because of human curiosity, the will to understand the universe, and sometimes as a matter of national pride. Presented here is every mission of exploration completed or under way that has a goal of studying the bodies of our solar system. These efforts are truly international, including private companies as well as governments. Recently, probes have touched down on a comet, visited distant dwarf planet Pluto, mapped the surfaces of planets and moons throughout the solar system, and delved into the mysteries of the sun.

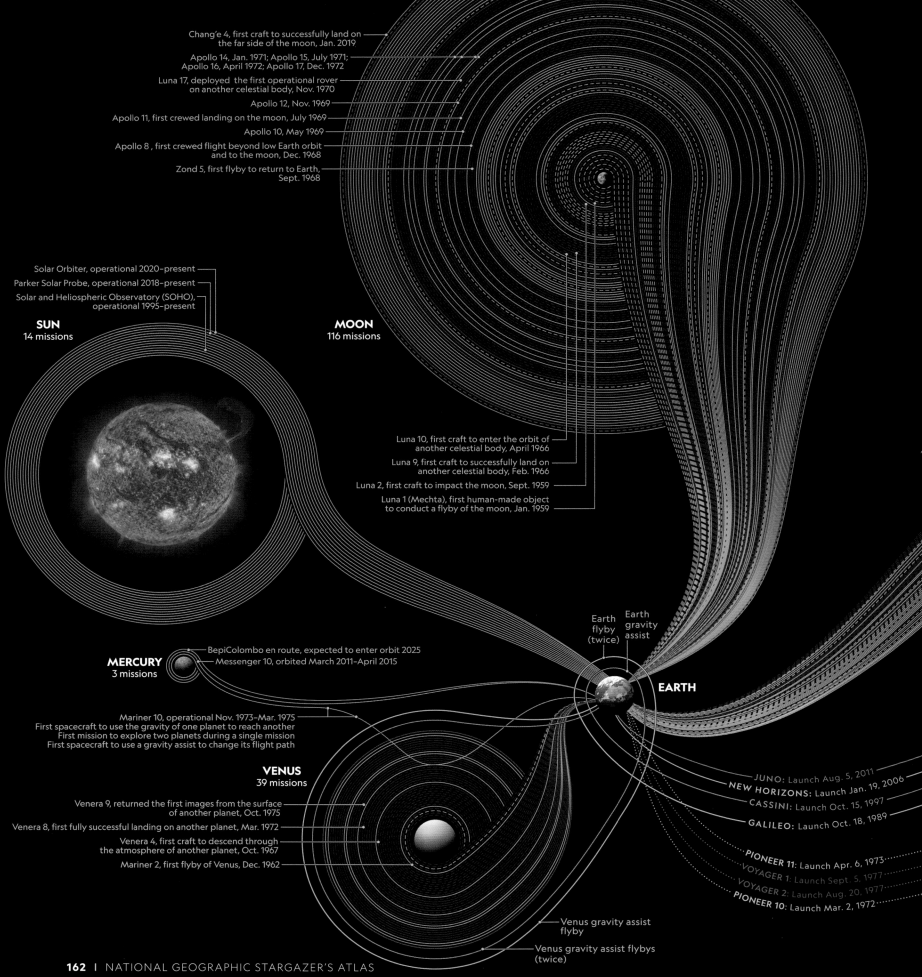

Chang'e 4, first craft to successfully land on the far side of the moon, Jan. 2019

Apollo 14, Jan. 1971; Apollo 15, July 1971; Apollo 16, April 1972; Apollo 17, Dec. 1972

Luna 17, deployed the first operational rover on another celestial body, Nov. 1970

Apollo 12, Nov. 1969

Apollo 11, first crewed landing on the moon, July 1969

Apollo 10, May 1969

Apollo 8 , first crewed flight beyond low Earth orbit and to the moon, Dec. 1968

Zond 5, first flyby to return to Earth, Sept. 1968

Solar Orbiter, operational 2020–present

Parker Solar Probe, operational 2018–present

Solar and Heliospheric Observatory (SOHO), operational 1995–present

SUN
14 missions

MOON
116 missions

Luna 10, first craft to enter the orbit of another celestial body, April 1966

Luna 9, first craft to successfully land on another celestial body, Feb. 1966

Luna 2, first craft to impact the moon, Sept. 1959

Luna 1 (Mechta), first human-made object to conduct a flyby of the moon, Jan. 1959

Earth flyby (twice)

Earth gravity assist

MERCURY
3 missions

BepiColombo en route, expected to enter orbit 2025

Messenger 10, orbited March 2011–April 2015

EARTH

Mariner 10, operational Nov. 1973–Mar. 1975
First spacecraft to use the gravity of one planet to reach another
First mission to explore two planets during a single mission
First spacecraft to use a gravity assist to change its flight path

VENUS
39 missions

JUNO: Launch Aug. 5, 2011

NEW HORIZONS: Launch Jan. 19, 2006

CASSINI: Launch Oct. 15, 1997

GALILEO: Launch Oct. 18, 1989

Venera 9, returned the first images from the surface of another planet, Oct. 1975

Venera 8, first fully successful landing on another planet, Mar. 1972

Venera 4, first craft to descend through the atmosphere of another planet, Oct. 1967

Mariner 2, first flyby of Venus, Dec. 1962

PIONEER 11: Launch Apr. 6, 1973

VOYAGER 1: Launch Sept. 5, 1977

VOYAGER 2: Launch Aug. 20, 1977

PIONEER 10: Launch Mar. 2, 1972

Venus gravity assist flyby

Venus gravity assist flybys (twice)

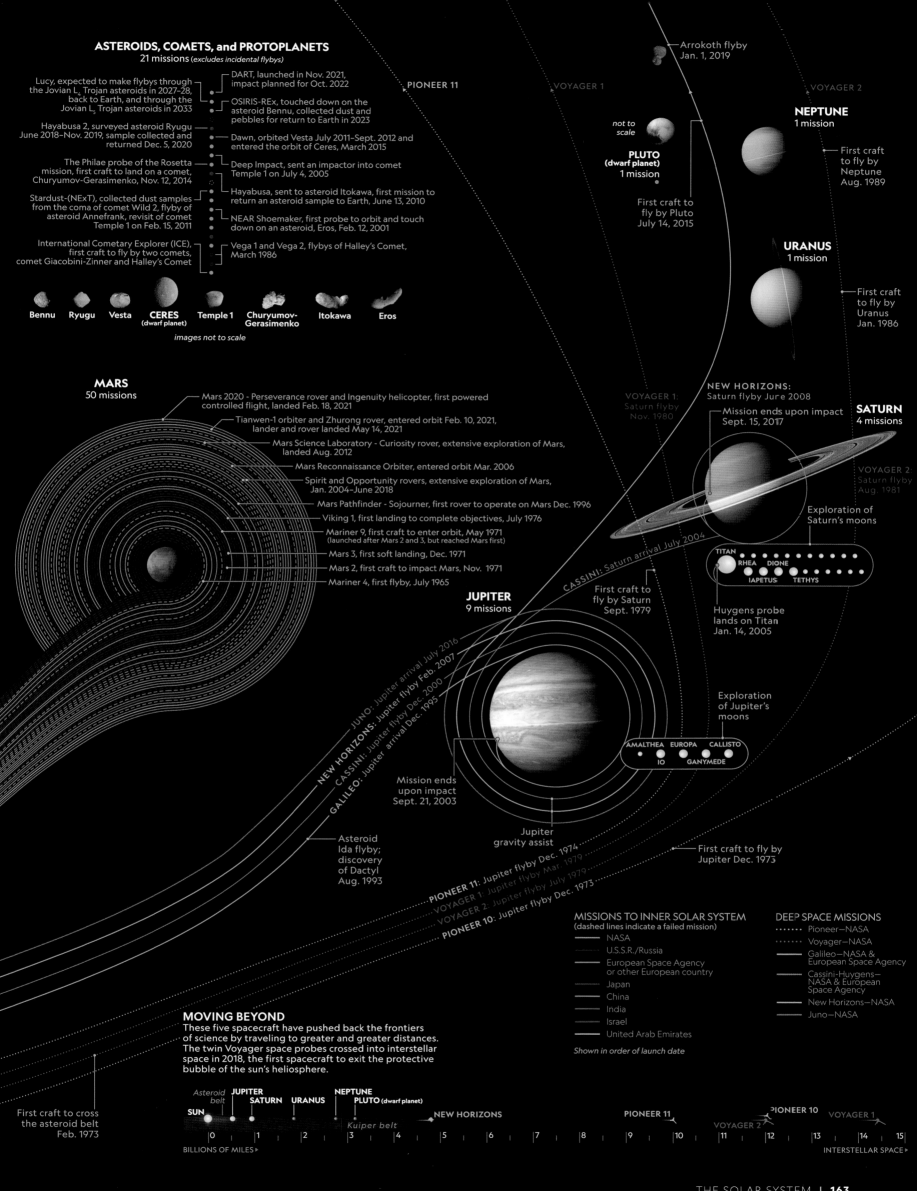

ASTEROIDS, COMETS, and PROTOPLANETS
21 missions *(excludes incidental flybys)*

Lucy, expected to make flybys through the Jovian L₄ Trojan asteroids in 2027–28, back to Earth, and through the Jovian L₅ Trojan asteroids in 2033

Hayabusa 2, surveyed asteroid Ryugu June 2018–Nov. 2019, sample collected and returned Dec. 5, 2020

The Philae probe of the Rosetta mission, first craft to land on a comet, Churyumov-Gerasimenko, Nov. 12, 2014

Stardust-(NExT), collected dust samples from the coma of comet Wild 2, flyby of asteroid Annefrank, revisit of comet Temple 1 on Feb. 15, 2011

International Cometary Explorer (ICE), first craft to fly by two comets, comet Giacobini-Zinner and Halley's Comet

DART, launched in Nov. 2021, impact planned for Oct. 2022

OSIRIS-REx, touched down on the asteroid Bennu, collected dust and pebbles for return to Earth in 2023

Dawn, orbited Vesta July 2011–Sept. 2012 and entered the orbit of Ceres, March 2015

Deep Impact, sent an impactor into comet Temple 1 on July 4, 2005

Hayabusa, sent to asteroid Itokawa, first mission to return an asteroid sample to Earth, June 13, 2010

NEAR Shoemaker, first probe to orbit and touch down on an asteroid, Eros, Feb. 12, 2001

Vega 1 and Vega 2, flybys of Halley's Comet, March 1986

Bennu **Ryugu** **Vesta** **CERES** (dwarf planet) **Temple 1** **Churyumov-Gerasimenko** **Itokawa** **Eros**

images not to scale

PIONEER 11   VOYAGER 1   VOYAGER 2

Arrokoth flyby
Jan. 1, 2019

NEPTUNE
1 mission

not to scale

PLUTO
(dwarf planet)
1 mission

First craft to fly by Pluto
July 14, 2015

First craft to fly by Neptune
Aug. 1989

URANUS
1 mission

First craft to fly by Uranus
Jan. 1986

MARS
50 missions

Mars 2020 - Perseverance rover and Ingenuity helicopter, first powered controlled flight, landed Feb. 18, 2021

Tianwen-1 orbiter and Zhurong rover, entered orbit Feb. 10, 2021, lander and rover landed May 14, 2021

Mars Science Laboratory - Curiosity rover, extensive exploration of Mars, landed Aug. 2012

Mars Reconnaissance Orbiter, entered orbit Mar. 2006

Spirit and Opportunity rovers, extensive exploration of Mars, Jan. 2004–June 2018

Mars Pathfinder - Sojourner, first rover to operate on Mars Dec. 1996

Viking 1, first landing to complete objectives, July 1976

Mariner 9, first craft to enter orbit, May 1971
(launched after Mars 2 and 3, but reached Mars first)

Mars 3, first soft landing, Dec. 1971

Mars 2, first craft to impact Mars, Nov. 1971

Mariner 4, first flyby, July 1965

VOYAGER 1:
Saturn flyby
Nov. 1980

NEW HORIZONS:
Saturn flyby June 2008

Mission ends upon impact
Sept. 15, 2017

SATURN
4 missions

VOYAGER 2:
Saturn flyby
Aug. 1981

Exploration of Saturn's moons

CASSINI: Saturn arrival July 2004

First craft to fly by Saturn
Sept. 1979

TITAN RHEA DIONE IAPETUS TETHYS

Huygens probe lands on Titan
Jan. 14, 2005

JUPITER
9 missions

JUNO: Jupiter arrival July 2016
NEW HORIZONS: Jupiter flyby Feb. 2007
CASSINI: Jupiter flyby Dec. 2000
GALILEO: Jupiter arrival Dec. 1995

Mission ends upon impact
Sept. 21, 2003

Exploration of Jupiter's moons

AMALTHEA EUROPA CALLISTO
IO GANYMEDE

Asteroid Ida flyby; discovery of Dactyl
Aug. 1993

Jupiter gravity assist

PIONEER 11: Jupiter flyby Dec. 1974
VOYAGER 1: Jupiter flyby Mar. 1979
VOYAGER 2: Jupiter flyby July 1979
PIONEER 10: Jupiter flyby Dec. 1973

First craft to fly by Jupiter Dec. 1973

MISSIONS TO INNER SOLAR SYSTEM
(dashed lines indicate a failed mission)

—— NASA
---- U.S.S.R./Russia
—— European Space Agency or other European country
—— Japan
—— China
—— India
—— Israel
—— United Arab Emirates

Shown in order of launch date

DEEP SPACE MISSIONS

········ Pioneer—NASA
········ Voyager—NASA
—— Galileo—NASA & European Space Agency
—— Cassini-Huygens—NASA & European Space Agency
—— New Horizons—NASA
—— Juno—NASA

MOVING BEYOND
These five spacecraft have pushed back the frontiers of science by traveling to greater and greater distances. The twin Voyager space probes crossed into interstellar space in 2018, the first spacecraft to exit the protective bubble of the sun's heliosphere.

First craft to cross the asteroid belt
Feb. 1973

Asteroid belt
SUN   **JUPITER**   **URANUS**   **NEPTUNE**
       SATURN   **PLUTO** (dwarf planet)

Kuiper belt   NEW HORIZONS

PIONEER 11   PIONEER 10   VOYAGER 1
VOYAGER 2

0 1 2 3 4 5 6 7 8 9 10 11 12 13 14 15
BILLIONS OF MILES ▶

INTERSTELLAR SPACE ▶

The remote Chajnantor Plateau in the Atacama Desert in Chile is home to ALMA, currently the largest radio telescope in the world.

CHAPTER FOUR

THE NIGHT SKY

VULPECULA et ANSER

CERBE

γ
δ
α
Scham
β
ε

ε

Scalocin
α
γ
δ
ζ
Rotanew

ζ
Deneb el okab

Deneb el Delphin

TAURU
PONIAT
SKI

γ

o
Tarazed
Athair
α

θ var.

μ

β
Alshain

SERPE

δ
C

γ var:

δ

I

I

H

γ

λ

G

i

SCUTO
SOBIES-

E

κ

B

QUARIUS

D

μ

ω

2 a
1 a
Dshabeh

β

E ——— W

CAPRICORNUS

SAGITTARIUS

S

A GUIDE TO THE CONSTELLATIONS

Through the ages, cultures around the world began to bring a certain order to the night sky by linking stars into eye-catching patterns that our human minds were immediately drawn to. Using our imaginations, we named these pictures after noteworthy characters from our societies, or deities, or animals, or stories, myths, and legends. We call these stellar figures constellations, which in Latin appropriately means "group of stars."

Though we are still struck with awe when we cast our eyes upward, thanks to centuries of advancements in science and technology our wonderment is now tied to the understanding that this view is a great journey that takes us deep into space and far back in time. Humanity's entire knowledge about the vast universe that surrounds our little blue planet comes from looking up and contemplating what it all means. So we build bigger and better tools to keep extending our vision back to the beginning of our universe. Today we live in the age of giant telescopes on the ground and in orbit, of computerized technology and space probes. Our collec-

tive understanding of the cosmos as a human species is unprecedented. Yet as individuals in our digital, cosmopolitan society, many of us have become disconnected from the wonders of the night sky. Throughout history, there has never been a substitute for going out under the stars and learning about them for yourself.

The majority of people in the world live within cityscapes, and for them many of the fainter stars are washed away by light pollution. Yet Earth's moon and the major planets are still within grasp of many urban dwellers, as are many of the brighter constellations, which are easily visible to most suburban skywatchers. In this chapter, we'll learn how to read star maps, so you can be oriented to the starry skies whether you are under bright city lights or in a dark countryside. And once you find your way among the constellations, it's time to peer deeper, learning about tracking down the many treasures buried within them, such as distant nebulae, star clusters, and galaxies. Remember to take your time, savor your moments under the stars, and enjoy your personal adventure each time you set out to navigate your night sky.

Part of a 32-card deck published in 1825 called "Urania's Mirror, or, A View of the Heavens," this illustration shows the constellations Delphinus, Aquila, Sagitta, and Antinous (now obsolete).

NAVIGATING THE NIGHT SKY

Learning constellations and their stars is the foundation for starting your own observations of the heavens.

Looking up at a night sky filled with stars can feel daunting and even a bit scary to some. But in reality making sense of the sky is like knowing how to read a simple road map. Finding your way around the constellations is all about becoming familiar with the placement and movement of celestial landmarks from night to night and through the seasons.

Astronomers have mapped the entire sky onto a giant celestial sphere, with Earth at the center. As our planet rotates around its axis every 24 hours, the sun, moon, stars, and planets all appear to move across the sky in sweeping arcs, rising in the east and setting in the west. And as Earth moves through its annual trip around the sun, we see the star patterns that we call constellations continually cycle, too. So for skywatchers, viewing the sky at the same time of night for a few weeks or months will slowly bring a parade of new groups of stars into view.

Easily recognizable stellar patterns—such as Orion and the Big and Little Dippers in the Northern Hemisphere, and Crux, the Southern Cross, in the Southern Hemisphere—are all great places to start learning how to navigate the heavens. Using these bright star formations as a starting point, you can easily and quickly jump from one region of the sky to another and begin learning about a handful of the brightest constellations that are visible each season of the year.

BRIGHT OR DIM

In addition to grouping stars into eye-catching patterns, astronomers classify them according to brightness. The ancient Greek astronomer Hipparchus is believed to have created the first stellar catalog that categorized stars into various classes of brightness, or magnitude, a system that is still used today. The brighter the star appears in the sky, the lower the magnitude number, with the brightest objects crossing over to negative numbers. A star's magnitude is shown on star charts by graded series of dot sizes. The fainter a star is in the sky, the tinier its counterpart dot is on a star chart, with each size representing a whole order of magnitude in brightness.

Skywatchers stuck within city suburbs can see stars down to around magnitude 4 on a clear moonless night. From a pristine dark location, far from any city lights, the unaided human eye can see stars down to about magnitude 6.5. However, large binoculars (ones with a 50-mm objective lens) will reveal stars as faint as magnitude 9 or 10, while a small backyard telescope can reach down to magnitude 12. The Hubble Space Telescope in orbit above Earth can record objects as faint as magnitude 30. This is about two billion times dimmer than the faintest object visible to the naked eye.

MEASURING THE SKY

Larger stars tend to shine brighter than smaller ones. Also, stars are not all at the same distance from us. So the brightness of the stars we see at night depends on both their size and how remote they are from us.

The distances between the stars and the galaxies themselves are so great that the conventional kilometers and miles we use on Earth become impractical on cosmic scales. Astronomical distances are expressed in astronomical units, light-years, and parsecs. Astronomical units, or AU for short, are used throughout the solar system, where one AU is the distance between the sun and Earth (about 150 million kilometers, or 93 million miles).

(Opposite) Attendees observe the night sky at the Saskatchewan Summer Party in the Cypress Hill Provincial Park Dark Sky Preserve in Alberta, Canada.

■ For more on dark-sky locations, see pages 341–347.

A beginner's telescope should be easily portable and provide bright, crisp views of objects in the night sky.

Binoculars are a great optical aid for beginner stargazers.

For interstellar distances, astronomers use the cosmic yardstick of light-years, the distance light travels in a one year—about 9.5 trillion kilometers (5.9 trillion mi) at a speed of 300,000 kilometers per second (186,000 mps). Then, for the truly gigantic scales used for large distances across the universe, astronomers use the parsec (parallax second) system, with one parsec being 3.26 light-years.

Skywatching measurements are different. The apparent sizes of objects and the distances that separate them in the sky are measured in angles— degrees, minutes, and seconds. For example, it is 90 degrees from the horizon to directly overhead. But it can be tricky to translate these measurements from handheld star charts to the real sky above. You can use your own hands and fingers held at arm's length (as noted in the sidebar on page 193) and the famous Big Dipper star pattern in the northern sky as convenient, portable angle-measurers.

These equivalencies will become second nature. Your outstretched hand is roughly equal to the 25-degree distance between the last star in the handle of the Big Dipper and the end stars in the bowl.

Your fist is about equal to the 10-degree width of the Dipper's bowl itself, while three middle fingers measure about the same as the five-degree height of the Dipper's bowl.

STARGAZING TOOLS

Once you are able to recognize the major, bright constellations with your unaided eyes, and when you become familiar with the general lay of the celestial land from night to night, it's the perfect time to graduate to binoculars and then telescopes. With optical aids you can get the magnified and more narrow field of views you need to begin exploring deep-sky treasures. As a rule of thumb, binoculars generally offer fields of view equal to about 10 to 20 full moons in the sky. Meanwhile, telescopes using the lowest magnifications generally give views equal to only about one full moon. These much narrower fields of view can make orientation in the night sky more challenging for the novice stargazer, so starting out learning the sky with just your naked eyes will lead to greater success as you get more serious about skywatching.

Telescopes with larger apertures are an excellent way to explore deep-sky targets.

Binoculars

For beginning skywatchers, binoculars offer a great stepping-stone between the naked eye and telescopes. They are perfectly suited for exploring discrete starry regions and larger objects within and between constellations. Binoculars offer splendid views of craters on the moon, hundreds of star clusters, and even the four largest moons of Jupiter. Optical quality, magnification, and light-gathering power are paramount. When trying out binoculars, look for pinpoint star views. For general viewing, a 7 x 50 configuration is a good place to start. The first number refers to magnifying power and the second refers to the aperture, or diameter, of the front lenses in millimeters. The higher the magnifying power, the greater the magnification of space objects, though it also magnifies the effects of shaky hands. And the wider the front lenses, the brighter the views (but this comes at the cost of weight).

Telescopes

When you train a telescope toward the starry skies, you begin to truly unlock the hidden beauty of the cosmos. Well-crafted backyard telescopes fall into two main types in terms of construction: lens-based refractors and mirror-based reflectors. Avoid the telescopes commonly found at department stores that are advertised by their maximum power. The main features to look for are quality optics and a rock-solid mount. Look for a telescope with a structure made of metal and wood, not plastic, and avoid models that make the image wobble every time you touch the 'scope. Keep in mind that the larger the main mirror or lens, the brighter and sharper the images—and the heavier the telescope.

Telescopes have also benefited from the digital age, with some models able to find and track tens of thousands of sky objects, from planets to distant galaxies, that are either loaded on an onboard computer or linked to your external digital device. Planetarium apps available for download on computers, handheld tablets, and smartphones can show millions of stars and realistic photographs of objects, and they can even guide naked-eye observers to specific sky targets.

Backyard astronomers have many options for instruments that will help bring the heavens into better view. When considering telescopes, prioritize optics and steady mounts.

USING
THE SKY MAPS

Just as we use terrestrial maps to find destinations on Earth, we use star charts to track down celestial treasures.

This chapter explores the entire night sky from both Northern and Southern Hemisphere locations. The maps feature the main celestial attractions in the sky, including constellations and their brightest stars, easy-to-find asterisms, star-hops, and eye-catching deep-sky objects.

On the following pages, the night sky for each hemisphere has been divided into four seasons, with each represented by a pair of semicircle maps looking north or south. The bottom edge is your horizon, and the top of the dome is the overhead point. The corners of the semicircles indicate due west and east. Each map shows mid-evening times

and mid-season dates for midlatitude Northern or Southern Hemisphere locations—however, each one can be used in a wide range of latitudes. Be aware, though, that the stars may then be lower or higher in your sky than is shown here.

To use the seasonal maps outside at night, first determine the direction you are facing. You can use a compass or simply note where the sun sets—that's west. To view the northern sky, face north and hold the "Looking North" map with the flat bottom (horizon) of your semicircle map toward you. To view the southern sky, use the "Looking South" counterpart with the flat horizon facing you. Once you're familiar with constellation locations, you can cross-reference these seasonal charts with the star-hopping maps (pp. 190–193) and with the individual constellation profiles later in the chapter.

(Opposite) The Milky Way, which stands out under clear and very dark skies, is a rich source of stargazing targets.

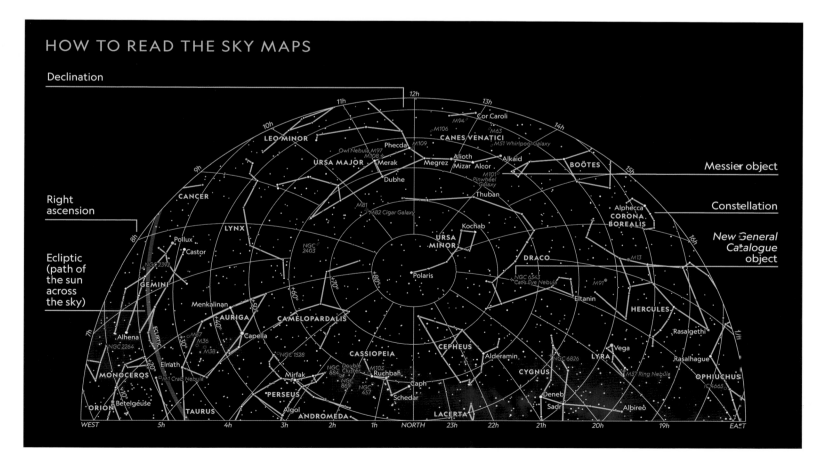

HOW TO READ THE SKY MAPS

NORTHERN HEMISPHERE

SPRING

T he bright Big Dipper asterism in the constellation Ursa Major dominates the near overhead northern skies, hanging upside down above Polaris and Ursa Minor. Its distinctive pattern of seven stars acts as a convenient pointer to neighboring bright stars and constellations. The southern skies offer a view into deep space—a window out of the Milky Way. The key springtime constellations looking toward the south are Leo and Virgo, with its brightest star Spica. Both constellations are rich with Messier objects, most of which are galaxies. Below Leo are the winding faint stars of Hydra and its bright star Alphard.

STELLAR MAGNITUDES

● -0.5 and brighter	● 2.1 to 2.5
● -0.4 to 0.0	● 2.6 to 3.0
● 0.1 to 0.5	● 3.1 to 3.5
● 0.6 to 1.0	· 3.6 to 4.0
● 1.1 to 1.5	· 4.1 to 4.5
● 1.6 to 2.0	· 4.6 to 5.0

DEEP-SKY OBJECTS

○ Open star cluster
⊕ Dark nebula
▫ Globular star cluster
▢ Bright nebula
✦ Planetary nebula
⬭ Galaxy

LOOKING NORTH

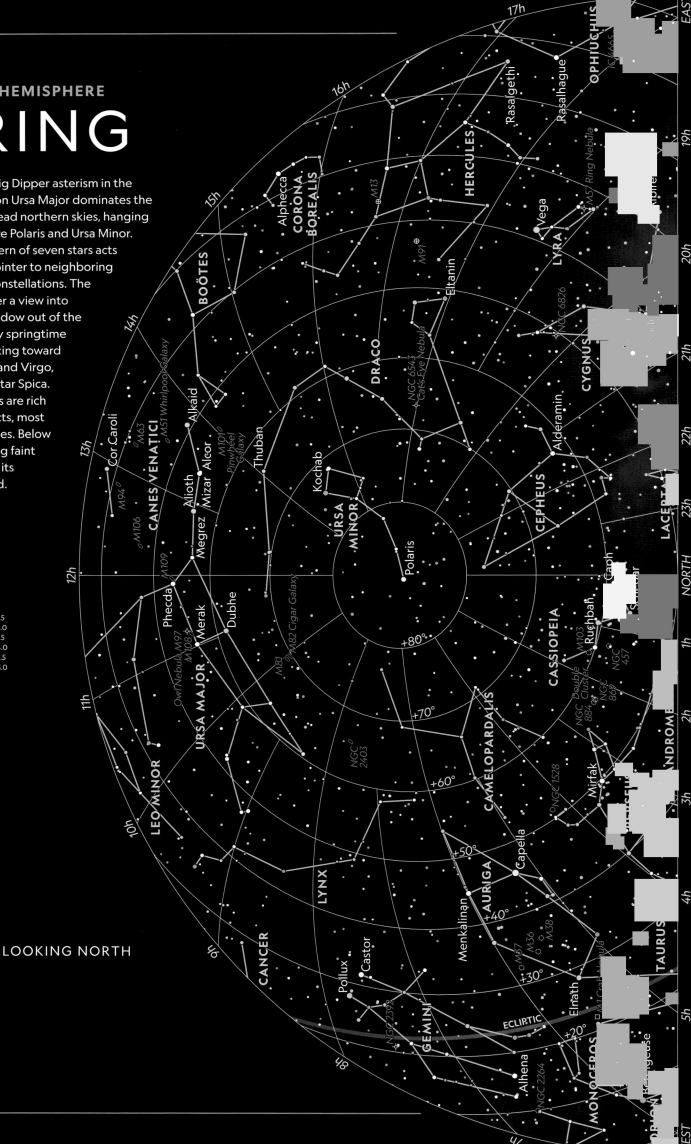

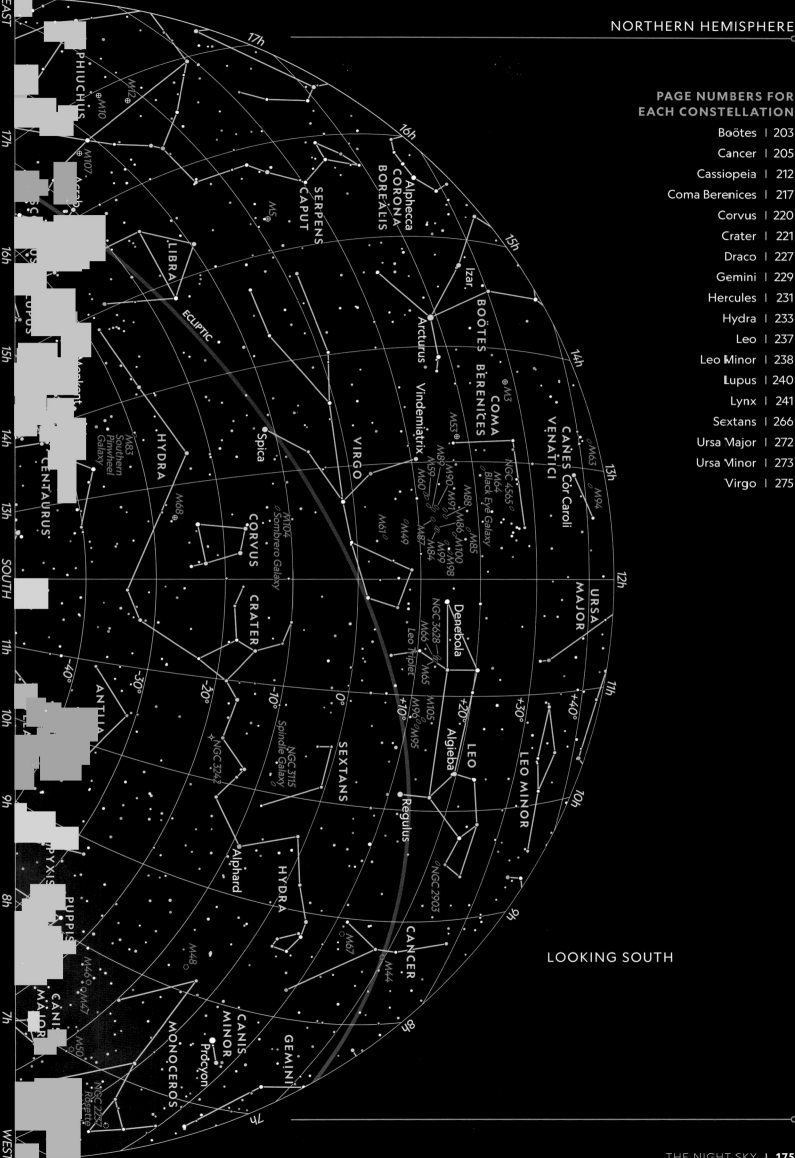

**PAGE NUMBERS FOR
EACH CONSTELLATION**

Boötes | 203
Cancer | 205
Cassiopeia | 212
Coma Berenices | 217
Corvus | 220
Crater | 221
Draco | 227
Gemini | 229
Hercules | 231
Hydra | 233
Leo | 237
Leo Minor | 238
Lupus | 240
Lynx | 241
Sextans | 266
Ursa Major | 272
Ursa Minor | 273
Virgo | 275

LOOKING SOUTH

NORTHERN HEMISPHERE
SUMMER

T he long, twisted constellation of Draco rides high in the northern sky above Polaris, with the three stars of the Summer Triangle asterism—brilliant Vega in Lyra high in the south, Deneb in Cygnus to its east, and Altair in Aquila— toward the low southern sky (see the star-hopping map on page 190). Bright orange Arcturus in Boötes is halfway down the western sky, tucked just underneath semicircular Corona Borealis. Between Arcturus and Vega is the wedge-shaped keystone asterism of Hercules. The Milky Way stretches from overhead down to the southern horizon, interacting with Sagittarius and Scorpius, whose eye is marked by brilliant orange Antares.

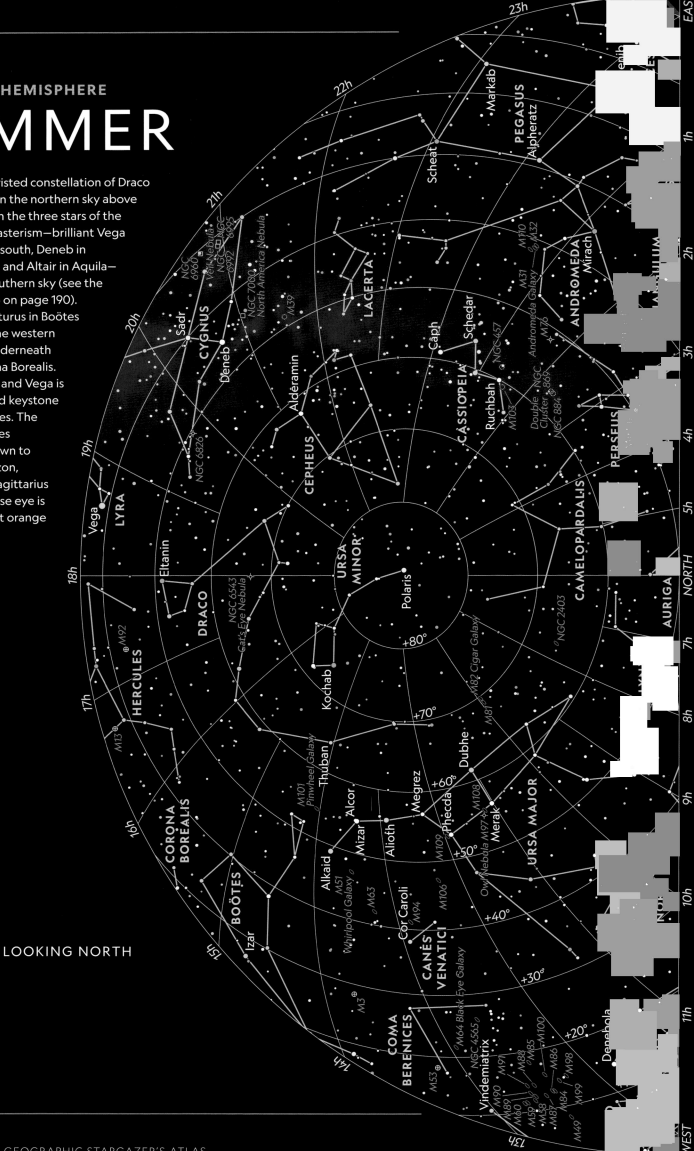

LOOKING NORTH

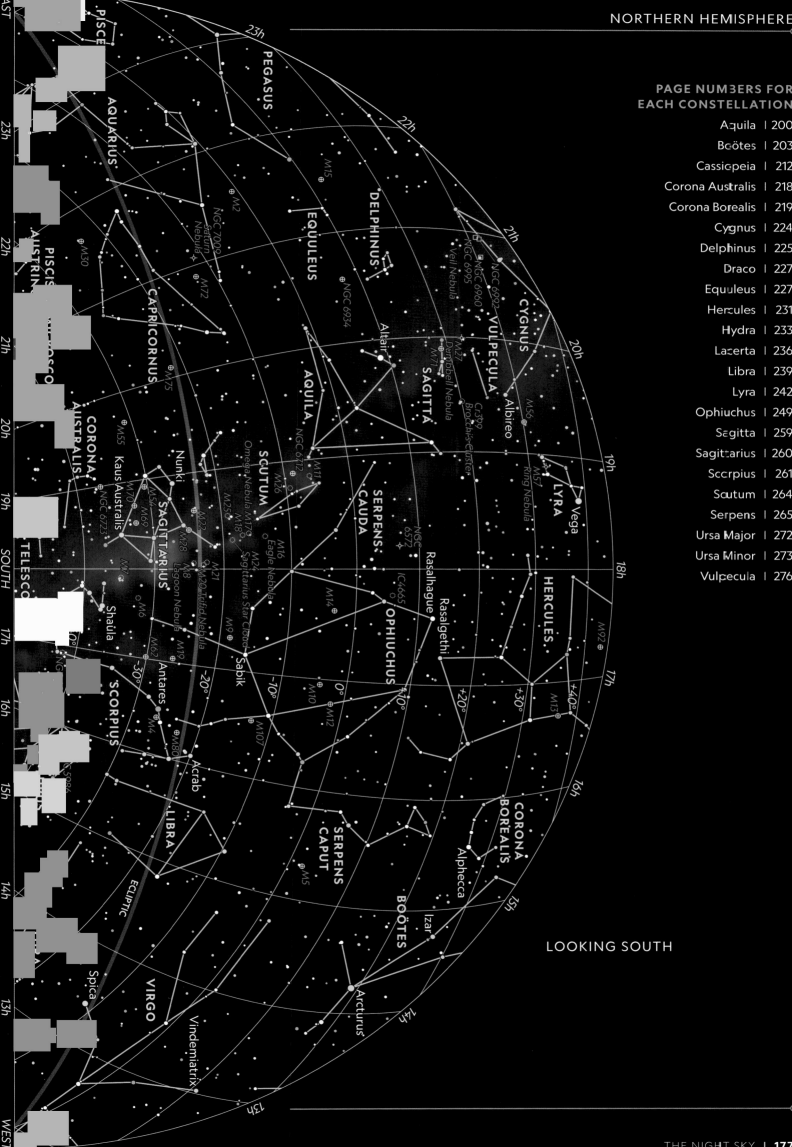

PAGE NUMBERS FOR
EACH CONSTELLATION

Aquila | 200
Boötes | 203
Cassiopeia | 212
Corona Australis | 218
Corona Borealis | 219
Cygnus | 224
Delphinus | 225
Draco | 227
Equuleus | 227
Hercules | 231
Hydra | 233
Lacerta | 236
Libra | 239
Lyra | 242
Ophiuchus | 249
Sagitta | 259
Sagittarius | 260
Scorpius | 261
Scutum | 264
Serpens | 265
Ursa Major | 272
Ursa Minor | 273
Vulpecula | 276

LOOKING SOUTH

NORTHERN HEMISPHERE
AUTUMN

The distinctive M-shape of Cassiopeia dominates the near overhead sky in the north. The Great Square of Pegasus, a bright asterism at the center of the constellation of the same name, stands out in the barren autumn sky, riding high in the south. Bright star Alpheratz is shared by Andromeda and the equine Great Square asterism: It points east, and then north to Perseus and the bright yellow-orange star Capella of Auriga. Below Pegasus, in the southeast, is the winding figure of Pisces, the faint stars of Aquarius, and the sprawling, monster whale constellation of Cetus. The star Fomalhaut pins down the low southern sky. The season also provides an opportunity to view galaxies, such as Andromeda (M31) in its namesake constellation, that are visible to the naked eye.

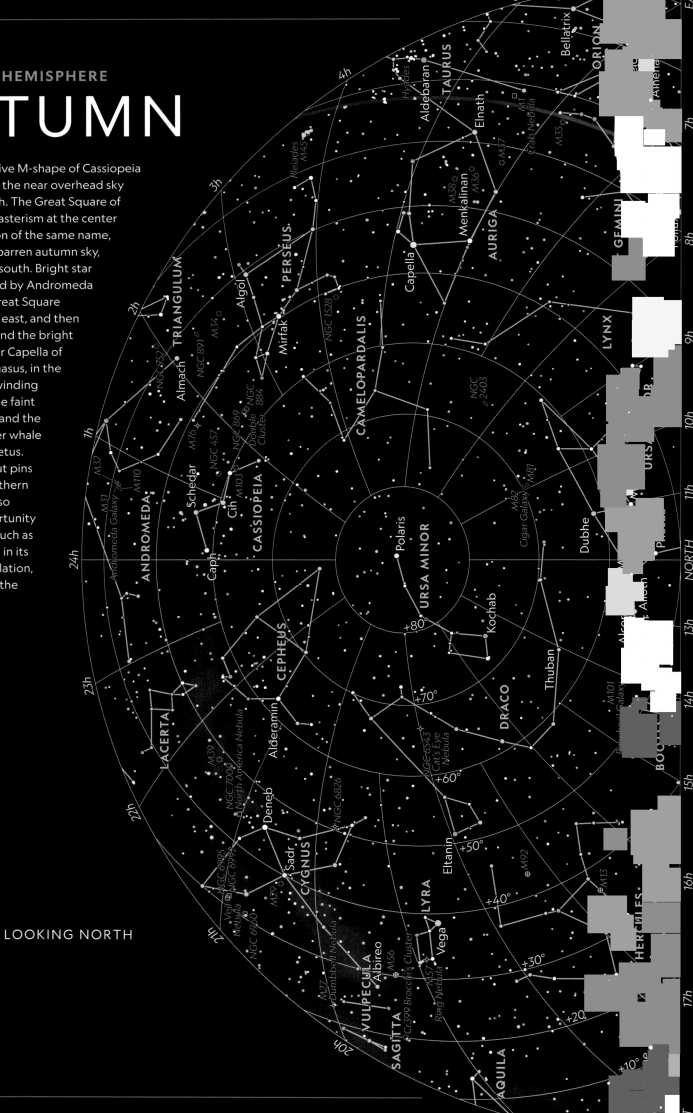

LOOKING NORTH

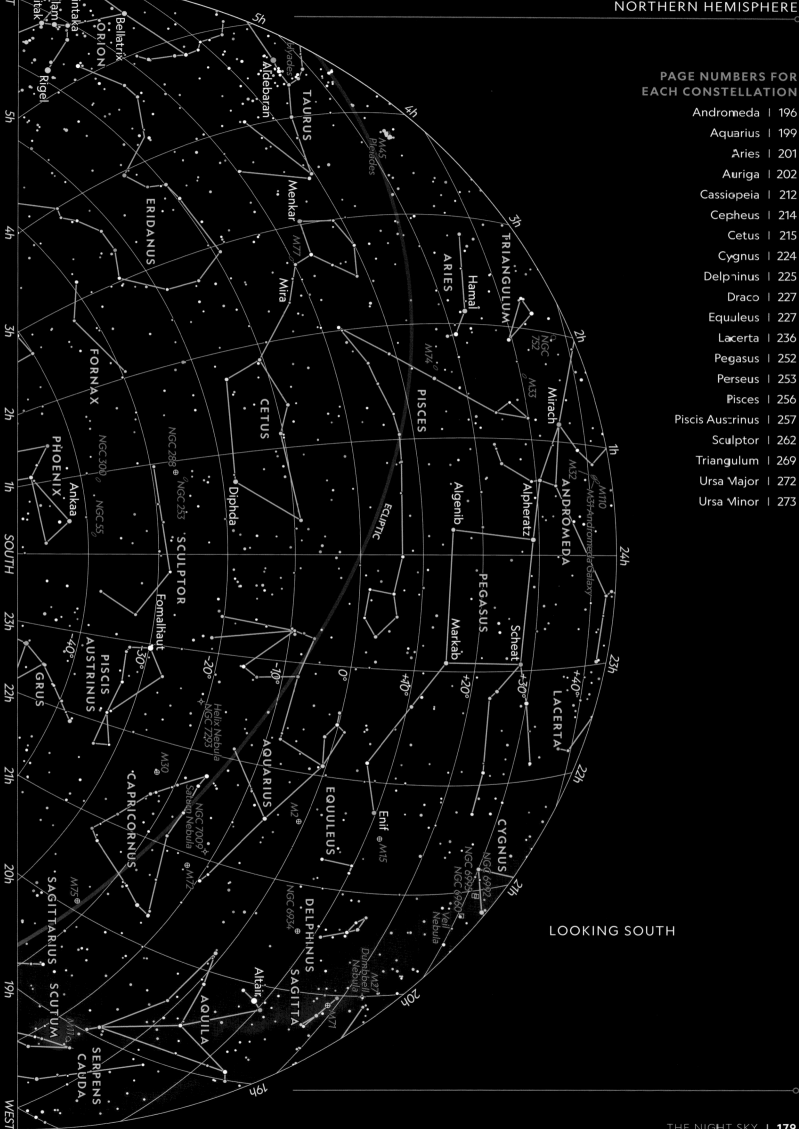

PAGE NUMBERS FOR
EACH CONSTELLATION

Andromeda | 196
Aquarius | 199
Aries | 201
Auriga | 202
Cassiopeia | 212
Cepheus | 214
Cetus | 215
Cygnus | 224
Delphinus | 225
Draco | 227
Equuleus | 227
Lacerta | 236
Pegasus | 252
Perseus | 253
Pisces | 256
Piscis Austrinus | 257
Sculptor | 262
Triangulum | 269
Ursa Major | 272
Ursa Minor | 273

LOOKING SOUTH

NORTHERN HEMISPHERE
WINTER

O rion dominates the view in the south. Bright orange Betelgeuse sits on Orion's shoulder (to the viewer's left) and blue-white Rigel forms the mythical hunter's foot (to the viewer's right). On the hanging sword below Orion's belt of three stars, you'll find that the middle "star" is in fact the Orion Nebula star factory. Just to the northwest of Betelgeuse is Taurus and two famed open star clusters, the Pleiades and the Hyades. Nearly overhead along the north-south sky you'll see Auriga and its bright yellow star Capella. Northeast of Orion lie the twin stars of Gemini, Castor and Pollux. To the southeast of Orion is brilliant white Sirius in Canis Major. Look northeast of Sirius to find its smaller companion, Canis Minor, and its lead star, Procyon.

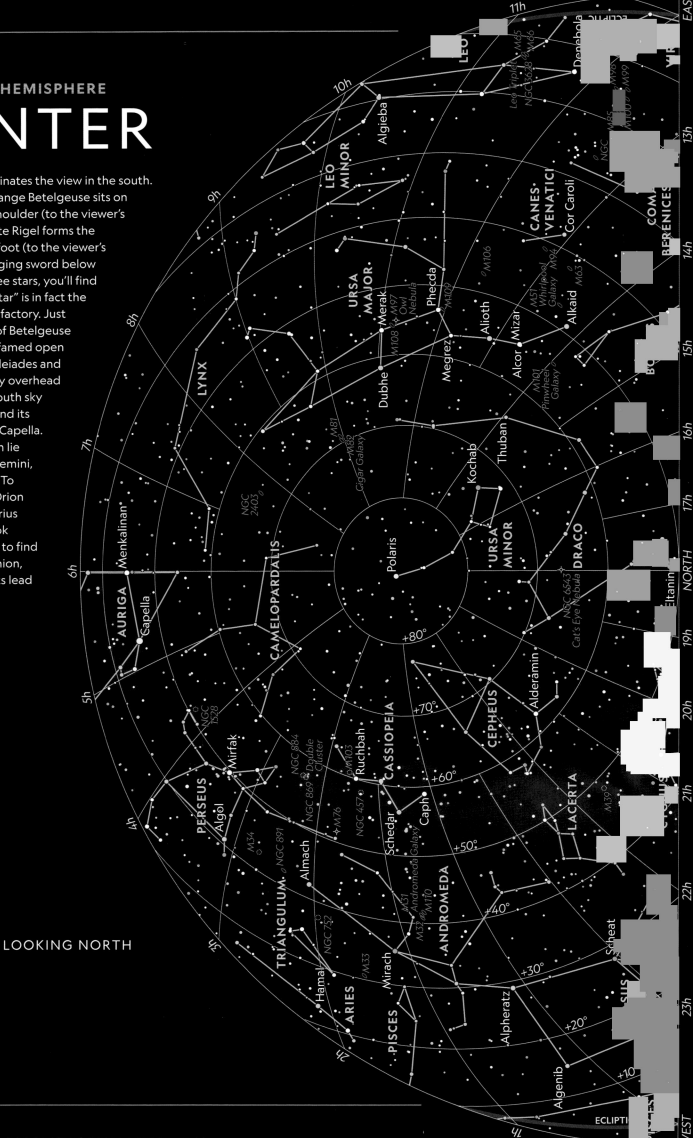

LOOKING NORTH

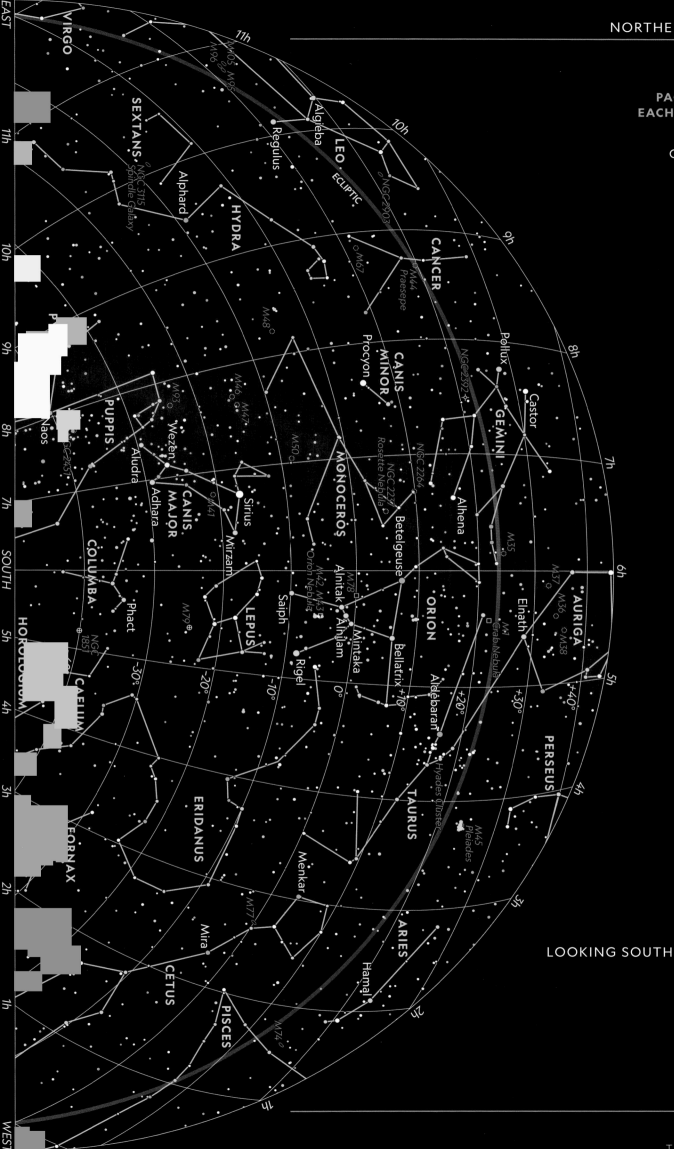

**PAGE NUMBERS FOR
EACH CONSTELLATION**

Auriga | 202

Camelopardalis | 204

Canis Major | 208

Canis Minor | 209

Capricornus | 210

Cassiopeia | 212

Columba | 216

Draco | 227

Gemini | 229

Lepus | 238

Lynx | 241

Monoceros | 245

Orion | 250

Puppis | 257

Pyxis | 258

Taurus | 267

Ursa Major | 272

Ursa Minor | 273

LOOKING SOUTH

SOUTHERN HEMISPHERE

SPRING

To the north, the sky appears dominated by the Great Square asterism of Pegasus with sprawling Cetus above it, nearly overhead, and faint Aquarius toward its west. To the east of the winged equine is the star Hamal, part of a stellar trio that identifies Aries (The Ram). Low in the north is bright star Deneb in Cygnus, with the most brilliant parts of the Milky Way running mostly parallel with the northern horizon. Centaurus and the Southern Cross in Crux have sunk to their lowest points, and winding Eridanus is high in the southern sky.

LOOKING NORTH

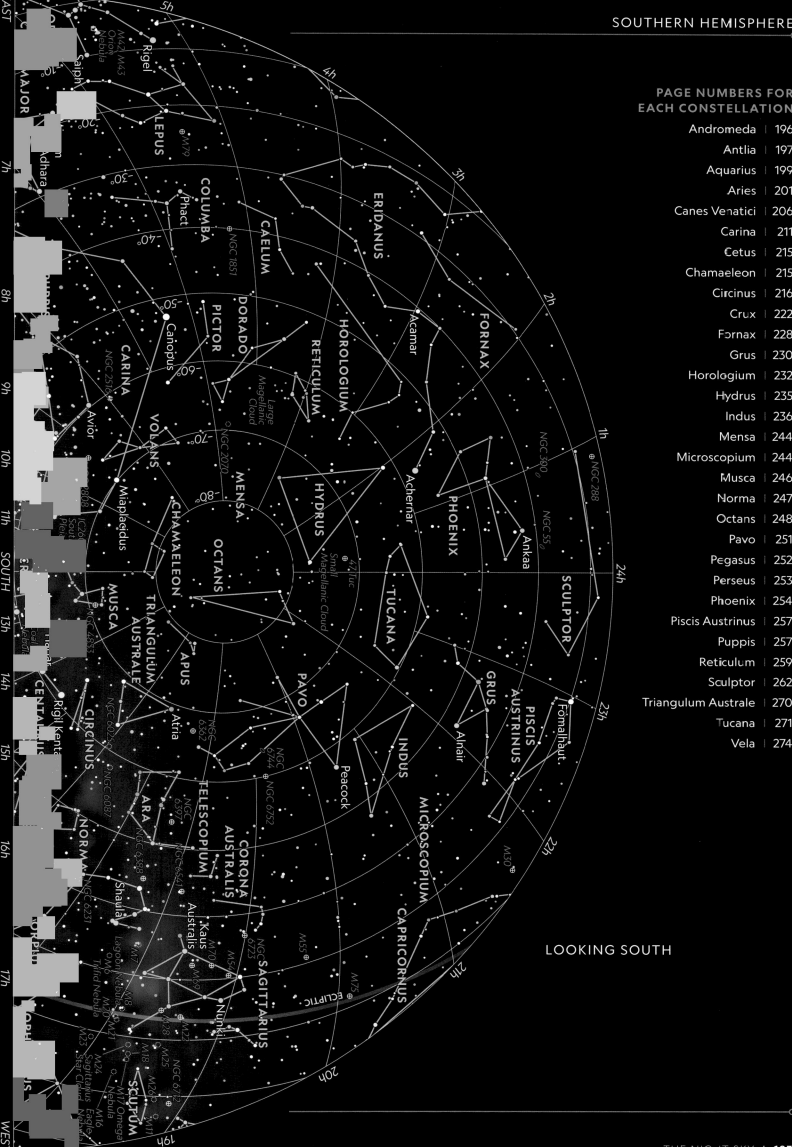

PAGE NUMBERS FOR
EACH CONSTELLATION

Andromeda | 196
Antlia | 197
Aquarius | 199
Aries | 201
Canes Venatici | 206
Carina | 211
Cetus | 215
Chamaeleon | 215
Circinus | 216
Crux | 222
Fornax | 228
Grus | 230
Horologium | 232
Hydrus | 235
Indus | 236
Mensa | 244
Microscopium | 244
Musca | 246
Norma | 247
Octans | 248
Pavo | 251
Pegasus | 252
Perseus | 253
Phoenix | 254
Piscis Austrinus | 257
Puppis | 257
Reticulum | 259
Sculptor | 262
Triangulum Australe | 270
Tucana | 271
Vela | 274

LOOKING SOUTH

SOUTHERN HEMISPHERE

SUMMER

Orion (The Hunter) rides high in the northern sky with a retinue of bright stars and constellations surrounding it—including the white beacon Sirius in Canis Major above it and bright orange Aldebaran in Taurus (The Bull) to its lower west. Auriga with its lead star Capella brushes along the northern horizon. Looking south, the brilliant yellow Canopus in Carina is dominant high in the sky, while lower down toward the east the Southern Cross is prominent, along with Centaurus and its main stars Alpha and Beta Centauri. Halfway up the sky in the southwest are both the Large and Small Magellanic Clouds.

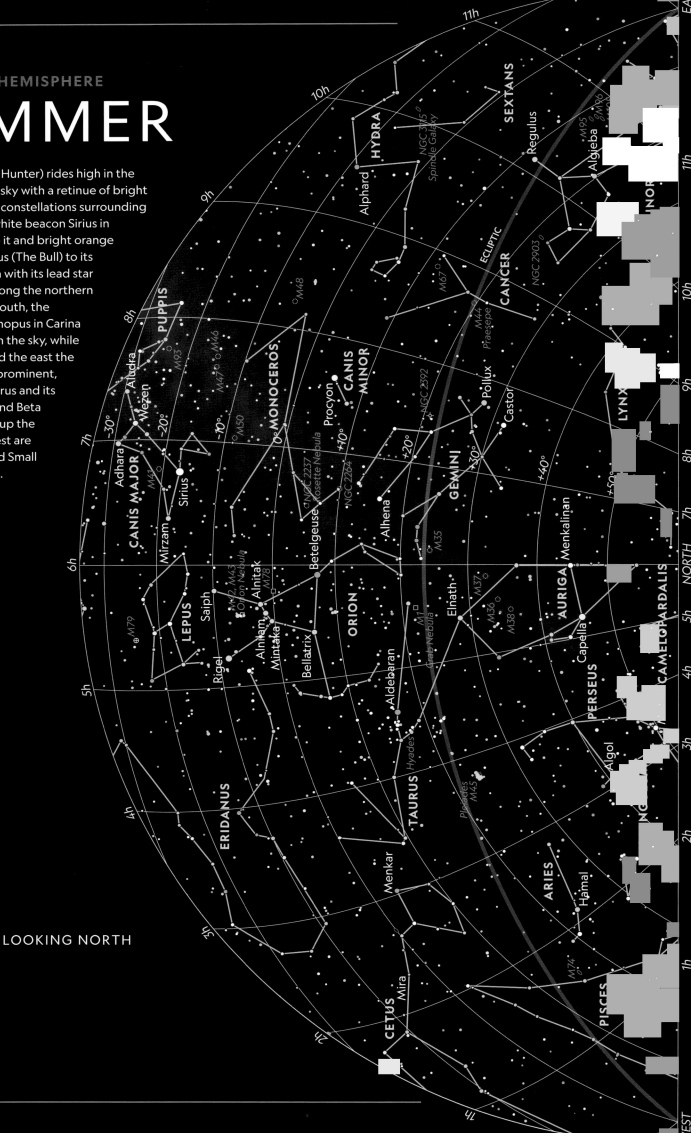

LOOKING NORTH

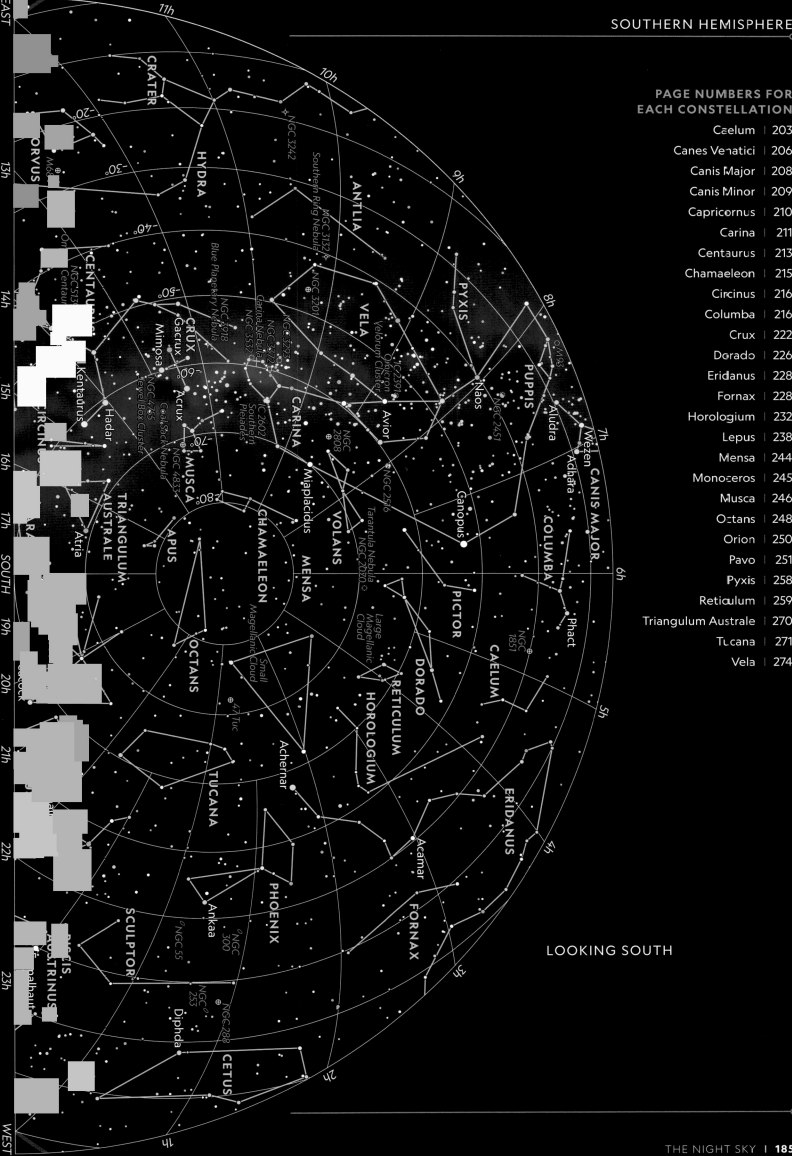

PAGE NUMBERS FOR EACH CONSTELLATION

Cælum	203
Canes Venatici	206
Canis Major	208
Canis Minor	209
Capricornus	210
Carina	211
Centaurus	213
Chamaeleon	215
Circinus	216
Columba	216
Crux	222
Dorado	226
Eridanus	228
Fornax	228
Horologium	232
Lepus	238
Mensa	244
Monoceros	245
Musca	246
Octans	248
Orion	250
Pavo	251
Pyxis	258
Reticulum	259
Triangulum Australe	270
Tucana	271
Vela	274

LOOKING SOUTH

SOUTHERN HEMISPHERE
AUTUMN

The northern sky's most prominent constellations are Leo with bright white Regulus and, to its east, Virgo with blue-white Spica. Fellow zodiacal constellations Gemini and Cancer lie to their lower west. The Crux constellation rises to its highest point in the south with the bright stars of Centaurus and Scorpius to its east. Even Pavo (The Peacock) and the Small Magellanic Cloud are visible in the low southern sky. Carina with bright yellow-orange Canopus, as well as superbright Sirius, are sinking in the southwest. Scorpius and orange-hued Antares are riding higher with each night in the east.

LOOKING NORTH

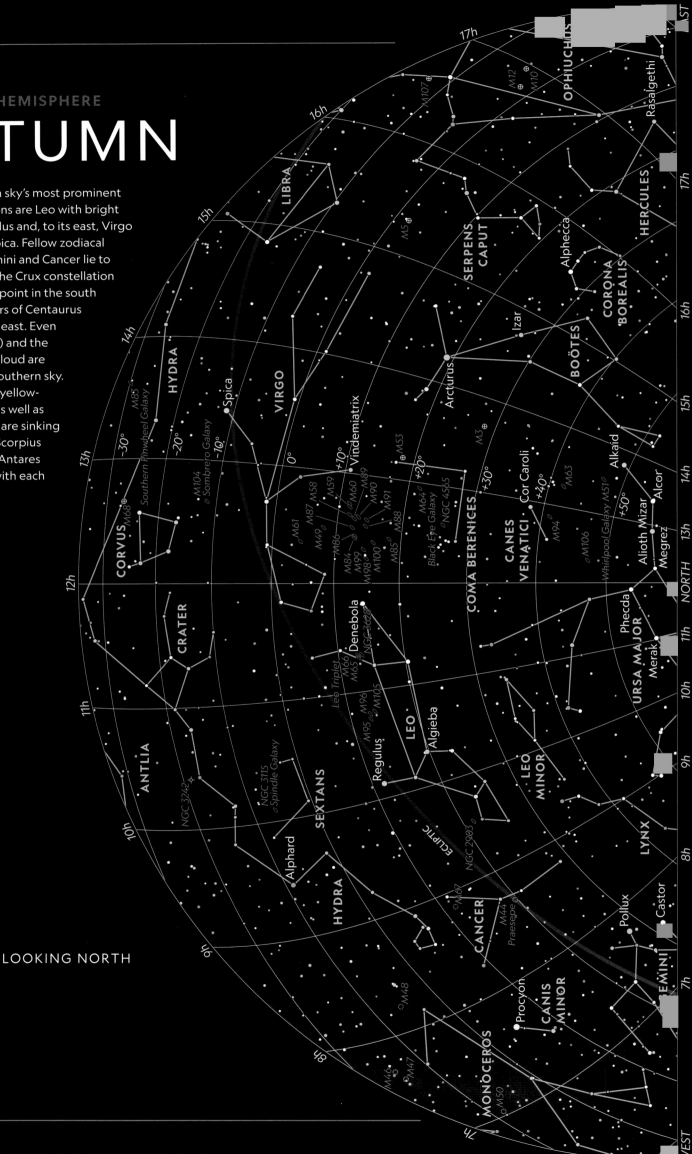

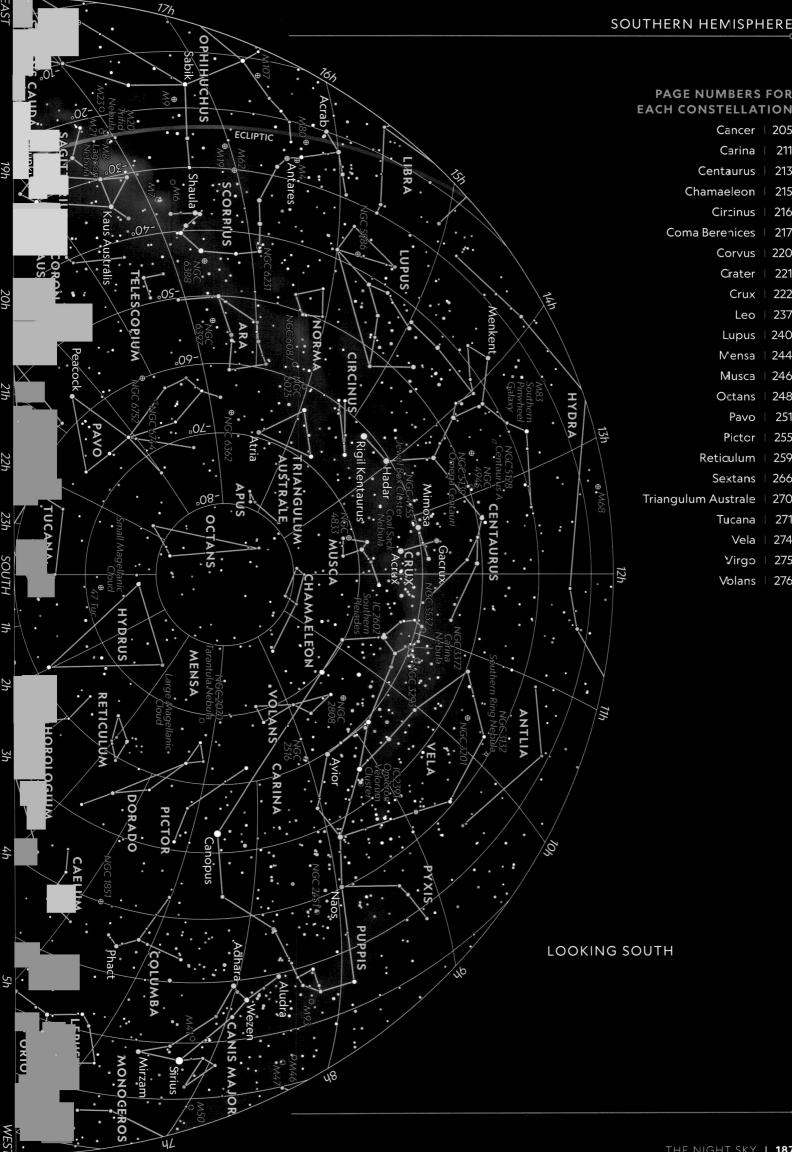

EAST

17h
OPHIUCHUS
Sabik
M107
M9
M23 M20 Trifid Nebula
M21 Lagoon Nebula
M8
M19
M62
16h
Acrab
M80
ECLIPTIC
Antares
M4
LIBRA
Shaula
SCORPIUS
M7
M6
Kaus Australis
NGC 6388
TELESCOPIUM
NGC 6231
15h
LUPUS
NGC 5986
ARA
NGC 6397
NGC 6087
NGC 6025
NORMA
CIRCINUS
14h
Menkent
M83 Southern Pinwheel Galaxy
HYDRA
Peacock
NGC 6752
NGC 6744
PAVO
NGC 6362
Atria
TRIANGULUM AUSTRALE
Rigil Kentaurus
Jewel Box Cluster
NGC 4755
Hadar
Coal Sack Nebula
NGC 5139 Omega Centauri
NGC 4945
Centaurus A
NGC 5128
13h
M68
APUS
OCTANS
Mimosa
MUSCA
NGC 4833
Acrux
CRUX
Gacrux
CENTAURUS
12h
SAGITT
19h
CORON AUS
20h
21h
22h
TUCANA
23h
SOUTH
1h
HYDRUS
47 Tuc
Small Magellanic Cloud
2h
MENSA
Large Magellanic Cloud
NGC 2020 Tarantula Nebula
CHAMAELEON
Southern Pleiades
IC 2602
VOLANS
NGC 3372 Carina Nebula
NGC 3532
NGC 3293
11h
ANTLIA
3h
RETICULUM
DORADO
PICTOR
NGC 1851
Canopus
NGC 2451
NGC 2808
NGC 2516
Avior
CARINA
IC 2391 Omicron Velorum Cluster
VELA
NGC 3201
Southern Ring Nebula
NGC 3132
10h
PYXIS
4h
CAELIUM
Naos
PUPPIS
9h
LOOKING SOUTH
COLUMBA
Adhara
Phact
Aludra
Wezen
CANIS MAJOR
M41
M93
M46
M47
5h
Mirzam
Sirius
MONOCEROS
M50
ORI
7h
6h
8h
WEST

PAGE NUMBERS FOR
EACH CONSTELLATION

Cancer | 205
Carina | 211
Centaurus | 213
Chamaeleon | 215
Circinus | 216
Coma Berenices | 217
Corvus | 220
Crater | 221
Crux | 222
Leo | 237
Lupus | 240
Mensa | 244
Musca | 246
Octans | 248
Pavo | 251
Pictor | 255
Reticulum | 259
Sextans | 266
Triangulum Australe | 270
Tucana | 271
Vela | 274
Virgo | 275
Volans | 276

SOUTHERN HEMISPHERE

WINTER

A bright triangle of stars prominent across the northern sky includes Vega and Altair in the mid-sky and Deneb near the horizon. Orange Arcturus in Boötes is still visible in the high northwest. Between Vega and Arcturus are the constellations Hercules and Corona Borealis. Meanwhile Antares, the lead star in the curvy constellation Scorpius, and next-door neighbor Sagittarius are in the overhead sky and filled with deep-sky treasures such as nebulae and clusters. The Milky Way band's brightest sections dominate winter skies, running through Crux and Centaurus in the southwest.

LOOKING NORTH

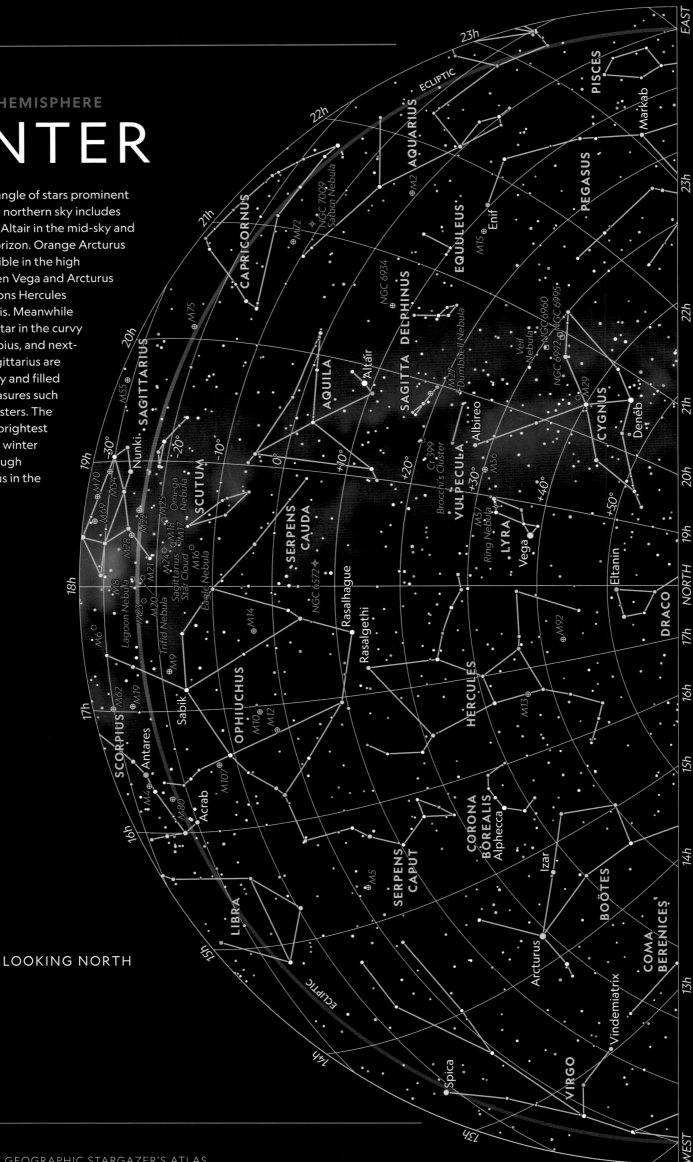

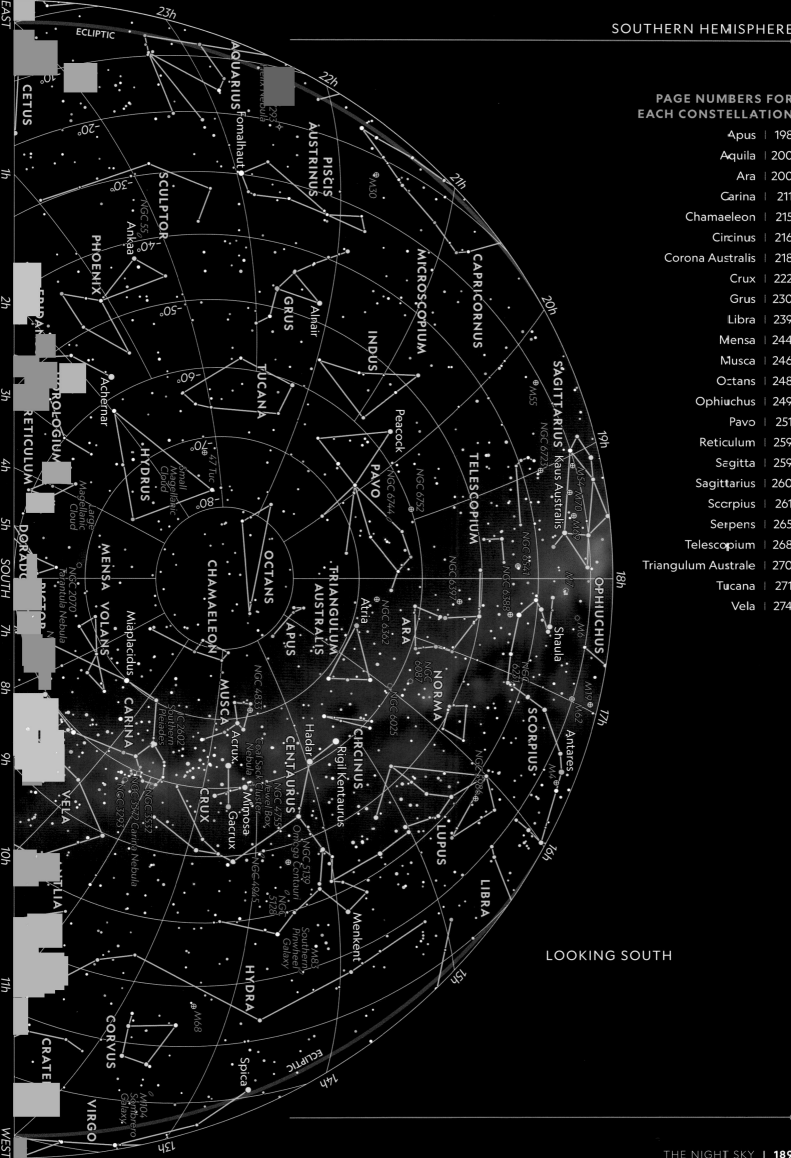

PAGE NUMBERS FOR
EACH CONSTELLATION

Apus | 198
Aquila | 200
Ara | 200
Carina | 211
Chamaeleon | 215
Circinus | 216
Corona Australis | 218
Crux | 222
Grus | 230
Libra | 239
Mensa | 244
Musca | 246
Octans | 248
Ophiuchus | 249
Pavo | 251
Reticulum | 259
Sagitta | 259
Sagittarius | 260
Scorpius | 261
Serpens | 265
Telescopium | 268
Triangulum Australe | 270
Tucana | 271
Vela | 274

LOOKING SOUTH

STAR-HOPPING NORTHERN HEMISPHERE

The night sky is full of wonders, and those wonders change over the course of the year as the constellations and other star patterns slowly make their way across the heavens. Finding specific targets in the inky black abyss can be a bit like reading a treasure map, using celestial signposts to reach dazzling destinations. One way to do this is by employing the technique known as "star-hopping," in which you start at the brightest objects in the sky and then navigate to the fainter ones. These seasonal maps provide a simplified view of the sky, focusing on easy-to-find asterisms (or star patterns) and the brightest constellations, with arrows to help you hop from one part of the sky to another.

SPRING STAR-HOPPING
NORTHERN HEMISPHERE

As the Great Bear prowls the sky overhead, the Big Dipper asterism points the way, in nearly every direction, to nearby constellations. Drawing an imaginary line down south from its bowl leads to blue-white star Regulus in Leo. Continuing down toward the horizon, this line goes on to point to Alphard of Hydra. By using other combinations of Big Dipper stars, observers can find Hercules to the east and Gemini to the west throughout the spring season. Follow the natural arc of the Big Dipper handle as it points to Arcturus of Boötes (The Herdsman), which marks one of the vertices of the Spring Triangle. Continue along this same line straight down south from Arcturus and you'll find Spica in Virgo. The third star in the Spring Triangle is Denebola, the tail star in the constellation Leo. The sickle or fishhook shape in Leo is an eye-catching marker for finding the lion.

SUMMER STAR-HOPPING
NORTHERN HEMISPHERE

Riding high in the southeast sky is the season's biggest asterism, the Summer Triangle, with each of its corners marking the starting point to a constellation. The second brightest star in this stellar pattern, Deneb, marks the tail of Cygnus, the swan constellation also known as the Northern Cross. The Milky Way appears to run up the spine of the swan, into the constellations of Aquila and Scutum. Extend an imaginary line from the mythical bird down to the southern horizon directly to the Teapot asterism within Sagittarius. Hanging low in the northwest is the Big Dipper asterism. By continuing a line out from the innermost handle stars straight out to the next brightest stellar pattern, we reach the Keystone asterism, which marks the mythical hero constellation of Hercules. Extending down south on the right side of the Keystone, we reach the bright star Antares, marking the heart of Scorpius. On moonless nights, sweep this entire region from Cygnus to Scorpius with binoculars and cruise the countless star clouds and deep-sky treasures.

AUTUMN STAR-HOPPING
NORTHERN HEMISPHERE

While the autumn sky lacks superbright stars, the four stars that form the recognizable square in Pegasus make a stellar signpost that helps skywatchers hop to several other seasonal constellations, including Andromeda. Extending one side of the Great Square straight south will point the way to Diphda, the brightest star of Cetus (The Sea Monster). Do the same with the other side of the Square and you arrive at Fomalhaut, the lead star in Piscis Austrinus. Stretch an imaginary line from the stars of the Great Square and Andromeda to the east and you will find fainter constellations such as Aries (The Ram) and Triangulum, the home of spiral galaxy M33. In the near center-northern sky is distinct, W-shaped Cassiopeia, the queen constellation, which also acts as a powerful guidepost in autumn. Extending stars of the giant W will help you track down Perseus (The Hero) to the east and Cepheus (The King) to the west.

WINTER STAR-HOPPING
NORTHERN HEMISPHERE

Low in the east, Leo—with its sickle-shaped pattern of stars—lies on its side. The Big Dipper of Ursa Major appears to be standing on its handle low in the northeastern sky. Extending imaginary lines from its "bowl" across the overhead sky leads to some of winter's brightest stars: brilliant-white Castor and yellow Pollux in Gemini and yellow-hued Capella of Auriga. Dominating the southern sky is the biggest asterism of the season, the Winter Hexagon. We start with Capella, then hop clockwise to trace out its seven stars: orange Aldebaran, then Rigel and Sirius, at which point moving up the eastern side takes us to Procyon, Pollux, and Castor, then back to Capella. Aldebaran is the brightest star in Taurus (The Bull), which is where Charles Messier sighted the first entry in his catalog: M1, the Crab Nebula.

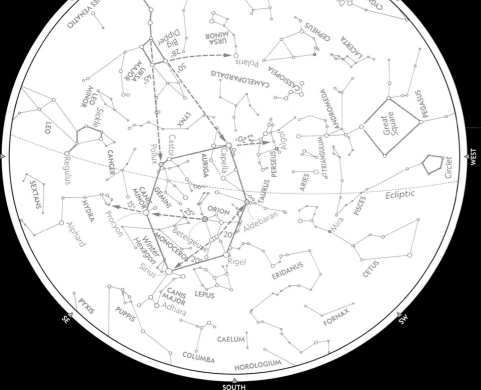

STELLAR MAGNITUDES

○ -0.5 and brighter	○ 2.1 to 2.5
○ -0.4 to 0.0	○ 2.6 to 3.0
○ 0.1 to 0.5	○ 3.1 to 3.5
○ 0.6 to 1.0	○ 3.6 to 4.0
○ 1.1 to 1.5	○ 4.1 to 4.5
○ 1.6 to 2.0	· 4.6 to 5.0
	◎ Variable star

LINES

→ 35° Star-hopping direction with angle of view
—— Asterism
—— Constellation

EXPRESSING DISTANCES BETWEEN STARS

The apparent distance between objects in the sky is described in units of angular measurements: degrees, minutes, and seconds. The distance from the horizon to the point on the celestial sphere just above your head is 90 degrees. This simplified graphic of the Big Dipper, for example, uses degrees to express the distances between one star and another within that constellation, as well as the distance from one corner of that constellation to the North Star. See "Handy Tips for Measuring Distances" on page 193.

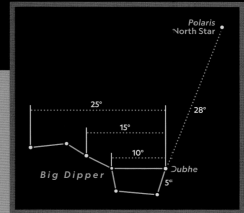

STAR-HOPPING SOUTHERN HEMISPHERE

The constellations of Orion and Scorpius, as well as the Great Square of Pegasus, shine brightly in southern skies and make good starting points as skywatchers navigate the jewels of the Southern Hemisphere. Two prominent stars, Alpha and Beta Centauri, are known as the Pointers, and they will guide you to the iconic Southern Cross. Several constellations lead you to brilliant Canopus, the second brightest star in the sky. In winter particularly, enjoy the sweeping view of the Milky Way available to southern observers. With Sagittarius as a guide, it is easy to make out the distinct shape of the bulge of the center of our galaxy and the spiral arms stretching across the sky—just by looking up.

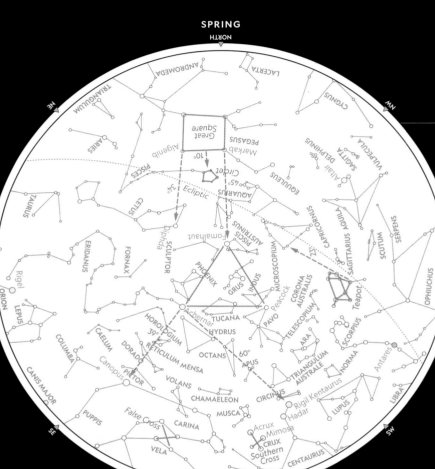

SPRING STAR-HOPPING
SOUTHERN HEMISPHERE

The winged horse Pegasus dominates the northern sky in spring. With relatively few bright stars, the Great Square of Pegasus asterism is easy to locate. Extending a line through the northeastern side of the square toward the south leads to Diphda, the brightest star in Cetus (The Whale). Use the other side of the square and extend south to find Fomalhaut in Piscis Austrinus (The Southern Fish).

Fomalhaut forms one point of an equilateral triangle, along with Peacock, the brightest star in Pavo, and Achernar, the brightest star in the long and winding river Eridanus. Using Achernar as a starting point, extend a line toward the southeast to find Canopus, the second brightest star in the sky, in Carina. A line southwest from Achernar finds Alpha Centauri. Start at the lid of the Teapot asterism, visible in the western sky, and follow a line through the top of the handle to find Capricornus (The Sea Goat).

SUMMER STAR-HOPPING
SOUTHERN HEMISPHERE

High in the northeastern sky is the Saucepan asterism, part of the constellation Orion. Orion's belt is the saucepan's base, a great starting point for finding other constellations. Extend the base toward the southeast and you find Sirius in Canis Major, the brightest star in either the Southern or Northern Hemisphere. Extend the base in the opposite direction and you spot Aldebaran, the brightest star in Taurus (The Bull). Nearby in Taurus's shoulder is the Pleiades star cluster, easily visible with the naked eye. From Orion's shield, draw a line through Betelgeuse to reach Canis Minor (The Lesser Dog) and star Procyon. Starting at the northwestern belt star, a line through Betelgeuse takes you to Castor, one of Gemini's twin stars. From the same belt star, travel in the opposite direction through Rigel and find Achernar, the brightest star in Eridanus. From the eastern belt star, a line through Saiph (Kappa Orionis) reaches bright Canopus in Carina (The Keel).

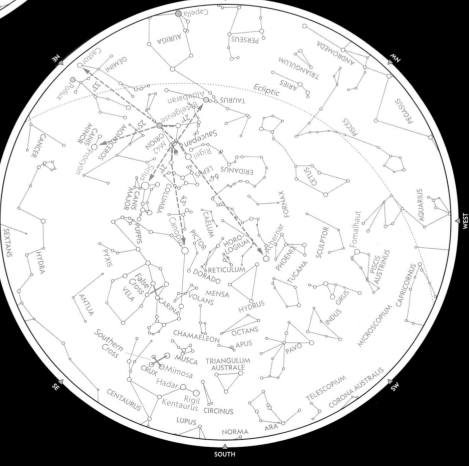

AUTUMN STAR-HOPPING
SOUTHERN HEMISPHERE

High in the southeastern sky is the famous constellation Crux (The Southern Cross). It is easily confused with the False Cross, which is larger and made up from two constellations—Vela and Carina. The best way to find the real Southern Cross is to look for the Pointers, Alpha and Beta Centauri: A line from Alpha through Beta Centauri will lead you to Gamma Crucis, the third brightest star in Crux. The South Celestial Pole is not easy to find, but it can be located by taking the long axis of Crux and extending this distance about 4.5 times toward the south: The point just outside of Octans that you reach is the South Celestial Pole. High in the eastern sky is the constellation Corvus. It looks more like an outstretched sail than a crow, and is sometimes known by its Sail asterism. Nearby is Virgo's Spica, one of three bright stars, along with Arcturus and Denebola, that form the Spring Triangle asterism.

WINTER STAR-HOPPING
SOUTHERN HEMISPHERE

Scorpius (The Scorpion) is easy to spot high overhead, as it's one of the few constellations that does look like its namesake. Extend a line from each end of Scorpius's head and you come to the two bright stars Zubenelgenubi and Zubeneschamali, whose names reflect their past as the northern and southern claws of the scorpion. To find Vega, the brightest star in Lyra, start at Beta Centauri and extend a line through Antares in Scorpius, toward the north-northeast horizon. Using the scorpion's sting as a starting point, zoom through Antares and continue toward the northwest horizon to reach the orange giant star Arcturus. Extending a line in the opposite direction, starting at Antares, leads to Alnair, the brightest star in Grus (The Crane).

This time of year sees the Milky Way stretching across the sky. From Scorpius's sting, head east to find Sagittarius, the Teapot asterism, and the center of our galaxy.

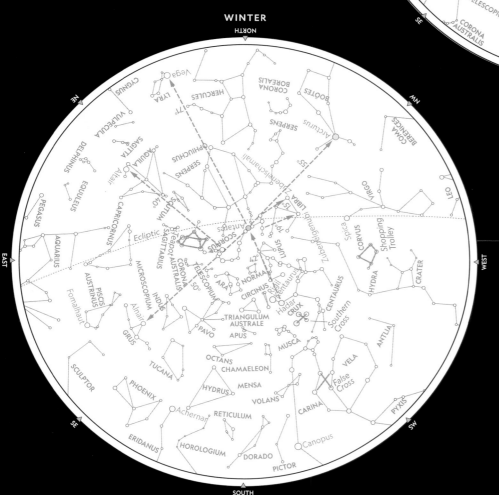

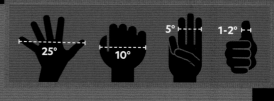

HANDY TIPS FOR MEASURING DISTANCES

An easy trick is to use your hands and fingers, held at arm's length, as convenient measurements of distance between stars in the sky. Your outstretched hand is about 20 to 25 degrees wide from the tip of the thumb to the tip of the little finger. Your fist is about 10 degrees across; your three middle fingers are about five degrees across; and your thumb and index finger are between one and two degrees wide, big enough to cover the sun or the moon. You'll find these handy measurement tools throughout this chapter. They will help you recognize the proportional size of constellations, as well as estimate distances between objects of interest.

STELLAR MAGNITUDES

◯ −0.5 and brighter	○ 2.1 to 2.5
◯ −0.4 to 0.0	○ 2.6 to 3.0
◯ 0.1 to 0.5	○ 3.1 to 3.5
◯ 0.6 to 1.0	○ 3.6 to 4.0
◯ 1.1 to 1.5	○ 4.1 to 4.5
◯ 1.6 to 2.0	○ 4.6 to 5.0
	◉ Variable star

LINES

- ←---35°--- Star-hopping direction with angle of view
- ——— Asterism
- ——— Constellation

目女八度至危十五度於辰在子屬玄枵者黑北方之色枵者龍青

之時陽氣下降陰氣上昇萬物幽死未有主者天地空虛督玄枵之次也

吴中
大津

太人

危

虛

造父

車府

將

兩

尖

離珠

敗臼

中

墳墓

蓋屋

泣

主公吏

天錢

北洛師門

天蠶

騰蛇

室

雷

電

鈎

雲雨

八魁

五月日會營室

三月日會婁胃中且之

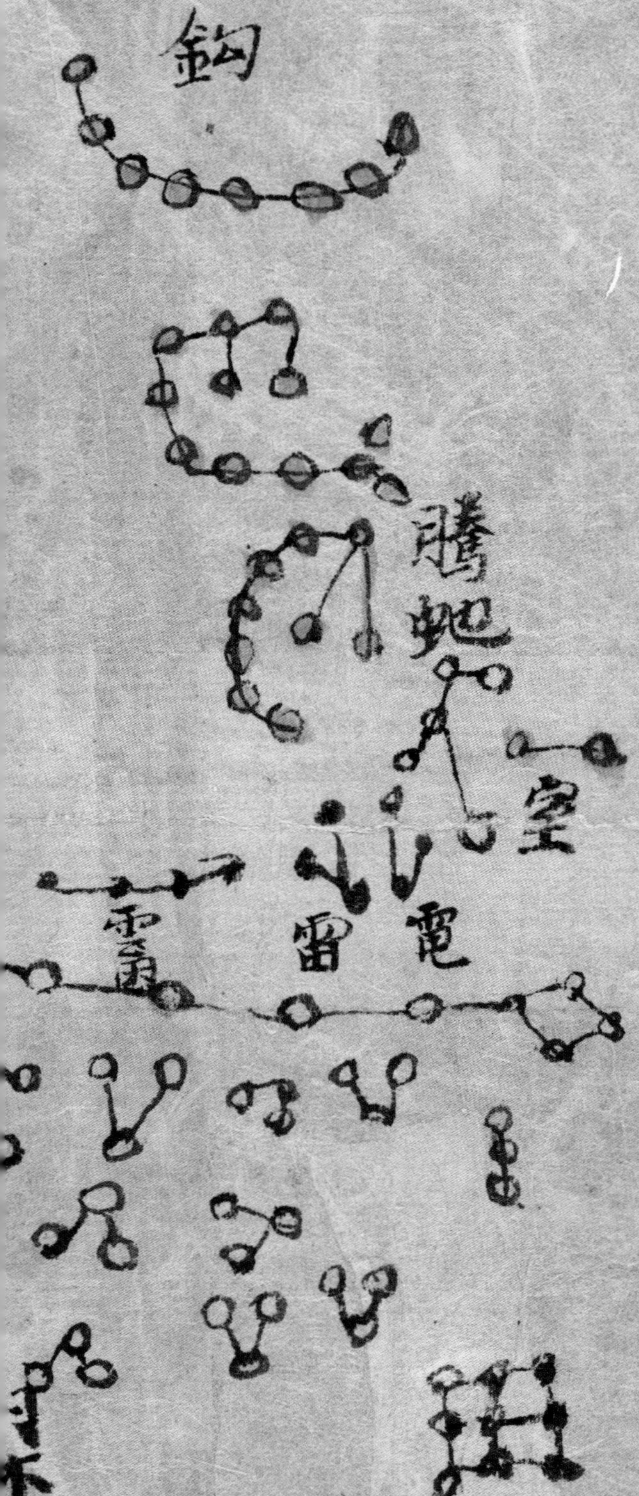

二月日會女虛

CONSTELLATIONS

The 88 constellations, each one with its own story,
make up our map of the night sky.

ook up on a clear, moonless night in the dark countryside away from the light pollution of the city, and how many stars can you see? While no one has actually counted every single star, it's estimated that at most about 6,000 stars are visible to the unaided eye under the most pristine skies around the world. For most of us, however, stuck under light-polluted domes, this number plummets to anywhere from about 200 to just a dozen or two in downtown skyscapes.

Ancient astronomers from cultures around the world divided the stars into their own groups called constellations: sky pictures that were important to their own people. They not only helped organize the sky but also helped create a calendar to time seasonal activities such as planting and harvesting, and they established guideposts to aid explorers in navigating across great distances.

The most widely recognized constellations today across both the amateur and professional astronomy communities are handed down from the ancient Greeks and Romans (with Latin names). It's important to remember that the star patterns we humans have assigned different star groups have no real scientific significance, as the stars themselves are located at vastly different distances. The patterns our minds have formed are simply due to the observational vantage point we have here on Earth.

In 1922, the International Astronomical Union—a global governing body of astronomers—officially adopted 88 constellations and standardized their names across the entire celestial sphere that surrounds Earth. Within a few years, the borders of each constellation were established. More than half of our constellations were first cataloged by Claudius

Ptolemy around A.D. 150, but there are hints that he included some stellar figures that dated back to at least the fifth century B.C. The modern constellations that were not part of the classical 48 are mostly faint and challenging to observe with the unaided eyes. They were added by European astronomers in the 16th and 17th centuries to fill in regions between established constellations, as well as to include previously unseen stars that were below the horizon for ancient Greeks and Romans.

In the following pages, you will find comprehensive, alphabetically listed descriptions of all 88 recognized constellations seen from the Northern and Southern Hemispheres. Each star pattern's profile offers details of mythological backstories from various cultures, the physical properties of its noteworthy stars, and a selection of its most stunning deep-sky targets for binoculars and telescopes.

(Opposite) A Chinese scroll map, created during the Tang dynasty (618–906), depicts the night sky as seen from the Northern Hemisphere.

HOW TO READ THE CONSTELLATION ENTRIES

Listed alphabetically, the entries that follow describe and map each of the 88 official constellations of the Northern and Southern Hemispheres. They include:

- The constellation's Latin name and English equivalent
- A fact box that tells you where and when to find the constellation, brightest stars, and significant deep-sky objects (such as galaxies and clusters)
- A description of the constellation, with information about its history and most interesting sights
- A map depicting the constellation's mythological outline, key stars (white), most major deep-sky objects (pink), the deep-sky object pictured on the page (yellow), asterisms (orange), and neighboring constellations. (An icon on each map shows the constellation's size on the sky as compared to an outstretched hand.)
- Often, sidebars with more information about the constellation's backstory or about particularly interesting targets for your binoculars or telescope

DEEP-SKY OBJECTS
- ○ Open star cluster
- ⊕ Globular star cluster
- ▫ Bright nebula
- ▫ Dark nebula
- ✧ Planetary nebula
- ⊘ Galaxy

ANDROMEDA (THE CHAINED MAIDEN)

Visible: Northern and Southern Hemispheres

Best Seen: Northern autumn and southern spring

Main Stars: 16

Brightest Star: Alpheratz, 2.07 mag.

Deep-Sky Objects: M31, M32, NGC 205, NGC 404, NGC 891

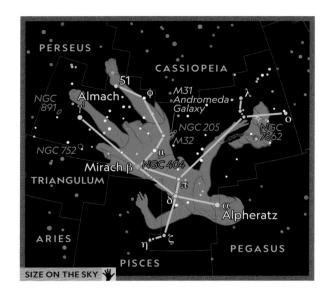

STAR STORIES

In classical Greek mythology, Andromeda is the daughter of Queen Cassiopeia and King Cepheus. According to legend, the hapless princess was chained to a seaside rock to be sacrificed to the sea monster Cetus. Fortunately, she was rescued by the young hero Perseus.

Andromeda, the chained maiden constellation, can be seen from August through February. This large constellation in the northern sky is conspicuous thanks to its attachment to the neighboring constellation Pegasus and that constellation's Great Square asterism (star pattern). The brightest star, Alpheratz, which represents the head of the maiden, happens to be shared with Pegasus. This blue giant star lies 97 light-years from Earth and shines 200 times brighter than our sun. Four main stars of Andromeda form a curved line toward the east, with Mirach representing her hips and Almach her chained foot. Although Mirach is nearly identical in brightness to Alpheratz, it is a larger red giant 197 light-years away. Meanwhile, Almach is a pretty color-contrasting triple star system that sits 355 light-years distant, with a primary star that appears orange and a secondary star glowing blue-green, making for a beautiful sight through backyard telescopes.

While Andromeda is considered a northern constellation, most of its stars can be glimpsed from the Southern Hemisphere, but always near the horizon, making it a bit trickier to observe its famous deep-sky objects. The most celebrated object within the constellation is the grand spiral called the Great Andromeda galaxy, or Messier 31. This, the closest large galaxy to our Milky Way, lies 2.5 million light-years away. To the surprise of many people, it can be spotted with the naked eye from a typical suburban backyard on clear autumn nights. While even binoculars will show the 3.4-magnitude central galactic core only as a faint, grayish oval smudge, the galaxy still represents one of the farthest objects the unaided human eye can see in the universe. And it's amazing to think that this ancient light left on its journey from the galaxy when parts of Earth were still covered by an ancient ice age and saber-toothed cats roamed North America. M31 can be tracked down by drawing a line from the stars Mirach and Mu Andromedae and then continuing an imaginary line the same distance north. Through a backyard telescope, more details in the galaxy's structures become visible, along with its two smaller companions, the faint elliptical galaxies M32 and M110.

Andromeda hosts a few additional galaxies, the brightest of which is 10th-magnitude NGC 891, only 3.5 degrees east of Almach. This edge-on spiral sports an eye-catching dark dust lane that sweeps across its entire length.

M31: the Andromeda galaxy

ANTLIA (THE AIR PUMP)

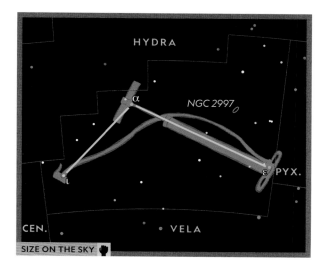

HYDRA

NGC 2997

α

ι

ε PYX.

CEN.

VELA

SIZE ON THE SKY

Visible: Southern Hemisphere

Best Seen: Spring

Main Stars: 3

Brightest Star: Alpha Antliae, 4.25 mag.

Deep-Sky Object: NGC 2997

Devoid of any bright stars, Antlia is one of the faintest and also one of the newer recognized star patterns in the sky. This tiny constellation's name means "pump" in Latin, and it celebrates the 17th-century invention that was a precursor to the modern vacuum chamber.

Antlia ranks 62nd in terms of constellation size and is completely visible as far north as 49 degrees latitude in the Northern Hemisphere. Its northern border is shared with Hydra, while Vela is to its south and Centaurus to its east. Antlia's brightest stellar member, Alpha Antliae, has an average magnitude of 4.25 and is faintly visible to the naked eye on dark nights, shining with a slight reddish hue. Located 320 light-years away, this orange giant is 41 times the size of our sun and is thought to be a variable star with 500 times the luminosity of our own star.

The constellation's next brightest star, Epsilon Antliae, is much farther from Earth at 590 light-years. It is also a stellar giant nearing the end of its life cycle, with a diameter that stretches 69 times wider than our sun. Antlia also hosts a few binary stars that are visible with small telescopes. Delta Antliae is 450 light-years away and consists of a primary blue-white star shining at a very faint naked-eye 5.6 magnitude and a much fainter 9.6 magnitude yellow-white secondary that requires at least binoculars.

Zeta Antliae is an optical double; its primary member is actually composed of two white stars shining at 6.2 and 7 magnitude, respectively, 410 light-years away. The second member is 386 light-

years distant and shines at 5.9 magnitude. S Antliae is an eclipsing binary star system in which one star regularly hides its companion from our line of sight, causing their combined light to appear to swing in brightness between 6.27 and 6.83 magnitude in just 15.6 hours. Astronomers believe the two stars are at least as old as our sun and that as they spin, they are being gravitationally pulled toward each other, eventually leading to a merger into a single star.

Antlia is home to multiple faint galaxies, with the brightest being NGC 2997, a magnitude-10 spiral galaxy that lies about 22.8 million light-years from Earth. With a small core, its disk and arms appear tilted nearly 45 degrees to our line of sight. Large-telescope views reveal spiral arms strewn with knots of ionized hydrogen that are star factories and the source of the majority of the galaxy's glow. NGC 2997 is thought to have a mass about 100 billion times that of our sun, but is probably less massive than our own Milky Way galaxy.

TELESCOPE TARGETS

Antlia is home to two star systems that host exoplanets. Macondo is a magnitude-8 star that is 93 light-years away and may host a terrestrial world within its habitable zone. Meanwhile, WASP-66 is a magnitude-11, younger version of our sun that hosts a hot Jupiter world.

Spiral galaxy NGC 2997

APUS (THE BIRD OF PARADISE)

Visible: Southern Hemisphere

Best Seen: Winter

Main Stars: 4

Brightest Star: Alpha Apodis, 3.8 mag.

Deep-Sky Objects: IC 4499, NGC 6101

STAR STORIES

Apus is one of the 12 new constellations Pieter Dirkszoon Keyser and Frederick de Houtman, Dutch navigators and stellar cartographers, placed in the sky during their first voyage to the East Indies from 1595 to 1597. Originally the constellation was known as Paradysvogel Apis Indica, with the first word meaning "bird of paradise" in Dutch and Apis Indica meaning "Indian bee" in Latin.

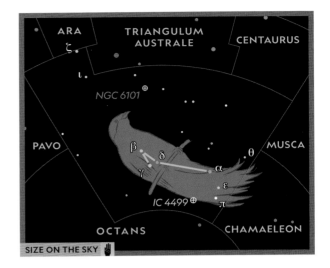

SIZE ON THE SKY

Apus is a faint, small constellation that represents the bird of paradise, but its name, from the Greek, erroneously implies it is "without feet" because early naturalists saw only a poorly preserved bird specimen. It ranks 67th in size out of all 88 constellations and has the more prominent constellation Triangulum Australe to its north, joined by Ara and Circinus. Skywatchers will also find Musca and Chamaeleon bordering it to the west, while Octans hangs to its south. It is home to only a few stars, none of which is brighter than magnitude 3. Three stars form its only eye-catching pattern—an elongated triangle.

Apus lies close to the South Celestial Pole, making this constellation almost completely exclusive to the Southern Hemisphere, save for observers up to about seven degrees latitude north of the Equator.

With an apparent magnitude of 3.8, its brightest star, Alpha Apodis, is an orange giant 430 light-years away that is nearing the end of its life. As its fuel has begun to run out, the star's diameter has ballooned to at least 48 times the size of our sun. Delta Apodis is a double star system that can be spotted with the naked eye under dark skies and is easily resolved with binoculars. The main component of this wide optical double star is an orange giant 545 light-years distant.

Three star systems within Apus have exoplanet discoveries. The star HD 131664 has a gas giant world that is 23 times larger than Jupiter. Sunlike HD 134606 owns three orbiting planets. And HD 137388, a star cooler than our sun, has been found to host a planet 79 times more massive than Earth. This giant circles its sun in a lopsided orbit every 330 days, a path that crosses the system's habitable zone. The planet's official name is Kererū, the Maori term for the New Zealand pigeon. The star itself is also known as Karaka, after a native tree in New Zealand, and at magnitude 8 is visible with binoculars.

Apus contains no Messier objects but is home to two prominent globular clusters. NGC 6101 is a small globular star cluster at magnitude 9, putting it within the reach of a small backyard telescope. This city of stars stretches 160 light-years across and lies some 47,000 light-years from Earth. It is thought to contain a large number of black holes. Another globular cluster, IC 4499, appears through larger backyard telescopes as the faint cotton ball that is typical of globulars, but its stars appear more scattered than usual. Shining at magnitude 8.5, it appears under low magnification to have an even, round glow.

Globular cluster IC 4499

AQUARIUS (THE WATER BEARER)

Visible: Northern and Southern Hemispheres

Best Seen: Northern autumn and southern spring

Main Stars: 10

Brightest Star: Sadalsuud, 2.91 mag.

Deep-Sky Objects: Helix Nebula, Saturn Nebula

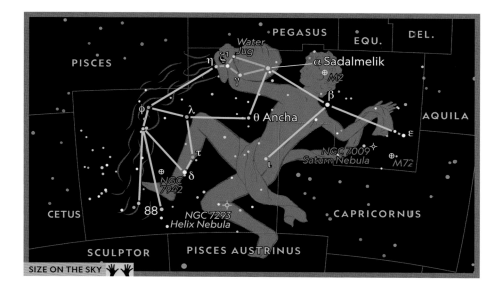

Across many cultures in Europe and North Africa, the stars of Aquarius have been connected to rain and the end of droughts, which is particularly relevant in regions with dry climates. For skywatchers, the constellation's stellar members are mostly faint and scattered, and drawing out the figure of a man kneeling and pouring water from a jar can be somewhat challenging. Yet parts of Aquarius do form distinctive patterns that draw the eye, particularly the water jar itself. To find it, draw an imaginary line from the stars Scheat and Markab in the Great Square of the nearby constellation Pegasus, and follow it out to a point below the circlet of stars in the neighboring constellation of Pisces. Then just scan to the right of the circlet of stars to find the Y-shaped stellar pattern that makes the water jar.

Two of the brightest stars in Aquarius are Sadalsuud and Sadalmelik, located 680 and 710 light-years, respectively, from Earth. For backyard telescopes, the constellation holds many deep-sky wonders, including two sparkly globular clusters, M2 and M72, which lie near Sadalsuud. M2 is a rich 6.5-magnitude cluster that lies five degrees

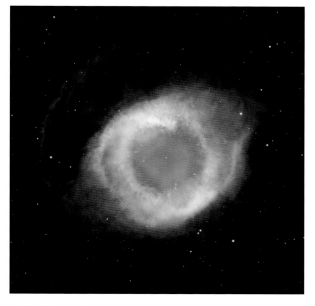

NGC 7293: the Helix Nebula

north of Beta Aquarii, making it an easy find using binoculars since both objects lie within the same field of view. The cluster itself measures about 175 light-years across and ranks as one of the largest of its kind known in our galaxy, while sitting 37,000 light-years away. Meanwhile, M72 is a much more modest magnitude-9 cluster that sits 56,000 light-years from Earth.

Tucked away in the southern part of the constellation is the largest of two planetary nebulae in Aquarius, the famous 700-light-years-distant Helix Nebula. On a dark night it shows off its overlapping ring-like structure, giving it an appearance similar to its famous cousins, the Ring and Cat's Eye nebulae. More recently, with high resolution imagery from space telescopes such as Hubble and Spitzer, the Helix Nebula has gained pop culture status with its alternate name of Eye of Sauron.

Meanwhile to the far south of Sadalsuud, near the faint naked-eye star Nu Aquarii, lies the Saturn Nebula. Under high magnification this 5,000-light-years-distant remnant of a dying sunlike star shows off an outer shell of green gas that it has blown into space, so that it resembles its namesake ringed planet. From end to end this expanding gas cloud is thought to stretch over a half light-year across.

In 2016, astronomers may, appropriately enough, have discovered potentially one of the most watery planetary systems beyond our solar system within the borders of Aquarius. The 40-light-years-distant, ultra-cool red dwarf star TRAPPIST-1 hosts seven Earth-size planets, three of which are in the habitable zone and could hold water.

STAR STORIES

Aquarius is one of the most global of all constellations, rooted in many myths, and its association with water crosses various cultures, including Greek and Indian One of the most ancient Middle Eastern legends dates back nearly six thousand years to the Sumerians and is associated with a story of a global flood that may be the seed of the Noah's Ark story in the Bible.

AQUILA (THE EAGLE)

Visible: Northern and Southern Hemispheres

Best Seen: Northern summer and southern winter

Main Stars: 10

Brightest Star: Altair, 0.76 mag.

Deep-Sky Objects: NGC 6781, NGC6804, Glowing Eye Nebula

Aquila, the great celestial eagle, soars high in the overhead skies across the summer months in the Northern Hemisphere. This prominent constellation lies just south of its brighter neighbor Cygnus, with the Milky Way band bisecting the mythical raptor's stellar form. According to ancient Greek legend, Aquila was the eagle that carried Zeus's bolts of lightning.

The constellation's lead star is Altair, the 12th brightest star in the sky. At only 16.8 light-years distant, it is also one of the closest bright stars to our solar system. Shining with a yellow-white light, it represents one of the three points in the Summer Triangle asterism. Because of its proximity to the rich Milky Way region, it pays off to scan Aquila with binoculars and telescopes for its many faint clusters of stars and nebulae. Planetary nebulae for moderate-size telescopes include NGC 6804 (12th magnitude), which takes on the shape of a comet with a tail. To find the 11th-magnitude Glowing Eye or NGC 6751 planetary nebulae, use the nearby faint naked-eye star Lambda Aquilae as a guidepost.

ARA (THE ALTAR)

TELESCOPE TARGETS

One of the youngest planetary nebulae known, the 18,000-light-years-distant Stingray Nebula first appeared in telescopes in the early 1970s, and the Hubble Space Telescope has seen the gas bubble continually expand. Today, the dying star's gaseous shell spans about half a light-year across, which is more than 260 times the size of our solar system—yet still smaller than most known planetary nebulae.

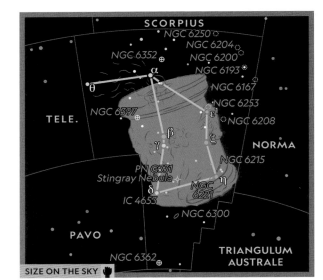

Visible: Southern Hemisphere

Best Seen: Winter

Main Stars: 8

Brightest Star: Beta Arae, 2.84 mag.

Deep-Sky Objects: IC 4653, NGC 6397, Stingray Nebula

One of the smaller constellations, Ara has an ancient heritage, having been described by Greek astronomer Ptolemy in the second century A.D. and included in his catalog of 48 constellations. Ara, "altar" in Latin, is associated with the altar on which the gods swore allegiance to Zeus before waging war with the Titans. The Milky Way, to the north, represented the smoke rising from the altar.

Ara is formed by two faint parallel lines of seven stars that straddle the Milky Way and lie just south of Scorpius, tucked underneath the bend in the scorpion's tail. Backyard telescopes will reveal a 12th-magnitude dwarf spiral galaxy, IC 4653, embedded within this dense region of interstellar dust clouds. Ara is also home to the second nearest globular cluster, NGC 6397, which lies 7,800 light-years away and is easily resolved with a small backyard telescope.

ARIES (THE RAM)

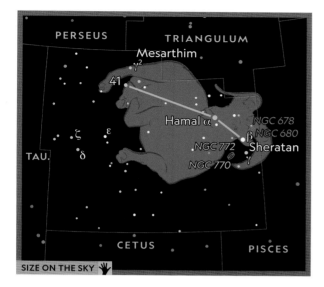

Visible: Northern and Southern Hemispheres

Best Seen: Northern autumn and southern spring

Main Stars: 4

Brightest Star: Hamal, 2.01 mag.

Deep-Sky Objects: NGC 772, NGC 678, NGC 680

About 4,000 years ago the sun was positioned within the pattern of stars known as Aries during the spring and signaled the start of that season in the Northern Hemisphere. Astronomically, this was the time when the sun appeared to be gliding through Aries as it passed from the southern to the northern celestial sphere for the year. Precession has since shifted this constellation's position with respect to the sun's equatorial crossing, but by tradition Aries remains where the zodiac begins. Starting with the Babylonians and then later with the ancient Greeks and Romans, astronomers assigned Aries as the first sign of the zodiac because it represented the start of a new year.

Many cultures around the world had their own celestial figures formed by several stars of Aries. In Chinese astronomy the stars were part of many different constellations representing a lasso, the emperor's hunting partner, and a marsh, among others. In some Polynesian cultures the constellation represented part of a porpoise, while the indigenous people of Peru called it the Kneeling Terrace, in reference to harvest time.

On evenings in late fall and early winter, Aries is high in the east between the Great Square of Pegasus and the Pleiades in Taurus. The two horns of the ram are marked by a pair of naked-eye stars: 66-light-years-distant Hamal (Arabic for "lamb"), which shines with a yellow glow, and blue-hued Sheratan (Arabic for "two-signs"), located 60 light-years away. These two stars, along with magnitude-4.6 Mesarthim, form a crooked asterism that has been used for navigation in the past. Mesarthim is in fact a nice double star with blue and white members that can be resolved with small telescopes. Another easy binary star target is fifth-magnitude Lambda Arietis.

A more challenging double to tease apart is Pi Arietis. While the brighter member of the stellar pair shines at fifth magnitude, its companion is much fainter at 8.8 magnitude and is separated by only three arc-seconds; you'll need high magnification and steady skies to resolve this pair.

Aries is home to a few galaxies that are faint and require at least a 150-mm (6-in) telescope. The brightest of them all is 10th-magnitude spiral NGC 772. Located 116 million light-years away, the galaxy is a monster, measuring twice the size of our Milky Way. Careful observation under high magnification reveals a concentrated, bright core surrounded by a faint halo filled with faint stars. A companion dwarf galaxy, NGC 770, appears to have a fine hazy band connecting the two galactic islands of stars.

STAR STORIES

To many cultures, Aries was associated with a ram, despite being made of only four stars. According to the Greeks, Aries was the source of the golden fleece. The ram was sent to rescue a king's children from an abusive stepmother. When the ram returned, the king sacrificed it, turned its fleece into gold, and left a dragon guarding the fleece, until it was stolen by Jason and his Argonauts.

Interacting galaxies NGC 772 and NGC 770 (top center)

AURIGA (THE CHARIOTEER)

Visible: Northern Hemisphere

Best Seen: Autumn and winter

Main Stars: 5

Brightest Star: Capella, 0.08 mag.

Deep-Sky Objects: Pinwheel Cluster, M37, Starfish Cluster, IC 405, IC 2149

TELESCOPE TARGETS

Astronomers have discovered that Capella actually consists of two binary star systems. One pair has two giant yellow stars in close orbit, while the other duo is made up of two feeble red dwarfs.

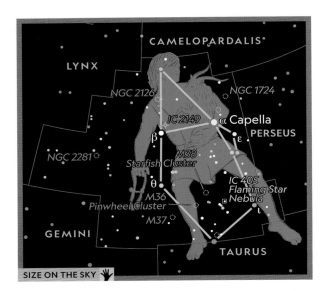

SIZE ON THE SKY

Auriga is the 21st largest constellation in the sky, and its brightest stars form a distinctive pentagon shape that stretches across a particularly rich part of the Milky Way. While not as well known to casual skywatchers, Auriga is a favorite among backyard astronomers for its many deep-sky treasures. One of the more ancient constellations, Auriga is classically portrayed as a charioteer, but for Romans and Greek legends he was shown without a chariot. Pinpointing Auriga in the sky is made easy thanks to its brightest star, Capella. For most northerners it rises in the northeast in early fall and as such has been

called the Harvest Star; it signals the coming winter as it heralds the approach of its brighter neighboring constellations Orion and Taurus. The sixth brightest star in the sky, Capella is relatively close to us, located only 42 light-years away. It has been a beacon in many cultures across the world, with mentions of it dating back to the 20th century B.C. in ancient Mesopotamia. It has represented the heart of Brahma in India, and in ancient Greece it was a mother goat that the charioteer carries on his back along with her three kids (the neighboring stars).

Auriga sports three beautiful, bright open star clusters found in a line down the center of the pentagon figure. All three are located roughly at the same distance from Earth, from 4,100 to 4,500 light-years away, and though they can be barely glimpsed with the naked eye under dark skies, they are easy targets for binoculars. A small telescope is, however, needed to begin resolving the individual diamond-like stars embedded within each cluster. The Pinwheel Cluster, also known as Messier 36, can be found just five degrees southwest of the star Theta Aurigae. It is a sixth-magnitude irregular concentration of about 60 stars that are considered adolescent, no more than 25 million years old. Arguably the most picturesque and brightest cluster of the cosmic trio is M37. Visible through binoculars as a fuzzy patch, it spans about the same area as the moon's disk. Views through small telescopes, however, have been likened by stargazers to a dark field strewn with sparkling gold dust. Finally, the Starfish Cluster, or M38, is the largest of Auriga's open clusters, but a bit fainter at seventh magnitude.

IC 405: the Flaming Star Nebula

BOÖTES (THE HERDSMAN)

Visible: Northern Hemisphere

Best Seen: Spring and early summer

Main Stars: 7

Brightest Star: Arcturus, -0.04 mag.

Deep-Sky Objects: NGC 5466, NGC 5248, NGC 5676

SIZE ON THE SKY

Boötes is a large, conspicuous northern constellation flanked by Canes Venatici, Coma Berenices, Corona Borealis, Draco, Hercules, Virgo, Serpens Caput, and Ursa Major. It is classically represented as the herdsman who invented the plough, but in Greek mythology Boötes was a bear chaser who followed Ursa Major and Ursa Minor around the polar star, Polaris. Today's skywatchers see the more modern form of an ice-cream cone.

The bottom of these modern forms is pinned down by the sky's fourth brightest star, Arcturus. Arcturus's name in Greek means "guardian of the bear," and it is an impressively brilliant orange star only 37 light-years from Earth. You can easily track down Arcturus by following the arc of the three stars that mark the handle of the Big Dipper.

CAELUM (THE ENGRAVING TOOL)

SIZE ON THE SKY

Visible: Southern Hemisphere

Best Seen: Summer

Main Stars: 4

Brightest Star: Alpha Caeli, 4.45 mag.

Deep-Sky Object: NGC 1679

After an expedition to southern Africa in the 1750s, the French astronomer Nicolas-Louis de Lacaille created 14 new constellations, one of which is the inconspicuous Caelum. Representing a sharp-edged tool, this small, faint group of stars takes the form of a chisel with a sharpened point. You can hunt it down by searching for an empty-looking part of the sky that is about 20 degrees northwest of the bright star Canopus and nestled adjacent to the tiny constellations Columba and Pictor to the west and Horologium to the east. Caelum's two brightest stars, Alpha and Gamma Caeli, are only fourth magnitude, making them barely visible to the naked eye.

CAMELOPARDALIS (THE GIRAFFE)

Visible: Northern Hemisphere

Best Seen: Winter

Main Stars: 8

Brightest Star: Beta Camelopardalis, 4.03 mag.

Deep-Sky Objects: NGC 2403, NGC 2146, NGC 1569

airly large in terms of the celestial area it takes up in the northern sky, Camelopardalis stretches from the North Star southward toward Cassiopeia and Ursa Major. Created by Dutch astronomer Petrus Plancius in 1612, it was first described as a camel. However, as the stellar figure clearly represents a giraffe, there is a bit of ambiguity regarding its origins.

Filled with a loose group of faint stars, Camelopardalis is considered a fairly barren northern constellation with no distinct stellar pattern, and with only four magnitude-4 stars it is barely visible to the naked eye from a typical suburban backyard. Its brightest member is Beta Camelopardalis, which is in fact a binary star system, with the primary star being a yellow supergiant. But it's the third brightest star in the Giraffe, 4.3-magnitude Alpha Camelopardalis, that holds special significance for astronomers. It ranks as one of the most distant bright stars visible to the unaided eye and is currently 6,000 light-years from our solar system. This blue supergiant is 32 times larger than our sun and is more than 670,000 times brighter, hence its relative brightness in Earth's sky despite its great distance. Another reason why Alpha Cam is so special is that it's thought to be a runaway star, gravitationally thrown out of the star cluster NGC 1502. Its expulsion must have been chaotic, since its current speed moving away from the cluster has been clocked at between 680 and 4,200 kilometers per second (1.5 and 9.4 million mph).

The Giraffe is home to a nice collection of deep-sky objects worth chasing down with a backyard telescope. One of the brightest and easiest galaxies for a small telescope in this long-necked constellation is the eighth-magnitude face-on spiral NGC 2403. Lying some 10 million light-years distant, it stretches 50,000 light-years across and is filled with star-forming regions marked by giant red spots along the spiral arms. The much smaller irregular galaxy NGC 1569, about a quarter the size of NGC 2403, has the same brightness. Its core is surrounded by a faint halo that shows mottling through larger telescopes. NGC 2146 is an 11th-magnitude barred spiral galaxy sitting 70 million light-years away. Its oval body stretches 80,000 light-years across, and high magnification shows off one of its unusually bent spiral arms, filled with dark dusty lanes, superimposed on its luminous white core in the background. This galactic contortion is thought to be a result of a gravitationally close encounter with a neighboring galaxy some 800 million years ago.

Small starburst galaxy NGC 1569

CANCER (THE CRAB)

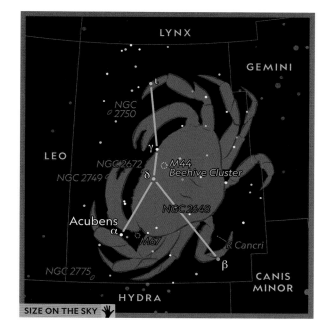

SIZE ON THE SKY

Visible: Northern and Southern Hemispheres

Best Seen: Northern spring and southern autumn

Main Stars: 5

Brightest Star: Tarf, 3.50 mag.

Deep-Sky Objects: M44, M67

The faintest of all zodiacal constellations, Cancer is devoid of any stars brighter than 3.5 magnitude and is best seen on moonless nights. Its scattered stars lie along the ecliptic and are sandwiched between Gemini to the west, Leo in the east, and Hydra to the south; it is found northeast of the superbright Orion's Belt asterism. Despite being so inconspicuous to the naked eye, it is one of the more famous constellations, dating back to ancient Greek times. In one myth, the goddess Hera hated Hercules the strongman and sent a giant crab to distract him while he battled the hydra, a multiheaded serpent. Hercules managed to crush the crab, and as a reward to the crustacean Hera placed it in the starry heavens.

The brightest star in Cancer is Tarf, which in Arabic means "the end"; the star lies at the end of one of the crab's legs. At 290 light-years from Earth, it shines with a luminosity 660 times brighter than our sun. It turns out that Tarf is actually a binary system with a very faint 14th-magnitude companion star. This star lies so far away from its larger companion that it takes 76,000 years for it to make one orbit.

Cancer is also home to one of the closest stars to our solar system. DX Cancri is a 14th-magnitude red dwarf located only 11.82 light-years away and ranks as the 18th closest star system to our sun. As of 2021, 10 planetary systems have been discovered around stars in Cancer. 55 Cancri, for instance, is a magnitude-6 binary star system 41 light-years from Earth. The main component star, known as Copernicus, hosts at least five exoplanets (one super-Earth and four gas giants) named after astronomers and telescope makers: Galileo, Brahe, Lipperhey, Janssen, and Harriot.

Cancer is home to two stunning star clusters that really shine through binoculars and telescopes. The Beehive, or M44, is just visible to the naked eye as a faint cotton ball but reveals itself through binoculars as a swarm of bees. Through a small telescope this 610-light-years-distant cluster shows at least 75 stars. Then there is one of the oldest known open clusters in the Milky Way, M67. This is a fainter sixth-magnitude open cluster that lies 2,700 light-years from Earth and is the size of the full moon in the sky. It looks rich through binoculars.

M44: the Beehive Cluster

CANES VENATICI (THE HUNTING DOGS)

Visible: Northern Hemisphere

Best Seen: Spring and summer

Main Stars: 2

Brightest Star: Cor Caroli, 2.90 mag.

Deep-Sky Objects: Whirlpool Galaxy, NGC 5195, M94, M63, M106, M3, Whale Galaxy

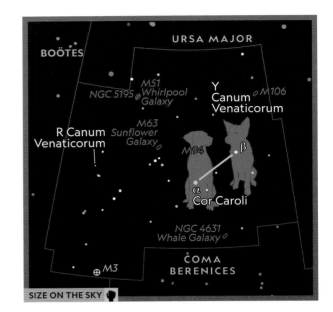

These hunting dogs were placed in the sky in 1687 by Johannes Hevelius, occupying a rather empty region between constellations Ursa Major to the north and Boötes to the east. The two greyhounds are held on a leash by Boötes, the herdsman, and are nicknamed Asterion (Starry) and Chara (Joy).

The brightest star is magnitude-2.9 Cor Caroli, which in Latin means "Charles's heart," a name popularized by famed comet hunter Edmond Halley in honor of England's King Charles I. Beta Canum Venaticorum, also known as Chara, is a yellow magnitude-4.2 star that sits 27 light-years away. Canes Venatici is also home to a couple of interesting variable stars worth following. Y Canum Venaticorum, or La Superba, is a magnitude-5 red giant star that varies over the course of 157 days from magnitude 4.8 to magnitude 6.6. R Canum Venaticorum is a Mira-class variable that fluctuates in brightness between magnitude 6.5 (a very easy bin-

ocular object) and magnitude 12.9 (needing at least a 200-mm/8-in aperture telescope) over a period of about 329 days.

To the naked eye these cosmic dogs appear to be in a barren part of the sky, but in fact it's an area rich with galaxies—a window out of the Milky Way. The dogs contain five Messier objects, with the most iconic and brightest galaxy being the Whirlpool Galaxy, or M51, which sits close to boundary of Ursa Major, just below the last star in the Big Dipper's handle. Located 31 million light-years away, it shows a bright core surrounded by prominent spiral arms, with smaller galaxy NGC 5195 next to it. Through a small telescope the two galaxies look like two connected fuzzy patches.

The other standout extragalactic target is the Sunflower Galaxy, or M63, located some 29 million light-years from Earth. It is an eye-catching magnitude-9 spiral whose loosely wound arms are filled with bright knots, visible through a small telescope under high magnification. Also worth hunting down with a scope is the barred spiral galaxy M94, at 16 million light-years away. Shining at magnitude 9, it is an easy target for binoculars, while 200- to 250-mm (8- to 10-in) aperture telescopes show its bright, circular core surrounded by a hazy halo and looking very much like a faint globular cluster. But larger telescopes are needed to reveal the knot-like textures within the halo that are actually part of the galaxy's spiral arms. Other standout spiral galaxies worth training a backyard telescope on include M106 and the edge-on Whale Galaxy.

M3, one of the largest and brightest globular clusters ever discovered

The majestic Whirlpool Galaxy displays its curving arms of gas, dust, and young stars.

CANIS MAJOR (THE GREAT DOG)

Visible: Northern and Southern Hemispheres

Best Seen: Northern winter and southern summer

Main Stars: 8

Brightest Star: Sirius, -1.46 mag.

Deep-Sky Objects: M41, NGC 2354, NGC 2362, NGC 2207

TELESCOPE TARGETS

Canis Major is home to a titanic collision between two spiral galaxies about 114 million light-years away. The larger of the two galactic monsters, NGC 2207, is in the early process of colliding and merging with its smaller neighbor IC 2163. In about 150 million years, both galaxies will have merged into one and become an even more massive elliptical galaxy.

Canis Major is the larger of the two faithful hunting dog companions to Orion, the hunter. Standing by the hunter's foot, the Great Dog is an easy-to-find constellation that is visible from most areas of the world, thanks to being positioned just south of the celestial equator. Canis Major's sparkling eye is the brightest star in the sky: Sirius. His gaze is fixed on neighboring Lepus, the hare constellation.

Only our sun, moon, and a handful of nearby planets outshine magnitude -1.46 Sirius. Thousands of years ago in ancient Egypt, priests eagerly waited for the first reappearance of the star, rising at dawn in late summer, to help predict when the Nile River would flood—a critical annual event for a desert nation that relied on the precious water to help grow their thirsty crops. Sirius also had special meaning for the ancient Greeks and Romans. The "dog days" of summer were named specifically for the Dog Star. In the Northern Hemisphere, skywatchers noticed the star would rise in the east, together with the

sun, and set in the west at sunset. The combined power of the two suns was thought to be the cause of the stretch of hot weather experienced annually from early July through mid-August. Today's astronomers, of course, know that Sirius is indeed the brightest star visible from anywhere on Earth because it ranks as one of the closest stars in our stellar neighborhood. But at 8.6 light-years away, equal to more than 81 trillion kilometers (roughly 50 trillion mi), it is much too distant for any of its heat radiation to affect us. Its annual meeting with the sun in our sky is due to Earth's travel along its orbit. We watch as the sun slowly appears to march across the same constellations each year, arriving near Sirius like clockwork in northern summer.

Sirius is about twice as massive as our sun and shines 25 times brighter. And hidden in that glow it has a magnitude-8.5 close companion: a white dwarf star that requires at least a 250-mm (10-in) aperture telescope for observation, because of its proximity to its larger companion.

For binoculars and telescopes, Canis Major is rich in star clusters, and leading the pack is the beautiful open cluster M41, also known as the Little Beehive (in reference to its more prominent namesake cluster, M44 in the constellation Cancer), which lies only four degrees south of Sirius. Containing some 80 stars, this fourth-magnitude cluster is 2,300 light-years away, yet is visible as a hazy patch to the naked eye on a dark night. It is impressive through a backyard telescope, with orange stars scattered within.

NGC 2362, a cluster containing the multiple star system known as Tau Canis Majoris

CANIS MINOR (THE LESSER DOG)

SIZE ON THE SKY ✋

Visible: Northern and Southern Hemispheres

Best Seen: Northern winter and southern summer

Main Stars: 2

Brightest Star: Procyon, 0.40 mag.

Deep-Sky Objects: NGC 2402, Abell 24

The smaller of the two hunting hounds of Orion, Canis Minor, the Lesser Dog, is small and mostly faint, save for its single bright beacon, Procyon. Besides being considered one of Orion's hunting dogs, Canis Minor is also depicted simply as a hound resting under a table, expecting scraps from the Gemini twins Castor and Pollux.

The Milky Way band appears to cut between Procyon and Sirius. According to an ancient Arabic legend, the two bright stars were once sisters who got lost in the wilderness. They encountered a wide river, the Milky Way. The older sibling, Sirius, took the plunge and crossed the river. Procyon was too scared and stayed behind, forever separated from her sister.

Orion's Betelgeuse, Canis Major's Sirius, and Procyon form the Winter Triangle asterism. Procyon is not only the eighth brightest star in the sky, but also the 14th nearest to our solar system at 11.4 light-years away. Procyon is about seven times more luminous than our sun, and like Sirius has a companion: It is in fact a binary star system that contains a faint 11th-magnitude white dwarf.

The second brightest star in the constellation is magnitude-2.9 Gomeisa. Shining with a distinct blue light, it lies 160 light-years away and, while fainter than Procyon, it is more powerful, with three times the mass of our sun and 250 times its inherent brightness. In most artistic representations of Canis Minor, Procyon marks the belly of the pooch, Gomeisa and Gamma its neck, and then Epsilon and Eta Canis Minoris its head and chest. S Canis

Minoris is the brightest variable in the constellation, with its brightness wildly swinging between a bright binocular-level magnitude of 6.5 to 13.7 over 332 days. The little dog is also home to a couple of the closest stars to our solar system. The red dwarf YZ Canis Minoris is estimated to be about 332 times the size of Jupiter and is only 20 light-years away, though it shines at a feeble 11.2 magnitude. Meanwhile Luyten's Star, another red dwarf, is brighter at 9.9 magnitude and is only 12.3 light-years from our solar system.

Even though Canis Minor runs through the edge of the Milky Way, it is curiously devoid of bright deep-sky targets for backyard astronomers, and it houses only a few very faint galaxies and a lone planetary nebula, Abell 24. Located four degrees southeast of Procyon, this stellar death shell has a rounded shape and shines at a very faint magnitude of 13.6.

Planetary nebula Abell 24

CAPRICORNUS (THE SEA GOAT)

Capricornus is part of the water-based set of zodiacal constellations that dominate the autumn skies in the Northern Hemisphere and include Pisces, Cetus, and Aquarius. With the head and front legs of a goat and the tail of a fish, Capricornus is the mythical monster that is the horned sea goat. Babylonian legends of this bizarre creature date back to 2100 B.C. and were inherited by the Greeks and then the Romans. In one story, the goat-legged god Pan leapt into the Nile River to escape a monster and the water transformed him. In an older tale, the constellation represents one of Zeus's warriors in the battle with the Titans. This soldier discovered conch shells, whose resounding call frightened the Titans into retreat. In appreciation, Zeus placed him in the sky with a fish tail and horns to represent his discovery.

Most skywatchers cannot make out the creature's stellar pattern, but instead see the constellation forming a child's drawing of an inverted hat. Capricornus lies southeast of bright Altair in Aquila and is bordered by Sagittarius, Piscis Austrinus, and Aquarius. The constellation is considered faint, with only the star Deneb Algedi rising above magnitude 3. This star's name is Arabic for "tail of the goat," and it is an eclipsing binary star located 39 light-years away. Meanwhile, Alpha Capricorni is an optical double star—a pair in which the component stars

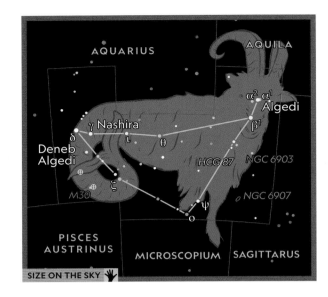

Visible: Northern and Southern Hemispheres

Best Seen: Northern winter and southern summer

Main Stars: 13

Brightest Star: Deneb Algedi, 2.85 mag.

Deep-Sky Objects: M30, HCG 87

are physically unrelated, appearing close together only because of our line of sight from Earth. The two stars can be separated with the unaided eye, with the brighter yellow-hued star shining at magnitude 3.6 and the other magnitude 4.3. The two stars lie 109 and 690 light-years distant, respectively.

Lying just east of the crowded star fields of the constellation Sagittarius along the Milky Way band, the Sea Goat sits along a relatively barren celestial landscape without many bright deep-sky objects. The most prominent object of note that backyard telescope users can hunt down is the 27,000-light-years-distant globular star cluster M30. This picturesque seventh-magnitude object can even be spotted with binoculars under dark skies. Its stellar swarm stretches about 100 light-years across and contains about 200,000 suns. The cluster's retrograde motion through our galaxy suggests that it was cannibalized from a hapless small satellite galaxy that wandered too close and got sucked up by the Milky Way's enormous gravitational forces. A very distant galaxy group is also nestled within Capricornus. Dubbed HCG 87, it consists of at least three galaxies 400 million light-years away that appear to be in the process of colliding; millions of years from now they may merge into a supergiant elliptical galaxy.

HCG 87, four galaxies bound by gravity

CARINA (THE KEEL)

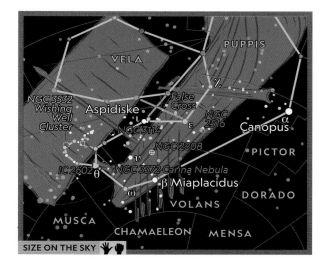

SIZE ON THE SKY

Visible: Southern Hemisphere

Best Seen: All year

Main Stars: 9

Brightest Star: Canopus, -0.72 mag.

Deep-Sky Objects: Carina, NGC 2516, NGC 3114, Wishing Well Cluster, IC 2602, NGC 2808

Carina is a large constellation in the far southern sky and is only partially visible up to mid-northern latitudes. With the Milky Way band flowing through this region, this constellation is a true treasure trove for observers in the Southern Hemisphere. Originally, Carina was part of a much larger stellar pattern known as Argo Navis, representing the ship sailed by Jason and his Argonauts. The old constellation was considered too large, however, and the 18th-century astronomer Nicolas-Louis de Lacaille carved it up into three smaller ones, including Carina.

Looking toward the extreme western part of Carina, you come across the second brightest star in the sky, Canopus, which can be glimpsed from the southernmost parts of United States in the winter sky below Sirius. It is bested in brightness only by Sirius, which is a half magnitude brighter. That is because, while Sirius is nearby at eight light-years away, Canopus is a stately 310 light-years distant, shining at least 13,000 times brighter than our sun. It is such a stellar beacon that many space probes exploring the solar system have used it as a navigation reference.

One of the most stunning sights in the southern heavens is the region around the giant star Eta Carinae, which appears to be shrouded inside an orange nebula, one of the brightest and largest in the sky. Binoculars reveal that the Carina Nebula is embedded in a rich star field. A telescope will show details in the gas cloud itself, notably a distinct L-shaped dark lane superimposed on the bright cloud. The nebula is an active stellar factory filled with young stars, while fledgling clusters pepper the surround-

ing area. The Eta Carinae star itself is barely visible today to the naked eye from a dark site at magnitude 6, but it surprised observers back in 1843 when it flared up to magnitude -1, brighter even than Canopus. Hubble Space Telescope observations indicate that it is actually a double star with one stellar member that is a monster, with a mass at least 100 times that of our sun, and the other also large at 30 solar masses or more.

Just northeast of Eta, and in the same binocular field, is the richest open star cluster of the region, 1,500-light-years-distant NGC 3532. Also known as the Wishing Well Cluster, because it has been likened to a batch of glittering silver coins at the bottom of a well, it is home to 400 stars, and looks amazing through a low-power telescope.

NGC 3532: the Wishing Well Cluster

CASSIOPEIA (THE SEATED QUEEN)

Visible: Northern Hemisphere

Best Seen: Autumn, but visible year-round for most northerners

Main Stars: 5

Brightest Star: Schedar, 2.24 mag.

Deep-Sky Objects: M52, M103, Caroline's Rose, Heart and Soul Nebulae, NGC 129, Bubble Nebula, Owl Cluster

STAR STORIES

While the official name of the middle star in the W-shaped Cassiopeia constellation is Gamma Cassiopeiae, it is also known affectionately as Navi. NASA astronaut Virgil "Gus" Grissom gave the star its nickname, which was his middle name spelled backward. After Grissom and two fellow astronauts perished in the 1967 Apollo 1 fire, this blue giant star 610 light-years away kept its new moniker as a celestial memorial.

For skywatchers across the Northern Hemisphere, Cassiopeia's distinct giant W- or M-shaped stellar pattern (depending on the time of year) is easy to distinguish and fun to tour. For many northern-latitude observers it is circumpolar, meaning it never sets below the horizon and is visible all year round—however, it rides highest in the evening sky in autumn. The celestial queen is bookmarked between the North Star and the constellation Andromeda and sits on the opposite side of Polaris from Ursa Major.

In Greek mythology, Cassiopeia and Cepheus were queen and king of Ethiopia and parents to the princess Andromeda. The queen is sitting on her throne in the sky, cursed to rotate eternally around the North Star as punishment for boasting that she was prettier than the daughters of Poseidon.

IC 1805 (right) and IC 1848 (left): the Heart and Soul Nebulae

The constellation's lead star Schedar, meaning "the breast" in Arabic, is actually a quadruple star system consisting of one giant and three very dim dwarfs, located 228 light-years from Earth. The brightest member of this system is an orange-hued giant that shines 771 times brighter than our sun and is estimated to be no more than 200 million years old. The second brightest star, Caph, meaning "hand" in Arabic, is a white star shining at magnitude 2.3 and is 55 light-years distant. Gamma Cassiopeiae, the center star in the W-shaped asterism, is a variable star 550 light-years away that varies in brightness unpredictably between 1.6 and 3 magnitude.

A dense field of stars with the northern section of the Milky Way running right through the constellation ensures that this region is packed with multitudes of deep-sky objects, including some of the sky's brightest open star clusters, colorful double stars, and planetary nebulae, as well as faint galaxies. In fact, Cassiopeia is so crowded that touring with binoculars shows off many tiny knots of stars scattered throughout the heart of the constellation, making some of them a challenge to pick out from the background curtain of stars. Train a telescope on the western corner of the constellation and enjoy the views of open cluster M52. Lying 5,100 light-years away, it contains more than 100 stars; meanwhile, M103 looks like a fainter version with about half the stars visible. Binoculars are perfect to show off the large and faint complex star system with an associated nebula pair called Heart and Soul. Both nebulae are lit up by star clusters buried within them and are nestled in the Perseus spiral arm of the Milky Way, 7,500 and 6,500 light-years away, respectively.

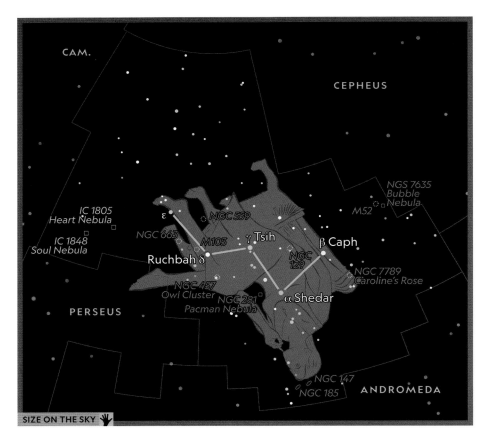

CENTAURUS (THE CENTAUR)

Visible: Southern Hemisphere

Best Seen: Late summer and autumn

Main Stars: 11

Brightest Star: Alpha Centauri, 0.01 mag.

Deep-Sky Objects: NGC 4945, Centaurus A, NGC 5253, NGC 5102, Pearl Cluster, NGC 3918, NGC 4622, Omega Centauri, Running Chicken Nebula

Despite being a prominent constellation for observers in the Southern Hemisphere, Centaurus has early origins farther north, with a centaur figure appearing in Assyrian art before 1200 B.C. In ancient times Centaurus was clearly visible across the Northern Hemisphere, but because of the movement of Earth's axis due to precession over the last few millennia, the constellation has sunk to a more southerly part of the sky. The classic centaur figure is a composite creature, with the upper part of its body human, while below its torso it is a horse. According to Greek mythology, it represents Chiron, a centaur talented in music, poetry, and medicine. Legends have Chiron as a tutor of Hercules and also as the inventor of constellations to help Jason and the Argonauts on their famed voyage. However, one of Hercules' poisoned arrows accidentally killed Chiron, who was rewarded by Jupiter with immortality in the stars.

Centaurus is mostly marked out by straggling lines of second- and third-magnitude stars that wind their way north and west without any easily recognized pattern. Its two leading stars, however, are unmistakable: Alpha and Beta Centauri, also known as the Pointers, as they direct observers to the nearby Southern Cross. Alpha Centauri's claim to fame, of course, is that it is the nearest bright star to our solar system at 4.36 light-years away, and it is the third brightest star in the entire sky. It appears similar to our sun but is part of a multiple star system. A small telescope will spot it as a double with a much fainter companion that orbits every 80 years. A third member of the system, the dim 11th-magnitude red dwarf Proxima Centauri, is a bit closer to us at 4.23 light-years away,

While the northern parts of the centaur are in a fairly empty part of the sky, its southern region crosses the Milky Way and so contains a rich deep-sky hunting ground. The most prominent globular cluster in the sky is actually found here. Omega Centauri is so bright that it is visible to the naked eye as a fuzzy fourth-magnitude object. A myriad of its stars can be resolved when looking through binoculars and small telescopes. Through larger backyard telescopes the views have been described as awesome. The amazing sights are probably thanks to Omega Centauri's sheer size: It contains over one million stars and is considered one of the closest globular clusters to Earth at 15,700 light-years.

NGC 5128: the Centaurus A Galaxy

CEPHEUS (THE KING)

Visible: Northern Hemisphere

Best Seen: Autumn

Main Stars: 5

Brightest Star: Alderamin, 2.45 mag.

Deep-Sky Objects: IC 1396, NGC 7510, NGC 7762, Bow-Tie Nebula, NGC 188, NGC 6946, NGC 7538, NGC 7023

STAR STORIES

Whatever the culture, this regal constellation has a lineage that dates back thousands of years. Instead of seeing this constellation as a king, though, other cultures envisioned very different pictures in this part of the sky. Arabian nomads saw a shepherd with his dog and sheep. Chinese know him as Zhao Fu, the famous charioteer of ancient times.

SIZE ON THE SKY

Cepheus is among the oldest constellations in the northern sky, and the king who inspired it played an important part in Greek mythology. In early Greek times Cepheus was known as the mythical king of Ethiopia, and Cassiopeia was his queen.

Five bright stars form a distinct stellar pattern reminiscent of a child's drawing of a house, with a roof that points roughly toward the North Star, Polaris. Errai, Cepheus's gamma star, 45 light-years distant, is both a binary star and a host to an orbiting planet estimated to be about 9.4 times the size of Jupiter. The star Mu Cephei is a stunning sight, known as the Garnet Star thanks to its gemlike color. Sitting at an enormous distance of 3,065 light-years, it is considered the reddest star visible to the naked eye in the Northern Hemisphere.

At the apex of the house's roof is Delta Cephei.

At 868 light-years away, it is considered the perfect model of a Cepheid variable. It has about a one-magnitude change in brightness, varying from 3.49 to 4.36 over the course of about 5.3 days, with a slow fading followed by rapid brightening.

Despite the Milky Way field running through Cepheus, the constellation lacks bright naked-eye deep-sky objects compared to neighboring Cassiopeia; however, there are plenty of fainter star clusters and nebulae to keep binoculars and telescopes busy. With an estimated age of 6.8 billion years, NGC 188 is one of the oldest open clusters known. Through backyard telescopes it appears faint and granular, sprinkled with yellow-hued stars. Located near the star Beta Cephei is the bright reflection nebula NGC 7023. Also known as the Iris Nebula, it holds an open cluster embedded within it, to which the NGC catalog number actually refers. This magnitude-6.8 gas cloud is located 1,300 light-years away, measures six light-years across, and is illuminated thanks to a nearby 7.4-magnitude star. The beautiful face-on spiral NGC 6946, nicknamed the Fireworks Galaxy, holds the record for producing more supernova explosions over the past century than any other observed galaxy.

Emission nebula IC 1396, holding the darker form of the Elephant's Trunk Nebula

CETUS (THE SEA MONSTER)

Visible: Northern and Southern Hemispheres

Best Seen: Northern autumn and southern spring

Main Stars: 14

Brightest Star: Diphda, 2.04 mag.

Deep-Sky Objects: M77, Skull Nebula

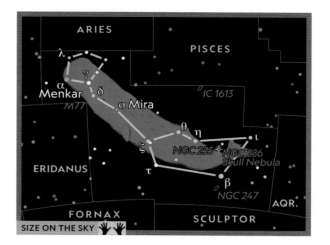

Cetus is the fourth largest member of the group of water-based constellations, with its vast form lying south of Pisces and east of Aquarius. In classic Greek myths Cetus was the sea monster that was sent to devour the princess Andromeda; it was turned to stone when young Perseus showed it Medusa's snake-haired head.

The constellation's second brightest star, called Menkar, represents the nose of the creature. It is a distinctly red, giant star that shines at magnitude 2.5 and sits 220 light-years away. Omicron Ceti, known commonly as Mira, is a prototype long-period variable star about 220 light-years away. Over a period of 331 days its brightness changes from magnitude 10 to 2, making it easy to track with the naked eye and binoculars as you follow its entire light cycle. The only Messier object in Cetus is the ninth-magnitude spiral galaxy M77. Sitting 45 million light-years away, the galaxy has an active core that has been found to be a strong radio source, indicating a hidden supermassive black hole. The creepy 1,600-light-years-distant Skull Nebula is a ninth-magnitude planetary nebula that looks like an eerie gray, round cloud with dark oval spots.

CHAMAELEON (THE CHAMELEON)

Visible: Southern Hemisphere

Best Seen: All year

Main Stars: 3

Brightest Star: Alpha Chamaeleontis, 4.06 mag.

Deep-Sky Objects: NGC 3195, IC 2631

Chamaeleon is a small, dim constellation nestled close to the South Celestial Pole, and it depicts the color-shifting lizard of the same name. First identified in the 16th century, this little celestial reptile has no stars brighter than fourth magnitude. (Gamma Chamaeleontis is the brightest at magnitude 4.1.) Its distinct central pattern of four stars is arranged in a diamond-shaped asterism, visible under dark skies, that may remind skywatchers of the animal it is supposed to represent.

Delta Chamaeleontis is a bright double star that can be separated with binoculars and small telescopes. The pair is particularly pretty, with the

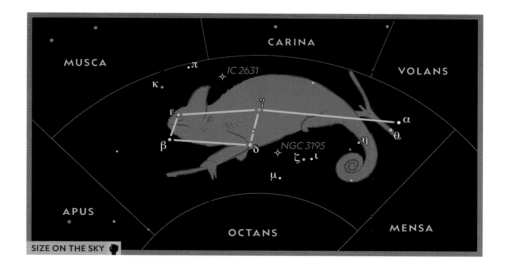

brighter one shining blue-white while its companion has a distinct yellow hue.

One of the brightest and best deep-sky objects in this part of the southern sky is the 6,440-light-years-distant planetary nebula named NGC 3195. Shining at a feeble magnitude 11, it can be spotted through backyard telescopes with at least 100-mm (4-in) apertures.

CIRCINUS (THE COMPASS)

Visible: Southern Hemisphere

Best Seen: All year

Main Stars: 3

Brightest Star: Alpha Circini, 3.18 mag.

Deep-Sky Objects: NGC 5315, Circinus Galaxy, NGC 5823

ircinus is a faint southern constellation that represents the drawing compass used by drafters and early carpenters and navigators. Its border lies east of the foot of Centaurus, with another tool, Norma the carpenter's square, to its northeast.

The fourth smallest of all star patterns in the sky, Circinus is fairly easy to track down using the lead stars in neighboring Centaurus: A line drawn through the stars Alpha and Beta Centauri points directly to the compass. Lying directly south of Alpha Centauri is Alpha Circini. With a magnitude of 3.2, it is the brightest star of this dim constellation and is in fact a great double star for small telescopes. While the brighter star appears white, its much fainter 8.6-magnitude companion appears distinctly red. One of the closest major galaxies to our Milky Way, the Circinus Galaxy is about 13 million light-years away. This spiral galaxy has an active nucleus, a supermassive black hole surrounded by a disk of matter that accretes onto it, generating immense energy. Through a large backyard telescope the galaxy appears as a 10th-magnitude round hazy spot.

COLUMBA (THE DOVE)

Visible: Northern and Southern Hemispheres

Best Seen: Northern winter and southern summer

Main Stars: 5

Brightest Star: Phact, 2.65 mag.

Deep-Sky Objects: NGC 1792, NGC 1851

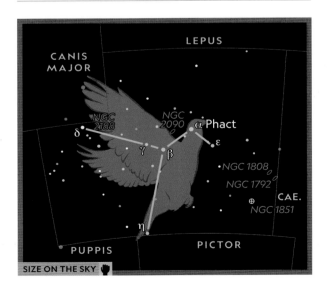

olumba is a tiny constellation south of Lepus that was created in the early 17th century, portraying the dove in the biblical story of Noah. Located south of Orion and to the west of the bright star Sirius, the dove is fairly easy to track down north of Carina and its superbright star Canopus. Columba's claim to fame is its star Mu Columbae, known as the Runaway Star for being ejected from the Orion Nebula in a possible stellar collision 2.5 million years ago. The fifth-magnitude star appears very faint to the naked eye under dark skies, but from light-polluted city suburbs it is an easy target with binoculars.

Columba contains a few deep-sky objects worth hunting down, including the bright globular star cluster NGC 1851. This eighth-magnitude object some 39,000 light-years distant is easily seen through binoculars and small telescopes as a tiny, pale cotton ball. Larger backyard telescopes will begin to resolve some of the brightest stars in its round structure.

COMA BERENICES (BERENICE'S HAIR)

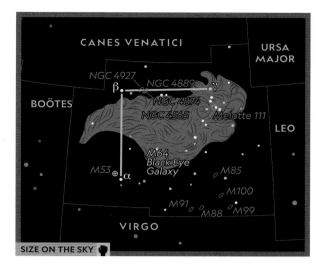

SIZE ON THE SKY 👊

Visible: Northern and Southern Hemispheres

Best Seen: Northern spring and southern autumn

Main Stars: 3

Brightest Star: Beta Comae Berenices, 4.26 mag.

Deep-Sky Objects: Melotte 111, M53, NGC 4874, NGC 4889, NGC 4927, M99, Black Eye Galaxy

Coma Berenices is a faint northern constellation that rides high in the sky during springtime in the Northern Hemisphere and can be seen in the Southern Hemisphere in the fall. Ranked as the 42nd largest constellation, it is located just north of Virgo and south of Canes Venatici and the Big Dipper. Coma Berenices was once actually considered part of Virgo and Leo, where it represented a wisp of the maiden's hair and even a tuft of the lion's tail. But the main legends have the constellation portraying the hair of Queen Berenice II of Egypt in stories that date back to the reign of the queen's husband, Ptolemy III. As this king went to war, his wife struck a deal with the goddess Aphrodite, promising her long, beautiful hair in exchange for the safe return of her husband. Upon his return, Berenice cut her long locks, placing them in a shrine. However, during the night the hair was stolen. To save the priests from execution, the royal astronomer convinced the king that a grateful Aphrodite had placed the queen's gift in the stars.

The best deep-sky feature within the constellation is the giant open star cluster called Melotte 111. Named after P. J. Melotte, the British astronomer who first cataloged it in 1915, this cluster is even easier to spot than its famous cousins, the Pleiades in Taurus and Cancer's Beehive. Filled with at least 40 bright stars, the cluster spans a massive five degrees across the sky—equal to 10 moon disks—making this a perfect target for binoculars. Lying some 280 light-years away, it ranks as one of the closest clusters to Earth, after the Big Dipper and Hyades groups. The Black Eye Galaxy (M64) is easily visible by telescope

between the two outermost stars in Coma Berenices. A dark dust cloud in front of its core creates the illusion of a black eye. A fine globular cluster (M53) sits 58,000 light-years from Earth and shines at magnitude 8. It is easily found next to the constellation's second brightest star, Alpha Comae Berenices. One of the brightest and largest spiral galaxies in the constellation's Virgo Cluster is 55-million-light-years-distant M100. While appearing as a faint magnitude-9 patch of light in small telescopes, it is famous for hosting seven visible supernovae explosions since the year 1900.

M64: the Black Eye Galaxy

CORONA AUSTRALIS (THE SOUTHERN CROWN)

Visible: Northern and Southern Hemispheres

Best Seen: Northern summer and southern winter

Main Star: 6

Brightest Star: Meridiana, 4.11 mag.

Deep-Sky Objects: NGC 6723, NGC 6729, NGC 6541

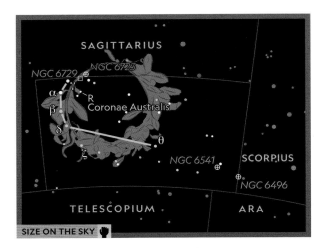

SIZE ON THE SKY

A semicircle of stars just south of Sagittarius and its Teapot asterism, Corona Australis, the Southern Crown, is the sister constellation to Corona Borealis, the Northern Crown. Its brightest stars—Alpha, Beta, Gamma, Delta, and Theta—form a distinct curving pattern in the sky. It was first mentioned by Greek writer Geminos in the first century B.C. and was frequently associated with the myth of Sagittarius, for centaurs were said to wear a crown-like laurel wreath. In Chinese mythology, this stellar pattern represents a giant turtle, while Arabic references include a tent or ostrich nest. To the San people of the Kalahari Desert it was called the House of Branches and represented a semicircle of people sitting around a campfire.

The constellation's brightest star, Alpha Coronae Australis, or Meridiana, is a middle-aged white star 2.3 times the mass of our sun. It sits 125 light-years distant, and measurements of the massive amounts of infrared radiation it is giving off indicate that it may be encircled by a dusty disk, similar to Vega in Lyra. Kappa Coronae Australis is a fine double star system nestled within the rich star field of the Milky Way. The blue-white stars shine at 5.6 and 6.2 magnitudes and sit at 695 and 692 light-years away, respectively, making them only an optical double and not physically connected to each other. Lambda Coronae Australis is another double that is easily resolved with a small telescope. Sitting at 202 light-years away, these two stars are truly part of the same star system. The primary member glows at magnitude 5.1, while its stellar partner is much fainter at only 9.7 magnitude.

Sitting on the edge of the Milky Way band, Corona Australis is a fairly easy constellation to locate thanks to its eye-catching pattern. While it is the ninth smallest constellation in the sky and has no bright stars, it is home to a small collection of deep-sky objects. The globular cluster NGC 6541 is the showpiece of these distant stellar sights. It can be found in the southwest corner of the constellation, nestled within a rich star field, and at magnitude 6 it is bright enough to be spotted easily with binoculars. Located 22,800 light-years from Earth, it lies near our galaxy's core. Its slightly fainter and more distant cousin NGC 6723 is a globular cluster 28,000 light-years away. A complex of three distinct and bright reflection nebulae—NGC 6726, 6727, and 6729—surrounds the variable star R Coronae Australis.

A star-forming region around the star R Coronae Australis

CORONA BOREALIS (THE NORTHERN CROWN)

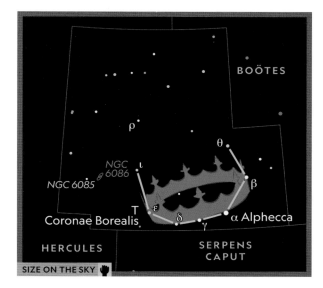

SIZE ON THE SKY

Visible: Northern Hemisphere

Best Seen: Summer

Main Stars: 8

Brightest Star: Alphecca, 2.22 mag.

Deep-Sky Objects: NGC 6085, NGC 6086

Corona Borealis's semicircle pattern of seven stars is quite eye-catching. Wedged between the kite-shaped Boötes and the keystone-patterned Hercules, the northern crown is easy to track down by tracing a line south between the bright stars Vega and Arcturus. This celestial crown was first listed by second-century astronomer Ptolemy and currently ranks 73rd in constellation size. According to Greek legends, Corona Borealis was the laurel wreath crown of the princess Ariadne, the daughter of King Minos of Crete. In ancient times, the laurel wreath was also traditionally awarded to winners of athletic competitions and poetry contests.

For skywatchers today the gem of the crown is the second-magnitude blue-white star Alphecca, which is an eclipsing binary 75 light-years away. It is similar in its physical properties to Vega and Sirius but appears fainter because of its much greater distance. Alphecca shines 67 times brighter than the sun and is nearly three times more massive. As an eclipsing binary system, it hosts a much fainter companion star that produces a very slight drop in brightness of 0.1 magnitude every 17.3 days. Large amounts of infrared radiation emitted from the star suggest it may be surrounded by a vast disk of dust that reaches out to 60 astronomical units.

Another star worth keeping an eye on with binoculars and telescopes is T Coronae Borealis. This is a recurring nova that last exploded in 1946, rising from magnitude 10 to 2 and then dropping back down again: It may explode again in the near future, and when it does it will become the brightest star

in the constellation. The star Rho Coronae Borealis has been found to host exoplanets. One planet larger than Jupiter has been discovered, and astronomers suspect it's not alone. At random intervals, as dark material erupts in its atmosphere, this variable star's magnitude drops sharply from 5.8 to 14.8.

The celestial wreath sits far away from the Milky Way and so is filled with galaxies rather than stars. NGC 6085 is a distant spiral lying 455 million light-years distant and is estimated to be twice the size of our Milky Way. Within the same telescope view is the brighter elliptical galaxy NGC 6086. One ultimate observing challenge for large-aperture telescopes (400 mm/16 in and more) is the Abell 2065 galaxy cluster. Under very dark and clear skies, observers can see a loose grouping of half a dozen 16th-magnitude and fainter galaxies located some one billion light-years from Earth.

STAR STORIES

Corona Borealis's prominent semicircle of stars has been noticed in cultures across the Northern Hemisphere. The ancient Chinese saw it as representing a cord, while Arab mythology called it the Bowl of the Poor People. The Cheyenne in the Americas knew it as the Camp Circle, since it resembled their traditional camp formations, and Australian Aboriginals likened it to the boomerang.

Spiral galaxy NGC 6085 (left) and elliptical galaxy NGC 6086 (right)

CORVUS (THE CROW)

Visible: Northern and Southern Hemispheres

Best Seen: Northern spring and southern autumn

Main Stars: 4

Brightest Star: Gienah, 2.58 mag.

Deep-Sky Objects: Antennae galaxies, NGC 4027

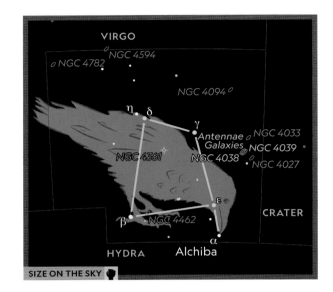

Known as a tent to Arabs, and part of a heavenly chariot in ancient China, the Corvus constellation is known today as a crow or raven. Four bright near-third-magnitude stars forming a trapezoid mark its pattern in a fairly empty part of the sky, making this asterism known as The Sail fairly easy to spot with the naked eye. Corvus ranks in size as the 70th out of 88 constellations and is bordered by Hydra and Virgo to its south and north, respectively. The crow can be glimpsed in its entirety by observers as far north as 65 degrees latitude.

Our association with this bird dates back to the ancient Greeks and the story of a white raven, the pet of the god Apollo. According to legend, the raven was tasked with looking after Apollo's lover. But after the raven reported that she had fallen in love with a mortal man, the furious Apollo turned the raven's feathers black. In another Apollo legend, the raven was sent to fetch him water, but the bird became distracted and stopped to eat figs. Fearful of Apollo's wrath, the raven picked up Hydra, the female water snake, and on his return claimed the snake had attacked him. Apollo knew the raven had lied and in revenge threw both creatures into the sky.

The constellation's brightest star Gienah, marking the bird's left wing, is an aging blue-white giant four times more massive than our sun. Lying 154 light-years away, it is actually a binary star system, hosting a tiny red dwarf partner that weighs no more than 0.8 solar masses and takes 158 years to complete one orbit. Delta Corvi, also known as Algorab, which means "the crow" or "the raven," marks the bird's right wing. It so happens that it is actually a wonderful color-contrasting double star lying only 85 light-years from Earth. A small telescope can easily separate the blue-white magnitude-2.7 primary from the much fainter magnitude-8.5 orange companion. Each star in this binary system has been discovered to host a ring of cosmic dust.

The deep-sky highlight is the pair of interacting galaxies found near the western boundary of Corvus, known as the Antennae galaxies, or NGC 4038 and 4039. With visual magnitudes of 10.5 and 10.3, the two colliding galaxies can be seen in small telescopes with the long, streaming tails of ejected stars, gas, and dust that have resulted from their entanglement. Another eye-catching deep-sky target is NGC 4027. A barred spiral lying some 83 million light-years distant, it shines at 11.1 magnitude and appears through 200-mm (8-in) telescopes under dark skies as a faint elongated core surrounded by a halo.

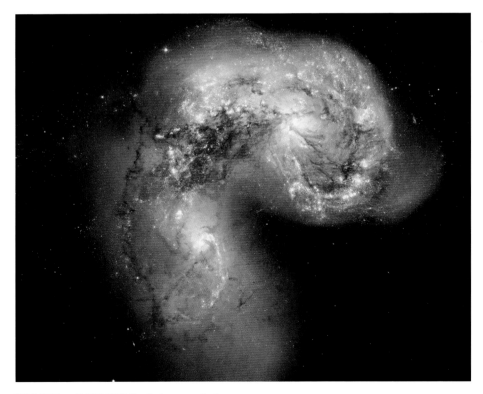

NGC 4038 and NGC 4039: the Antennae galaxies

CRATER (THE CUP)

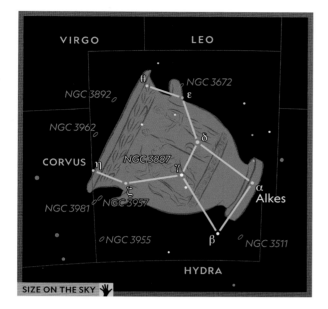

Visible: Northern and Southern Hemispheres

Best Seen: Northern spring and southern autumn

Main Stars: 4

Brightest Star: Delta Crateris, 3.56 mag.

Deep-Sky Objects: NGC 3887, NGC 3511, NGC 3981

Crater, the Cup constellation, represents the goblet in which Corvus the crow brought water for the god Apollo. With their legends eternally intertwined, when the crow falsely implicated Hydra, the female water snake, for his delay in carrying out his god's mission, he was cast into the sky with the cup forever just out of reach.

Crater is bordered by Corvus to its east, Leo and Virgo the north, and Hydra to the west and south. While none of its stars is brighter than magnitude 3.5, this faint naked-eye constellation consists of six stars that form a pattern resembling a footed goblet. The brightest member star is Delta Crateris, or Labrum, which sits 163 light-years from Earth. An older orange giant, it has only about 1.5 solar masses, but as its fuel has started to run out in its core, its outer atmosphere has puffed out, expanding its size to about 23 times the diameter of our sun. The second brightest star in the cup is magnitude-4.4 Beta Crateris. This binary system is made up of a giant star and a white dwarf sitting 296 light-years from our solar system. Gamma Crateris is an easy double star for backyard telescopes, with a brighter 4.1-magnitude primary and a 9.6-magnitude companion. The cosmic pair takes over 1,150 years to circle each other and is located 85 light-years from Earth. Meanwhile, Epsilon Crateris marks one side of the rim of the celestial goblet. Magnitude-4.8 Epsilon lies 366 light-years away, has a diameter nearly 45 times larger than our sun, and shines nearly 400 times brighter, too. On the other side of the cup's rim lies Zeta Crateris, which is a bit fainter at magnitude 4.7.

Though it is devoid of any bright deep-sky objects, Crater is far enough away from the Milky Way band that it contains many faint galaxies visible to telescopes with at least 100- to 150-mm (4- to 6-in) apertures. NGC 3887 is a 10th-magnitude spiral galaxy that appears to be half the size of our Milky Way and is 59 million light-years from Earth. Another spiral galaxy, NGC 3981, is 62 million light-years away. Shining at 11th magnitude, it appears under high magnification as a faint gray oval disk with tapered ends. Crater also holds a powerhouse quasar galaxy an astounding six billion light-years away. In 2014, astronomers were able to measure the supermassive black hole at the center of this galaxy, RX J1131, and determined that it was spinning at more than half the speed of light.

Barred spiral galaxy NGC 3887

CRUX (THE SOUTHERN CROSS)

Visible: Southern Hemisphere

Best Seen: All year

Main Stars: 4

Brightest Star: Acrux, 0.87 mag.

Deep-Sky Objects: NGC 4755, Coalsack Nebula

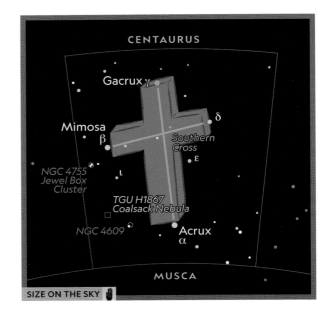

SIZE ON THE SKY

STAR STORIES

Petroglyphs showing the clear stellar pattern of the Southern Cross have been found carved in stone at Machu Picchu, Peru. Some archaeoastronomers believe this cross was an important symbol to the Inca people, who believed it marked the center of the universe and called the stars of Crux Chakana, meaning "to cross" or "to bridge."

One of the most famous of all southern constellations, Crux is the smallest constellation in the sky, yet among the most easily recognizable. Its distinctive cross asterism is marked simply by four bright stars. In ancient times before Earth's precession shifted the stars toward the south, the stars of Crux were visible from Europe. Ancient Greek astronomers used the constellation's stars to mark the back legs of Centaurus. To the Maori in New Zealand the cross was called Te Punga, meaning "the anchor." It wasn't until the 17th century, when European navigators sailing south recognized the cross-like pattern, that Crux became its own official constellation.

While considered a celestial jewel best explored from south of the Equator, keen-eyed skywatchers as far north as the southern tip of Florida can glimpse the Southern Cross. Just as novice skywatchers learn to use the pointer stars in the bowl of the Big Dipper to point to the North Celestial Pole, marked by nearby Polaris, the base stars of Crux are used to show the way to the South Celestial Pole—

but without the luxury of a bright star marking its place in the sky.

The lead star in Crux, Alpha Crucis, or Acrux, shines at magnitude 0.87 and is a wonderful starting point in exploring the constellation. It is in fact a blue-white double star 320 light-years away that can be separated through small telescopes, and with the light from its partner it is the 14th brightest star in the sky. Shining at magnitude 1.3, Beta Crucis, or Mimosa, is the 20th brightest star in the sky, located 350 light-years from our solar system. Then there is ruddy Gamma Crucis, shining at magnitude 1.6 and sitting only 88 light-years away. It is a pretty optical double star with the brighter member being a red giant, and a much fainter, magnitude-6.5 star that is also considerably farther at 264 light-years away. The final star that forms the cross asterism is Delta Crucis, or Imai, a blue-white star shining at magnitude 2.8 that sits 345 light-years away.

The Cross straddles a rich section of the southern Milky Way band and so is filled with deep-sky treasures of all kinds. Just east of the Cross lies what appears to be a hole in the sky—a dark region devoid of stars known as the Coalsack Nebula. At about 600 light-years away, it is clearly visible to the naked eye as a strikingly large, dark gas cloud silhouetted against a bright star-studded Milky Way. Train binoculars or a telescope near the northern edge of the Coalsack and spot the Jewel Box. This 6,400-light-years-distant star cluster contains about 60 stars that are mostly young supergiants no more than a few million years old.

The Coalsack Nebula (Caldwell 99), a dark cloud of interstellar dust

The Running Chicken Nebula (IC 2944) gained its name from the birdlike shape of its brightest region.

CYGNUS (THE SWAN)

Visible: Northern Hemisphere

Best Seen: Summer and early autumn

Main Stars: 9

Brightest Star: Deneb, 1.25 mag.

Deep-Sky Objects: NGC 7000, M39, NGC 6910, NGC 6826, Veil Nebula, Cygnus X

STAR STORIES

In Chinese myths the constellation Cygnus contains Que Qiao, meaning "the magpie bridge." The story tells of a fairy falling in love with a mortal man, only to be separated forcefully by a goddess placing a white, raging river (the Milky Way) in the sky between them. According to legend, magpies around the world converge one night a year to form a bridge so that the star-crossed lovers can briefly unite.

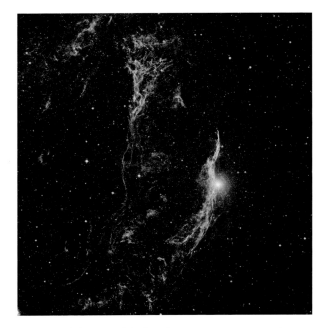

The Veil Nebula is one of the best known supernova remnants.

Cygnus, the swan constellation, dominates the nighttime overhead views during the summertime. Its prominent birdlike figure, with long neck and outstretched wings, appears to fly through the dense star fields of the Milky Way band, heading southwest toward Aquila (The Eagle). Because of its distinct arrangement of six stars, Cygnus is more commonly recognized as the Northern Cross—the larger cousin of the famous Southern Cross, seen best from the Southern Hemisphere. Several Greek myths may link to the constellation: In one, Cygnus was a great warrior who was tragically killed in battle. As he drew his last breath on the battlefield, Poseidon, his father, transformed him into a beautiful swan and carried him to the starry heavens.

Leading this mythical constellation is the 19th brightest star in our sky, blue-white Deneb. As the Arabic name Tail implies, it marks the tail of the swan (and the head of the cross). Together with the neighboring bright stars of other constellations, Vega of Lyra and Altair of Aquila, Deneb forms the Summer Triangle asterism—a guidepost for beginner stargazers. While it may appear as the faintest member of this famous trio, 1.25-magnitude Deneb is in fact one of the most powerful and remote naked-eye stars in the entire sky. Burning at least 70,000 times brighter than our own sun, far-off Deneb is a blue supergiant 1,500 light-years from Earth.

Nestled right under Deneb is a legendary cloud of gas and dust nicknamed the North American Nebula, NGC 7000. Binoculars will show a gray, irregular glow, but long-exposure photographs reveal an uncanny similarity to the North American continent, complete with the state of Florida and the Gulf of Mexico. With low-power binoculars, the ghostly shape of this huge nebula can be glimpsed only from dark locations. Locating this sprawling star factory is not difficult since it lies close to Deneb and covers three times the width of a full moon in the sky.

An easier target is a fifth-magnitude open cluster just northeast of Deneb called M39. Containing about 20 stars, it lies some 800 light-years distant. Cygnus is also home to a magnificent supernova remnant visible with binoculars from a dark location. The 2,400-light-years-distant Veil Nebula really shines through backyard telescopes, which begin to reveal delicate filaments and streamers of gas.

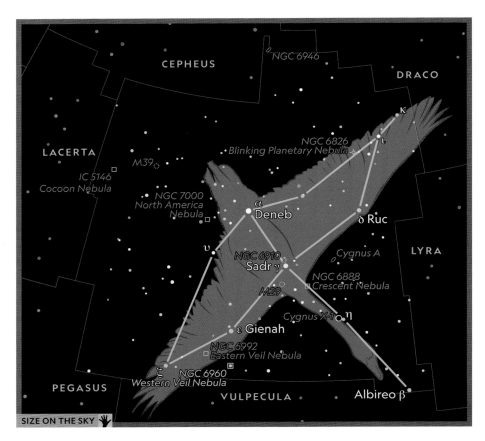

CEPHEUS
NGC 6946
DRACO
LACERTA
IC 5146
Cocoon Nebula
M39
NGC 6826
Blinking Planetary Nebula
NGC 7000
North America Nebula
Deneb
δ Ruc
Cygnus A
LYRA
υ
NGC 6910
Sadr γ
NGC 6888
Crescent Nebula
M29
Cygnus X-1
η
ε Gienah
NGC 6992
Eastern Veil Nebula
ζ
NGC 6960
Western Veil Nebula
PEGASUS
VULPECULA
Albireo β

SIZE ON THE SKY

DELPHINUS (THE DOLPHIN)

VULPECULA

NGC 6905

SAGITTA

NGC 7006

γ α Sualocin

δ ζ

Rotanev β

θ

NGC 6891

ι ε

κ

NGC 6934

AQUILA

EQUULEUS

SIZE ON THE SKY

Visible: Northern Hemisphere

Best Seen: Summer and early autumn

Main Stars: 5

Brightest Star: Rotanev, 3.63 mag.

Deep-Sky Object: NGC 6905

Delphinus may be one of the smallest constellations, but it is easily recognizable as a flattened, curved, diamond pattern just east of Aquila and west of Aquarius. It has been associated with a dolphin since ancient times—ancient Greeks considered dolphins to be messengers of the ocean god, Poseidon. While some Polynesian cultures may also have seen a dolphin-like figure, Chinese astronomers placed these same stars within their celestial symbol for the Black Tortoise of the North.

To many skywatchers, the constellation's tight kite structure of bright stars, with a few finer stars forming the dolphin's tail, may look like a star cluster, even reminiscent of the Pleiades. This is just an optical illusion, however, since its stars vary their distance from Earth across hundreds of light-years. The four brightest stars form a well-defined rhombus-shaped asterism known as Job's Coffin. The two lead stars, Alpha and Beta Delphini, are called Sualocin and Rotanev, respectively, but their odd sounding names are grounded in a practical joke. The 18th-century Italian astronomer Niccolò Cacciatore was assisting in building a star catalog for the Palermo Observatory. He latinized his first and last names (Nicolaus Venator), spelled them backward, and somehow included them as the star names in the catalog. In 2016, the International Astronomical Union officially recognized the star names.

The nose of the dolphin is marked by Gamma Delphini, which is in fact a beautiful contrasting double star that is easily split with a small telescope. Lying 114 light-years away, the brighter 4.5-magnitude star appears white, while the fainter star, magnitude 5.5, glows with an eye-catching bluish or greenish hue. Sitting above the dolphin's back is the bizarre remnant of a dying star, NGC 6905, also known as the Blue Flash Nebula. This 11th-magnitude planetary nebula lies some 8,800 light-years from our solar system and can be glimpsed as a blue disk with a telescope with at least a 100-mm (4-in) aperture. NGC 6934 is a magnitude-9.7 globular cluster 52,000 light-years away. It's an easy target for small telescopes, appearing small and rounded with a concentrated center. A much fainter globular cluster, NGC 7006, is located 137,000 light-years away at the very outskirts of our galaxy. One of the farthest globulars from Earth, it shines at a feeble 10.5 magnitude and is a modest challenge for small telescopes.

NG 6934, a globular cluster home to some of the most distant stars in our galaxy.

DORADO (THE SWORDFISH)

Visible: Southern Hemisphere

Best Seen: Summer

Main Stars: 3

Brightest Star: Alpha Doradus, 4.3 mag.

Deep-Sky Objects: N44, Tarantula Nebula, NGC 2080

TELESCOPE TARGETS

More than three decades ago, the brightest supernova seen since the invention of the telescope was spotted by astronomers as it erupted in the Large Magellanic Cloud. Supernova 1987A exploded with the power of 100 million suns for many months, appearing at its peak as a moderately bright star visible to the naked eye.

As its name in Spanish suggests, Dorado indeed represents a golden celestial region filled with deep-sky treasures to explore. Taking on the figure of a swordfish or dolphinfish, the constellation's faint line of third- and fourth-magnitude stars is sandwiched between the constellations Reticulum and (appropriately) Volans, the flying fish. This southern summer constellation is fairly easy for skywatchers to track down since it lies between the bright stars Achernar and Canopus.

Along Dorado's southern border with the constellation Mensa is one of the greatest celestial prizes for skywatchers: the Large Magellanic Cloud. This cloud is actually an irregular galaxy that is a close companion to our Milky Way. Lying 160,000 light-years away, it stretches 14,000 light-years across and is filled with 40 billion stars, with a multitude of star clusters and nebulae. While it was recorded

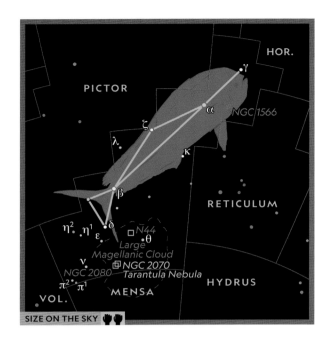

by Persian astronomers as far back as A.D. 964 and called the White Ox, the Large Magellanic Cloud became known more widely in western culture when its namesake explorer Ferdinand Magellan recorded it in 1519. Shining with a collective magnitude of 0.1, this giant gray cloud is clearly visible to the naked eye, looking like a broken piece of the Milky Way, and is bright enough to even be spotted on moonlit nights. Sitting along its eastern edge is one of the sky's largest and brightest emission nebulae, the Tarantula Nebula, visible to the naked eye under dark skies at magnitude 8. Long-exposure photography reveals that it is filled with colorful streamers and knots expanding out from its white-hot core region, indeed making it resemble a spider. The radiation of the young stars at its core makes the surrounding hydrogen gas glow and expand, rendering it 100 times larger than its famous cousin, the Orion Nebula.

Another amazing emission nebula within this satellite galaxy is N44. With a bubble-like appearance, it is shaped by strong stellar winds emitted by about 40 baby stars huddled within its core region. This colorful star factory expands out to about 1,000 light-years across. Meanwhile, scanning along a chain of eye-catching stellar nurseries to the southern end of the Large Magellanic Cloud, you'll find the Ghost Head Nebula, or NGC 2080. Two bright bubbles of white-hot gas mark the eyes of this cosmic phantom, formed within the last 10,000 years by clutches of massive stars hidden inside.

The Tarantula Nebula, a star-forming region in the Large Magellanic Cloud

DRACO (THE DRAGON)

Visible: Northern Hemisphere

Best Seen: All year

Main Stars: 14

Brightest Star: Eltanin, 2.24 mag.

Deep-Sky Objects: NGC 5879, NGC 5907

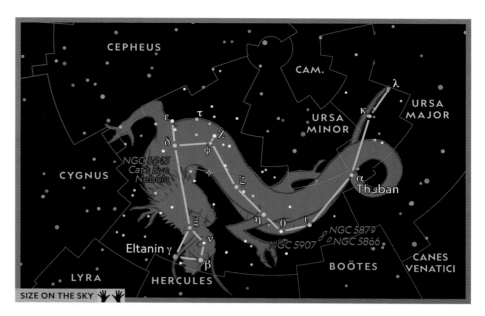

Riding high in the northern sky, Draco (The Dragon) is the eighth largest constellation. It covers a large chunk of celestial real estate as it coils around the North Star and the small constellation of Ursa Minor (The Little Bear). Draco is a very ancient grouping. The Sumerians saw these stars as representing the dragon Tiamat. Later it became one of the creatures that Hercules, the strongman, was sent to kill. In the heavens, we now see Hercules standing (albeit upside down) on Draco's head.

The ancient Egyptians had a special connection to this constellation, too. A faint star in the dragon's tail (the third from last), known as Thuban, used to point to true north as the North Star about 4,600 years ago. Egyptians revered this star, as the night sky appeared to revolve around it. Draco is also home to the beautiful eighth-magnitude globular cluster NGC 5879, while NGC 5907 is a 10th-magnitude, warped spiral galaxy more than 50 million light-years away, with a central dust lane visible even in small telescopes.

EQUULEUS (THE LITTLE HORSE)

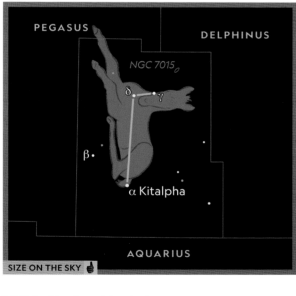

Representing a small horse, Equuleus is the second smallest constellation in the entire sky. Faint stars form a relatively easy-to-find trapezoid between Delphinus and the head of Pegasus. Greek legends say it represents the foal Celeris, gifted by the god Mercury to Castor, one of the Gemini twins. In Chinese this asterism is appropriately dubbed the Emptiness.

The brightest star at the tip of the elongated triangle marking the constellation is called Kitalpha, meaning "part of a horse" in Arabic. Epsilon Equulei is a triple star system that offers a pretty color contrast when using high magnification with a telescope. The two brightest stars are too close to separate, but they appear collectively as yellow while the third member is distinctly blue. When it comes to deep-sky objects, Equuleus is quite barren, save for a few very faint galaxies. For larger backyard telescopes the 11th-magnitude spiral galaxy NGC 7015 offers a pretty face-on view, despite being 203 million light-years from Earth.

Visible: Northern Hemisphere

Best Seen: Late summer and autumn

Main Stars: 3

Brightest Star: Kitalpha, 3.93 mag.

Deep-Sky Object: NGC 7015

ERIDANUS (THE RIVER)

Visible: Southern Hemisphere

Best Seen: Summer

Main Stars: 24

Brightest Star: Achernar, 0.45 mag.

Deep-Sky Objects: NGC 1535, NGC 1300

SIZE ON THE SKY

Meandering across the night sky, Eridanus is much like a lazy river in the heavens. Since biblical times Eridanus has been associated with the Nile and Euphrates. One of the largest constellations, its faint stars wind across the sky southwest of Orion, passing the hunter's foot marked by Rigel, crossing under Cetus, and then flowing deep into the southern sky until reaching an end next to the Phoenix and Hydrus constellations. As such, this makes Eridanus immensely long, and in fact it is the second longest constellation in the sky.

It is here in the deep southern sky that Eridanus's brightest star, Achernar, pins down the end of the constellation. This blue-white stellar beacon is the ninth brightest star in the entire sky, shining at magnitude 0.45. In Arabic its name appropriately means "the river's end." Other fainter targets that really shine for medium-size telescopes include the 10th-magnitude blue planetary nebula NGC 1535, known also as Cleopatra's Eye, as well as the 11th-magnitude barred spiral galaxy NGC 1300.

FORNAX (THE FURNACE)

Visible: Southern Hemisphere

Best Seen: Late spring and summer

Main Stars: 2

Brightest Star: Alpha Fornacis, 3.85 mag.

Deep-Sky Objects: NGC 1097, NGC 1316, Fornax dwarf galaxy, Fornax Cluster

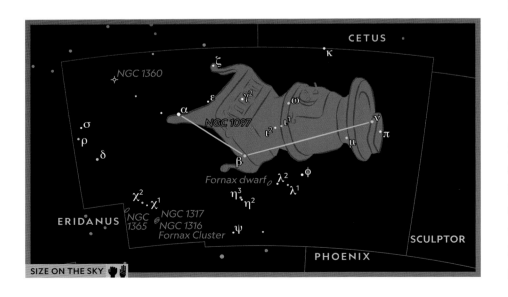

SIZE ON THE SKY

Fornax is a modern constellation, having been invented by Nicolas-Louis de Lacaille in 1752 and representing a chemical furnace. Marked by a dim, flattened triangle of stars, the constellation is bordered by Eridanus, Cetus, and Phoenix. Fornax's brightest member is no more than magnitude 3.9, making it just visible from a typical suburban sky.

While it may appear barren to the naked eye, Fornax is strewn with galaxies, including a tight grouping of about two dozen that are part of the Fornax Cluster, 55 million light-years from our solar system. NGC 1097 is a face-on, barred spiral galaxy that is an easy target for small telescopes at only magnitude 9. At the edge of the galactic cluster is NGC 1316, a small elliptical galaxy only 60,000 light-years across. Shining at a relatively bright eighth magnitude, it appears to have a bright, condensed core. Larger telescopes reveal a dark dust lane cutting across its length. Astronomers believe this feature marks the remains of an ancient collision with a spiral galaxy.

GEMINI (THE TWINS)

Visible: Northern Hemisphere

Best Seen: Winter and early spring

Main Stars: 8

Brightest Star: Pollux, 1.15 mag.

Deep-Sky Objects: Medusa Nebula, NGC 2355

Marked by two bright stars huddling together, the Gemini constellation is easy to find thanks to this super-bright first-magnitude pair, representing the heads of the mythical twins known as Castor and Pollux. Even from light-polluted suburbs it is possible to trace out the loose lines of stars that form the hands and feet of the pair, though they are much fainter and are more of a challenge to track down. The twin stars are not exactly alike. Pollux is noticeably yellow, and it is brighter and closer to Earth at 36 light-years than white Castor at 53 light-years distant.

Gemini is one of the 12 zodiacal constellations and represents the sons of Leda (the queen of Sparta) and the brothers of Helen of Troy. Positioned beside the Taurus constellation, Gemini's feet appear to dip into the rich Milky Way band that runs through the star-studded Orion constellation, while the twin's heads are in a relatively more barren region. As such, Gemini contains a fine selection of star clusters, nebulae, and galaxies. And because the ecliptic runs through this constellation, there are times when the moon or planets disrupt the familiar pattern of stars.

A fainter cousin to the more famous M35 star cluster is NGC 2355. It shares the same binocular field of view with Alkibash, the "knee" star of the twin Pollux. Located some 6,400 light-years from Earth, this cluster has 20 brighter stars; more than 40 fainter ones behind them form an S-shape. Train a telescope toward Gemini's border with Canis

Minor to find the Medusa Nebula. A more modest version of the Clown Face Nebula, Medusa is shaped like a half moon. The expanding gas cloud from this old planetary nebula has a very low contrast, making it a great target for larger backyard telescopes.

Finally, Gemini holds an ancient drama in the form of a long-dead neutron star known as Geminga. Located only 800 light-years from Earth, it is thought to have produced one of the closest supernova explosions to the planet 360,000 years ago. At its peak brightness Geminga may have even rivaled the full moon for a few weeks. Our planet's ozone layer may have been damaged temporarily by the rush of radiation from the supernova, which might even have sunburned any ancient human ancestors. But now its celestial footprint is an Earth-size, spinning pulsar that is one of the sky's brightest sources of gamma rays.

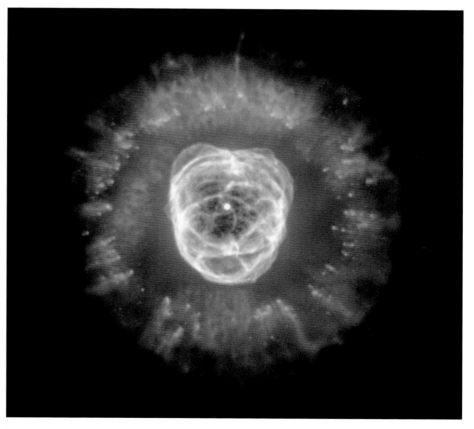

NGC 2392: the Clown Face Nebula

GRUS (THE CRANE)

Visible: Southern Hemisphere

Best Seen: Winter and early spring

Main Stars: 8

Brightest Star: Alnair, 1.73 mag.

Deep-Sky Objects: IC 5148, NGC 7213, NGC 7410, NGC 7424

ook deep into the southern sky for a flock of faint celestial birds, with Grus (The Crane) being the most prominent. A long line of stars represents its long neck and beak, with the stars straddling either side of the body forming its flapping wings. The crane is surrounded by other avian constellations in the far southern sky: Apus, Pavo, Phoenix, and Tucana.

Grus is one of a dozen new constellations invented by Dutch voyagers Pieter Dirkszoon Keyser and Frederick de Houtman during their trip to the East Indies at the end of the 16th century. The brightest stellar member, Alnair, meaning "the bright one" in Arabic, is easy to spot as it is the 31st brightest star in the sky.

Its estimated diameter is three times that of our sun, and it is a powerful beacon 380 times brighter than our own star. This 101-light-years-distant sparkling point of light pins the end of one of the crane's

wings and shines with a beautiful bluish white color. For an eye-catching contrast, look in the lower body of the crane for 170-light-years-distant Tiaki, the constellation's second brightest star. This red giant star, the 59th brightest in the sky, shines with a warm orange hue and is so large that if it were to replace the sun in our solar system it would swallow Venus. Tiaki is easy to track down if you draw a line from the Great Square of Pegasus to Fomalhaut on to Alnair and then to Tiaki. On closer inspection, naked-eye observers will notice that the stars Tiaki and Delta Gruis appear to be impressive, wide double stars—however, their double star appearance is just an optical illusion due to our line of sight.

In 2019, astronomers detected within the constellation the speediest star on record. Dubbed S5-HVS1, the star is traveling at 1,755 kilometers a second (1,090 mps) and is currently 29,000 light-years away, heading out of the Milky Way thanks to a close encounter with the supermassive black hole at the center of our galaxy. Affectionately dubbed the Spare Tyre Nebula, IC 5148 is a planetary nebula sitting 3,000 light-years from Earth.

Grus is also home to a few galaxies within reach of backyard telescopes. The brightest is the barred spiral NGC 7424, 37 million light-years away. This face-on galaxy shines at magnitude 10.4 and is similar in appearance and size to our own Milky Way. The next brightest, at a feeble 11.7 magnitude, is the barred spiral galaxy NGC 7410, some 122 million light-years from our solar system.

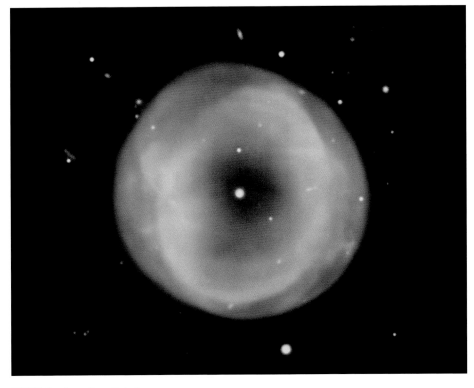

IC 5148: the Spare Tyre Nebula

HERCULES (HERCULES)

Visible: Northern Hemisphere

Best Seen: Late spring and summer

Main Stars: 14

Brightest Star: Kornephoros, 2.78 mag.

Deep-Sky Objects: Abell 39, NGC 6210, Abell 2151

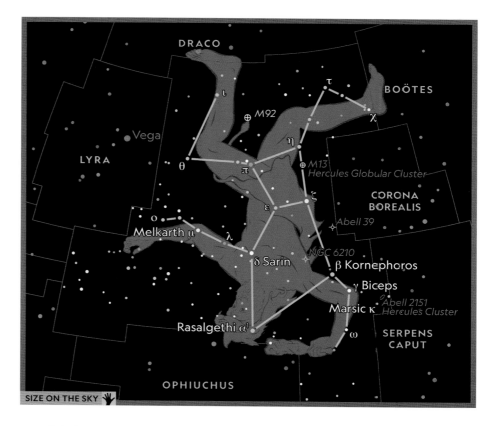

A sprawling constellation that ranks as the fifth largest in the entire sky, Hercules is easy to track down since it lies between the brilliant stars Vega and Arcturus.

Hercules is truly an old figure in the sky, predating even the ancient Romans and the Greeks before them. Earlier civilizations such as the Babylonians, as far back as 2500 B.C., associated it with Gilgamesh and the legendary flood. The original strongman of Greek myths, Hercules is the superhero on which all others are based. This classical hero was said to be the son of the king of all gods, Zeus, and a mortal mother. Legends have him slaying a magical lion, a nine-headed snake, and monster crabs, and stealing flesh-eating horses, among other labors. At the end of Hercules' hectic life, Zeus honored his weary son by making him a god and placing him in the sky. His celestial pose has him kneeling with an upraised club in one hand and a foot on the head of the mythical dragon Draco.

M92, one of the brightest globular clusters in the Milky Way

While his stars are not the brightest, four do make up a distinctive pattern that is pretty easy to identify: a wedge-shaped trapezoid of stars marking his torso known as the Keystone asterism. While its star clusters may get all the fame, Hercules is also home to an amazing planetary nebula and numerous galaxies. A line drawn northeast through Gamma Herculis and the brightest star in Hercules, Kornephoros, will point to NGC 6210, a 10th-magnitude planetary nebula 5,400 light-years from Earth. Small telescopes under dark skies show it as an elliptical, green-hued gas cloud with regular borders. A much fainter planetary nebula is 6,800-light-years-distant Abell 39. It holds the record for the largest perfectly spherical stellar cloud, stretching five light-years across.

For a real telescope-observing challenge, try spotting Abell 2151, also known as the Hercules galaxy cluster. With an average magnitude of 13, this is a collection of more than 200 galaxies with an average distance of 500 million light-years. Also running through the constellation is the Hercules–Corona Borealis Great Wall, which is the largest structure seen in the universe. Composed of a meandering cluster of gamma ray bursts that stretches 10 billion light-years across, this superstructure is about one-tenth as long as the observable universe.

TELESCOPE TARGETS

The Hercules constellation is home to two of the finest globular clusters in the northern sky. Both M92 and M13 can be glimpsed with binoculars, but the latter is bright enough to be just visible to the naked eye as a "fuzzy star" under a dark sky. A small telescope reveals both of their mesmerizing ball-shaped structures to be composed of hundreds of thousands of stars.

HOROLOGIUM (THE CLOCK)

Visible: Southern Hemisphere

Best Seen: Late spring and summer

Main Stars: 6

Brightest Star: Alpha Horologii, 3.85 mag.

Deep-Sky Objects: NGC 1261, NGC 1512

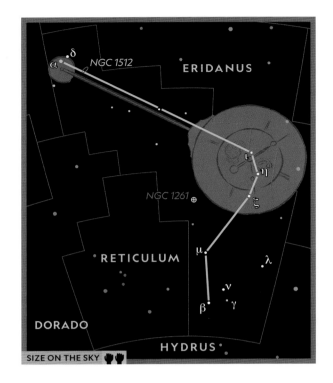

Horologium is one of the obscure, faint constellations in the Southern Hemisphere that were created to celebrate technological advances of the 17th and 18th centuries: in this case, the clock. It was first known as Horologium Oscillatorium, or Pendulum Clock, in honor of that clock's inventor, Christiaan Huygens. Located just east of the bright star Achernar, in a fairly barren part of the sky, it consists of mostly fifth- and sixth-magnitude stars, making it a challenge to see unless under a dark sky. Look for the clock to run parallel to the winding river constellation of Eridanus, blocked in by Hydrus, Reticulum, Dorado, and Caelum.

The brightest star is the near-fourth-magnitude Alpha Horologii, an orange giant 115 light-years from Earth that marks the end of the clock's pendulum. This elderly star is reaching the end of its life and has swollen to 11 times the diameter of our sun.

Meanwhile, just under Achernar is the face of the clock, near the southern end of the constellation. Some artists' interpretations have Beta and

Zeta Horologii positioned as the end of the clock's two hands and Mu Horologii marking the center of the timepiece's face. R Horologii is a long-period variable that is fun to watch. With a period of 404.8 days, it dips in brightness from a faint naked-eye star shining at 4.7 magnitude down to 14.3 magnitude. This 1,000-light-years-distant object has one of the widest swings in brightness of any known star.

Horologium lies far from the star fields of the Milky Way and so has no open clusters or nebulae worth noting, yet it is home to a nice variety of galaxies and one picturesque globular cluster. NGC 1261 is a small eighth-magnitude globular cluster about 53,000 light-years away. It appears as a faint ball through smaller telescopes, but larger instruments begin to resolve the individual stars around the cluster's outer halo. For larger backyard telescopes the 10th-magnitude barred spiral galaxy NGC 1512 makes for a pleasing target. High-resolution photos using the Hubble Space Telescope have revealed an ethereal double ring structure between the nucleus and the main disk. This 38-million-light-years-distant island of stars appears as a bright, elongated oval, with a tinier elliptical galaxy, NGC 1510, lying just to its southwest. The two galaxies recently begin merging, and the process likely won't finish for another 400 million years, when the larger galaxy has completely gobbled up the hapless galactic dwarf.

Barred spiral galaxy NGC 1512

HYDRA (THE FEMALE WATER SNAKE)

Visible: Northern Hemisphere

Best Seen: Spring and early summer

Main Stars: 17

Brightest Star: Alphard, 1.99 mag.

Deep-Sky Objects: Abell 33, M48, M83, M68, NGC 2835, NGC 3242, NGC 5694, NGC 3314

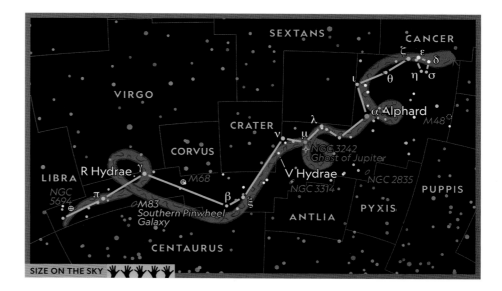

SIZE ON THE SKY

Hydra (The Female Water Snake) holds the record as the largest of all the constellations in the sky. It appears to meander across a quarter of the celestial sphere and is associated with mythology surrounding the god Apollo, and also with the ancient Greek legend of the multiheaded monster killed by Hercules. The constellation appears to be sandwiched between constellations Cancer to its north and Centaurus and Libra to its south.

Despite its enormous size, Hydra surprisingly has relatively few prominent stars, with the brightest, Alphard, marking the snake's heart and just about punching above the second-magnitude level. The star lies in a relatively barren part of the sky, and so its name appropriately means "the solitary one" in Arabic. Alphard is an orange giant 85 light-years away, and it is easy to track down with the bright twin Gemini stars Pollux and Castor pointing directly to it.

M83: the Southern Pinwheel Galaxy

Hydra contains three good Messier object targets for backyard telescopes. Sitting along Hydra's western border, within the snake's head, is the 5.5-magnitude open cluster M48. The 1,500-light-years-distant cluster lies very close to the boundary the constellation shares with Monoceros, and in fact the Gamma Monocerotis star acts as a great guide. Just visible with the naked eye as a faint smudge, the cluster will be revealed in a small telescope as about 60 stars huddled together within an area 24 light-years across. Astronomers estimate the stars are fairly young at about 300 million years old. Lying between Gamma and Beta Hydrae, the globular cluster M68 is not hard to find with binoculars. Sitting 33,000 light-years from Earth, this 7.8-magnitude cluster is one of the more distant globulars known, embedded within the halo around our Milky Way galaxy. Stretching about 106 light-years across, about 100,000 stars call this stellar city home.

An even more distant globular cluster within Hydra is NGC 5694. More than 114,000 light-years away, it is estimated to be heading out of our galaxy's gravitational pull. Known affectionately as the Southern Pinwheel, M83 is one of the closest and brightest large spiral galaxies to our own Milky Way and a real showpiece for backyard astronomers. At 7.6 magnitude, this face-on spiral is visible through binoculars with a bright core and bar running through it and spiral arms encircling it. Only about 15 million light-years away, M83 is not far out from our Local Group of galaxies. For a time it held the record for harboring the most observed supernovae, with six seen within the 20th century.

TELESCOPE TARGETS

The Ghost of Jupiter (NGC 3242) is a ninth-magnitude planetary nebula that through backyard telescopes resembles its namesake gas giant planet. High magnification through a small telescope shows it off as a relatively bright blue-green oval with a faint central star. Larger telescopes, however, reveal this tiny nebula to have a brighter inner disk, enveloped by a faint halo.

NGC 602 in the Small Magellanic Cloud gives off x-rays in this composite image from three orbiting telescopes.

HYDRUS (THE MALE WATER SNAKE)

Visible: Southern Hemisphere

Best Seen: Spring

Main Stars: 3

Brightest Star: Alpha Hydri, 2.85 mag.

Deep-Sky Objects: NGC 602, NGC 1466, NGC 1511

SIZE ON THE SKY

Hydrus (The Male Water Snake) was created in 1603 by Johann Bayer, who intended it to be a companion to a more ancient constellation—the female water snake, Hydra.

Hydrus is visible only to observers in the Southern Hemisphere, and it is not hard to find since it sits in between the constellation Reticulum and the Large Magellanic Cloud to its east, and the constellation Tucana and the Small Magellanic Cloud to its west. The head of Hydrus touches Octans around the South Celestial Pole. Also making it easy to locate is the fact that its lead star, Alpha Hydri, lies only four degrees southeast of the constellation Eridanus's stellar beacon Achernar. Hydrus's bright yellow star is also considered the closest bright star to the South Celestial Pole, located about 12 degrees away. When finding Hydrus, look for the constellation's three brightest stars, all of which shine at about third magnitude, forming a distinct triangle.

Hydrus is home to two fine optical double stars within reach of binoculars. Pi Hydri is composed of two fifth-magnitude stellar giants, Pi1 and Pi2, sitting 476 and 488 light-years away, respectively. Meanwhile, Eta Hydri consists of a very faint blue star, 700 light-years distant, and a 4.7-magnitude yellow giant 218 light-years away. In 2005, this yellow star was discovered to host a gas giant planet at least six times the mass of Jupiter. The Hydrus constellation has been a fruitful hunting ground for exoplanet hunters. To date four planetary systems have been identified, with the most notable hosted by the seventh-magnitude star HD 10180, located 127 light-years away. The sunlike star holds the record of having a total of nine planets ranging in size from Earth to Neptune.

Hydrus is considered a bit of a desert in terms of deep-sky objects, but it is home to a few targets for backyard telescopes, including the fairly bright star cluster NGC 602, which is surrounded by a diffuse nebula. Estimated to be lying on the periphery of the Small Magellanic Cloud, some 200,000 light-years away, the cluster is composed of dozens of newborn stars. These stellar youngsters belch out copious amounts of radiation and shock waves that carve out undulating shapes in the surrounding cloud of gas and dust.

NGC 1466 is an 11th-magnitude globular cluster that is estimated to be around 160,000 light-years away, nestled within the outskirts of the Large Magellanic Cloud. Astronomers calculate that it contains as many as 140,000 suns and is truly a cosmic fossil. With an age around 13.1 billion years, NGC 1466 is nearly as old as the universe.

TELESCOPE TARGETS

According to legend, this water snake had to take a long journey to find its mate, swimming up the winding Eridanus river, sliding below Orion, and then crossing the Milky Way river.

NGC 1466, an ancient globular cluster in the Large Magellanic Cloud

INDUS (THE INDIAN)

Visible: Southern Hemisphere

Best Seen: Spring

Main Stars: 3

Brightest Star: Alpha Indi, 3.11 mag.

Deep-Sky Objects: NGC 7090, NGC 7049

Indus is meant to commemorate the American Indians encountered by European explorers to the Americas and is one of a set of constellations created by German astronomer Johann Bayer in 1603. Its faint stars, none brighter than third magnitude, lie south of the constellations Grus and Microscopium and run down to Octans near the South Celestial Pole. The lead star, Alpha Indi, shines bright orange at third magnitude some 98 light-years from Earth. The brightest three stars, Alpha, Beta, and Delta, form a recognizable right-angled triangle pattern in the southern sky. And the constellation contains Theta Indi, a nice binary system 98 light-years from Earth, visible through small telescopes.

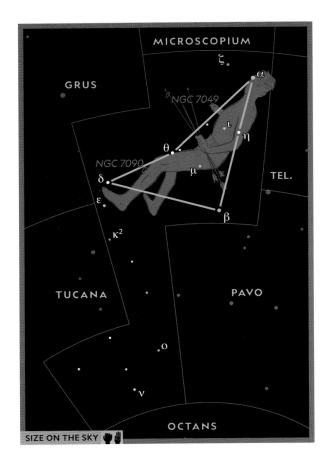

SIZE ON THE SKY

LACERTA (THE LIZARD)

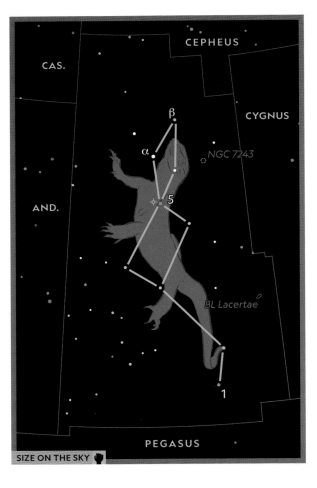

SIZE ON THE SKY

STAR STORIES

While the lizard constellation is fairly new and is missing from ancient star maps across European and Middle Eastern cultures, its stars are part of a larger celestial figure in China. Ancient Chinese astronomers included it in their Tianshe, or Flying Serpent, formed by 22 stars that include Lacerta and the eastern half of the neighboring constellation Cygnus.

Visible: Northern Hemisphere

Best Seen: Summer and early autumn

Main Stars: 5

Brightest Star: Alpha Lacertae, 3.76 mag.

Deep-Sky Objects: NGC 7243, BL Lacertae

Nestled within a fairly barren region of the northern sky, Lacerta (The Lizard) is a lackluster assemblage of stars encircled by some of the sky's brightest constellations. The celestial reptile is sandwiched between Cepheus to its north and Pegasus in the south, and it is flanked by Andromeda and Cassiopeia to its east and Cygnus to its west. Credited to Polish astronomer Johannes Hevelius in 1687, it represents a Mediterranean creature known to some locals as the star lizard. Lacerta is filled with dim fourth- and fifth-magnitude stars; the brightest member of the constellation is the young, blue-white, near-3.8-magnitude Alpha Lacertae, 102 light-years away.

LEO (THE LION)

Visible: Northern and Southern Hemispheres

Best Seen: Northern spring and southern autumn

Main Stars: 15

Brightest Star: Regulus, 1.35 mag.

Deep-Sky Objects: M65, M66, M95, M96, M105, NGC 3628, NGC 2903

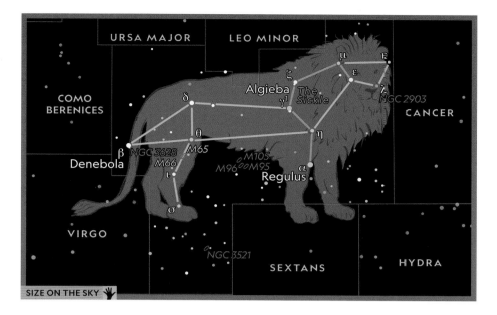

The legendary figure of a lion has been associated with the constellation Leo for many millennia, representing one of the most ancient of all stellar patterns. With a distinctive and bright figure, it is one of the easier constellations to construct mentally, as it resembles the classic image of the Sphinx and lions in general. One gruesome ancient Greek legend has the mighty feline landing on Earth as a shooting star and menacing the countryside, savagely attacking people and livestock. The task of slaying this brute fell to the Greek hero Hercules. As the story goes, upon finding the lion, Hercules wrestled with it and then killed it with his bare hands. Using stars as nails, the gods placed Leo in the sky in memory of this dreadful battle.

Leo takes up a large tract of the night sky, but it is devoid of many very bright stars. It is one of the 12 zodiac constellations that the sun glides through each calendar year. The formidable feline is sandwiched between fellow zodiacal figures Cancer and Virgo. Leo is easily found by novice skywatchers, thanks to a form consisting of a small triangle connected to a giant sickle or hook. This giant sickle, or backward question mark, is an eye-catching asterism that marks the lion's head, mane, and chest. Meanwhile the stars Zosma, Chertan, and Denebola trailing to the east mark Leo's hindquarters and tail.

The period in the celestial punctuation is marked by the constellation's lead star, Regulus. Shining at first magnitude, it is the 21st brightest star in the sky and dominates the northern spring skies with its blue-white brilliance. Observations of this star date back to Babylonian tablets. Better known in antiquity as Cor Leonis, the Lion's Heart, the star's current name is taken from the Latin word meaning "little king," and it is often known as the Royal Star. Regulus is approximately 84 light-years away and shines some 160 times brighter than our own sun, with a diameter five times that of our own puny star. When Regulus is viewed through binoculars or a small telescope, a very dim companion star is revealed, making for a great observational challenge.

Look at the third brightest star in Leo, Algieba, which means "lion's mane," as it shines with an orange color. Algieba is a giant, more than 10 times larger and 80 times brighter than our own star. It's a beautiful example of a double star, and binoculars will reveal a strikingly different colored stellar companion. Binoculars and telescopes scanning the belly of Leo will also find faint fuzzy patches of light, each representing far-off galaxies. One of the brightest groups of galaxies includes M65, M66, M96, and M105.

STAR STORIES

The annual rise of the Nile River was believed by ancient Egyptians to occur when the sun rose in the part of the sky occupied by the constellation Leo. This link to the Nile may explain why later Greek and Romans often placed the familiar lion's head at springs and fountains, just as Egyptians placed theirs at canal gates.

(Left to right) NGC 3628, M65, and M66: the Leo Triplet galaxies

LEO MINOR (THE LESSER LION)

Leo Minor is a minor northern constellation that sits between its larger companion, Leo, to its south and Ursa Major to its north. Officially a 17th-century invention, the little lion was seen by ancient Arab cultures as a leaping gazelle, while to others, affectionately, as the paw prints of Ursa Major, the great bear. For observers the faint constellation consists of a bent line turning east to west made up of only faint fourth-

Visible: Northern Hemisphere

Best Seen: Spring

Main Stars: 3

Brightest Star: Praecipua, 3.83 mag.

Deep-Sky Objects: NGC 3432, NGC 3003, NGC 3344, NGC 3486, NGC 3021

magnitude stars. Its brightest star, Praecipua, or 46 Leonis Minoris, is an orange giant more than eight times larger than our sun, lying 95 light-years away.

For a tiny constellation, Leo Minor is rich with galaxies to explore with backyard telescopes. The brightest ones at 10th magnitude are the 22-million-light-years-distant barred spiral NGC 3344 and the face-on spiral NGC 3486, 27 million light-years from Earth. Fainter and suited for larger backyard telescopes with 250- to 300-mm (10- to 12-in) apertures are edge-on spirals NGC 3003, 63 million light-years away, and NGC 3432, which appears to be undergoing a collision with a dwarf galaxy 30 million light-years from us.

LEPUS (THE HARE)

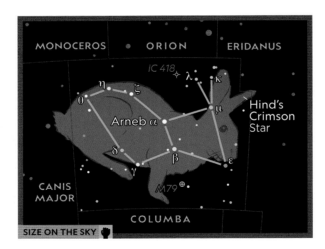

Visible: Northern and Southern Hemispheres

Best Seen: Northern winter and southern summer

Main Stars: 8

Brightest Star: Arneb, 2.58 mag.

Deep-Sky Object: M79

Most often shown as a hare or rabbit being hunted by Orion and his dogs, Canis Major and Canis Minor, Lepus is a small but distinct constellation just below the bright star Rigel. The celestial rabbit is best seen from December to March around the world outside of the Arctic Circle. Its four main stars form a quadrilateral pattern that was one of the 48 constellations cataloged by Ptolemy in the sec-

ond century A.D. According to some Arabic legends, the stars of Lepus either represented four camels crossing the desert or were thought to form Al 'Arsh al Jauzah, meaning "the throne of Orion." Meanwhile, ancient Egyptians imagined that these same stars formed the boat of Osiris, the god of the dead. Lepus's brightest star is white Arneb, which is from the Arabic, appropriately meaning "the hare."

Lepus contains only one deep-sky object of interest for backyard stargazers, the globular cluster Messier 79. Located near the constellation's southern border, this 42,000-light-years-distant ball of stars shines with a combined eighth magnitude, making it a great showpiece in small telescopes.

LIBRA (THE SCALES)

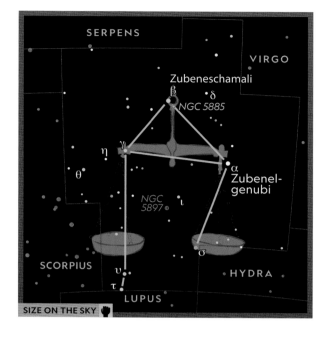

SERPENS

VIRGO

Zubeneschamali
β • δ
NGC 5885

η γ

θ α
Zubenel-
genubi

NGC
5897 ⊕ ι

SCORPIUS

υ
τ HYDRA

LUPUS

SIZE ON THE SKY

Visible: Northern and Southern Hemispheres

Best Seen: Northern summer and southern winter

Main Stars: 6

Brightest Star: Zubeneschamali, 2.61 mag.

Deep-Sky Objects: NGC 5897, NGC 5885

Hiding within Libra is a faint 8.5-magnitude globular cluster named NGC 5897. Through a telescope this 41,000-light-years-distant star city is a large, poorly concentrated, hazy glow that appears elongated and has a span about as wide as a third of the full moon's disk. Astronomers have examined the chemical fingerprints of the stars of this cluster and have determined that it is quite old, born before the Milky Way even had any spiral arms. NGC 5885 is a beautiful 12th-magnitude face-on spiral galaxy 100 million light-years from Earth. Large backyard telescopes show it as a grayish oval with a tight star-like core. About 20 light-years away, Gliese 581 is a red dwarf at the center of a planetary system that has gained much attention for having ideal conditions for habitable environments.

TELESCOPE TARGETS

Zubenelgenubi, or Alpha Librae, the second brightest star in Libra, is a bit of an observing mystery. Many observers over the centuries have pointed out that it's the only star in the sky to have a distinctive green hue. What color do you see?

L ying between the constellations Virgo and Scorpius, little Libra—the golden balance or scales—is, like them, a zodiacal constellation. It was first outlined in the heavens by the ancient Romans, who thought the founding of their empire was tied to this celestial sign. In more ancient times the Greeks made its two brightest stars part of Scorpius, representing the arachnid's claws. But carvings from ancient Mesopotamia dating back to 2200 B.C. depict the sky pattern taking the form of a balance scale, used to weigh the souls of the departed. Other ancient astronomers in the Middle East saw Libra's stars as the lamp of the great lighthouse of Alexandria, one of the Seven Wonders of the Ancient World, held within the grasp of the scorpion's claws.

Libra is a faint constellation, and its three main stars form a triangular pattern that represents the top of the figure, while its scales appear toward the west in the direction of the orange stellar beacon Antares, the brightest star in Scorpius. The names of Libra's two brightest stars harken back to the constellation's ancient association with the scorpion. Zubeneschamali is the brightest star and Zubenelgenubi the second brightest at 2.75 magnitude; their names are from the Arabic, meaning "northern claw" and "southern claw," respectively. Delta Librae, found just west of the northern end of the balance's beam, is an eclipsing variable whose 2.3-day cycle ranges from magnitude 4.9 to 5.9, perfectly visible to the naked eye on a clear night.

Spiral galaxy NGC 5885

LUPUS (THE WOLF)

Visible: Northern and Southern Hemispheres

Best Seen: Northern spring and southern autumn

Main Stars: 9

Brightest Star: Alpha Lupi, 2.30 mag.

Deep-Sky Objects: IC 4406, NGC 5986, NGC 5882, NGC 5927, NGC 5822, NGC 5824, SN 1006

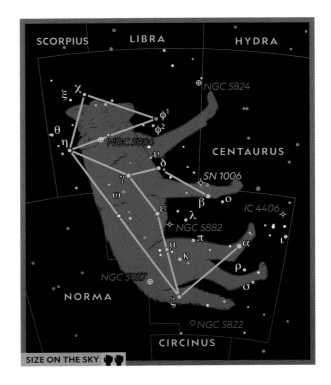

SIZE ON THE SKY

Lupus is a well-defined constellation in the southern sky located south of Libra and east of the prominent constellation Centaurus. While Lupus is best positioned in the sky for observers in the Southern Hemisphere, northern skywatchers up until the midlatitudes can see the top half of the constellation pop above the southern horizon, just below Scorpius, during the spring season. The stars of Lupus were associated with a wolf figure going back to the Babylonians. Later on the Greeks inherited it and called it Therium, meaning "the wild dog." Their legends had this creature intertwined with Centaurus, representing a sacrifice carried by the hoofed half-monster. Meanwhile to some Arab cultures, it was known as Al Asadah, meaning "the lioness," while others called it al-Shamareekh, meaning "branch of the date palm." In Chinese astronomy the same stars are part of a much larger pattern of stars known as Qíguān, meaning "the imperial guards."

Lupus is positioned such that the northeast edge of the band of the Milky Way runs along its back, through neighboring Centaurus. In many artistic interpretations, the bright stars Alpha and Beta Centauri lie just behind the wolf's tail, and Antares is in front of its head. Alpha Lupi is a powerful blue-giant shining at magnitude 2.3 despite being 460 light-years from Earth. Epsilon Lupi, a multiple star, has a blue-white primary of magnitude 3.4 and a wide ninth-magnitude companion that is visible in small telescopes. Kappa Lupi is an easily visible double star for small telescopes; it consists of blue-white stars of magnitudes 3.7 and 5.7.

Since it is bordering Milky Way territory, Lupus is filled with star clusters and other deep-sky treasures perfect for exploring with binoculars and telescopes. A beautiful, large 10th-magnitude open cluster, NGC 5822, holds more than 120 stars loosely scattered across the field of view of a small telescope. Two globular clusters hold court in Lupus's western and northern parts. NGC 5824 is near Centaurus, and telescope views show the ninth-magnitude cluster as having a very distinct condensed core surrounded by a wide scattering of stars that span more than 200 light-years. However, with its immense distance of 104,000 light-years, NGC 5824 appears only as a tiny cotton ball in high-magnification views with larger telescopes. NGC 5986 is brighter at eighth magnitude and is closer to us at 33,900 light-years, making this globular cluster an easy target for binoculars.

Supernova remnant SN 1006

LYNX (THE LYNX)

Visible: Northern Hemisphere

Best Seen: Winter and early spring

Main Stars: 4

Brightest Star: Alpha Lyncis, 3.14 mag.

Deep-Sky Objects: NGC 2419, NGC 2537, IC 2233, NGC 2541, NGC 2500

Lynx is a faint northern constellation that occupies a part of the sky that is devoid of bright stars, situated between major constellations Auriga and Ursa Major, with Gemini to its south. Lynx owes its existence to Polish astronomer Johannes Hevelius, who officially placed it in his sky charts in 1687 and chose this secretive feline as the perfect representation of a star pattern that can be challenging to find. Under dark skies, its crooked line of brightest stars appears sandwiched between Leo and Camelopardalis. The lead star is the 3.1-magnitude Alpha Lyncis, which is an orange giant 203 light-years away. It has twice the mass of our sun and is 55 times its diameter.

The most famous deep-sky object in Lynx is one of the most distant globular clusters known to be associated with the Milky Way. NGC 2419 is located about seven degrees north of the bright twin star Castor in Gemini. Often called the Intergalactic Wanderer, this city of stars lies around 300,000 light-years from Earth and is thought to be in a highly elliptical orbit around our galaxy. Even though it shines at magnitude 10, a telescope with at least a 200- to 250-mm (8- to 10-in) aperture is needed to begin resolving the cluster's outer halo of individual stars and make it appear as more than just a fuzzy star in the eyepiece. The fact that it can even be spotted with a small telescope, despite its great distance, tells astronomers that it has a very high luminosity. Some of its brightest members are stellar powerhouses, including yellow and red giants that intrinsically shine hundreds of times brighter than our own sun.

Located about four degrees northwest of the star, 31 Lyncis is a compact dwarf galaxy known as the Bear Paw, or NGC 2537. Lying about 20 million light-years away, it appears in even large backyard telescopes as a faint, blue, round disk that resembles a planetary nebula. Next door to the Bear Paw is an edge-on spiral galaxy, IC 2233, that for a long time was thought to be interacting with it, but that we now know is 10 million light-years more distant. A great observing challenge for larger telescopes is a triplet of distant spiral galaxies—NGC 2541, NGC 2500, and NGC 2552. Dubbed the Modest Galaxy, UGC 3855 is a highly tilted spiral galaxy. The Hubble Space Telescope has been able to look past the blinding glare of the nearby superbright foreground star and resolve intricate dark dust lanes that follow its spiral structure.

Spiral galaxy UGC 3855

LYRA (THE LYRE)

Visible: Northern Hemisphere

Best Seen: Summer

Main Stars: 5

Brightest Star: Vega, 0.03 mag.

Deep-Sky Objects: M56, M57

STAR STORIES

An appealing tale of the star Vega comes from Chinese lore, dating back to the Zhou dynasty in the sixth century B.C. Telling of a romantic relationship between the beautiful daughter of the sun god—the weaving princess (Vega)—and a poor herdsman (Altair), one version of the story says that the two starstruck lovers neglected their heavenly duties. As punishment they were condemned to remain on opposite sides of the great celestial river in the sky (the Milky Way), never to cross.

According to Greek mythology, Lyra represents a tiny lyre, or harp, belonging to Orpheus, the gifted music maker capable of taming wild beasts and charming Olympian gods. Resembling a small equilateral triangle latched onto a parallelogram, Lyra is one of the smaller yet more easily recognized classical stellar patterns in the sky. It appears wedged between two prominent constellations, Cygnus and Hercules. But what makes it particularly easy to spot is its brilliant star Vega, part of the famous Summer Triangle asterism. Vega forms one of the corners, with Deneb and Altair the others, making this giant triangle a handy reference point to three separate constellations in the summer. One of the brightest stars visible across the Northern Hemisphere, Vega is the fifth brightest star in the sky and is relatively nearby at only 25 light-years away. Three times the size of our own sun and producing 50 times as much light, blue-white Vega may be young and newly formed, only about 400 million years old.

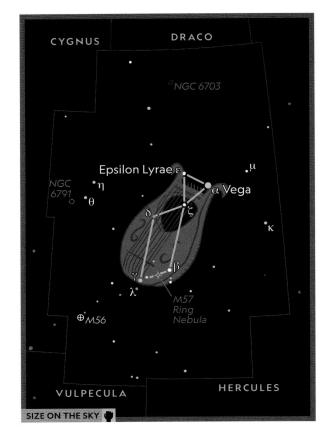

SIZE ON THE SKY

Those with keen eyesight will notice that a star near Vega is actually an eye-catching pair of stars known as Epsilon Lyrae. Orbiting each other around their common center of gravity, they lie 175 light-years from Earth. Seen through a small telescope, these celestial twins will surprisingly reveal that both stars are themselves double, forming a spectacular quadruple star system known popularly as the Double Double.

Lyra lies on the edge of the Milky Way band and so is rich with deep-sky treasures. The Ring Nebula, or M57, is 2,300 light-years away and visible with at least an 80-mm (3-in) aperture telescope, appearing as a faint doughnut-shaped cloud. This famous planetary nebula is formed by gas emitted by a dying star. The nebula looks circular from our perspective, but scientists have concluded that it is actually cylindrical. Large backyard scopes may reveal a faint star, a white dwarf, in the center of this celestial smoke ring.

M56 is an eighth-magnitude globular cluster lying halfway between Gamma Lyrae and Albireo in Cygnus. Considered one of the fainter Messier globulars, it is 33,000 light-years distant and so small that telescopes will reveal it only as a fairly bright, unresolved fuzzy ball.

Globular cluster M56

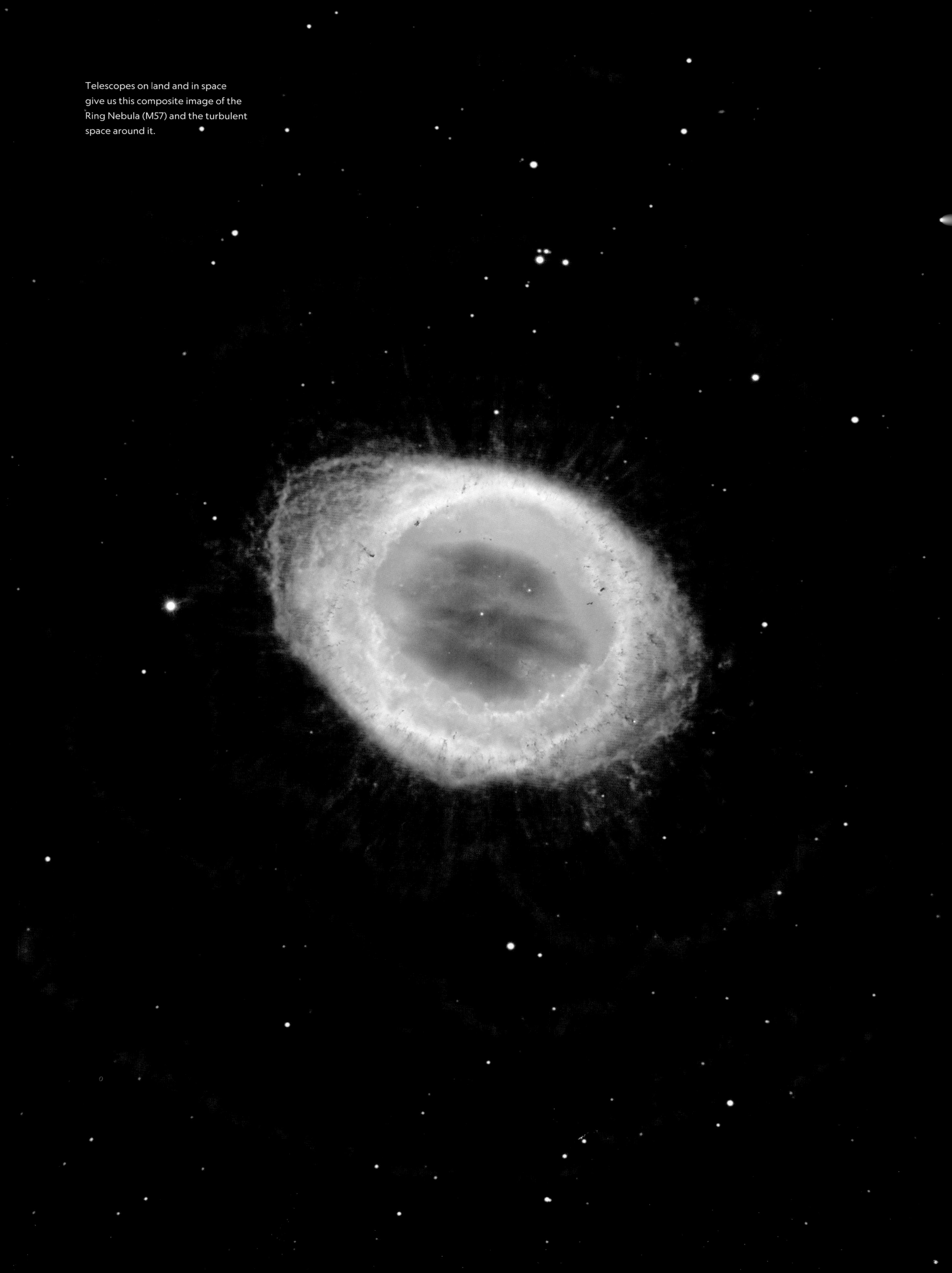

Telescopes on land and in space give us this composite image of the Ring Nebula (M57) and the turbulent space around it.

MENSA (THE TABLE MOUNTAIN)

Visible: Southern Hemisphere

Best Seen: All year

Main Stars: 4

Brightest Star: Alpha Mensae, 5.09 mag.

Deep-Sky Objects: NGC 1987, IC 2051

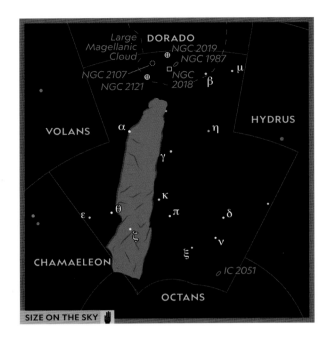

One of the faintest constellations in the sky, Mensa holds no bright stars: Even its lead member, Alpha Mensae, is barely visible to the naked eye under dark skies. The constellation's name literally means "table," and the stellar shape was created in the 18th century to honor the famous table mountain that dominates the skyline of Cape Town in South Africa. It is in a fairly barren part of the southern sky bordered by Hydrus, Dorado, and Volans; as a circumpolar constellation it is visible across the entire Southern Hemisphere.

Part of the Large Magellanic Cloud from neighboring Dorado crosses into Mensa, making it a great guidepost for finding the constellation. Along the edge of the satellite galaxy is a mysterious star cluster that exhibits both open and globular traits. This object, NGC 1987, appears to be a very loose cluster with some concentration at its center, and it is filled with dying red-giant stars estimated to be no more than 600 million years old.

MICROSCOPIUM (THE MICROSCOPE)

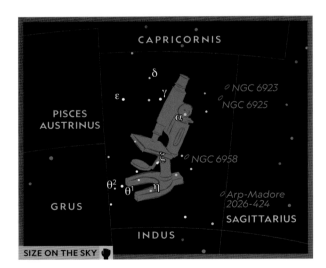

Visible: Southern Hemisphere

Best Seen: Spring

Main Stars: 5

Brightest Star: Gamma Microscopii, 4.67 mag.

Deep-Sky Object: Arp-Madore 2026-424

Named in the 18th century in honor of the revolutionary instrument used to explore the microcosm, Microscopium is a very faint southern constellation. Its dim fifth-magnitude stars make finding it with the unaided eyes from city limits all but impossible, and even under dark skies it's a bit challenging. However, the constellation lies directly south of Capricornus and east of Sagittarius, so it is just barely visible to observers in the tropical Northern Hemisphere.

Microscopium's brightest star is Gamma Microscopii, representing the instrument's eyepiece. It is a yellow giant 2.5 times the mass of our sun, located 223 light-years from Earth. Astronomers estimate that the star may have glided as close as 1.14 light-years past the sun 3.9 million years ago, gravitationally affecting objects in the outer solar system. Alpha Microscopii, an older yellow giant star, is one part of a nice optical double star system easily separated with a 100-mm (4-in) aperture telescope.

MONOCEROS (THE UNICORN)

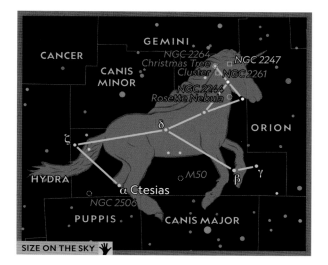

SIZE ON THE SKY ✋

Visible: Northern and Southern Hemispheres

Best Seen: Northern winter and southern summer

Main Stars: 4

Brightest Star: Beta Monocerotis, 3.74 mag.

Deep-Sky Objects: M50, Rosette Nebula, NGC 2264, NGC 2261

Straddling the celestial equator is the mythical figure of Monoceros, the unicorn. Depending on the artist, most of its stellar pattern lies within the triangle of bright stars Sirius, Betelgeuse, and Procyon, with its magical horn lying closest to Orion's shoulder star. While the constellation is considered modern, having appeared on sky globes by Dutch astronomer Petrus Plancius only in 1613, legends of horned steeds and their healing powers are ancient.

While the unicorn lacks any truly bright stars, it is home to a great triple star system perfect for exploring with a small telescope. Beta Monocerotis has components that shine at magnitudes 4.7, 5.2, and 6.1 and form a beautiful blue-white stellar triangle 690 light-years from Earth. It was discovered in 1781 by British astronomer Sir William Herschel, who described the triplet system as "one of the most beautiful sights in the heavens."

A portion of the Milky Way band appears to run through Monoceros, offering skywatchers a great variety of deep-sky targets, from clusters to nebulae. The cluster M50 is one of the brightest in this region at sixth magnitude. Containing more than 80 stars, it is located on the southern border with Canis Major. The Rosette Nebula is a stunning, shell-like emission nebula more than 100 light-years across. Shining at sixth magnitude, this 4,900-light-years-distant nebula has a young star cluster nestled at its center that offers eye-catching views when using telescopes with wide field and low power. A similar combination of a star cluster with a nebula can be found with NGC 2264, also known as the Christmas Tree Cluster. This bright, 2,400-light-years-distant open cluster resembles its namesake holiday tree and is associated with a dark conical nebula appropriately named the Cone Nebula. Located just off the tip of the "Christmas tree," it looks amazing in photographs but is a challenge to observe. And just south of the Cone Nebula lies another enigmatic ghostly glow, Hubble's Variable Nebula, or NGC 2261. Visible in larger telescopes, it's a triangular gas cloud that regularly changes brightness and shape thanks to a variable star embedded inside the nebula.

Monoceros also happens to be the home of one of the nearest recorded black holes to Earth. Located about 3,300 light-years away, A0620-00 is a binary star system consisting of a main sequence star and black hole companion estimated to weigh 6.6 times the mass of our sun. The cosmic predator, no bigger than Earth, gravitationally pulls gaseous material off its neighboring star and devours it on a regular basis. In the process it emits large amounts of detectable visible light and x-rays.

STAR STORIES

On June 15, 2018, a signal containing a message of peace and hope was beamed from a radio antenna in Spain to the black hole A0620-00 in memory of the late astrophysicist Stephen Hawking, who had died three months before. Once it arrives in the year 5475, the signal may become the first human interaction with a black hole.

Cone Nebula (left) and Rosette Nebula (right)

MUSCA (THE FLY)

Visible: Southern Hemisphere

Best Seen: All year

Main Stars: 6

Brightest Star: Alpha Muscae, 2.70 mag.

Deep-Sky Objects: Coalsack Nebula, Hourglass Nebula

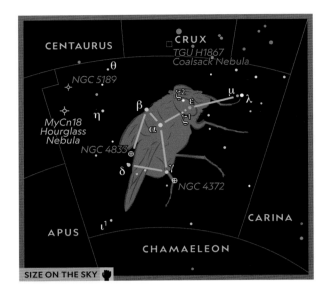

SIZE ON THE SKY

Musca is a small but distinct constellation in the far southern sky, and it is easily found thanks to being embedded in the rich jewel-studded Milky Way band, east of Carina and south of Crux, the Southern Cross. Originally the constellation was named Apis, "the bee," back in 1603, but more than a century later it was dubbed Musca Australis, meaning "southern fly." This made sense when there was a northern counterpart, Musca Borealis, which was formed out of the stars on the back of Aries (The Ram). When the Northern Fly fell out of favor and was removed from star maps, Musca Australis dropped the southern suffix altogether.

Musca's tiny stellar pattern crudely resembles the Little Dipper found in Ursa Minor, complete with a bowl and handle. Alpha Muscae is the brightest member of the constellation and lies just off Acrux, the brightest star in neighboring Crux. A third-magnitude subgiant star, it sits 310 light-years from Earth and has a diameter five times larger than that of our sun. Beta Muscae is a bright and very close pair of white stars 341 light-years away: It offers a great observing challenge for small telescopes. The fly's tail is marked by Gamma Muscae, a pulsating blue-white variable star that swings in brightness between magnitude 3.84 and 3.86 over 2.7 days. Delta and Epsilon Muscae mark the fly's left and right wings (from the viewer's perspective).

The famous Coalsack Nebula, the most prominent dark nebula in the sky, mainly resides in Crux and a bit in Centaurus, but it also reaches into the Musca constellation's boundary. The faint globular cluster NGC 4833 near Delta Muscae sits 21,200 light-years away. Another globular cluster near Gamma Muscae, NGC 4372, is somewhat closer at 18,900 light-years away, but our telescope views appear dimmed because it lies partially hidden behind the Milky Way's cosmic dust clouds. In Earth's skies it appears to have an angular size equal to two-thirds the size of the moon's disk. Astronomers have examined the chemical fingerprints of its stars and found they contain little metallicity, indicating the cluster may be one of the oldest known at 12.5 billion years of age. Located just over two degrees east of Eta Muscae is a faint 12th-magnitude planetary nebula made famous by the Hubble Space Telescope. Known as the Hourglass Nebula, this expanding cloud of gas and dust some 8,000 light-years from Earth cuts a distinctive form against the dark backdrop of space. The main eye-catching structure of this nebula consists of multiple vividly colored, overlapping shells of gas. They are the result of a dying sunlike star expelling its outer atmosphere into space.

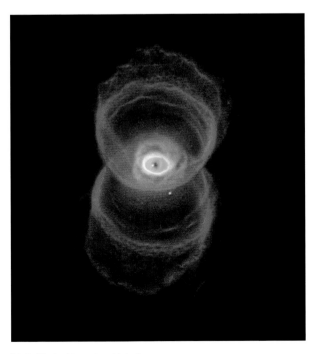

MyCn18: the Hourglass Nebula

NORMA (THE CARPENTER'S SQUARE)

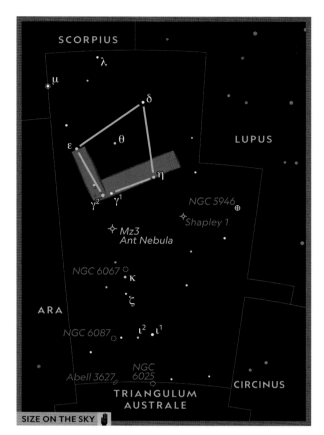

SIZE ON THE SKY ✋

Visible: Southern Hemisphere

Best Seen: Spring

Main Stars: 4

Brightest Star: Gamma² Normae, 4.02 mag.

Deep-Sky Objects: Ant Nebula, NGC 6087, NGC 6067, Shapley 1, Abell 3627

Norma is a small and faint southern constellation embedded within the thick star fields of the Milky Way. It lies east of Lupus and southwest of the prominent stinger star pattern of Scorpius. The four brightest stars of Norma—Gamma, Delta, Epsilon, and Eta—form a celestial square pattern, but an observer will need fairly dark skies away from light pollution to track it down, as Norma ranks 74th out of 88 constellations in size.

Astronomer Nicholas-Louis de Lacaille in 1751 carved Norma (celebrating the carpenter's tool used during early ocean voyages) out of undesignated stars found in the Lupus and Scorpius constellations. As a result it has only one star close to the magnitude 4 mark, Gamma² Normae (the exponent in the name refers to the double star nature of this system). High magnification will reveal it as an optical double, with the brighter star a yellow giant 10 times the diameter of our sun and located 129 light-years away. Its 10th-magnitude companion is just an optical illusion, as it sits at a much greater distance of 1,500 light-years. A small telescope will resolve the true double star Epsilon Normae, at 530 light-years away, with its eye-catching contrasting yellow and blue colors. Just visible to the naked eye at about

4.6 magnitude is Mu Normae, a blue supergiant star that ranks as one of the most luminous stars in the Milky Way galaxy. Astronomers suspect that it could be nearly a half million times more luminous than our sun and 40 times its mass.

Straddling the Milky Way ensures that Norma, despite its small size, is chock-full of deep-sky targets to explore. NGC 6067 and NGC 6087 are sixth- and fifth-magnitude open clusters perfect for binoculars. Better known as the Fine Ring Nebula, Shapley 1 is a 12th-magnitude planetary nebula 4,900 light-years away that looks like a round blue bubble. Astronomers believe that it was formed less than 9,000 years ago. Located about 200 million light-years away is a treasure trove of galaxies called the Norma Cluster, or Abell 3627, thought to be one of the most massive collections of galaxies in the universe.

In 2008, the Hubble Space Telescope revealed the intricate beauty of the Ant Nebula and showcased why it resembles its insect namesake. Located some 8,000 light-years from Earth, this planetary nebula displays tenuous columns and rays of expanding gas reaching out of two interconnected lobes.

Mz3: the Ant Nebula

OCTANS (THE OCTANT)

Visible: Southern Hemisphere

Best Seen: All year

Main Stars: 3

Brightest Star: Nu Octantis, 3.73 mag.

Deep-Sky Objects: NGC 2573, NGC 7098, Melotte 227, Collinder 411

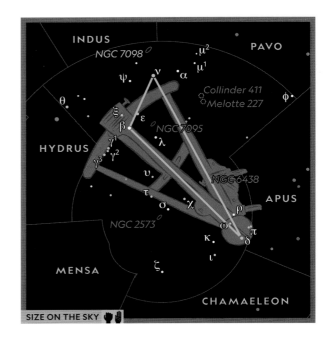

Octans is a small constellation, devoid of bright stars, that straddles the South Celestial Pole. It is therefore circumpolar and visible throughout the year in the Southern Hemisphere. The constellation honors the octant, an important navigational invention and a forerunner to the sextant that was used to determine the position of stars during ocean voyages. While the South Celestial Pole doesn't have a bright star marking its position, such as the Northern Hemisphere's Polaris, the closest visible star to the pole belongs to this constellation—the orange 5.5-magnitude Sigma Octantis. It sits just over one degree from the southern pole, but because it is so faint to the naked eye it is not a practical marker for navigation. Its official name, Polaris Australis, was given in the 1700s, and it holds the distinction of being the southernmost star that is named. (Incidentally, another, much fainter star—BQ Octantis at magnitude 6.8—is located even closer to the southern pole.) As an easier alternative, observers can use the stars of the Southern Cross in the constellation Crux to draw a line to the South Celestial Pole.

The brightest member of this constellation is Nu Octantis, a 3.7-magnitude giant star that sits 643 light-years from Earth. It is actually a binary system with its primary star being an orange giant and its companion a dwarf. Octans lacks major deep-sky targets, but skywatchers still get a star cluster and a couple of faint galaxies to tour. The closest NGC-class galactic object to the southern pole that is within reach of backyard telescopes is NGC 2573, also dubbed Polarissima Australis. It is a barred spiral galaxy that lies 117 million light-years from our solar system. Only 40 arc minutes from the South Celestial Pole, it shines with a feeble 13th magnitude. As a result, it's a real observing challenge, being just barely visible with 200-mm (8-in) aperture telescopes under dark skies using averted vision. Under high magnification in larger instruments with at least a 300-mm (12-in) aperture, it appears as an extremely faint, diffuse, and elongated halo with a starlike core. Another galaxy that's a bit brighter at 11th magnitude and closer to us at 95 million light-years is NGC 7098, a bizarre double-barred spiral that stretches 152,000 light-years across.

Two lonely star clusters call Octans home: Melotte 227 and Collinder 411. Melotte 227 is the brighter one at 5.3, but it is very loose and is scattered well over one degree across, making it difficult to identify as a distinct cluster. To do so, it's best to use lowest-power views with a telescope.

Spiral galaxy NGC 7098

OPHIUCHUS (THE SERPENT BEARER)

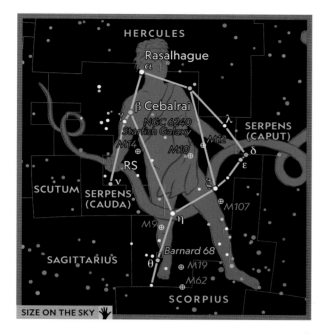

SIZE ON THE SKY

Visible: Northern and Southern Hemispheres

Best Seen: Northern summer and southern winter

Main Stars: 10

Brightest Star: Rasalhague, 2.08 mag.

Deep-Sky Objects: M9, M10, M12, M14, M19, M62, M107, Starfish Galaxy, Barnard 68

A gigantic constellation that crosses the celestial equator, Ophiuchus is wedged between fellow major stellar figures Hercules to its north and Scorpius to the south. It is the 11th largest constellation. The ecliptic appears to run through the southern portion of the constellation.

Ophiuchus and its neighboring constellation Serpens were often depicted in the past as one constellation, but in the 20th century they were officially split. However, their mythologies remain as intertwined as their stars. In ancient times Ophiuchus was seen as the serpent holder, and in both ancient Greek and Roman legends Ophiuchus is thought to honor Asclepius, the god of medicine. The snake wrapped around Asclepius's staff taught the healer about the properties of plants, and he soon became skillful enough to raise the dead. Hades, god of the dead, became concerned and convinced Zeus to kill Asclepius. Now the god, known as Ophiuchus, and his serpent watch over us from the sky.

The summer Milky Way band runs through Ophiuchus and so the region is filled with 22 globular clusters, with the brightest—M9, M10, M12, M14, M19, M62, and M107—visible through binoculars. A most unusual object, Barnard 68, is a dark molecular cloud that sits at a distance of only 410 light-years. It is so close to our solar system that there are no stars between us and its half-light-year-wide inky form. Astronomers have found that the nebula is actually vibrating, an indication that a supernova shock wave may have slammed into it in the recent past.

Binoculars and telescopes offer great tours of one of the closest star-forming regions to our solar system, the Rho Ophiuchi cloud complex. Located 460 light-years away, it is a collection of many dark and emission nebulae with filamentous and colorful structures. The entire complex stretches about six degrees across the sky—an area equal to 12 full moons! Finally, the Starfish Galaxy, or NGC 6240, is a strange object that showcases the wounds of a titanic multigalactic pile-up. Located 400 million light-years away, the 12th-magnitude object we see today is the result of a collision between three galaxies that has resulted in a monster, a butterfly-shaped galaxy with two visible cores. Each of these cores hides a supermassive black hole, with the two separated by only 3,000 light-years.

TELESCOPE TARGETS

The most famous of the stars in Ophiuchus is a ninth-magnitude red dwarf known as Barnard's Star. The second closest star to our solar system, it is visible through binoculars and appears to move very quickly relative to background stars, traveling about the width of the full moon every 200 years.

Dark absorption nebula Barnard 68

ORION (THE HUNTER)

Visible: Northern and Southern Hemispheres

Best Seen: Northern winter and southern summer

Main Stars: 7

Brightest Star: Rigel, 0.12 mag.

Deep-Sky Objects: M42, M43, M78, NGC 1999, Horsehead Nebula, NGC 2174, Barnard's Loop, Flame Nebula, IC 2118

STAR STORIES

The seven bright stars that form Orion's distinctive figure in the sky are known to cultures around the world. In India the constellation was known as Mriga the deer. In Hungarian folklore it took the form of the Reaper, while the Ojibwe of North America called it the Winter Maker.

Barnard 33: the Horsehead Nebula

Visible across the world, Orion is one of the most identifiable as well as one of the oldest constellation figures, crossing cultures and thousands of years. Orion straddles the celestial equator, so it is well known to observers in both the Northern and Southern Hemispheres and holds the record for containing the most bright stars in one stellar pattern. Orion has his trusted canine companions—Canis Major and Canis Minor—with him, both following him near his feet.

Many stories are attached to Orion, a great hunter in Greco-Roman myths that date back to the seventh and eighth century B.C. The most famous legend says that he was stung by a scorpion (Scorpius) in an epic battle, which is why the two figures have been placed in opposite parts of the sky. Orion boasts two first-magnitude stars, with Betelgeuse marking the shoulder to the viewer's left and Rigel his foot on the right. Betelgeuse is a monster-size red supergiant star located about 548 light-years from Earth. Its diameter is at least 400 times larger than that of our sun. It is a variable that pulsates, irregularly changing its brightness over several years. Sitting between Rigel and Betelgeuse is Orion's stellar line of three stars, Alnitak, Alnilam, and Mintaka, marking the mythical hunter's belt. The trio are not physically linked but are a distinct asterism that also acts as a great guidepost for navigating to nearby constellations.

Orion includes an area of the Milky Way that features intense star production. Beneath the hunter's belt, in the middle of three stars that form Orion's sword, is the Orion Nebula (M42). Visible to the naked eye as a fuzzy, faint patch, backyard telescopes will begin to show delicate wreathlike structures in this 1,300-light-years-distant gas cloud, where star formation is taking place at a furious pace. Focus on the bright core of the emission nebula using high magnification and four newborn stars become evident. Known as the Trapezium, their strong winds of radiation light up the Orion Nebula like a neon sign. Hanging down from the belt star Alnitak is the famous dark Horsehead Nebula with the Flame Nebula nearby. Meanwhile, the sky's brightest reflection nebula, M78, lies just northwest of Alnitak. This 1,300-light-years-distant patch of cold dust, lit up by a nearby star, is easily glimpsed through binoculars. Finally, surrounding a great portion of the Orion region is Barnard's Loop, a titanic arc-shaped nebula.

PAVO (THE PEACOCK)

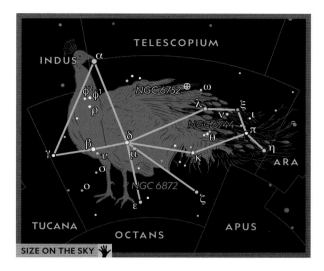

SIZE ON THE SKY 🖐

Visible: Southern Hemisphere

Best Seen: All year

Main Stars: 7

Brightest Star: Peacock, 1.91 mag.

Deep-Sky Objects: NGC 6752, NGC 6744, NGC 6872

Pavo is considered a modern constellation, having been officially listed in the 17th century, but it honors the classical Greek myths related to the builder of the ship Argus, the name of the constellation in ancient times. As one legend goes, the goddess Hera changed this builder into a peacock and placed him in the sky. To Australian Aboriginal people, the stars of Pavo represented one of two flying foxes in combination with the stars of the Ara constellation. In western astronomy, Pavo is one of five celestial birds flocking across the southern skies, along with Apus, Grus, Phoenix, and Tucana.

A fairly large constellation, Pavo is circumpolar and is found next to Apus and Telescopium and just north of Octans. The constellation's brightest stellar member, Alpha Pavonis, is also called the Peacock. It appears as a moderately bright blue-white star, but is actually a very tight binary star about 180 light-years distant. The two stars of Alpha Pavonis are separated by just half the distance between Mercury and our sun in our solar system. The celestial bird's fancy tail feathers are drawn out by the stars Beta Pavonis and Gamma Pavonis. Six stars in Pavo are known to host planetary systems, one of which, HD 172555, is thought to have been a scene of a titanic collision between two planets, one the size of Earth's moon and the other the size of Mercury, only a few thousand years ago.

Pavo hosts the third brightest globular cluster in the sky, after Omega Centauri and 47 Tucanae. NGC 6752 is estimated to be about 100 light-years across and looks impressive through backyard telescopes, with its dense core shining at fifth magnitude. Located less than two degrees east of fifth-magnitude star Omega Pavonis, this stunning cluster is home to over 100,000 stars lying relatively close to Earth compared to other globulars, at only 13,000 light-years away. With a slight pan of a telescope only three degrees south you will see a beautiful spiral galaxy that is thought to be the most Milky Way–like in structure of all nearby galaxies.

Making it even more similar to our home galaxy, NGC 6744 also hosts its own satellite companion galaxies, much like our own Milky Way's Magellanic Clouds. High-magnified views of NGC 6744 show the ninth-magnitude galaxy as a small, bright core surrounded by an extended halo that appears elongated. Larger scopes show the 30-million-light-years-distant object has hints of fluffy spiral arms that wrap around its core. In 2005, astronomers spotted a supernova explosion there that brightened up to 16th magnitude for a few nights.

Globular cluster NGC 6752

PEGASUS (THE WINGED HORSE)

Visible: Northern and Southern Hemispheres

Best Seen: Northern autumn and southern spring

Main Stars: 9

Brightest Star: Enif, 2.38 mag.

Deep-Sky Objects: M15, NGC 7331, NGC 7742, Stephan's Quintet

Spiral galaxy NGC 7331

TELESCOPE TARGETS

In 1995, astronomers discovered the first exoplanet found around a stable star. This hot Jupiter gas giant was orbiting close to its sunlike star, 51 Pegasi. Since then, 11 other stars in Pegasus are now known to host worlds. These include HD 209458b, which provided the first evidence of water vapor in an exoplanet atmosphere, and the world's first directly imaged exoplanets, captured orbiting HR 8799.

A mythical winged steed, Pegasus is an impressively large northern constellation that is the dominant stellar figure in the autumn skies. The pattern of a magical horse with wings dates back to Babylonian times, but it was made famous through Greek legends, which told of how Pegasus was born from the blood of the snake-headed Medusa after Perseus vanquished her. In another tale, Athena gave the Greek hero Bellerophon a special bridle in order to tame the flying horse and the pair went on to fight the Chimera monster. Pegasus's final act of service was to carry Zeus's lightning bolts, after which the horse was rewarded with a place in the sky.

The mythical winged steed is found in the southern sky right by Andromeda, with whom Pegasus shares the star Alpheratz (Alpha Andromedae). This shared star joins with the horse's three brightest stars to form the Great Square of Pegasus, an especially useful asterism that serves as a signpost for star-hopping to constellations in the regions close by. The constellation's brightest star is Enif, which means "the mane" in Arabic. It is an orange supergiant lying some 670 light-years from Earth. The second brightest is 2.4-magnitude Beta Pegasi, or Scheat, which means "shoulder," a dying red giant some 670 light-years away. Two variable stars can be found in the Great Square asterism. Phi and Psi Pegasi are red giants that are both faintly visible to the naked eye under dark skies. The stars Zeta, Xi, Rho, and Sigma Pegasi all delineate the equine's neck, while Theta Pegasi marks the horse's eye.

The M15 globular cluster is one of the brightest and most impressive visible in the Northern Hemisphere, and it is located on the far eastern border of the constellation near Enif. The 32,000-light-years-distant cluster is easy to spot even with binoculars and represents a pesky fly just off the tip of the steed's nose. You need a telescope to see the main attraction of this compact cluster—some 100,000 stars in a tight, sparkling ball. Astronomers have discovered that the cluster belches out copious amounts of x-rays, indicating that a black hole may be lurking at its core. Point a small telescope at the northwest corner of Pegasus, and you'll find the spiral galaxy NGC 7331. Similar to our Milky Way, this spiral is tilted at a steep angle from our line of sight. Despite being some 38 million light-years away, views even with 150- to 200-mm (6- to 8-in) aperture telescopes offer hints of the galaxy's cloudy spiral arms. Stephan's Quintet is an eye-catching cluster of five galaxies visually close to one another near Pegasus's northern border, lying an average distance of at least 300 million light-years away. Four are in the process of colliding, while the fifth may be a foreground object closer to the Milky Way.

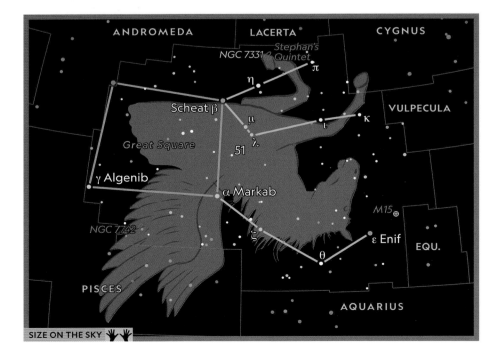

PERSEUS (THE HERO)

Visible: Northern and Southern Hemispheres

Best Seen: Northern autumn and southern spring

Main Stars: 19

Brightest Star: Mirfak, 1.79 mag.

Deep-Sky Objects: Perseus Double Cluster, M34, M76, California Nebula, NGC 1333, NGC 1260

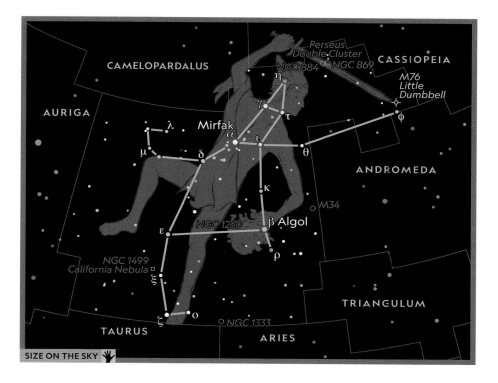

Considered one of the landmark autumn constellations in the Northern Hemisphere, Perseus is intertwined with neighboring star patterns made famous in ancient Greek legends. According to Greek myths, Perseus was the half-mortal son of Zeus who slew the snake-haired monster Medusa. The hero used his polished shield as a mirror to avoid her direct gaze, which could turn any creature into stone. When Perseus lopped off Medusa's head, the winged horse Pegasus was born of her blood. In his second great feat, Perseus used Medusa's head as a weapon to help him rescue Andromeda from the sea monster Cetus.

One of the constellation's naked-eye stars, called Algol, represents the eye of Medusa. Its name means "the demon's head" in Arabic; in Hebrew, the star is known as "Satan's head." Shining as the second brightest star in Perseus, Algol (Beta Persei) is a well-known eclipsing variable that dims by a full magnitude from magnitude 2.1 to 3.4 for about 10 hours every three days. The constellation itself is fairly easy to locate, with its bright stars forming a V-shaped pattern and its brightest six all magnitude

M76: the Little Dumbbell Nebula

3 or greater. Its brightest star is Mirfak, meaning "elbow" in Arabic, and is a blue-white supergiant located 510 light-years from Earth.

Skywatchers will find that the lavish star fields of the Milky Way band cross Perseus, which is appropriately bookmarked between Cassiopeia and Andromeda. Perseus contains a fascinating deep-sky object, the famous Perseus Double Cluster (NGC 869 and 884), which was even seen by ancient Greeks, being fairly bright at 4.4 and 4.7 magnitudes, respectively. Known as Perseus's sword-handle, the double cluster presents the illusion of a large, connected field but is in fact two separate, unassociated swarms of stars. Each cluster spans about half a degree in the sky—equal to about the size of the full moon disk. Best views are with binoculars or through small telescopes that will provide a wide field of view for this area, rich in double and multiple stars.

Located 2,500 light-years from Earth, the Little Dumbbell Nebula (M76) takes its name from its bigger, brighter planetary nebula cousin. This magnitude-10 fuzzy spot is fairly easy to hunt down with small telescopes, since it lies near the border with Andromeda and just south of Cassiopeia. While visible to the naked eye as a very faint fuzzy patch, binoculars begin to show the California Nebula as a faint rectangular object. Using nebula filters will bring out filamentous details in this 1,000-light-years-distant emission nebula.

TELESCOPE TARGETS

Perseus is the namesake constellation from which the annual Perseid meteor shower appears to radiate. Like clockwork, the shower is visible every summer starting in late July, peaking between August 11 and 13. The meteors are debris shed by Comet Swift-Tuttle, first spotted in 1862 and with an orbit of 133 years. It last passed by Earth in 1992.

PHOENIX (THE PHOENIX)

Visible: Southern Hemisphere

Best Seen: Spring

Main Stars: 4

Brightest Star: Ankaa, 2.40 mag.

Deep-Sky Objects: NGC 625, Robert's Quartet, Phoenix Cluster

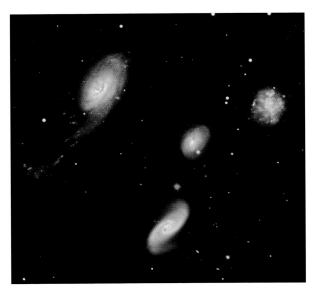

(Left to right) NGC 92, NGC 89, NGC 88, and NGC 87: the Robert's Quartet galaxies

Phoenix is a diminutive constellation visible throughout the Southern Hemisphere and low in the sky north of the Equator, though entirely invisible north of 40 degrees in latitude. It represents the legendary bird that is reborn from its own fiery ashes every 500 years or more. It is a modern constellation officially recognized on sky maps in 1603, but its stars were important to various cultures around the world. To the ancient Egyptians the phoenix was closely tied to their sun god and was a symbol of immortality and of sun and fire worship. Ancient Chinese astronomers also saw a bird in this group of stars, calling it Ho-Niao after a bird in Chinese lore, Gu Huo Niao. But to Arabs this small stellar shape was Al Zaurak, meaning "the boat," named after the small vessels that used to sail the Euphrates River.

Containing only three stars brighter than magnitude 4, Phoenix is nestled south of the constellations Fornax and Sculptor, north of Tucana, and west of Eridanus. Use the bright star Achernar in Eridanus as an easy guide to finding the tail of Phoenix, about 10 degrees to its northwest—a separation equal to the width of a person's fist held at arm's length. The brightest member of Phoenix is named Ankaa, Arabic for "the phoenix," but an appropriate alternative moniker for this star is Nair al-Zaurak, which means "the bright one of the boat." Ankaa is a dying yellow-orange giant that is on its way to becoming a full-blown red giant. It lies 77 light-years from our solar system, and it has already ballooned to be 13 times wider than our sun. Astronomers have discovered that Ankaa has a mysterious, smaller stellar companion that orbits seven times farther from its primary partner than Earth does from the sun, so that it takes 10 years to make one orbit.

Phoenix holds the record for hosting what might be the oldest star yet discovered, located some 36,000 light-years away in our own Milky Way galaxy. Known only as HE0107-5240, its poor metal content (1/200,000 that of the sun) indicates that it may be in the second generation of stars born after the big bang more than 13 billion years ago. While appearing faint, Phoenix does host many galaxies, with 11th-magnitude NGC 625 being the brightest. This barred spiral lies about 11 million light-years distant and appears in larger backyard telescopes to have a distinctly mottled core surrounded by a fainter oval halo. Robert's Quartet is a tight little group of faint colliding galaxies 160 million light-years away. This constellation is also home to its namesake galaxy cluster, located some 5.7 billion light-years from Earth. It is the most powerful producer of x-rays ever seen in a galaxy cluster.

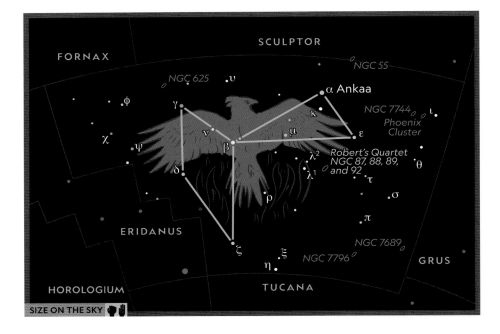

PICTOR (THE PAINTER'S EASEL)

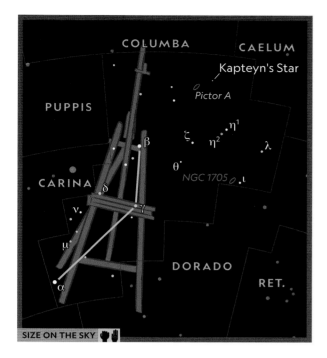

SIZE ON THE SKY

Visible: Southern Hemisphere

Best Seen: Autumn

Main Stars: 3

Brightest Star: Alpha Pictoris, 3.30 mag.

Deep-Sky Objects: NGC 1705, Pictor A

The faint constellation Pictor is a modern invention first published in a sky catalog in 1763. Surrounding brighter constellations that help track the dim stars of the painter's easel include Carina and Dorado; the Large Magellanic Cloud lies just south of Pictor. Only six degrees southeast of Pictor's three brightest stars, superbright stellar beacon Canopus also makes for a great guidepost.

The star Alpha Pictoris is the constellation's brightest member and lies 97 light-years away, but it's the second brightest star, 3.8-magnitude Beta Pictoris, that has gained fame. Located 63 light-years from Earth, it is younger, more massive, and brighter than our sun and holds court to a wide disk of dust with two exoplanets embedded within it. Both of these alien worlds have been confirmed with direct imagery. The constellation also has five other stars hosting their own planets. One of these, the 42-light-years-distant orange dwarf HD 40307, may have six super-Earth-size planets, at least one of which is believed by some astronomers to be orbiting within the star's habitable zone.

The ninth-magnitude Kapteyn's Star is part of the elite club of nearest stars to Earth, at just 12.7 light-years distant. In 2014, two super-Earths were discovered orbiting this red dwarf star. But its pedigree appears intriguing, too—when we trace the star's motion through the Milky Way, it appears to be traveling with a renegade pack of stars in the oppo-site direction from our galaxy's other stars. This has led astronomers to believe that Kapteyn's Star may have originated in an ancient dwarf galaxy that was gravitationally sucked in and cannibalized by the Milky Way in the distant past.

One of the most active galactic star factories known in the local universe is found only 17 million light-years away within Pictor. NGC 1705 is a 12th-magnitude elliptical galaxy located under one degree east of Iota Pictoris—a separation less than the width of two full moon disks. Observations have shown most of the star formation is happening in its core. In a structure just 500 light-years across, there may have been over a half million sun-size stars born in the last 10 million years. Pictor A is another extreme galaxy in the constellation. Classified as a double-lobed galaxy 485 million light-years away, it has a supermassive black hole no bigger than our solar system hiding within its core. The black hole powers a jet of particles that is visible as a strong x-ray source.

Pictor A, a galaxy whose spectacular jet emanates from a black hole at its center

PISCES (THE FISHES)

Visible: Northern Hemisphere

Best Seen: Autumn

Main Stars: 18

Brightest Star: Eta Piscium, 3.61 mag.

Deep-Sky Objects: M74, NGC 488, NGC 520

Two merging galaxies known as NGC 520, or the Flying Ghost Galaxy

TELESCOPE TARGETS

The westernmost star in the Circlet asterism, dubbed TX Piscium, is most famous for being a highly irregular variable. It is known as the reddest star visible, and it varies erratically in brightness from magnitude 4.5 to 6.2. This puts this 900-light-years-distant star within range of binoculars, and even visible to the naked eye from a dark site when at its peak brightness.

Pisces is one of the 12 zodiacal constellations, and according to the ancient Greeks it takes the shape of two fish bound at their tails by a long cord. Their stellar patterns together form a giant V-shape in the autumn sky, with one fish facing west, tucked underneath Pegasus and its Great Square asterism, while its counterpart faces north with Aries nearby. The length of cord that holds the two celestial fishes together bends at alpha star Alrescha (Arabic for "rope"), just outside the constellation Cetus. Despite its designation, this star is actually only the third brightest in Pisces at 3.94 magnitude. It lies 151 light-years from our solar system and is a binary system, with the two stars taking over 3,000 years to orbit each other. Beta Piscium, also known as Fumalsamakah, which comes from the Arabic word for "mouth of fish," shines at magnitude 4.4 and is 410 light-years away.

The ancient Greeks believed the goddess Aphrodite and her son Eros changed themselves into fish to escape the sea monster Typhon, and the cord holds mother and son together as they swim. In Roman stories, the fish are Venus and Cupid, but legends of these stellar fish date further back in history to the time of the Babylonians, to whom fish sometimes embodied sacred gods of water and wisdom. Pisces is the location of the vernal equinox (the first day of spring), where the sun is within the constellation from March 13 to April 19, and is the same location in the sky where the ecliptic crosses the celestial equator, entered on March 21.

Five faint stars in the south of the constellation form the asterism known as the Circlet, representing the head of the larger fish. Zeta Piscium is a stunning double star shining at magnitudes 5 and 6. Van Maanen's Star, at magnitude 12, is a rare white dwarf—the hot, naked core of a dead star—that is within reach of a 200-mm (8-in) telescope.

While Pisces is a large constellation, since it is not near the Milky Way it is a sparse field when it comes to bright deep-sky objects. However, M74 is a picturesque spiral galaxy about 32 million light-years away that appears face-on, with its beautiful, yet faint, spiral arms loosely coiling around its tiny, bright core. Shining at 10th magnitude, M74's structural details begin to resolve with at least 200- or 250-mm (8- or 10-in) aperture telescopes. Three supernovae explosions have been spotted in this 30-million-light-years-distant galaxy, and that's just in the 21st century. Much farther away at 90 million light-years is another 10th magnitude face-on spiral galaxy, NGC 488. Its multiple arms appear very tightly wrapped around the galaxy's nucleus. The Flying Ghost Galaxy, known officially as NGC 520, is actually a pair of colliding galaxies 105 million light-years away that began merging some 300 million years ago.

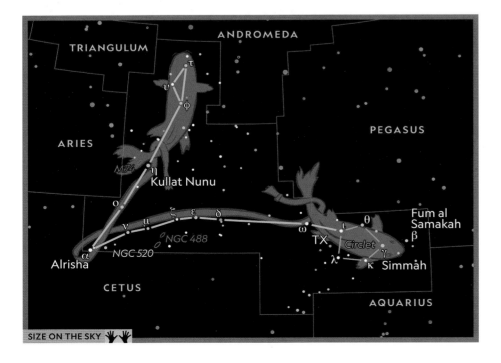

ANDROMEDA

TRIANGULUM

ARIES

PEGASUS

M74

Kullat Nunu

NGC 488

NGC 520

Alrisha

Fum al Samakah

TX

Circlet

Simmah

CETUS

AQUARIUS

SIZE ON THE SKY

PISCIS AUSTRINUS (THE SOUTHERN FISH)

Visible: Northern and Southern Hemispheres

Best Seen: Northern autumn and southern spring

Main Stars: 7

Brightest Star: Fomalhaut, 1.16 mag.

Deep-Sky Object: NGC 7314

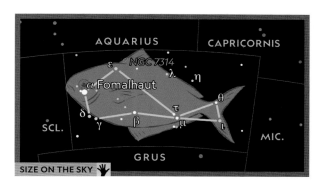

SIZE ON THE SKY

Piscis Austrinus is a small group of stars that first the Babylonians, and then the ancient Greeks, saw in the form of a fish. It is located just east of Capricornus and south of Aquarius, where the fish has been traditionally depicted drinking water from the water-bearer's jar.

The most prominent point in this part of the sky is the fish's brightest star, Fomalhaut, which marks its mouth. The 18th brightest star to the naked eye and just 25 light-years from Earth, Fomalhaut is a relatively young star no more than 300 million years old. It is twice as large as the sun and burns between 14 and 17 times as bright. Several expanding disks of dust surround the star, which the latest observations suggest may contain two super-Jupiter-size exoplanets. The second brightest star in this fishy constellation is Epsilon Piscis Austrini, which is a blue-white star shining at magnitude 4.2 and located 400 light-years from Earth. Estimates are that it is four times more massive than our sun and about 660 times brighter.

Ranked as the 11th closest star to our solar system is the red dwarf Lacaille 9352. Shining at 7.34 magnitude, it is too dim to observe with the unaided eye but is easy to spot with binoculars. In 2020, astronomers discovered two super-Earth-candidate exoplanets, and possibly a third, orbiting the red dwarf.

PUPPIS (THE STERN)

Visible: Northern and Southern Hemispheres

Best Seen: Northern winter and southern summer

Main Stars: 9

Brightest Star: Naos, 2.25 mag.

Deep-Sky Objects: M46, M47, M93, NGC 2451, NGC 2477, NGC 2438

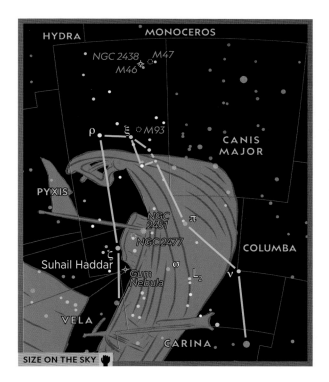

SIZE ON THE SKY

Puppis is a southern constellation between Hydra and Canis Major. It represents the stern of the one-time constellation Argo Navis, a ship divided into three parts in the mid-18th century.

Puppis contains no bright stars, but since the Milky Way band runs through this part of the sky there are plenty of clusters to explore. M47 is easy to find with the naked eye under dark skies, looking like a hazy patch, and is home to only about 50 stars. Look closely through binoculars and you'll get a cosmic two-for-one-deal with a second open cluster, M46. Photobombing the cluster is a puffball planetary nebula, NGC 2438, superimposed in front. Look just to the south and you'll find two more open clusters, Messier 93 and NGC 2477.

PYXIS (THE COMPASS)

Visible: Northern and Southern Hemispheres

Best Seen: Northern winter and southern summer

Main Stars: 3

Brightest Star: Alpha Pyxidis, 3.67 mag.

Deep-Sky Objects: NGC 2818, NGC 2627, NGC 2613

TELESCOPE TARGETS

Located about four degrees northeast of Alpha Pyxidis—less than the width of a person's middle three fingers held at arm's length—is the 3,100-light-years-distant star T Pyxidis. Astronomers have spied on this recurrent nova as it explodes in brightness every few years, skyrocketing from magnitude 15.5 to 7 in the years 1890, 1902, 1920, 1944, 1966, and 2011.

SIZE ON THE SKY

Pyxis is a southern constellation that was located near the large and ancient stellar figure known as Argo Navis, representing the ship of the Argonauts from Greek legend. In the mid-17th century the constellation was named after a nautical compass used by mariners. Pyxis is easy to track down from the Southern Hemisphere, where it is seen high in the overhead sky, while for those in the north it remains a challenging figure to observe, never wandering far from the southern horizon.

The star fields of the Milky Way run through Pyxis, making it a great region of the sky to tour with a pair of binoculars, despite it ranking as only the 65th in size out of 88 constellations. Pyxis is bordered by Vela to the south, Hydra in the north, Puppis to the west, and Antlia to its east. The mariner's compass sports three main stars appearing in a straight line. Its brightest, Alpha Pyxidis, is a blue-white giant that sits 880 light-years away. It has a light output 10,000 times brighter than our sun and is 10 times our sun's mass, meaning it is expected to go out with a bang when it dies as a supernova. In Chinese mythology Alpha Pyxidis is part of an asterism referred to as the Celestial Dog, and the star itself is referred to as the Fifth Star of Celestial Dog. The three brightest stars in Pyxis together formed part of a figure of a temple called Tianmiao, which honored an emperor's ancestors. Beta Pyxidis is the second brightest star at 3.9 magnitude and is a yellow giant that lies 420 light-years away. It has a magnitude-12 companion star that is difficult to observe even in large telescopes.

A stunning planetary nebula, NGC 2818, appears superimposed in front of a more distant open cluster in the background. Lying some 10,400 light-years from Earth, this nebula shows beautiful layers of glowing shells of gas ejected from a dying star at its center. NGC 2627 is an eighth-magnitude open cluster with about 15 stars visible through binoculars. The 78-million-light-years-distant edge-on spiral galaxy, NGC 2613, shines at magnitude 10.5 and through a small telescope appears spindle shaped. Two space telescopes, Hubble and Chandra, have revealed a dwarf galaxy dubbed Henize 2-10 in Pyxis, some 34 million light-years from our own Milky Way. It is undergoing an intense period of star formation due to a giant black hole with 300,000 times the mass of our sun hidden within its core. Because of this combination of features, some experts believe we are witnessing a young galaxy in its earliest stages of evolution.

Planetary nebula NGC 2818

RETICULUM (THE RETICLE)

Visible: Southern Hemisphere

Best Seen: All year

Main Stars: 4

Brightest Star: Alpha Reticuli, 3.33 mag.

Deep-Sky Objects: Topsy-Turvy Galaxy, NGC 1559

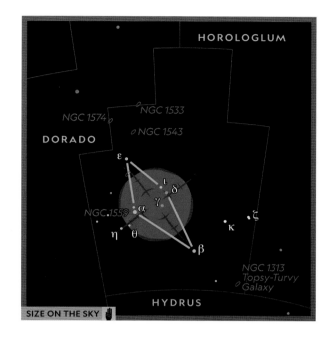

Reticulum is a faint southern circumpolar constellation invented in the mid-18th century to represent the scale or grid used in eyepieces to measure positions of stars. However, it was only recognized by the International Astronomical Union as an official constellation in 1922.

Reticulum is wedged between constellations Dorado and Horologium, north of the Large Magellanic Cloud. Only two of its stars are brighter than magnitude 5, making it best viewed from dark skies away from light pollution. Other than two faint galaxies, it doesn't have many deep-sky targets worth hunting down. Scanning the southwest part of Reticulum, look for the ninth-magnitude barred spiral dubbed the Topsy-Turvy Galaxy, cataloged officially as NGC 1313. Estimated to be about half the diameter of our own Milky Way, this lopsided galaxy in magnified views is filled with starburst activity, meaning it is showing an unusually high rate of new stars being born. Another barred spiral in Reticulum is 52-million-light-years-distant NGC 1559, which shines at a feeble 11th magnitude.

SAGITTA (THE ARROW)

Visible: Northern and Southern Hemispheres

Best Seen: Northern summer and southern winter

Main Stars: 4

Brightest Star: Gamma Sagittae, 3.51 mag.

Deep-Sky Objects: M71, NGC 6886

Nestled within an abundant part of the Milky Way band in the northern sky, the tiny, ancient constellation Sagitta represents the arrow of the archer constellation Sagittarius. Keen-eyed observers under dark skies will notice that our Milky Way's Great Rift—dark dust clouds that hide the bright stars behind them—stretches across Sagitta. The brightest star in the little constellation, Gamma Sagittae, is an aging red giant 275 light-years from Earth that has ballooned into a titanic monster with a diameter 54 times larger than our sun. Alpha Sagittae, also known as Sham, an Arabic word meaning "arrow," is a yellow giant of magnitude 4.4 sitting 475 light-years away.

While there are no bright galaxies in Sagitta, it does host some eye-catching galactic treasures. M71 is an eighth-magnitude globular cluster filled with a visually scattered group of stars. Through small telescopes it shows as a small, diffuse patch, but larger ones will help resolve some of the individual stars. IC 4997 is an 11th-magnitude planetary nebula in the southeastern corner of Sagitta.

SAGITTARIUS (THE ARCHER)

AQUILA
SCUTUM
SERPENS (CAUDA)
NGC 6822
M17 Omega Nebula
NGC 6537 Red Spider Nebula
M18
Great Sagittarius Star Cloud
M25
M23
NGC 6445 Little Gem
M21
M20
Trifid Nebula
M8 Lagoon Nebula
M22
M28
OPH.
M75
CAP.
Teapot
NGC 6528
M54
M55
M69
Alnasl
M70
Kaus Australis ε
Rukbat
CORONA AUSTRALIS
SCORPIUS
Arkab Prior
Arkab Posterior
INDUS
TELESCOPIUM
ARA

SIZE ON THE SKY

Visible: Northern and Southern Hemispheres

Best Seen: Northern summer and southern winter

Main Stars: 4

Brightest Star: Epsilon Sagittarii, 1.85 mag.

Deep-Sky Objects: Lagoon Nebula, Omega Nebula, Trifid Nebula, Red Spider Nebula, NGC 6445, NGC 6522, M54, NGC 6528

While there are no outstandingly bright stellar beacons in Sagittarius, the pattern formed by its stars is eye-catching. It appears dim but distinctive to the naked eye even from light-polluted suburban skies, and it is located at the widest band of the Milky Way, offering a window to the center of the galaxy. The eight central stars form the shape of a teapot, complete with a spout, lid, and handle, and you can think of the Milky Way as steam coming from the teapot's spout.

Bordered by zodiac neighbors Scorpius and Capricornus, Sagittarius appears in mid to late northern summer in the southern sky as he pursues his prey, Scorpius. Sagittarius, described by the ancient Greeks and Romans as a four-legged archer, aims his arrow at the heart of the scorpion directly to his west. When a poisoned arrow accidentally wounded this immortal half-man, half-horse archer, a son of Saturn, he renounced his immortality. But after the centaur's death, the gods took pity and raised him to the starry skies for eternity. In some legends he is connected to the centaur Chiron.

Eight of the most prominent stars in Sagittarius

are brighter than magnitude 3. But at magnitude 4 the alpha star, Rukbat, is only the 14th brightest in the entire constellation. Sagittarius is oriented toward the core of the galaxy, which is some 28,000 light-years away. In this part of the sky, binoculars or a telescope will get you cruising through a half dozen or more bright deep-sky objects. One of the most striking deep-sky targets in the region is the 10,000-light-years-distant Large Sagittarius Star Cloud above the lid of the teapot; it is visible with binoculars, but telescopes will show countless stars. On one side of the teapot's lid is an amazing group of bright Messier objects all within reach of binoculars and small telescopes: the 4,000-light-years-distant Lagoon Nebula (M8), which is arguably on par with the grand Orion Nebula; the Omega Nebula (M17) at 4,890 light-years away; and the famous Trifid Nebula (M20), a colorful stellar nursery 4,100 light-years from Earth. The Trifid takes its name from the three dark dust lanes that split up the bright emission and reflection nebula. The pretty globular cluster Messier 54 is easily found close to Zeta Sagittarii, the third brightest star in the constellation.

M17: the Omega Nebula

STAR STORIES

Every year, from December 18 to January 19, we can find the path of the sun crossing within the celestial borders of Sagittarius. This location in the sky is where our sun reaches its southernmost position in its travels during the year. The moment it reaches this point is the winter solstice, celebrated with holidays around the world.

SCORPIUS (THE SCORPION)

Visible: Northern and Southern Hemispheres

Best Seen: Northern summer and southern winter

Main Stars: 18

Brightest Star: Antares, 1.0 mag.

Deep-Sky Objects: M4, M6, M7, M80, NGC 6302

NGC 6302: the Butterfly Nebula

Filled with brilliant stars, Scorpius is one of the most beautiful and easily recognized constellations in the sky. Even so, observers in the Northern Hemisphere may find it a challenge to trace out the entire constellation, since its most striking part, a winding chain of stars that forms the scorpion's curved tail and stinger, may be hidden behind low-lying haze, or even lie below the local horizon.

In ancient Greek myths, this cosmic beast represents the scorpion that killed Orion the hunter after he bragged that no living creature could ever hurt him. Zeus wanted to make sure the scorpion was as far from the hunter as possible, so he placed the scorpion here, on the opposite side of the celestial sky from Orion.

Pinning down the heart of the giant arachnid is the brilliant orange star Antares, the 15th brightest star in the sky. A red supergiant with a diameter about 830 times larger than our sun, Antares shines 60,000 times brighter than our star and is a distant 554 light-years away. Its name means "rival of Mars," because its conspicuous red color resembles that of the planet Mars (and Ares is the Greek name for Mars).

The Milky Way passes through the lower portion of Scorpius, and even from light-polluted suburbs you can see that the region is chock-full of clusters and clouds of stars. First, though, look for a beautiful blue double star that sits in the northern portion of the stick figure. Acrab is the brightest star at the top of a north-south line of stars that forms the scorpion's head. Also known as Graffias or Beta Scorpii, this 540-light-years-distant binary star system is easily split with binoculars or a small telescope using at least 50x magnification.

M4 is one of the easiest globular clusters to track down in the sky as a result of its proximity to Antares. Simply point your telescope, using a low-power eyepiece, at the bright star, and nudge your view only one degree westward to catch this giant ball of stars. Located just 7,000 light-years from Earth, M4 is considered one of the closest globulars and is bright and distinctly round through binoculars. The Scorpion's second globular cluster, M80, is located halfway between Antares and Acrab, but it is four times farther away from us than M4. About 4.5 degrees northwest of Antares and more than 32,000 light-years away, it shines at 7.2 magnitude. Small telescopes show it only as a tiny, circular patch of light with a bright core.

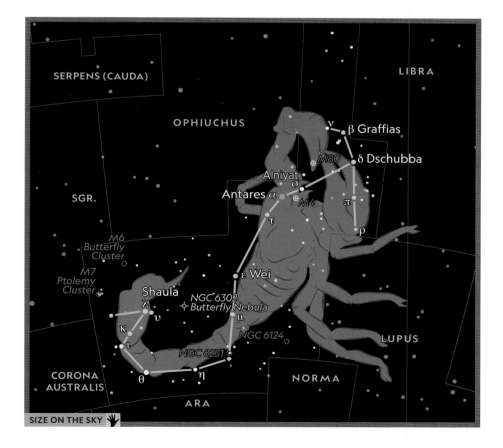

SCULPTOR (THE SCULPTOR)

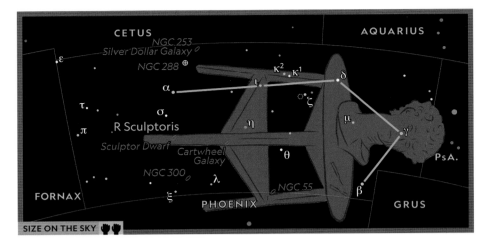

SIZE ON THE SKY

Visible: Northern and Southern Hemispheres

Best Seen: Northern autumn and southern spring

Main Stars: 4

Brightest Star: Alpha Sculptoris, 4.30 mag.

Deep-Sky Objects: Cartwheel Galaxy, R Sculptoris, Silver Dollar Galaxy, NGC 55

TELESCOPE TARGETS

R Sculptoris is a variable star system about 1,500 light-years from Earth that consists of an aging red giant star and at least one very dim companion star. The main star appears to pulsate in brightness from 9.1 to 12.9 over a period of 370 days. Recent observations have revealed a spiral structure and ring around the star that are believed to be caused by the hidden companion.

Sculptor is a minor constellation filled with fourth-magnitude and fainter stars that pays homage to a sculptor's studio. It was invented in the mid-18th century, along with 13 other diminutive constellations, by French astronomer Nicolas-Louis de Lacaille after his expedition to South Africa. Sculptor lies south of the larger constellations of Cetus and Aquarius. It is most famous for being the location of the south galactic pole (SGP), where the axis of the Milky Way passes perpendicular to star fields of our galaxy's spiral arms. (The north galactic pole is visible 180 degrees away in the northern constellation of Coma Berenices.) Since this region of the sky is looking "south" away from the plane of the Milky Way into the universe at large, observers will see few bright stars, but many faint galaxies. You can track down the location of the SGP thanks to the brightest star Alpha Sculptoris, which lies 780 light-years away. The other main stars, barely visible to the unaided eye from bright suburban skies, include Beta Sculptoris at magnitude 4.4 and Gamma Sculptoris, a yellow giant of magnitude 4.4. An interesting variable star is Eta Sculptoris, a red giant that varies in magnitude from 4.8 to 4.9 over periods ranging from 22 to 158 days. It is estimated to be over 1,000 times more luminous than our sun and is located 460 light-years away. Another well-known variable is R Sculptoris, which is a red giant that hosts bizarre spirals of matter around itself. Astronomers believe that the elderly star released this material about 1,800 years ago.

Sculptor is also home to a few faint galaxies that lie just beyond our Local Group within an average distance of eight million light-years, all within reach of common backyard scopes. The Silver Dollar Galaxy (NGC 253) is the brightest of the lot. An eighth-magnitude edge-on, barred spiral galaxy, it appears in 150- to 200-mm (6- to 8-in) aperture telescopes as a large, unevenly lit oval with a dark dust lane cutting across its core. The second brightest object in this Sculptor galaxy group is the eighth-magnitude NGC 55. This is another barred spiral, which many telescope observers have said looks like a miniature version of the Large Magellanic Cloud.

At a distance of 500 million light-years from our solar system, the Cartwheel Galaxy is one of the most dramatic galaxies in the entire sky. Too faint at 15th magnitude to be spotted by most backyard telescopes, this ring galaxy was put on full display by the Hubble Space Telescope. Images show the result of a violent smash-up after a smaller galaxy plowed through a large spiral galaxy millions of years ago, creating ripples of gas and dust and forming a ring 1.5 times the size of our own Milky Way.

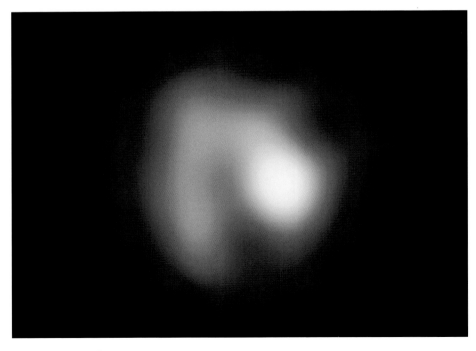

Red giant star R Sculptoris

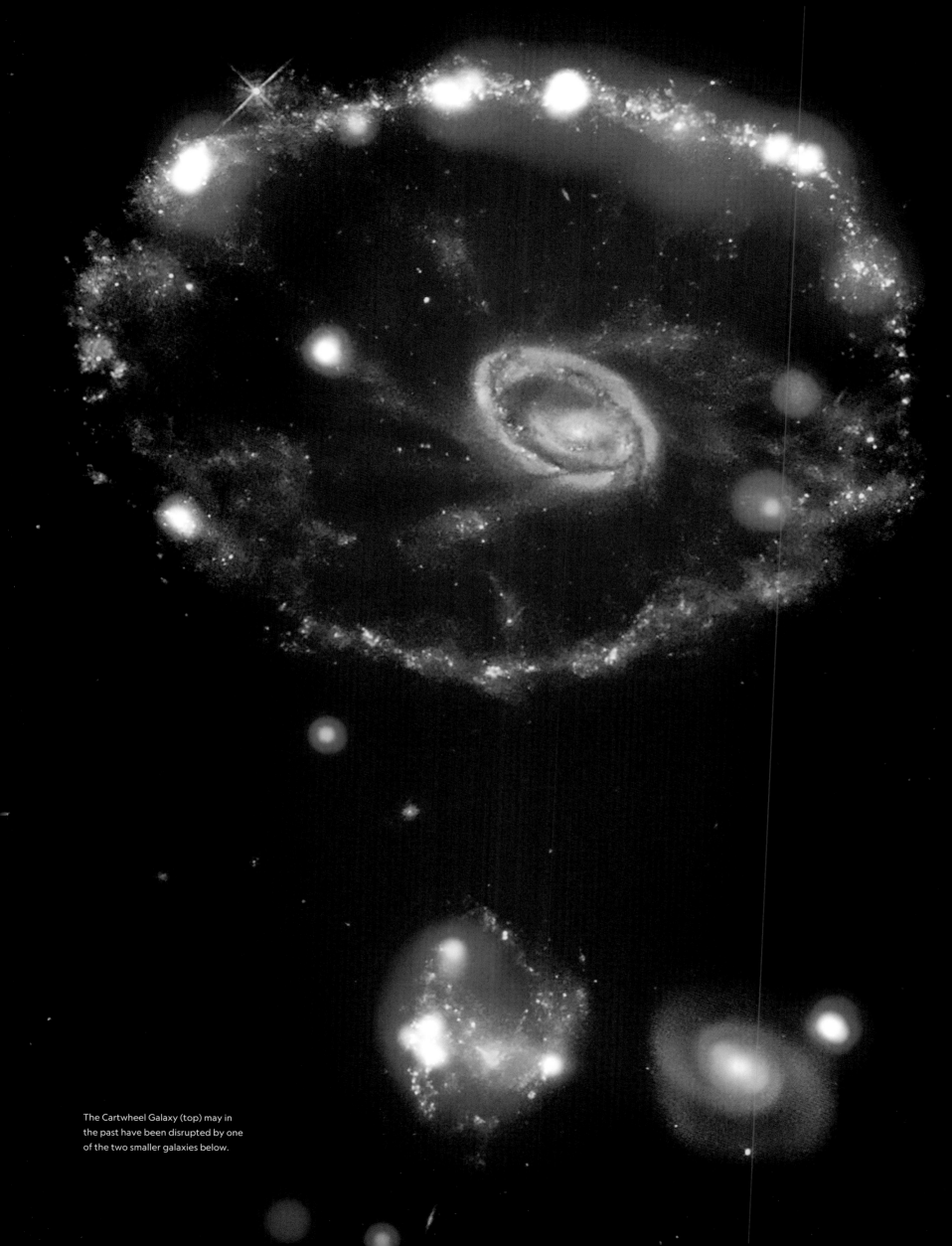

The Cartwheel Galaxy (top) may in
the past have been disrupted by one
of the two smaller galaxies below.

SCUTUM (THE SHIELD)

Visible: Northern Hemisphere

Best Seen: Summer

Main Stars: 2

Brightest Star: Alpha Scuti, 3.85 mag.

Deep-Sky Objects: M11, M26, M1-63, NGC 6712, IC 1295

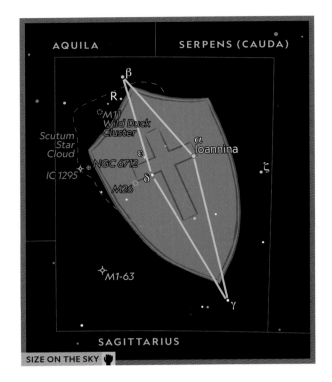

TELESCOPE TARGETS

The Wild Duck Cluster (M11) is definitely a deep-sky gem for backyard astronomers. Binocular views show it looking much like a fan-shaped flock of star-studded waterfowl, flying southwest of the constellation's northern-most star (Beta Scuti). Just a 100-mm (4-in) aper-ture telescope will start to show the 6,200-light-years-distant cluster's compact V-pattern, spanning 25 light-years.

Scutum is a small northern constellation best seen during the summertime. Looking like a simple rectangular pattern of stars, it sits just north of the major constellation Sagittarius, with the dense star fields of the Milky Way running through it. Scutum represents a shield and is a rather newly invented constellation, created by astronomer Johannes Hevelius in the late 17th century. He created this star pattern in honor of the king of Poland at the time, John III Sobieski, commander of the forces that defeated the Ottoman Empire in the critical 1683 Battle of Vienna. The original name for the constellation was Scutum Sobiescanum, which translates as "shield of Sobieski." The Shield appears sandwiched between Sagittarius, Aquila, and the lower half of Serpens.

Scutum is home to a pulsating variable star called R Scuti. While this 3,600-light-years-distant yellow supergiant spends much of its time shining at around magnitude 5, it can plunge down to magnitude 8 every four to five months. For backyard skywatchers, R Scuti is a great variable to follow—it's always within reach of binoculars but can at times be bright enough to spot with the unaided eye.

Just slightly southeast of the star Delta Scuti sits the open cluster M26. Estimated at 5,000 light-years from Earth, it's an easy eighth-magnitude target for binoculars; however, it may be a bit of a challenge to identify among the surrounding star field because the cluster's members are so scattered. A much easier sight is the Scutum Star Cloud, tucked away in the northeast part of the constellation, between Alpha and Beta Scuti. Both binoculars and small telescopes show it as a mass of scintillating stars.

A challenging globular cluster to observe from light-polluted skies is eighth-magnitude NGC 6712. Located 26,000 light-years from Earth, through small telescopes it looks like an unresolved circular haze. Astronomers believe that the globular may have lost multitudes of its stars over billions of years as it repeatedly swept too close to the Milky Way galaxy and its strong gravitational pull, which is why the cluster is so dim despite its relative proximity to us. Keen-eyed observers may notice the planetary nebula IC 1295 in the same field of view. The stellar puff cloud lies much closer, at only 4,700 light-years away. Astronomers are particularly interested in the shell of supernova G21.5-0.9. We have known about the supernova for years, but between 1999 and 2004 the Chandra X-ray Observatory was able to highlight its outer shell of ejected x-ray material.

Bipolar planetary nebula M1-63

SERPENS (THE SERPENT)

Visible: Northern and Southern Hemispheres

Best Seen: Northern summer and southern winter

Main Stars: 11

Brightest Star: Unukalhai, 2.63 mag.

Deep-Sky Objects: M5, Eagle Nebula, NGC 5921, NGC 6118, Hoag's Object, Seyfert's Sextet

Serpens is an ancient constellation that holds the unique distinction of being cut into two parts. It represents a snake that coils around Ophiuchus (The Serpent Bearer), whose body separates the reptile into two apparently unconnected sections named Serpens Caput, "the head of the snake," and Serpens Cauda, "the tail of the snake." The medicinal power of the snake is legendary and was said to lie in the serpent's venom, which could kill or cure depending on how it was used. The shedding of a snake's skin was also said to be associated with rebirth.

Serpens Cauda arcs in the space between bright stars Altair and Antares. Just within the limits of the naked eye is the globular cluster Messier 5, which is quite stunning through a small telescope. You can hunt it down about one-third of the way between Arcturus and Antares. Sitting 25,000 light-years distant, this city of a half million stars is one of the largest clusters, spanning 165 light-years. Astronomers estimate its age at 13 billion years, making it one of the oldest globulars of those scattered in a halo around the Milky Way.

A hidden, not-to-be-missed gem in Serpens is Graff's Cluster (IC 4756). Taking up more space in the sky than the moon's disk, this fourth-magnitude cluster sits 1,300 light-years away. Binoculars will reveal a bewildering number of stars across the entire field of view.

Another notable object is Seyfert's Sextet, which is relatively faint though it contains several galaxies. These galaxies are packed so closely together that their gravitational pulls have started to rip stars away from each other. In billions of years they may completely merge into a supersize galaxy. The Eagle Nebula (M16) in Serpens Cauda, 7,000 light-years away, is a combination nebula and star cluster. Hot, energetic young stars no more than six million years old are embedded inside the nebula and light it

The Pillars of Creation within the Eagle Nebula (M16)

up. Two can be spotted with a 200-mm (8-in) telescope or larger. However, even binoculars will give glimpses of the 20 brightest stars that make up the cluster and the faint, wispy hints of the gas cloud that surrounds them. Dark pillars of dense material rise in the center of the nebula; these can be seen with a 300-mm (12-in) telescope.

STAR STORIES

The snake in this constellation is supposed to have taught medicine to the Greek physician-god Asclepius, who then used his skills to restore the dead to life. The other gods disapproved, so they killed the physician and sent him and the serpent to the sky. Asclepius's rod, with a serpent winding around it from bottom to top, has become a modern symbol associated with healing. This staff is featured in the seal of the World Health Organization.

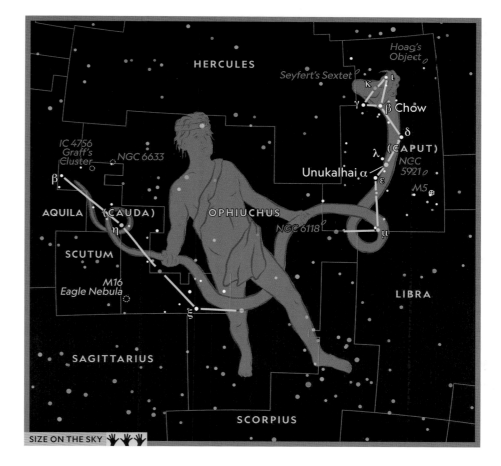

SEXTANS (THE SEXTANT)

Visible: Northern and Southern Hemispheres

Best Seen: Northern spring and southern autumn

Main Stars: 3

Brightest Star: Alpha Sextantis, 4.49 mag.

Deep-Sky Object: Spindle Galaxy

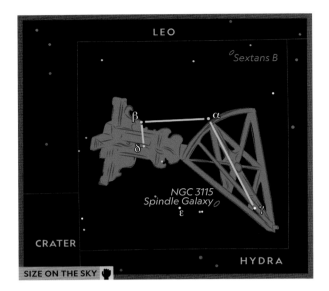

Sextans is one of the faintest and most obscure constellations in the heavens, with no stars brighter than fifth magnitude. This modern constellation was invented by Polish astronomer Johannes Hevelius to honor the instrument used by observatories of the time to measure star positions. Considered an equatorial constellation, Sextans sits along the celestial equator between the much brighter stellar patterns of Leo and Hydra. While its stars are best visible during dark, moonless nights with minimal light pollution, it can be tracked down by looking directly south of Leo's brightest star, Regulus, and northeast of the brightest member of Hydra, Alphard.

Sextans's brightest star, Alpha Sextantis, is about 280 light-years from our solar system. It is considered by many skywatchers to be the unofficial "equator star" since it currently sits only a quarter-degree south of the celestial equator—the extension of Earth's Equator out onto the celestial map of stars. But because of the slow wobble of Earth's axis, in 1923 the star appeared to actually cross the celestial equator, moving from the north.

Despite being in a sparse patch of sky, Sextans has its fair share of galaxies, but they are mostly visible to larger backyard telescopes with at least a 200-mm (8-in) aperture. However, the constellation is home to a fairly bright dwarf galaxy called Sextans B. Located some four million light-years from our Milky Way, it is a distant member of our Local Group of galaxies. Shining at a feeble 11th magnitude, it appears as an elongated, faint patch of light using high magnification and is thought to be more than 6,000 light-years across, just over 5 percent the size of our Milky Way. At the other extreme in terms of galactic distances is the galaxy cluster known only as CL J1001+0220. In 2016, astronomers used the Chandra X-ray Observatory to determine that its 17 members were the most distant group of galaxies known, at 11.1 billion light-years away.

Sextans also contains one of the oldest, most distant individual galaxies ever observed. Known simply as Cosmos Redshift 7, the galaxy has been measured at an incredible distance of 12.9 billion light-years—meaning we are seeing it only 800 million years after the big bang.

For backyard skywatchers the one standout deep-sky target is the Spindle Galaxy (NGC 3115), viewed edge-on as a magnitude-9 flattened disk. Visible even through binoculars on dark, moonless nights with no light pollution, this lens-shaped galaxy is 32 million light-years away. Many times larger than our Milky Way, it hides at its core one of the largest, closest supermassive black holes ever discovered beyond our galaxy, weighing in at a whopping two billion times the mass of our sun.

NGC 3115, which contains what is considered to be the nearest billion-solar-mass black hole to Earth

TAURUS (THE BULL)

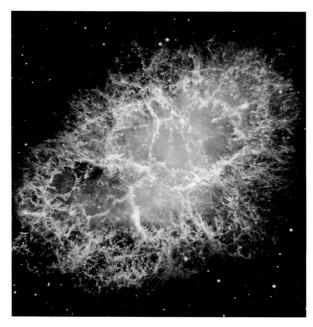

M1: the Crab Nebula

Visible: Northern Hemisphere

Best Seen: Winter

Main Stars: 19

Brightest Star: Aldebaran, 0.85 mag.

Deep-Sky Objects: Crystal Ball Nebula, Crab Nebula, NGC 1817, NGC 1647, Pleiades Cluster

Taurus is a large, ancient zodiacal constellation that is one of the most prominent across the Northern Hemisphere on winter evenings. It is easy to locate because Orion's belt line of three bright stars points right toward the bull's lead star Aldebaran, the celestial bovine's red eye.

Many myths are associated with Taurus, some dating back as far as the Bronze Age around 3000 B.C. To the ancient Egyptians, the bull had magical ties to Osiris, god of life and fertility, while in Greek mythology the bull is one of Zeus's disguises, used to capture Europa and bring her across the sea to the continent that now carries her name.

The stellar beacon of Taurus, of course, is magnificent Aldebaran, which in Arabic means "follower," so named because it follows the bright Pleiades star cluster in the same constellation. This red giant star is ranked the 13th brightest in the sky and is 65 light-years away. It appears to pin down one side of the giant V-shaped star pattern known as the Hyades Cluster. While Aldebaran just happens to be in front of the cluster, this gathering of stars forming the face of the bull lies about 150 light-years from Earth. Only about a dozen stars within the Hyades Cluster can be spotted with the naked eye on a dark, moonless night, but hundreds of members of this grouping are visible through binoculars. Hyades spans about five degrees—equal to the width of the middle three fingers held at arm's length.

Farther to the northeast, the 444-light-years-distant Pleiades star cluster marks the animal's shoulder. Also known as the Seven Sisters, it is one of the more easily identified deep-sky objects, visible even with the naked eye from typical suburban skies as a tiny, fuzzy patch of light. A line traced from Betelgeuse (in Orion) through Aldebaran will bring the dazzling group into view. Myths of many cultures have been spun around the seven bright stars in this cluster. A Native American story tells of seven young sisters who took a walk and lost their way: Now they stay in the sky as a reminder to children not to stray too far from home. Binoculars will help reveal the cluster's 500-plus stars.

With a fairly dim magnitude of 11, the Crystal Ball Nebula (NGC 1514) is a little planetary nebula, 1,000 light-years distant, with a dim outer gaseous shell protecting a brighter mottled inner bubble of gas and dust.

TELESCOPE TARGETS

The supernova that produced the Crab Nebula, also known as Messier 1, appears in A.D. 1054 in records worldwide, from ancient Chinese writings to rock art in the American Southwest. The 6,500-light-years-distant supernova explosion is estimated to have been brighter than the planet Venus and visible even during the daytime for a few weeks, before taking nearly two years to fade from sight. Today it takes a small telescope to view it as a tiny gray cloud.

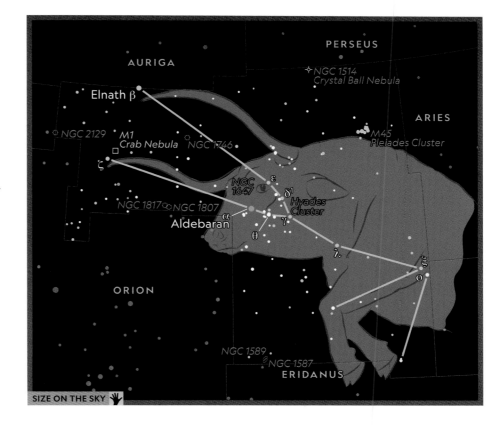

SIZE ON THE SKY

TELESCOPIUM (THE TELESCOPE)

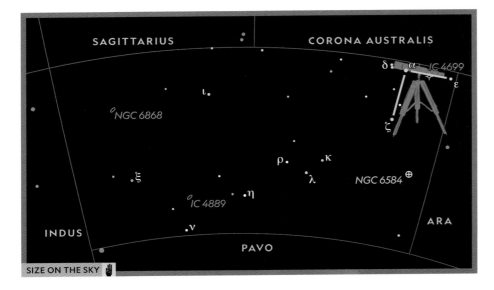

SAGITTARIUS

CORONA AUSTRALIS

δ α IC 4699
ε
ι
°NGC 6868
ζ
ρ κ
λ NGC 6584 ⊕
ξ
°IC 4889 η
ARA
ν
INDUS
PAVO

SIZE ON THE SKY

Visible: Southern Hemisphere

Best Seen: Winter

Main Stars: 2

Brightest Star: Alpha Telescopii, 3.49 mag.

Deep-Sky Object: NGC 6584

Telescopium is a faint southern constellation created in 1752 to commemorate the telescope, just as many other obscure constellations were named at the time for pivotal scientific instruments. Located between Ara and Indus, Telescopium sports three main stars that form a triangular pattern just southeast of neighboring Scorpius's stinger and southwest of Corona Australis.

Its brightest star, Alpha Telescopii, is a blue-white giant at least three times the diameter of our sun, at 250 light-years away. No exoplanets have been detected around the stars in Telescopium, but there have been a few brown dwarf companions and dusty debris disks confirmed around some of its stars. Most are young stars, with the brown dwarfs weighing in at least a couple dozen times the mass of Jupiter.

Telescopium is also home to a few interesting variable stars whose swings in brightness can be followed with just the unaided eye and binoculars; they include irregular variables Xi Telescopii (magnitude 4.89–4.94), RX Telescopii (magnitude 6.45–7.47), eclipsing binary BL Telescopii (magnitude 7.09–9.08), and eruptive variable RS Telescopii (magnitude 9.6–16.5). BL Telescopii is a rare type of supergiant that is believed to have been formed by the gravitational merger of two white dwarfs, with its wild swings in brightness occurring randomly over thousands of days. Fewer than 100 such stars have been recorded.

The ninth-magnitude globular cluster NGC 6584 sits near the border with Ara and its star Theta Arae,

TELESCOPE TARGETS

Astronomers have discovered what may be the closest known black hole to Earth at 1,000 light-years away, residing in a visible double or triple star system in the southwest corner of the constellation Telescopium. While the brightest member of QV Telescopii is just visible to the unaided eye at magnitude 5.3, its invisible companion was found thanks to its gravitational pull on neighboring stars.

making it easier to track down. Views through backyard telescopes show the 45,000-light-years-distant globular to have a concentrated and bright core surrounded by a halo of unresolved stars. As an observing challenge for larger telescopes, look at the midway point along a straight line between the Alpha and Epsilon Telescopii stars for the 13th-magnitude planetary nebula IC 4699. Spanning only three degrees of sky in the northeastern part of the constellation—a part of the sky equal in size to about six full moon disks—are 12 galaxies all about 120 million light-years from Earth known collectively as the Telescopium Group. The brightest member of this galactic group is the elliptical galaxy NGC 6868. Lying just to its west is the second brightest deep-sky object in the constellation, the spiral galaxy NGC 6861. Telescopes with at least 250-mm (10-in) apertures will begin to show it as an oval haze with a bright, elongated central region. Larger telescopes will also reveal that its core is partially obscured by a ring of dust.

Globular cluster NGC 6584

TRIANGULUM (THE TRIANGLE)

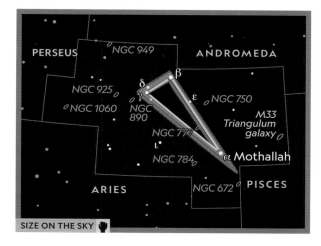

Visible: Northern Hemisphere

Best Seen: Autumn

Main Stars: 3

Brightest Star: Beta Trianguli, 3.00 mag.

Deep-Sky Objects: Triangulum galaxy, NGC 672

Triangulum is a small but distinct northern constellation, named appropriately for its long and narrow triangular pattern of naked-eye stars. In ancient Babylonian texts dating back to 1000 B.C., the stars of Triangulum were part of the constellation known as the Plough. The rising of the Plough at dawn each February marked when it was time to begin preparing fields throughout Mesopotamia. Later on, ancient Greeks called this star pattern Deltoton since it looked so much like their letter delta. In Chinese astronomy the same trio of stars formed part of the stellar pattern known as Teen Ta Tseang Keun, meaning "heaven's great general."

Triangulum appears between Andromeda to its north and Aries to its south. The base of the triangle is formed by a white star, third-magnitude Beta Trianguli, 125 light-years away, and fourth-magnitude Gamma Trianguli at a distance of 118 light-years.

While the constellation ranks as the 11th smallest in the sky, it still contains plenty of deep-sky objects worth exploring. Its definite showpiece is the Triangulum galaxy (M33). Shining at magnitude 5.4, it is bright enough to be glimpsed with the naked eye under truly dark-sky conditions with no light pollution. At a distance of 2.7 million light-years, M33 is the second closest spiral galaxy after the Andromeda galaxy (M31), and as such ranks as the most distant object the unaided human eye can spot. M33 and M31 plus its satellite galaxies form one unit, while the Milky Way and its Magellanic Clouds are a second unit, together forming what is called the Local Group of galaxies.

The Triangulum galaxy is spread over 70 x 45 arc minutes of the sky, meaning it takes up as much space as four full moons. It therefore has low surface brightness, and so it's best to use low magnification when viewing it. Telescopes with around 150-mm (6-in) apertures under dark skies show it as a faint oval glow with a small bright core, its loose spiral arms filled with knots and clumps of stars. Some of the knots are bright enough to have been cataloged with their own numbers, such as NGC 588, 592, and 595. Just beyond Alpha Trianguli—with the Triangle asterism actually pointing to it—is 11th-magnitude spiral galaxy NGC 672. Small telescopes show this 23-million-light-years-distant galaxy as having an elongated body with even surface brightness. Larger instruments will begin to reveal a bar-shaped core that shows signs of mottling.

M33: the Triangulum galaxy

TRIANGULUM AUSTRALE (THE SOUTHERN TRIANGLE)

Visible: Southern Hemisphere

Best Seen: All year

Main Stars: 3

Brightest Star: Atria, 1.91 mag.

Deep-Sky Objects: NGC 5979, ESO 69-6, NGC 6025

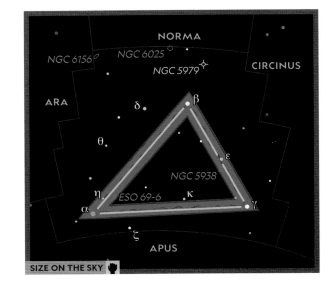

Triangulum Australe is a small constellation that appears circumpolar for most observers across the Southern Hemisphere and is the southern counterpart to Triangulum in the north. Lying between the constellations Ara and Circinus, its three brightest stars form a roughly equilateral triangle pattern and first appeared on sky charts back in 1603, when they were considered useful for ocean navigators. Famous explorer Amerigo Vespucci, who explored the Americas in the preceding century, described a stellar triangle in the southern skies that is likely Triangulum Australe.

The constellation's brightest star, Atria, is ranked as the 42nd brightest in the sky and is an orange giant. Its diameter is so big that if it were to replace our sun in the center of our solar system, its edge would reach the orbit of Venus. Despite being at a stately distance of 391 light-years from Earth, the star is prominent enough to have been included on the flag of Brazil. Beta Trianguli Australis lies 40 light-years away and is a magnitude-2.85 double

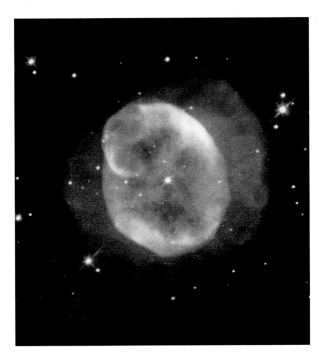

Planetary nebula NGC 5979

star. However, the companion is extremely faint at magnitude 13. The final member of the constellation's main triangular pattern is Gamma Trianguli Australis. With a magnitude of 2.8 it is a white star lying some 180 light-years from Earth. Near Beta lies the fourth brightest member of this small constellation, Delta Trianguli Australis. Shining at magnitude 3.8, this yellow star is 606 light-years away.

The constellation is easily identifiable within the Milky Way band, and while it is relatively sparse when it comes to deep-sky objects, Triangulum Australe is home to a pretty open cluster, NGC 6025. With just over 60 stars with an average magnitude of 7, through binoculars it appears as an elongated, bright, hazy patch of light. Examine the cluster through a 200- or 250-mm (8- or 10-in) aperture telescope and the cluster's intricate details begin to come to life: Long twisted strands of stars curve out from its central area. The cluster stands out well in front of black background sky and appears to be as wide as the quarter moon's disk.

A round fuzzy disk shows off in a 200-mm (8-in) aperture telescope when observing the blue-green planetary nebula NGC 5979. The 9,000-light-years-distant nebula readily brightens when using a light-pollution filter in the eyepiece.

The Hubble Space Telescope created a stunning portrait of two interacting galaxies, ESO 69-6, located about 650 million light-years from Earth, that look very much like musical notes. Long filaments of gas and dust appear to be pulled and torn away from their host galaxy's spiral arms by strong gravitational forces.

TUCANA (THE TOUCAN)

Visible: Southern Hemisphere

Best Seen: All year

Main Stars: 3

Brightest Star: Alpha Tucanae, 2.87 mag.

Deep-Sky Objects: NGC 104, NGC 121, NGC 362, NGC 346, Small Magellanic Cloud

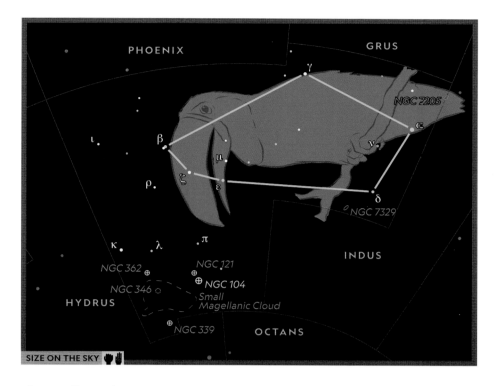

Tucana is a deep southern constellation near the South Celestial Pole, making it visible year-round for most Southern Hemisphere observers. It is named after the impressive bird with a large, colorful beak that Dutch explorers came across during their voyages to the East Indies at the end of the 16th century. Tucana is depicted between fellow avian constellations Phoenix and Grus to its north, with Indus to its west and the serpent Hydrus to its east.

The constellation's stars are no brighter than about third magnitude, except for Alpha Tucanae. This star is estimated to be at a distance of 184 light-years from Earth and is part of a binary system. The primary member is a cool giant 424 times more luminous and 37 times larger in diameter than our sun.

Tucana is ranked as the 48th constellation in size, but its claim to fame is hosting the Small Magellanic Cloud (SMC), on which it appears to be perched.

Globular cluster 47 Tucanae

This satellite galaxy orbits the Milky Way and lies 210,000 light-years away. It can be glimpsed with the unaided eye lying in the southern border of Tucana, and many observers describe it as looking like a small hazy patch of cloud, or as a piece broken off from the Milky Way band. Along with spiral galaxies M31, M33, and the Large Magellanic Cloud, the SMC is considered one of the farthest objects the unaided human eye can see. It measures about three degrees across—equal to six full moon disks in the sky. In reality, the Milky Way's small companion galaxy stretches across some 7,000 light-years. And because its masses of stars are spread out, it has low surface brightness and so is best viewed from dark locations with no light pollution.

Within the SMC itself are deep-sky treasures to explore, similar to those that we find in our own Milky Way, such as clusters and nebulae. NGC 121 is a globular cluster within the SMC that shines at 11th magnitude and looks like a tiny cottontail through larger backyard telescopes.

Just northwest of the SMC is a stunning globular cluster that is much closer. NGC 104, also known as 47 Tucanae, is one of the most eye-catching of all globulars in the sky, second only to Omega Centauri. It is visible to the naked eye as a tiny, hazy point of light. Through a small telescope, its large halo of stars measures 25 arc minutes across—nearly equal in size to the full moon's disk.

TELESCOPE TARGETS

Tucana is the home of one of the Hubble Space Telescope's farthest views into the cosmos, an area known as the Hubble Deep Field South. In 1998, the space instrument spied for two continuous weeks at one small patch of seemingly empty sky within this constellation, seeing galaxies down to the 30th magnitude and out to 12 billion light-years away.

URSA MAJOR (THE GREAT BEAR)

Visible: Northern Hemisphere

Best Seen: All year

Main Stars: 20

Brightest Star: Alioth, 1.77 mag.

Deep-Sky Objects: M81, M82, M97, M101, NGC 2787, NGC 3079, NGC 3310, NGC 4013

TELESCOPE TARGETS

The galaxy M101 is a face-on spiral that forms a perfect equilateral triangle with the two last stars in the Big Dipper's handle (Alkaid and Mizar). Sitting at a distance of 25 million light-years from Earth, it appears large but very faint in small telescopes, due to low surface brightness—so low-magnification eyepieces are best to use.

Ursa Major, also known as the Great Bear, is one of the most prominent northern constellations, and its main claim to fame is that it contains the Big Dipper, a highly recognizable star pattern, or asterism, engraved across human cultures since prehistory. The easily identifiable Big Dipper represents the bear's rear torso and tail, with the other stars of the constellation mapping out its long nose and legs.

While to the ancient Greeks this stellar group represented a large bear, other cultures saw everything from a chariot to a horse and wagon, a team of oxen, and a hippopotamus (by Egyptians, who had likely never seen a bear). Some Native American tribes believe the cup of the dipper represents a bear and the stars in the handle represent warriors that pursue it.

While the seven-star asterism itself is popularly called the Big Dipper in North America, it is also

Distant galaxies within Ursa Major

known as the Plough in England (not to be confused with the ancient Plough constellation, now part of Triangulum). Not surprisingly, in France, where astronomy meets gastronomy, the same group of seven is called the Casserole. Before the Civil War in the United States, the Big Dipper was known by enslaved people as the Drinking Gourd. In songs, it pointed the way north toward freedom. For most observers in the Northern Hemisphere, the Great Bear is close enough to the North Celestial Pole that it never sets below the horizon and it rotates around the North Star once a day.

Ursa Major ranks as the third largest constellation and is full of amazing deep-sky targets. Can you spot the double stars Mizar and Alcor in the Big Dipper's handle? Visible as two stars to a sharp naked-eye observer, they are not actually bound in a binary star system. However, looked at through a small telescope, second-magnitude Mizar, which is 78 light-years away, reveals itself as a true double as well. The Owl Nebula (M97) is a fine planetary nebula that takes the form of a very faint circular patch of light about 2.5 degrees southeast of the Big Dipper star Merak. Shining at only magnitude 9.9, it has a low surface brightness; averted vision is useful to bring out details when using smaller 100- to 150-mm (4- to 6-in) aperture telescopes. The bear also hosts a pair of galaxies 12 million light-years away, the spiral galaxy M81 and the starburst galaxy M82, that can be spotted together with binoculars.

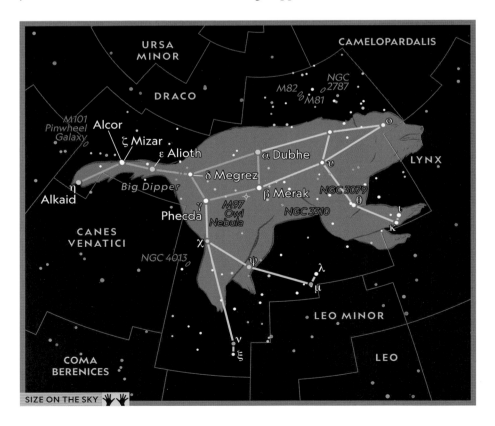

SIZE ON THE SKY

URSA MINOR (THE LITTLE BEAR)

SIZE ON THE SKY

Visible: Northern Hemisphere

Best Seen: All year

Main Stars: 7

Brightest Star: Polaris, 1.97 mag.

Deep-Sky Object: NGC 6217

Located near the North Celestial Pole, Ursa Minor is the northernmost constellation visible year-round throughout the Northern Hemisphere. This "little bear" is a faint group of stars surrounded by Camelopardalis, Draco, and Cepheus. None of its members is brighter than second magnitude.

Ursa Minor was given its name by ancient Greek skywatchers in the sixth century B.C. One of many Greek legends about the two celestial bears tells of the god Zeus being saved from his murderous father by two bears that helped him hide out in a cave. In return for their good deed, Zeus swung them up to the sky by their tails—hence, both Ursa Major and Ursa Minor have long tails. In earlier times the Little Bear was part of the next-door constellation Draco and was known simply as the asterism Dragon's Wing.

The constellation's value as a navigational tool was recognized long ago. Travelers still look to the constellation's famous asterism, the Little Dipper, to locate Polaris, the North Star: It is the last star in the tail of the Little Bear or in the handle of the Little Dipper. Earth's axis of rotation happens to point almost directly toward the North Star, so that it remains nearly motionless in the sky for observers across the Northern Hemisphere.

All the other stars rotate around the celestial pole, making Polaris and the Little Dipper reliable constants in the night sky. Earlier cultures even used the constellation as a clock, marking time as the Little Dipper swung around the seemingly fixed point of the North Star. Over thousands of years the position of the North Celestial Pole in the sky has shifted. It continues to approach 433-light-years-distant Polaris and will reach its closest around the year 2100. The pole will then move past it and through a succession of new pole stars, first in Cepheus, then in Cygnus, and eventually in Vega about 12,000 years from now.

Ursa Minor lies away from the Milky Way band, so it is mostly devoid of deep-sky objects for skywatchers, except for a few faint galaxies. A 200-mm (8-in) telescope will begin to reveal structural details in the magnitude-11 barred spiral galaxy NGC 6217. At 67 million light-years away, the center nucleus can be spotted with 100-mm (4-in) telescopes.

TELESCOPE TARGETS

Look for the annual Ursid meteor shower to peak on the nights of December 21 and 22, with average rates of about 10 to 15 shooting stars per hour and rare outbursts of 30 or more. The meteors, which are debris shed from the parent comet 8P/Tuttle, appear to shoot out from a region of sky just above the bowl of the Little Dipper.

Barred spiral galaxy NGC 6217

VELA (THE SAILS)

Visible: Southern Hemisphere

Best Seen: All year

Main Stars: 5

Brightest Star: Gamma Velorum, 1.83 mag.

Deep-Sky Objects: Eight-Burst Nebula, NGC 2547, NGC 3201, NGC 2899

STAR STORIES

While Gamma Velorum may seem lost among the many stars, it does have a standout place in the history of space exploration. Unofficially, the star is also known as Regor, which is Roger spelled backward. It honors the legacy of the late Apollo astronaut Roger Chaffee, who died in the 1967 Apollo 1 launchpad fire. The name was originally placed onto NASA star charts by Chaffee's crewmate Gus Grissom as a joke, and the star was used by Apollo astronauts as a navigational aid.

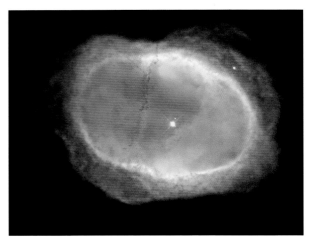

NGC 3132: the Eight-Burst Nebula

Embedded almost entirely in the swath of the southern Milky Way, Vela is a southern constellation that lies south of the bright star beacons of Sirius and Canopus. Its figure represents canvas sails, one of three parts of the Greek mythological ship *Argo*. The original constellation of this legendary ship occupied such a large portion of the sky that French night sky cartographers in 1756 decided to divide it into smaller sections: Puppis (stern), Carina (hull or keel), and Vela (sails).

Ranked as 32nd in constellation size, Vela has Antlia and Pyxis to its north, Carina to its south, and Centaurus to its east. Vela's brightest star, Gamma Velorum, may look like a single bright star to the naked eye, but in fact it is a quadruple star system about 1,200 light-years distant, composed of two pairs of stars. The second brightest star, Delta Velorum or Alsephina, is one of the points that helps form the diamond-shaped asterism known as the False Cross.

Vela holds the distinction of hosting the third closest star system to Earth. In 2013, astronomers discovered a binary brown dwarf system named Luhman 16, only 6.5 light-years distant. And two stars in the constellation hold court to super-Earth-size exoplanets. Located 142 light-years away, HD 93385 is similar to our sun and hosts a pair of worlds that are eight and 10 times Earth's mass. HD 85390 is an orange dwarf 109 light-years distant; it hosts a planet one-seventh the size of Jupiter.

The Eight-Burst Nebula (NGC 3132) is a stunning, bright eighth-magnitude planetary nebula that lies an estimated 2,821 light-years from Earth, one of the closest of its kind. Its doughnut-like appearance closely resembles that of its northern counterpart, the Ring Nebula in the constellation Lyra. The Eight-Burst Nebula appears as a little oval disk, about the same size as Jupiter, through small telescopes. In reality the nebula is expanding at speeds of 24 kilometers per second (15 mps) and measures half a light-year across. A much more distant, fainter (12th magnitude), and reddish-hued planetary nebula is NGC 2899. NGC 2547 is a bright fifth-magnitude loose open cluster that is about as large as the quarter moon disk in the sky and lies 1,177 light-years from Earth. It makes for a pretty target even for binoculars. NGC 3201 is a sixth-magnitude loosely structured globular cluster that shows off well in 150-mm (6-in) or larger telescopes. Keen-eyed observers may notice its slightly less than round shape, and because it is partially covered by the dark dust lanes of the Milky Way, the light from some of its member stars twinkles with a reddish glow.

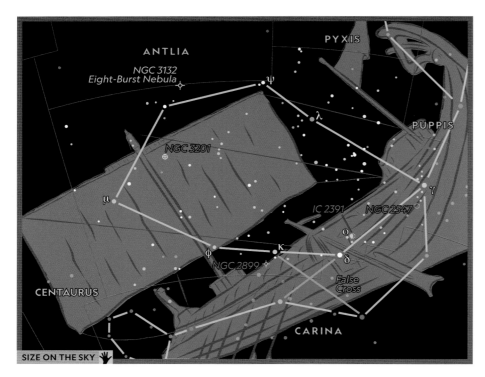

ANTLIA
PYXIS
NGC 3132
Eight-Burst Nebula
ψ
λ
PUPPIS
NGC 3201
μ
IC 2391
NGC 2547
γ
o
φ
κ
δ
NGC 2899
CENTAURUS
False Cross
CARINA

SIZE ON THE SKY

VIRGO (THE MAIDEN)

Visible: Northern and Southern Hemispheres

Best Seen: Northern spring and southern autumn

Main Stars: 9

Brightest Star: Spica, 0.98 mag.

Deep-Sky Objects: M49, M58, M59, M60, M61, M84, M86, M87, M89, M90, M104, NGC 4639, NGC 4380

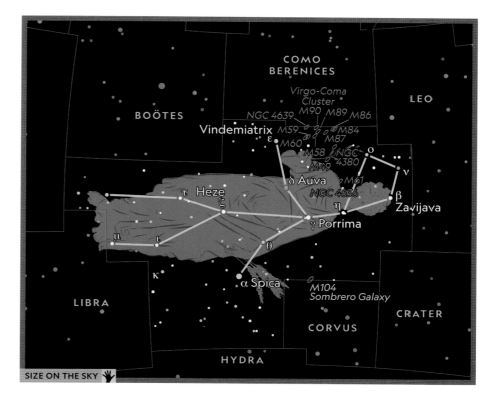

D epicting a female figure, Virgo the maiden is an ancient zodiacal constellation that straddles the ecliptic and is a wonderland of galaxies to explore. Virgo ranks as the second largest constellation in the sky and is depicted clutching a stalk of wheat, a traditional symbol of harvest across the ancient world. This association with agriculture is believed to have been cemented when astronomers figured out that the sun's path would pass through this constellation during the time of harvest. While to the Romans Virgo may have represented the goddess of justice and all the laws of nature, for the Babylonians she is said to have represented their ancient goddess of fertility: Ishtar, queen of the stars and lover of the fertility god Tammuz, whose death she mourns every autumn when the grain is milled.

Shining brilliant blue-white, Virgo's brightest star, Spica, is about 260 light-years away and is the 16th brightest star in the heavens. This blue giant is about 14 times the mass of our own sun and 2,000 times more luminous. While it may look like a single star, it is in fact a double. Both are hot blue giants that orbit each other only 18 million kilometers (11 million mi) apart.

While Virgo appears empty to the naked eye, it is a galactic treasure trove for telescope users. Spanning more than five degrees of the sky, an area about 10 times that of the full moon, the Virgo Cluster is nestled within the constellation's borders and represents a monster collection of thousands of spiral and elliptical galaxies huddled together in a 15-million-light-year diameter.

With its center about 53 million light-years away, the giant cluster is one of the closest such groups to our own Milky Way. Even with casual trolling with small telescopes with wide-field views, dozens if not hundreds of galaxies can be spotted. The bright, fuzzy core of elliptical M49 is one of the easiest to observe, as is the twin sight of M84 and M86 in the same field of view. M87, a supergiant elliptical, appears as a cotton ball, but it is home to many trillions of stars. Look east of Spica for the famous Sombrero Galaxy (M104), marked by a delicate dark dust lane that cuts across its core and spiral arms, providing its distinctive look. Shining at magnitude 9, the 32-million-light-years-distant Sombrero is one of the brightest spiral galaxies visible through backyard telescopes. Stretching some 130,000 light-years across, it is slightly larger than our Milky Way.

M104: the Sombrero Galaxy

VOLANS (THE FLYING FISH)

Visible: Southern Hemisphere

Best Seen: Autumn

Main Stars: 6

Brightest Star: Gamma Volantis, 3.62 mag.

Deep-Sky Object: Meathook Galaxy

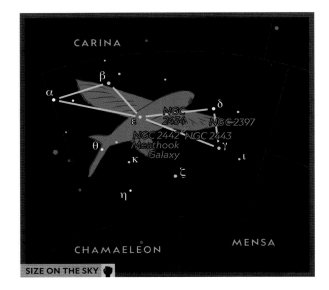

SIZE ON THE SKY

Volans is a deep southern constellation visible mostly from the Southern Hemisphere; it represents a tropical flying fish. It was placed on celestial maps during the late 16th century by navigators exploring the southern seas—the winged fish appears on old sky maps to be following the ship *Argo,* while also being hunted by fellow fish constellation Dorado.

Volans is a small and faint star pattern, with most stars being fourth magnitude and dimmer. But it can be easily found thanks to being wedged between the superbright star Canopus, second-magnitude star Beta Carinae, and the Large Magellanic Cloud. Volans is home to three great double stars. The constellation's brightest star, Gamma Volantis, is a pretty gold and yellow stellar pair located at the tip of one of the fish's wings. Epsilon Volantis is a blue and yellow pair, while Zeta Volantis is an easily separated yellow and orange double. The brightest galaxy within reach of backyard telescopes is the 10th-magnitude spiral Meathook Galaxy (NGC 2442), an oddly asymmetric, snakelike structure located about 50 million light-years from Earth.

VULPECULA (THE FOX)

SIZE ON THE SKY

Visible: Northern Hemisphere

Best Seen: Summer

Main Stars: 5

Brightest Star: Anser, 4.44 mag.

Deep-Sky Objects: Dumbbell Nebula, Coathanger, NGC 7052

This little fox is a faint constellation nestled between the bright stars Altair in Aquila and Vega in Lyra. It lies within the lavish star fields of the Milky Way with Cygnus to its north and Sagitta and Delphinus to its south. As such, Vulpecula is filled with deep-sky targets for binoculars and small telescopes, including some stunningly beautiful star clusters. For binoculars the most famous find is the asterism known as the Coathanger (Collinder 399) on the border of Sagitta. On a dark night, it is visible to the naked eye, but binoculars and small telescopes resolve it into individual stars—a straight line of six with a "hook" of four in the middle.

The magnitude-7 Dumbbell Nebula (M27) is one of the sky's best and brightest examples of a planetary nebula. A small telescope using high magnification will reveal its distinctive shape. It can even be spotted as a faint, fuzzy patch through binoculars just north of the brightest star in Sagitta. The most distant target in Vulpecula for small scopes would be the 12th-magnitude elliptical galaxy NGC 7052. At the center of this 214-million-light-years-distant galaxy is a supermassive black hole surrounded by a disk of dust, resembling a giant hubcap.

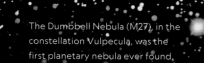

The Dumbbell Nebula (M27), in the
constellation Vulpecula, was the
first planetary nebula ever found.

The Perseids meteor shower lights up
the sky over Mount Shasta, California,
in this composite image.

CHAPTER FIVE

MORE CELESTIAL SIGHTS

THE NIGHT SKY IS
A BUSY PLACE

Step outside and look up at the star-filled night sky, and at first glance it may look the same as last night's sky. Nothing could be further from the truth. Astronomically speaking, no two nights are ever alike, even from the same location. Gaze up at the clear sky for an hour or two a night, week to week, and month after month, and it becomes easily noticeable that there is a changing cycle in the heavens above. These daily and seasonal rhythms were seen by ancient skywatchers thousands of years ago.

Fixed patterns of stars slowly glide across our sky in unison as Earth spins on its axis every 24 hours and moves in its orbit around the sun. Keen-eyed onlookers can see planets slowly parade along their ecliptic pathways, appearing as untwinkling stars, patiently moving across the fixed backdrop of constellations over weeks and months. Two or more worlds, sometimes including Earth's moon, will pass each other in our sky, appearing strikingly close to each other. Special visitors chronicled through the ages include comets and their debris trails—meteor showers—which regularly grace our skies for only a few weeks at a time. These create sky shows that delight both the novice and the seasoned backyard astronomer. On special occasions, skywatchers in the right place and at the right time witness space weather in the form of late-night colorful auroras, displays that appear to dance and shimmer across large tracts of the overhead sky. On certain times of the year at twilight, far away from light-polluted cities, one can even catch the ghostly glows from sunlight reflecting off countless grains of space dust.

And then there are human-made objects that appear to silently sail in the overhead sky. These spacecraft range from spent rocket boosters tumbling and twinkling as they orbit Earth, to long trains of communication satellites, to crewed space stations that may appear for a few fleeting moments as bright as the brightest star.

Finally, buried within the constellations lies a treasure trove of hidden objects far beyond our solar system. These deep-sky objects include giant star factories, enormous stellar islands, and individual stars that are ticking time bombs ready to explode. It's a veritable cosmic zoo ready for skywatchers to explore with their binoculars and telescopes.

Depicting the Battle of Hastings, this segment of the 11th-century Bayeux Tapestry (exhibited in Bayeux, France) is the first known illustration of Halley's comet (top left).

METEORS AND COMETS

Ancient cosmic relics made of iron, stone, and ice represent the birth and death of worlds in the solar system.

On any given night we can look up at a clear sky and see a falling star—a meteor—blazing a quick trail across the sky. One of the more impressive sights in nature is when a flurry of meteors begins to rain down in the course of a few hours, a phenomenon we call a meteor shower.

METEOR SHOWERS

These streaks of light can brighten the entire sky. The "falling stars" are not actually stars of course, but bits and pieces of debris left behind by icy comets and a few asteroids. These meteoroids, most of which are no bigger than a grain of sand, hit Earth's atmosphere at tens of thousands of miles an hour. While most are faint and burn up in a fraction of second, a few bigger stones can be mixed in, perhaps the size of a baseball to a basketball, producing unusually bright meteors known as fireballs, or bolides. These can last many seconds as they race dramatically across the sky, producing a super-bright light show and even, rarely, a sonic boom.

Meteor showers occur several times a year and are predictable, happening annually around the same dates on the calendar. For example, the Perseids peak on August 12 and 13, while the Leonids reliably peak in performance on November 17 and 18. That's because around these dates, our planet's orbit intersects with clouds of debris deposited behind comets as they swing around the sun. Over the course of several nights, this swarm of debris hits our planet's atmosphere like bugs on a speeding car's windshield, producing a fireworks show of shooting stars. It's important to note that a meteor shower's peak activity usually happens over the course of just a few hours on a single date, when our planet is in the densest part of the cloud. However, Earth can take many days, even weeks, to plow through the entire length of these clouds in space. So for skywatchers this means that meteors appearing at lower hourly rates can be seen for many days on either side of the peak dates.

In some years, meteor showers can put on a better-than-usual performance, with displays of

Syrian refugees search for meteorite fragments in the eastern province of Bingol, Turkey, in 2015.

several hundred or more shooting stars per hour at peak times. And rarely, jaw-dropping displays of thousands of shooting stars, known as a meteor storm, appear in just a few hours. These special cosmic surprises occur when Earth encounters particularly dense or freshly deposited debris clouds. In 1966, the legendary Leonids meteor storm was a true fireworks show, producing thousands of shooting stars per minute for nearly an hour.

VIEWING TIPS FOR METEOR SHOWERS

Every meteor shower is named for the constellation from which its meteors appear to radiate. For example, the Geminids originate from Gemini, while the

---

(Opposite) In 2020, the comet NEOWISE sailed through the skies above the mountain Suchý Vrch in the Czech Republic (Czechia).

Meteor Crater in Arizona came to scientific attention in 1891 and is thought to be approximately 50,000 years old.

Perseids stream out from Perseus. So when you're observing a meteor shower, locating its namesake constellation is a great guidepost for knowing where to direct your gaze.

Remember that a moonless, dark sky will offer the best views of even the faintest meteors. So plan ahead by finding out where the moon will be during peak viewing hours of the shower and try to time your observing window to times before or after the moon has risen or set below your local horizon. If the moon does happen to be in the sky, turn your back to it to maintain your dark-adapted vision. The best times to watch meteor showers are usually between local midnight and the predawn hours. That's when the sky is at its darkest and a shower's radiant (the point from which the meteors appear to originate) will be at its highest point in your local sky.

Finally, it's worth timing meteor watching, if possible, to the predicted absolute peak time (refer to the International Meteor Organization website). That's when Earth will enter the densest part of the debris cloud that forms the meteor showers.

The best locations from which to catch meteor showers—and even the faintest shooting stars—are in the countryside, far away from all the light pollution surrounding cities. If you can't get away, seek out a dark urban park and remember that the process of adapting your eyes to the dark can take up to a half hour. Simple stray light from a car or even a cell phone can ruin your night vision in seconds. Try watching for meteors streaking at least halfway up the sky and position yourself comfortably, perhaps in a reclining lawn chair or on a blanket facing the shower's radiant. You don't need binoculars or a telescope: Meteors zip across large areas of the sky,

A FIELD GUIDE TO SKY ROCKS

The solar system is riddled with debris left behind from its formation and from the collisions between celestial bodies that have gone on for billions of years since. The resulting space rocks range in size from dust grains to behemoths nearly as big as the state of Texas. The largest ones, which can be many miles across, are called asteroids and comets (left, top two objects). The vast majority of asteroids circle the sun in the asteroid belt, between the orbits of Mars and Jupiter. Comets also orbit the sun, but they are mostly found in elongated paths, where they spend most of their days far past Neptune and Pluto at the outer edges of the solar system.

Smaller pieces that break off these giant objects in space are called meteoroids (left, third from top). When they slam into a planet's atmosphere at high speeds and burn up, we call that flash of light a meteor, or "shooting star" (left, fourth and fifth from top). When a meteoroid survives its trip through the atmosphere and hits the planet's surface, that fragment is called a meteorite (left, bottom).

An illustration of a large asteroid colliding with Earth, thought to be the event that led to the extinction of the dinosaurs some 66 million years ago

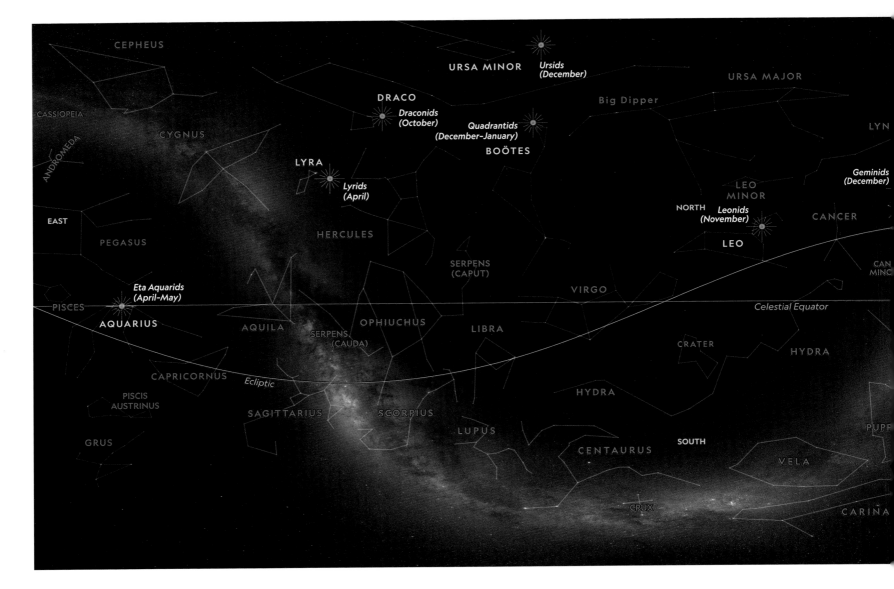

Each meteor shower is named for the constellation from which it seems to originate, though its tracks then spread across the sky.

so unaided eyes are best to see wide-angle views. Make sure you have an unobstructed view of at least 80 percent of the overhead sky—and you'll be ready to make a bunch of wishes.

COMETS

Comets fascinate both professional and amateur astronomers alike because they are visually mesmerizing when they appear in Earth's skies, while

also representing the ultimate time capsule— frozen bits and pieces left behind from the birth of the planets, including Earth.

The vast majority of comets we can observe in our skies were born in the frozen outer regions of the solar system beyond the orbit of Neptune, from the debris left behind after the planets had coalesced and formed. Numbering in the billions, most are invisible to even the most powerful telescopes as

RAINBOW SPECTACLE

Every meteor that streaks across the sky brings a unique experience to skywatchers. For one thing, these flashes of light can come in a rainbow of colors. These colors depend on the chemical composition of the meteors and the speed at which they enter the atmosphere. For instance, meteors made mostly of magnesium will give off a green light, while those that have an abundance of calcium glow purple. The Geminids, known for producing different-colored meteors, shine in orange and yellow because of their high sodium and iron content. Speed adds to the spectacle, with faster meteors reportedly burning blue, while slower ones give off a distinct orange or red light.

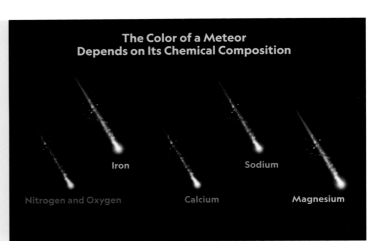

The Color of a Meteor Depends on Its Chemical Composition

Iron

Sodium

Nitrogen and Oxygen

Calcium

Magnesium

only weeks or perhaps a few months at most. As a comet approaches the sun, observers under dark skies can watch for its tail to grow in brightness and increase in length over the course of many days and weeks.

To hunt down comets and see them in all their glory, a transparent, dark sky with as little light pollution as possible is essential. Expect the comet to usually appear in the lower half of the eastern sky before dawn or in the western sky after dusk. Snap photos, or use binoculars or low-power telescopes to see details in the dust- and gas-filled tails. Wide-angle views through binoculars are perfect for scanning the skies to locate a comet and get a look at its sweeping tail.

SKYLIST

METEOR SHOWER CALENDAR

Meteor showers are a skywatcher's delight, not least because they arrive at the same time every year and start from the same spot ("radiant point") in the sky. Below, you'll find a list of major meteor showers and the dates at which they are at their heights. Moon phases will vary, so check the lunar calendar for prime viewing nights when moonlight won't interfere with your night vision.

■ **QUADRANTID: JANUARY 3–4**
Radiant point: Just before the handle of the Big Dipper "arcs to Arcturus," follow a line at approximately a right angle to the Quadrantid radiant point in the constellation Boötes.

■ **LYRID: APRIL 21–23**
Radiant point: Near the star Vega in the constellation Lyra.

■ **ETA AQUARID: MAY 6–7**
Radiant point: The constellation Aquarius.

■ **PERSEID: AUGUST 12–13**
Radiant point: Just above the constellation Perseus.

■ **ORIONID: OCTOBER 20–21**
Radiant point: Between the red supergiant star Betelgeuse and the constellation Gemini.

■ **LEONID: NOVEMBER 17–18**
Radiant point: Within the constellation Leo, from the center of the lion's head.

■ **GEMINID: DECEMBER 13–14**
Radiant point: The constellation Gemini, closest to the star Castor.

frozen debris, until they approach the sun, experience its warmth, and begin to awaken.

Most visible comets appear to the naked eye and even through binoculars and telescopes as faint, tiny cotton balls, but on special occasions one of these icy visitors brightens enough to put on a real sky show. Some of the more recent famous ones that have painted our skies include Hyakutake (1996), Hale-Bopp (1997), and McNaught (2007). Their journeys have taken hundreds of thousands to millions of years to make it into the inner solar system. And when they reach the orbit of Jupiter or closer and swing around the sun they begin to glow with enough light for small telescopes and binoculars to begin tracking them. On average, every decade or so a comet grabs the attention of the world by rising in brightness and becoming easily visible to the naked eye. Every two to five years there is a comet that brightens enough to be an easy target for binoculars. Observers have to be ready as comets are fast travelers, meaning their peak performance lasts

A meteor (captured in a time-lapse photo) burns bright over a forest in Kustavi, Finland.

DIVING INTO THE DEEP SKY

Stargazers peering out of our solar system will find deep-sky delights, from jewel-like clusters to star islands.

Beyond our solar system lies a beautiful universe filled with cosmic treasures known as deep-sky objects. These targets for observation range from well-known star clusters and nebulae to obscure galaxies. Their exploration is a perennially favorite pastime for both novice and experienced backyard stargazers using the naked eye, binoculars, and telescopes.

The main challenge in catching a glimpse of these treasures—say, a bright knot of stars across a galaxy's spiral arms, or dark dust lanes cutting across its central core—is not that they appear too small through a telescope, but that they are just so dim. However, with patience, training, and the right optical aids, a surprising amount of detail can be teased out of your live views, even ones of a cluster of galaxies hundreds of millions of light-years away. When exploring the solar system, it's important to remember that a telescope is used specifically to magnify details such as craters and mountains on Earth's moon, and storms and rings on neighboring planets. When it comes to star clusters, clouds of gas and dust, and especially galaxies that can lie billions of light-years away, the telescope's main purpose is simply to augment our insensitive eyes' ability to see in near-total darkness.

The human eye's direct, real-time view of deep-sky objects can never match the richly detailed photographs created at the telescope, whether amateur or professional. The light-gathering power of the human eye is simply not adapted as well as a digital camera to seeing the dark universe. Optical aids allow us to collect as many of these ancient, feeble photons of light as possible, creating bright, detailed views of the distant cosmos that otherwise would remain invisible. However, we can improve our night vision by following a few tips when planning a deep-sky adventure under the stars.

DEEP-SKY VIEWING TIPS

The human eye is an amazing light detector, but it needs time—at least 20 to 30 minutes—to adapt to the darkness and let the necessary chemical changes occur in the retina. And once outside, at all costs avoid being exposed to any bright lights. Aviators have a trick of wearing sunglasses for at least a

half hour before heading out to fly at night. Doing something like this gives your eyes a head start in being able to see much fainter celestial details. But to see really distant, super-dim objects such as nebulae and galaxies, as well as their structural details, stargazers also rely on averted vision: a technique of looking out of the corner of your eye instead of directly at an object. Our peripheral vision appears to be more sensitive to picking up very faint objects in the night, doing wonders for bringing a bit of clarity to our views of night sky.

While boosting your night vision does help reveal many hidden cosmic gems, light pollution is the biggest factor limiting deep-sky observing,

Illustrations of the three main types of galaxies—elliptical, spiral, and irregular—are matched with images of actual galaxies.

(Opposite) A composite, color-boosted image shows the Andromeda galaxy's size if it were as bright as the moon.

The Seven Sisters, or Pleiades, are an open cluster of about 1,000 stars located in the constellation Taurus.

especially with galaxies. The best views will always be had in the dark countryside, at least an hour or more away from any metropolis. Using higher magnification at the telescope will help somewhat to darken the background sky when observing within bright city limits. Timing observations to the hours when the deep-sky objects are highest in the late-night sky will help cut through some of the light pollution, too. There is a reason why stargazers affectionately call these distant islands of stars "faint fuzzies." These ghostly, tiny patches of light that we seek out with binoculars and telescopes do suffer the most from bright artificial lights.

WHAT'S IN A NAME?
CATALOGING THE UNIVERSE

Humans have been keeping records of the night sky and its treasures for thousands of years across many cultures. For instance, one of the first known catalogs of the brightest stars may have been collected by the ancient Greek astronomer Hipparchus around 129 B.C. But it took the invention of the telescope in the early 17th century for the first nebulae—

■ For more on Hipparchus, see page 20.

gas- and dust-filled clouds in space—to be discovered. Soon thereafter astronomers meticulously scanned the skies with ever bigger telescopes and began to hunt down, chart, and catalog a growing number of deep-sky objects.

Flash forward to the 21st century, and today's backyard skywatchers have the ultimate luxury of using computerized, robotic telescopes with onboard libraries of the names and locations of tens of thousands of deep-sky objects. But at the heart of these electronic databases are a handful of core catalogs of the deep sky's best and brightest objects, many of them visible from the backyard.

The most famous of all inventories, filled with some of the brightest and easiest-to-locate deep-sky objects, is the Messier catalog. In the late 18th century, French astronomer Charles Messier tallied a list of more than 100 of the brightest star clusters, nebulae, and galaxies visible in the northern two-thirds of the sky. Messier, who was a comet hunter, compiled this list of fixed "fuzzy" objects so that in his future searches he and his colleagues would not be fooled by these comet look-alikes.

Centuries later, many stargazers still see it as a challenge to find everything on the Messier list (known by Messier numbers such as M33 or M57), and for those new to the hobby it is the first guide they turn to when looking through their new telescopes or binoculars. Some targets on the list, though, are visible to the naked eye, and several of these also are known by their common names. For example, the Pleiades, or Seven Sisters, are M45. The nebula in Orion is M42, and the Andromeda galaxy is M31.

The New General Catalogue of Nebulae and Clusters of Stars (NGC), published in the late 19th century by Johan Dreyer, was the first major collection that went way beyond Messier's, as it compiled a list of 7,840 deep-sky objects. Unlike Messier's seemingly random scattering of objects across the sky, all the NGC objects were conveniently numbered in order of right ascension, starting with 0 hours and increasing in number as we sweep west to east.

Soon, newly discovered objects needed to be cataloged, and so the NGC was expanded with the *Index Catalogue,* adding thousands of additional objects. Revisions, corrections, and updates have continued into the 21st century. Meanwhile, other deep-sky lists abound, such as the Herschel 400 and the Caldwell catalog, which includes some objects visible only in the Southern Hemisphere.

The Crab Nebu a, or M1, is a supernova remnant that can typically be seen using a small telescope.

DEEP-SKY TREASURES

As the seasons change, so does our view of the night sky and hence our window out to the universe at large. In the following section, we take a cosmic journey beyond the edge of our solar system, into the deep space filled with stargazing wonders of various shapes and sizes. Through a series of carefully curated tours of discrete parts of the night sky, our explorations take us through the star-cluster and nebula-packed spiral arms of our own Milky Way, into the most distant realm of the legions of galaxies within reach of binoculars and backyard telescopes.

M94, the Cat's Eye Galaxy, in Canes Venatici

DEEP-SKY TREASURES: TOUR 1

Visible: Northern Hemisphere

Season: All year

Location: Ursa Major/Ursa Minor

Featuring: Galaxies

Ursa Major and Ursa Minor are two of the perennially visible constellations that pin down a region visible across the Northern Hemisphere all year round. These two mythical bear constellations and the surrounding star fields represent a rich hunting ground for various deep-sky objects, but galaxies abound. That's because we leave behind the stellar clutter of the Milky Way and can see out into the depths of the greater universe, filled with these great islands of stars. Surprisingly, many of these cosmic won-

ders are visible with binoculars and backyard telescopes of all sizes.

M106

Look to the south of the handle of the Big Dipper for a great variety of galaxies offering hours of celestial browsing. First on our tour is the large, bright spiral galaxy M106. Its saucer-shaped nucleus can be spotted even with binoculars and the smallest telescopes, looking like a faint smudge just within the border of the faint neighboring constellation Canes Venatici. Lying some 24 million light-years away, this oddly shaped spiral shows off two spindly, faint arms with a distinct S shape that comes to light in at least 100-mm (4-in) telescopes. Larger telescopes show that M106 appears to have rich star-forming clouds strewn chaotically across its arms. Astronomers have found the culprit in a supermassive black hole at the galaxy's core.

M94

Sitting just off the line between the two brightest stars of Canes Venatici is the picturesque galaxy M94. At 13.7 million light-years distant, this face-on spiral presents an intensely bright center contrasted with very faint and diffuse arms. With a visual magnitude of 9 it's an easy target for small telescopes, even from light-polluted suburbs. Larger telescopes reveal it to have a bright ringlike system of spiral arms wrapped tightly around its central core. This brightness indicates massive star-forming regions. M94 also holds a cosmic mystery: Unlike many other galaxies we have examined across the universe, this one comes up short on dark matter, leaving astronomers scratching their heads as to why.

SUNFLOWER GALAXY (M63)

About two-thirds of the way along an imaginary line between Alkaid, the last star in the handle of the

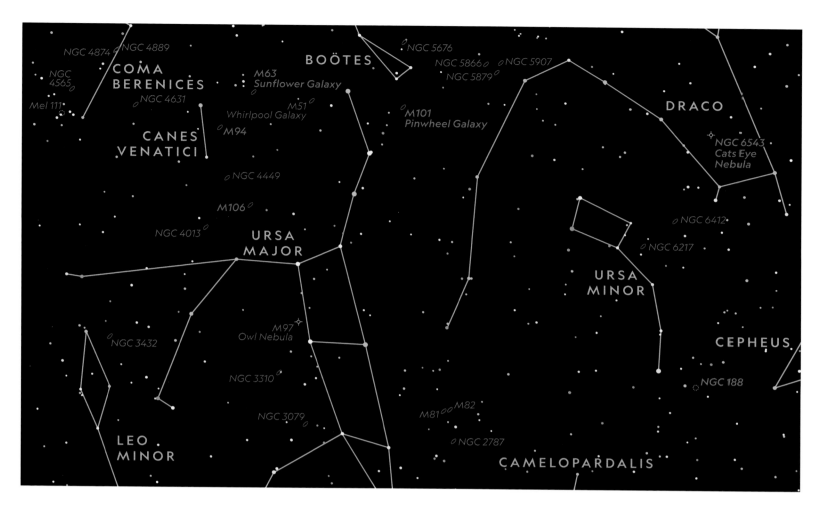

Big Dipper, and Cor Caroli, the brightest star in the constellation Canes Venatici, lies the famous Sunflower Galaxy, or M63. Shining at 9.3 magnitude, it looks like a tiny oval in small telescopes. It seems to have no well-defined spiral arms, but instead possesses knotlike clouds of stars scattered in a loose, disjointed formation, similar in appearance to its namesake flower's overlapping petals. Located 27 million light-years away, its faint, mottled spiral pattern is revealed only with 250-mm (10-in) and bigger telescopes.

PINWHEEL GALAXY (M101)

Star-hopping to the great spiral galaxy M101 couldn't be easier, thanks to a chain of four easy-to-follow stars leading away from the middle star in the handle of the Big Dipper. At 7.9 magnitude and about 24 million light-years away, it can be glimpsed with binoculars, but a telescope will offer better views. However, despite offering a bright core, the face-on view of its spiral arms means that the majority of the galaxy has a very low surface brightness. Best views of this galaxy, nearly twice the size of our own, will be achieved with at least 150-mm (6-in) telescopes using low magnification under dark skies.

NGC 188

Located near the North Star is an open star cluster that may hold the record for being the most ancient of its kind, at around 6.8 billion years old. NGC 188, also known as Caldwell 1, is located in the constellation Cepheus (The King), above the plane of the Milky Way galaxy at about 5,000 light-years from Earth. The cluster is easy to find, being less than five degrees away from Polaris: a distance equal to the separation between the two end stars in the Big Dipper's bowl.

CAT'S EYE NEBULA (NGC 6543)

Glide into the large and winding constellation Draco (The Dragon) and spy on one of the most famous planetary nebulae in the entire sky. Made famous by the Hubble Space Telescope, the Cat's Eye Nebula, or NGC 6543, is a bubble-like structure created when a sunlike star gently ejected its outer layers of gas into space, forming clouds in intricate shapes. Through small telescopes this eighth-magnitude object resembles an out-of-focus star, but with a distinctive light blue-green hue. Use larger telescopes to reveal its central star—the naked white-hot core remnant called a white dwarf.

■ For more about the Draco constellation, see page 227.

DEEP-SKY TREASURES: TOUR 2

Visible: Northern Hemisphere

Season: Autumn

Location: Cassiopeia/Cepheus/Perseus region

Featuring: Nebulae, clusters, star clouds

As crisp autumn nights settle in across the Northern Hemisphere, stunning celestial gems are scattered across the northern sky, perfect for binoculars and telescopes. The center of attraction this time of the year is the area around the giant W-shaped constellation Cassiopeia (The Queen). This bright, easy-to-find stellar pattern dominates the high northeastern sky and holds court with neighboring constellations Cepheus (The King) and Perseus (The Hero). All three figures are conveniently placed nearly overhead during fall evenings, with the Milky Way's glowing band spilling across this entire region.

ELEPHANT'S TRUNK NEBULA (IC 1396)

A great two-for-one treat for keen skywatchers is the Elephant's Trunk Nebula, located within IC 1396, some 3,000 light-years away in the Cepheus constellation. A companion star cluster is visible with binoculars or even with the unaided eye under truly dark-sky conditions. With a small telescope, the surrounding nebula looks like a hazy, circular glow.

The best times to look for this dim nebula are on moonless nights using wide-field binoculars and low-power telescopes. The entire cloud of gas and dust stretches out for 20 light-years and spans nearly three degrees across our sky—equal to about six moon disks. The nebula is a stellar nursery filled with baby stars estimated to be at least 100,000 years old. Its distinctive dark, condensed gas formation, known as a dark globule, is shaped by intense winds of stellar radiation from the young stars and is the feature that gives the nebula the appearance of an elephant's trunk.

M52

One of the best open star clusters for binoculars in the entire sky is M52. Star-hopping takes you straight to the seventh-magnitude cluster: Draw a line from the two brightest stars on the side of the W-shaped Cassiopeia constellation and continue that line for the same distance to get to M52. Sitting over 5,000 light-years away, it looks like a glowing

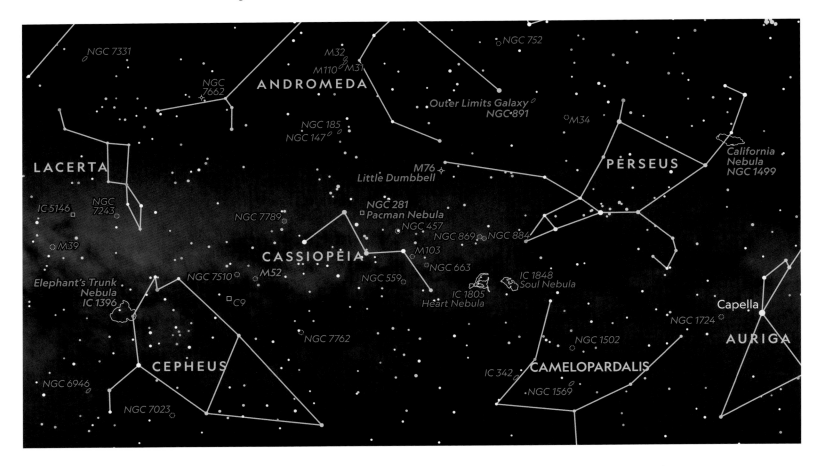

patch of light and is filled with young stars estimated to be about 20 million years old. Meanwhile, a telescope begins to show off the cluster's true beauty as an irregular pattern of shimmering stars that spans 24 light-years in space. Small telescopes can resolve about 50 stars in the cluster, while larger ones may reveal as many as 200 under ideal sky conditions.

PACMAN NEBULA (NGC 281)

Resembling the iconic video game character, this giant star factory in Cassiopeia is a real challenge for a backyard telescope. Because the nebula (also known as NGC 281) has a low surface brightness, dedicated skywatchers will need to use a wide field of view with a special nebula filter under dark skies to spot it. Lying less than 1.5 degrees (about three moon disks) southeast of the star Eta Cassiopeiae, the nebula's position is given away by a tiny star cluster (IC 1590) near its center. The Pacman's estimated distance is 9,200 light-years, which puts it in the neighboring spiral arm of our galaxy, within a much larger region rich in hydrogen gas that is primed to give birth to a new generation of stars.

M34

Gaze at the constellation Perseus and a faint fuzzy patch of light will catch the unaided eye. This is the open cluster M34, more than 1,800 light-years away from Earth. A small telescope reveals nearly 50 member stars loosely spread over 10 light-years. A couple dozen of its stars are easily resolved with

NGC 281, the Pacman Nebula, in Cassiopeia

binoculars and are amazing sights through small telescopes, even from light-polluted backyards.

OUTER LIMITS GALAXY (NGC 891)

Located only three degrees to the west of M34 is the pretty edge-on spiral galaxy NGC 891, affectionately known as the Outer Limits Galaxy because of its appearance in the end credits of the TV series of the same name. Sitting within the border of the constellation Andromeda (The Chained Maiden), this 10th-magnitude galaxy is about 30 million light-years away and has a diameter more than 1.2 times that of our own Milky Way. Through a 150- to 200-mm (6- to 8-in) telescope, the galaxy has a distinct silvery, needle-like shape with a broad dark dust lane that appears embedded in the galaxy's white glow, running its entire length.

CALIFORNIA NEBULA (NGC 1499)

A great challenging visual target for binoculars and small telescopes is this beautiful California-shaped nebula. Its large cloud glows with red hydrogen gas that measures more than one moon disk wide by almost six moon disks long along the side of the Perseus constellation. With an actual distance from Earth of more than 1,000 light-years, this emission nebula glows so strongly because of the radiation from the nearby bright star Menkib. The best bet for seeing it unaided is to observe it from a dark sky and patiently scan the stellar region once your eyes have become dark-adapted.

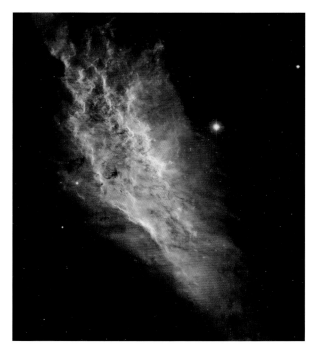

NGC 1499, the California Nebula, in Perseus

Collinder 399, the Coathanger, in Vulpecula

DEEP-SKY TREASURES: TOUR 3

Visible: Northern Hemisphere

Season: Summer

Location: Cygnus region

Featuring: Nebulae, clusters, star clouds

The onset of summer across the Northern Hemisphere heralds short nights, but also amazing views of the Milky Way arching overhead late at night. Superimposed on this galactic tapestry are some of the most stunning cosmic gems in the sky. A focal point in the near-overhead, late-night sky this time of the year is the distinctive stellar pattern of Cygnus (The Swan), also known as the Northern Cross. The stately figure of the mythical swan provides a rich hunting ground for multitudes of bright star clusters and eye-catching nebulae.

■ For more about the Veil Nebula, see page 224.

MILKY WAY BAND

The Northern Cross star pattern is actually the simplified version of the full constellation of Cygnus, complete with the swan's outstretched wings and its long neck. On a clear night during midsummer and even into early autumn, a hazy band of light from the Milky Way stretches across the sky. This faint glow we see under a dark countryside sky is actually the edge-on view into the disk of our galaxy. When we look up at Cygnus we are seeing our galaxy's disk as it runs down the swan's neck and body. It's amazing to think we are seeing hundreds of millions of individual stars within our home galaxy. On any clear night it's worthwhile to use binoculars, scan the sky through Cygnus, and enjoy the jam-packed views of countless star clouds and clusters.

VEIL NEBULA COMPLEX

Discovered in 1784 by astronomer William Herschel, the Veil Nebula in Cygnus is one of the largest supernova remnants. Located just three degrees south of the bright star Aljanah, the Veil spans 110 light-years and is roughly 2,400 light-years away. Astronomers estimate that it was born when an ancient giant star, at least 10 times more massive than our sun, exploded 10,000 to 20,000 years ago in a neighboring arm of our galaxy. When the explosion was at its brightest, the supernova may have been as bright as Venus in our skies. Today, the Veil Nebula is shaped like a fragmented loop, a form composed of two distinct, bright, narrow arcing sections and a fainter triangular patch of gas with feathery tendrils. The entire complex stretches more than three degrees—equal to six moon disks in the sky. The best views through a telescope will come from using a wide-field, low-power eyepiece, perhaps with a nebula filter, under dark skies.

M56

Just off the side of the main band of the Milky Way in Lyra (The Lyre) lies a small but bright globular cluster, M56. It's easy to find using the star Albireo, the head of the Cygnus swan, as a guidepost. A giant city of stars located 33,000 light-years distant, M56 shines at eighth magnitude, an easy target for small

Globular cluster M56 in Lyra

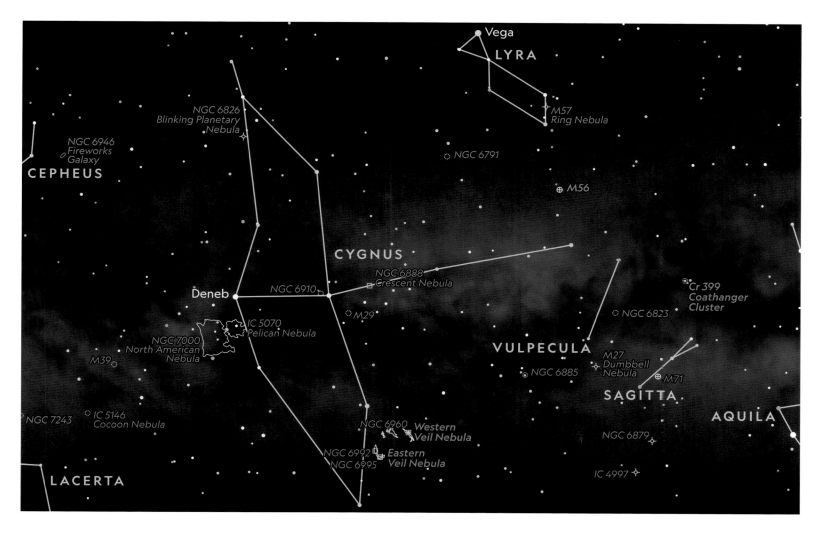

telescopes. Nestled within a rich stellar field, the globular cluster appears quite compact, making it difficult to resolve since it is packed so tightly with stars. Using the Hubble Space Telescope, astronomers have studied the chemical ingredients inside the cluster and found it devoid of elements heavier than helium, suggesting that its stars were born when the universe was young—and that it may be an alien to the Milky Way, gravitationally stolen from a satellite galaxy.

THE COATHANGER

A group of 10 stars huddled together in the shape of a common coat hanger resides in the constellation Vulpecula (The Fox). To find it, star-hop from Albireo in Cygnus to Anser, the brightest star in Vulpecula, continuing onward for a bit more. The Coathanger is finely visible to the naked eye and was first spotted as a tiny, hazy patch of light by Persian astronomers back in A.D. 964. It was not until 1997, when stellar distance measurements were carried out on the member stars of the Coathanger, that it was discovered they are not physically associated, and are in fact scattered across a large span of space ranging from 200 to 1,000 light-years from our solar system. Their eye-catching formation is simply a chance alignment from our vantage point here on Earth. To fit the entire Coathanger formation in a single field of view, it's best to use low-power binoculars.

The Milky Way seen in infrared light

DEEP-SKY TREASURES: TOUR 4

Visible: Northern Hemisphere

Season: Spring

Location: Virgo/Leo region

Featuring: Galaxies

Late spring and early summer are the prime times for hunting galaxies. While these islands of stars are scattered everywhere across the cosmos, in many parts of our night sky our own Milky Way blocks our view of the rest of the universe. This is especially true anywhere around the hazy band of light that marks the plane of the galaxy. However, there are a few parts of the sky that offer great open windows to those far-off galaxies. These include a large area that holds the Virgo and adjoining Leo and Coma Berenices constellations.

SOMBRERO GALAXY (M104)

Virgo contains many of the sky's brightest galaxies, featuring all types, including ellipticals and spirals. This galactic richness stems from the Virgo cluster of galaxies, a rich concentration of thousands of galaxies in our part of the universe, found within the borders of the namesake constellation. Our Milky Way is in fact a part of this same giant galactic group. Far south of the cluster's main group is M104, a beautiful edge-on spiral more than 28 million light-years away. Known also as the Sombrero Galaxy because of its distinctive hat-brim shape—a dark lane that cuts across the edge of the bright saucer-shaped disk—it shines at eighth magnitude and is easy to observe with any size of telescope, even under moderate light pollution. While the dark dust lane signifies areas of star formation, the galaxy's center is thought to be home to a supermassive black hole, one at least one billion times the mass of our sun.

NGC 4636

Travel northeast of the bright star Spica in the heart of the Virgo constellation and this is where the giant elliptical galaxy NGC 4636 makes its home. This gas-rich, ball-shaped galaxy lies about 50 million light-years away and sits at the edge of the Virgo cluster. The galaxy shines with the power of 30 billion suns and stretches some 100,000 light-

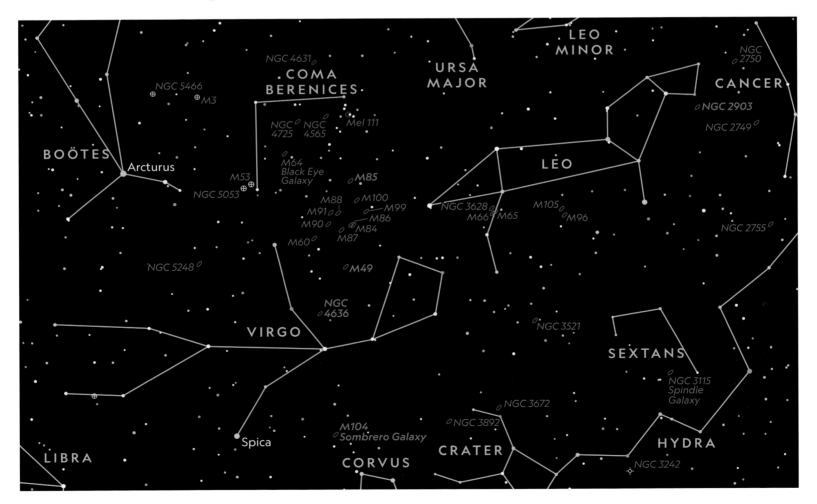

years across. Unlike spiral galaxies, ellipticals like NGC 4636 appear featureless for the most part, but they tend to be easier to spot because of their compact cores. Through the telescope this galaxy has a bright, starlike, condensed core that is surrounded by a diffuse outer disk.

M49

Lying near the center of the 20-million-light-year-wide concentration of galaxies in the Virgo cluster is M49. This giant elliptical is about 56 million light-years from Earth and at magnitude 9.4 is bright enough to be an easy target even for binoculars. Its well-defined, elongated appearance is evident in small telescopes under high magnification. Recent observations have revealed that a halo of nearly 6,000 globular clusters envelops M49 (compared to about 200 such clusters in our Milky Way), indicating that this galaxy may be filled with ancient stars that formed not too long after the big bang.

M85

Within the border of Coma Berenices, but still part of the giant Virgo cluster of galaxies, lies M85. In small telescopes it looks like a fuzzy cotton ball with a bright, condensed core, and it is officially classified as a lenticular galaxy, appearing to have characteristics of both elliptical and spiral galaxies. Despite being magnitude 9.2, which is fairly bright for galaxies in this region, it is quite diffuse, making

it a bit of a challenge to observe with binoculars and small telescopes.

NGC 2903

A true outlier of the Virgo cluster of galaxies is NGC 2903, a lonely, beautiful, loosely wound spiral galaxy hanging off the chin of Leo (The Lion). Shining at magnitude 9, this is a fairly bright galaxy that is easy to track down, being only 1.5 degrees south of the star Lambda Leonis. Through small telescopes it shows up as a patchy oval ball with a hazy halo around it. Larger 150- to 200-mm (6- to 8-in) telescopes can begin to reveal some of its mottled appearance, which comes from the clouds of gas and dust that form the spiral arms. These are the places where stars are born and die at high rates. Astronomers estimate that NGC 2903 lies some 30 million light-years away, measures 80,000 light-years across, and is home to about 80 billion suns.

Infrared view of M104, the Sombrero Galaxy, in Virgo

Spiral galaxy NGC 2903, in Leo

IC 2118, the Witch Head Nebula, in Orion

DEEP-SKY TREASURES: TOUR 5

Visible: Northern and Southern Hemispheres

Season: Northern winter and southern summer

Location: Orion/Taurus region

Featuring: Nebulae, clusters

Gaze up on a winter night and you will see a stunning collection of some of the brightest stars in the entire sky. A group of constellations dominates, and leading the pack are Orion and Taurus. This part of the sky contains some of the finest celestial attractions our Milky Way galaxy has to offer, from multitudes of ghostly nebulae to sparkling star clusters.

CHRISTMAS TREE CLUSTER

Visible to the naked eye as a fuzzy star off to the side of the right arm of the bright Orion constellation is a strange grouping of 20 brighter stars and more than 100 fainter ones, called the Christmas Tree Cluster, which many skywatchers consider a classic hidden cosmic gem. Located in the tiny Monoceros (The Unicorn) constellation, it begins to show off nicely

■ For more about Orion, see page 250.

in binoculars and small telescopes as a tight scattering of stars almost as wide as the full moon disk. Adding to its splendor is the hazy Milky Way as the backdrop. The cluster belongs to a star-forming region that also hosts the Cone and Fox Fur nebulae in the same Orion spiral arm of the Milky Way that holds Earth.

Careful examination through a telescope shows that the 2,500-light-years-distant cluster's main tree formation is outlined by the six brightest members arranged in pairs, forming its eye-catching triangular shape. The Christmas Tree spans about 30 light-years, with its stars considered young at roughly three million years old.

BARNARD'S LOOP

Orion (The Hunter) dominates the winter evening sky and is a veritable celestial garden filled with all manner of nebulae in different shapes, sizes, and colors. But there is one gaseous formation that beats all the rest when it comes to sheer size and mysterious origin. Barnard's Loop is a long arc of red nebulosity that is centered more or less around Orion's belt of three stars. Its namesake is astronomer E. E. Barnard, who was the first to record it on long-exposure photographs of the Orion region back in 1894. This great bubble of ionized gas stretches some 14 degrees, equal to 28 full moons side by side in the sky. Speculation is that Barnard's Loop may have formed when an expanding shock wave from an ancient supernova explosion swept up large clouds of gas and dust and spread them through space. For observers, one of the biggest challenges in catching the loop is having enough of a low-power view to encompass all its sections in one look. There have been reports of skywatchers being able to directly spot the entirety of Barnard's Loop just by looking through a light-pollution filter under dark skies. Individual fragments of the loop can be spotted through telescopes as bright twisty filaments of warm hydrogen gas.

WITCH HEAD NEBULA (IC 2118)

Tucked away just northeast of the bright blue-white star Rigel in the Orion constellation is a spooky glowing cloud called the Witch Head Nebula. As the name implies, the very faint reflection nebula looks like a selfie of an old sorceress from fairy tales and Halloween legends. While located at the end of the toes of Orion, the Witch Head Nebula is actually within the borders of the constellation Eridanus (The River). Astronomers suspect the bizarrely shaped cloud may have been formed when a giant star exploded, throwing its outer layers of atmosphere into space. Currently, the nebula is 900 light-years distant and is physically near enough to the powerful supergiant star Rigel that the star lights up the nebula's dust particles, which happen to reflect a blue color, giving the nebula its wonderfully creepy appearance. When you use binoculars under a dark, pristine sky, the nebula looks like a ribbon, while larger backyard telescopes will begin to reveal faint glowing tendrils in a thin, rectangular shape the size of the moon's disk.

Newborn stars in the Christmas Tree Cluster, Monoceros constellation

HIND'S VARIABLE NEBULA (NGC 1555)

One of the oldest and strangest deep-sky objects in the winter heavens can be found in the constellation of Taurus (The Bull). Hind's Variable Nebula, also known as NGC 1555, is a variable nebula stretching about four light-years across and located some 400 light-years away from Earth. Incredibly, this space cloud regularly changes its brightness, size, and even shape. The secret to its magical powers lies with the variable star embedded inside, a star of the T Tauri variety that fluctuates in brightness between magnitudes 8.5 and 13.5.

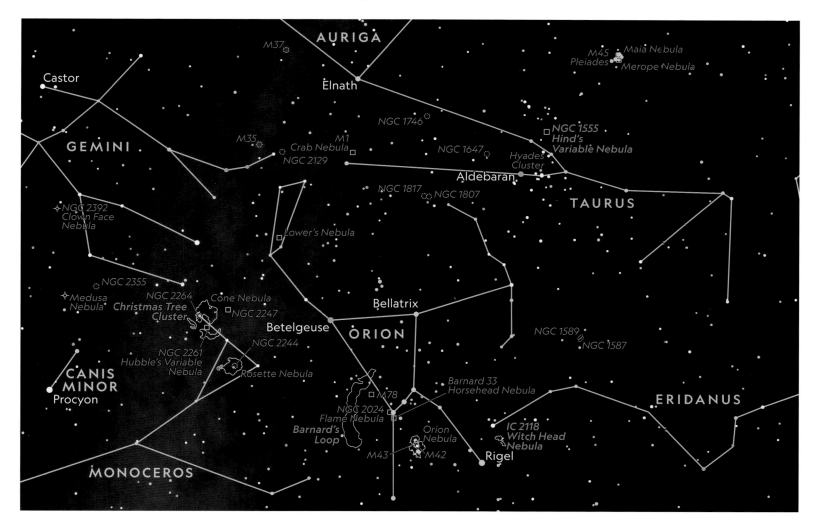

DEEP-SKY TREASURES: TOUR 6

Visible: Southern Hemisphere

Season: All year

Location: Crux/Centaurus/Lupus region

Featuring: Star clusters

Any skywatcher touring celestial objects of the southern sky will find a panorama of deep-sky wonders, filled with exotic names and forms, that can be seen only from places such as Australia, Chile, and southern Africa. Centaurus, Crux, and Carina are three southern circumpolar constellations visible from most locations in the Southern Hemisphere. They offer some of the most fascinating touring grounds south of the celestial equator.

GEM CLUSTER (NGC 3293)

While Carina's claim to fame is holding court to the second brightest star visible from Earth—Canopus—and the grand Eta Carinae Nebula, it also is home to one of the most stunning open star clusters, NGC 3293, also known as the Gem Cluster. Binoculars show a rectangular formation of spar-

The Gem Cluster, NGC 3293, in Carina

kling stars that is reminiscent of a very loose globular cluster. Look closely, and three bright stars in a line jump out, with the southern one shining with a deep orange color. The Gem Cluster sits 8,000 light-years away and is filled with dozens of stars approximately 10 million years old.

PEARL CLUSTER (NGC 3766)

Scan the sky along the Milky Way and along the back leg of the Centaurus constellation, and the unaided eye can easily spot a fuzzy patch of light that marks the Pearl Cluster, or NGC 3766. Point binoculars at the 5,800-light-years-distant cluster and the light blossoms into multiple chains of stars, all packaged in a tight triangular formation. Each of the three points is pinned by a brilliant star, two of which are distinctly orange-red in color. Through a backyard telescope the cluster begins to resolve into at least 30 stars, with a myriad of much fainter stars in the background. NGC 3766 is thought to be a typical open cluster, in that its stellar members are still quite young—about 30 million years old.

BLUE PLANETARY NEBULA (NGC 3918)

About 3,200 light-years from us in the constellation Centaurus sits the secretive Blue Planetary Nebula, or NGC 3918. Spanning only 0.3 light-years, this faint ghostly glow has a central red giant star that has pushed out colorful gaseous envelopes, giving it an eye-catching elongated form. For the type of object it is, the Blue Planetary Nebula is quite bright at magnitude 8, but still deceptively diffuse, taking up nearly as much space in the sky as the full moon. This object is one of the best examples of a planetary nebula: the end product of a sunlike star that in its death throes produces gas clouds in some of the most dramatic shapes and combinations of colors visible in the sky. In astronomical time frames, planetary nebulae like this Centaurid beauty are gone in a blink of a cosmic eye—dissipating into deep space in a few tens of thousands of years.

NGC 4945

Located relatively nearby in galactic terms, NGC 4945 is a beautiful edge-on spiral galaxy roughly 12 million light-years away. Nestled between two faint naked-eye stars within the Centaurus constellation and shining at 8.8 magnitude, it is a fairly easy target

for most backyard telescopes. Its shape is a flattened pinwheel—however, since we see it at a steep angle, it appears elongated. Even through binoculars this beautiful galaxy shows its distinctive cigar shape. With patience and practice, southern skywatchers using backyard telescopes can begin to see mottling all along its spindly length, as well as hints of a dark dust lane sweeping across the 77,000-light-year length of the galaxy. The mottled clumps represent

NGC 3766, the Pearl Cluster, in Centaurus

sites of intense star formation, and astronomers recently discovered what may be the driving force for this stellar activity. Waves of x-rays have been found to be pouring out from the galaxy's central core, indicating that a supermassive black hole is actively feeding on the gas and dust that is being gravitationally drawn to it.

CALDWELL 84 (NGC 5286)

A beautiful globular cluster, Caldwell 84—also known as NGC 5286—appears as a hazy seventh-magnitude ball through binoculars and is a very easy deep-sky object to track down, thanks to its proximity to the faint naked-eye star M Centauri. Observers will immediately notice the striking color difference between the greenish hue of the cluster stars and the intense orange-red glow from M Centauri. Larger scopes will show a bright core and several dozen stars along its edge being resolved. Orbiting just above the plane of the Milky Way, at 35,000 light-years from Earth, this globular may be an alien. Astronomers have found that it belongs to a stream of stellar material that may have originated from an ancient dwarf galaxy cannibalized by the Milky Way about 10 million years ago. The results of this ancient merger left behind star clusters with unusual orbits.

SUPERNOVAE

The most massive stars in the universe end their lives in a spectacular physical transformation.

Some of the brightest stars we see with the naked eye are ticking time bombs in our cosmic backyard. They are nearby giant and binary stars primed to undergo some of the most spectacular death throes the universe has to offer. And skywatchers are standing at the ready to witness what promises to be a once-in-a-lifetime event.

Over 5,000 supernova explosions have been witnessed in other galaxies, some occurring as far as billions of light-years away. These exploding stars can outshine their entire home galaxies for weeks at a time before they fade from view forever. It is estimated that a supernova goes off once a century in most galaxies. Backyard astronomers with at least a 150-mm (6-in) telescope can visually, or with camera systems, search for these exploding stars by surveying brighter galaxies. Some of the best candidate galaxies to watch include those from the famous deep-sky Messier list, such as M51, M60, M61, M64, M65, M66, M81, M82, M84, M85, M101, and M104.

What will the next nearby supernova look like?

The first signal astronomers expect to notice from a star about to explode nearby (within a few thousand light-years) will be a flood of high-energy neutrino particles inundating observatories around the world. The tiny particles erupt from the core of a dying stellar giant, cascade through its layers, and escape into space at the speed of light. A few hours later, the visual light counterpart of the explosion should hit Earth, gradually brightening over many days and then plateauing for a few weeks before fading from view over a period of a few months.

Soon skywatchers will have a dedicated alert system in place for potential supernovae as far away

(Opposite) Chandra X-ray Observatory image of Cassiopeia A supernova remnant

as the center of the Milky Way galaxy, about 26,000 light-years distant. SuperNova Early Warning System (SNEWS) will catch the rise in neutrinos from a star, providing a more reliable, precise, and timely notice of any star going supernova in the Milky Way.

Which star will explode next? Here are a few candidates.

BETELGEUSE

At 725 light-years distant, Betelgeuse is one of the nearest red supergiant stars to Earth. Expectations are that it is ripe to explode as a supernova. In September 2019, the star began getting noticeably fainter. And with the star's distance, this dimming would have actually occurred around the year 1300. This dimming of Betelgeuse, created thanks to a large dark spot in its atmosphere, caused some to believe an explosion was imminent. But since then, this beaconlike star in the constellation Orion has gone through a roller-coaster ride in brightness, apparently now returning to a calmer state. Evidently, more volatility could happen at any time. Skywatchers therefore should keep an eye on Betelgeuse. It's easy to spot as the second brightest star in Orion's right shoulder, and it is perfectly placed to be observed around

■ For more about supernovae, see pages 44–48.

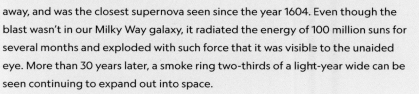

THE 1987 SUPERNOVA

The last bright supernova visible from Earth occurred in 1987. It was located in the Large Magellanic Cloud, about 168,000 light-years away, and was the closest supernova seen since the year 1604. Even though the blast wasn't in our Milky Way galaxy, it radiated the energy of 100 million suns for several months and exploded with such force that it was visible to the unaided eye. More than 30 years later, a smoke ring two-thirds of a light-year wide can be seen continuing to expand out into space.

Betelgeuse, a red supergiant star in the constellation Orion, baffled astronomers when it began to mysteriously dim in late 2019.

the world. When will Betelgeuse explode? Astronomers estimate it could happen anytime in the next 100,000 years. When it does, it's not expected to pose any danger to life on Earth, but it will be quite a spectacular sight.

Some people fear that an exploding star could threaten our planet by way of debris from the blast, or that the resulting radiation and gamma rays could destroy Earth's ozone layer, in turn triggering a mass extinction. It has been found, however, that a supernova needs to be within 40 to 50 light-years of Earth to cause any sort of harm, and there are no known candidates within that distance.

ETA CARINAE

Another candidate for explosion is Eta Carinae, located about 7,500 light-years from Earth. This stellar monster may hold a record as one of the largest stars in the Milky Way galaxy—it is estimated to be one hundred times more massive and five million times brighter than our sun. It nearly annihilated itself in a giant outburst seen in 1843, when it became one of the brightest stars in the southern sky. Though the star released as much visible light as a supernova, it somehow survived the outburst. The explosion produced two lobes moving outward at about 2.4 million kilometers an hour (1.5 million mph). Scientists say that it could explode tomorrow or in the next few thousand years. This binary star is found in the southern constellation Carina (The Keel), embedded within the bright Carina Nebula.

The nebula itself shines at third magnitude, bright enough to be seen with the naked eye. Using at least 150-mm (6-in) telescopes, the 5.5-magnitude Eta Carinae star is easily visible, with definite hints of the explosive double lobes.

V SAGITTAE

V Sagittae is a more exotic celestial bomb, one known as a cataclysmic variable. In this kind of variable, a white dwarf star is in a binary star system, and it gravitationally steals hydrogen from its neighboring star. As the gaseous material reaches the surface of the white dwarf, the hydrogen accumulates until it ignites in a burst of nuclear fusion, which causes the star's brightness to skyrocket. V Sagittae has been getting brighter rapidly over the decades. Astronomers have found evidence that the two stars in the V Sagittae system are spiraling in

toward each other at an ever increasing rate. Sometime between the years 2067 and 2099 they may finally merge, creating a mega-burst flash, called a nova, that will light up Earth's skies as the brightest star in the night. Today, V Sagittae is an inconspicuous star nestled within the tiny constellation Sagitta (The Arrow), which is between the Vulpecula and Aquila constellations. The eclipsing binary star system has brightened by a factor of 10 in the past century and now shines at a faint 11th magnitude, making it a challenging target for midsize backyard telescopes.

The binary stars of V Sagittae (here in an artist's rendering) are expected to explode as a nova in this century.

Astronomers expect that the transformation of volatile double star system Eta Carinae to supernova is not far off.

PASSING WORLDS

Every once in a while, lucky skywatchers spot eye-catching close encounters between two or more worlds.

As we watch the planets wandering in the sky in front of a backdrop of stars, on some nights or mornings a celestial traffic jam can occur. When two or more planets or the moon are grouped closely together, it is called a conjunction.

These striking events occur fairly regularly, on nearly a monthly basis, and offer amazing views with the naked eye, binoculars, and telescope. From our vantage point here on Earth, they may look like they are passing close to each other, but of course this is just an optical illusion. Planets that look snug to our eyes can actually be hundreds of millions of miles apart.

WHAT'S IN A NAME?

Conjunctions can take on different meanings in the astronomy world. Among amateur astronomers, a conjunction is when worlds appear near each other for days, perhaps about only one degree apart: a separation equal to the width of two full moon disks in the sky. But officially, a conjunction occurs when two celestial bodies appear to converge in the sky along the same right ascension, which is equivalent to the longitude coordinate system used on Earth. Conjunction also has a stricter astronomical meaning. If a planet lies exactly on the opposite side of the sun from Earth, that planet is at superior conjunction. When a planet lies between Earth and the sun, it is at inferior conjunction.

The most frequent kind of conjunction involves various planets and the moon, with some of the most dramatic occurring when the moon is at its crescent phase in the western sky at dusk or eastern sky at dawn. Other conjunctions involve the moon or planets and a bright star. We see the pair slowly converge from night to night, only to split apart and move away in their respective orbits.

Lineups of three or more celestial objects are known as alignments, and they can be quite eye-catching, especially when they involve the five classical naked-eye planets: Mercury, Venus, Mars, Jupiter, and Saturn. Some encounters can be incredibly close, with two celestial bodies appearing in the same telescopic field of view. In solar transits, the planets closer to the sun than Earth—Mercury and Venus—move as dark silhouettes across the bright face of the sun. (To view these, make sure you use only specially designed solar viewing glasses, not just sunglasses, to protect your vision.) "Great conjunctions" happen when the bright planets Jupiter and Saturn have very close encounters, an event that occurs roughly every two decades.

CATCHING A CONJUNCTION

Close encounters like conjunctions, especially involving the bright and easy-to-find moon, offer a chance for novice skywatchers to learn how to locate our neighboring planets more easily. And for those new to telescopes, these bright events can be inspirational in terms of exploring these worlds up close through the eyepiece.

A great observing challenge is to look for planets and stars involved in conjunctions during the day. These events, particularly if they target the moon and the brightest planets such as Venus and Jupiter, are the easiest to spot. Training binoculars or telescopes on the moon and then scanning the surrounding sky can help skywatchers locate the planet despite the glare from reflected sunlight. Conjunctions also offer wonderful photographic opportunities, even with smartphone cameras, allowing for beautifully framed wide-angle shots with landscape in the foreground.

(Opposite) To the right of a crescent moon above the Cerro Armazones Observatory in Chile, four planets are visible. From top to bottom: Venus, Mercury, Jupiter, and Mars

TRACKING SATELLITES

Earthbound observers can watch satellites parade across our skies—if they know where and when to look.

Spend enough time under a clear night sky any time of the year and soon enough a tiny dot of light will appear, sailing across the heavens. Some lights appear solitary, while others move in long chains, crossing the sky in a short few minutes from horizon to horizon. They may resemble moving stars, but these are human-made satellites, and hundreds are visible even from light-polluted suburbs.

Since the 1957 launch of Sputnik 1, thousands of spacecraft have been launched into Earth orbit, with more than 27,000 objects currently being tracked by the U.S. Department of Defense. The objects can range from bolts and paint chips to old rocket boosters, and of course include active and dead satellites used for communication, weather, research, Earth reconnaissance, and spying services. The larger the object, the easier it is to spot, with the largest ones easy to track with unaided eyes. Satellites are visible, just like planets and the moon, thanks to sunlight bouncing off their highly reflective metallic surfaces and mirrorlike solar panels.

HOW TO SPOT SATELLITES

With some patience, you can easily spot between 10 to 20 satellites per hour on any clear night of the year. But timing your hunt will be critical. Satellites are best seen within 90 minutes after sunset or before sunrise, when the sun is a few degrees below your local horizon. While it may be hidden from earthbound skywatchers, the sun will still be shining at altitudes where satellites orbit. Wait too long into the overnight hours, however, and the same satellites will be blanketed in the shadow of Earth, making them invisible to skywatchers.

(Opposite) Made with a long exposure near Salgótarján, Hungary, this image shows the movement of satellites as shining tracks across the sky.

The brightest and easiest are the low-earth orbit satellites, orbiting between 150 and 1,200 kilometers (93 to 746 mi) above Earth. One of the brightest and most famous is the International Space Station—the largest human-made object in space. Orbiting at an altitude of about 402 kilometers (250 mi), it is the size of a football field and circles Earth once every 90 minutes. Its huge array of solar panels is highly reflective, making the orbiting laboratory appear sometimes as bright as the planet Venus. But unlike that planet, the ISS moves across the sky over the course of many nights.

THE INTERNATIONAL SPACE STATION

In orbit since 1998, the ISS is operated by 15 countries and usually has six astronauts on board from all over the world, sharing as much living space as available in a Boeing 747. It travels at more than 28,000 kilometers an hour (17,500 mph) and circles Earth 15 times a day, encountering about 16 sunrises and sunsets every day. When the sun hits its large, reflective solar panels at just the right angle, the ISS can appear to ground observers as bright as magnitude -6, much brighter than Venus, and about the same apparent brilliance as the crescent moon.

Satellites are highly reflective. As more are launched, parent companies are looking into strategies to reduce their glare.

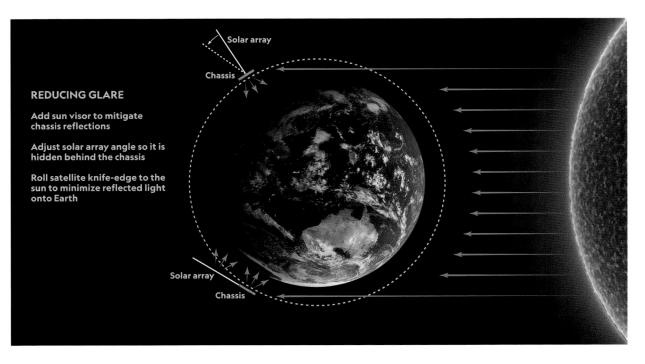

REDUCING GLARE

Add sun visor to mitigate chassis reflections

Adjust solar array angle so it is hidden behind the chassis

Roll satellite knife-edge to the sun to minimize reflected light onto Earth

The ISS, like other satellites, will appear to travel quickly, and it shines with a steady light, unlike the blinking lights of an airplane. Its orbit is inclined by 51 degrees, making it visible from a wide swath of Earth, ranging from 63 degrees north to 63 degrees south latitude.

Some satellites may appear to zip across the sky in a minute, while others can take several minutes to glide silently across the fixed backdrop of stars. The higher their orbit, the slower their track in the sky. Most will appear to travel from west to east, while others, particularly Earth reconnaissance and spy satellites, loop our planet from pole to pole, so they appear to move from north to south or south to north. Some satellites can offer more visually dramatic events. When a satellite appears to be moving toward the eastern sky, it can brighten dramatically as its metallic body or solar panels get hit

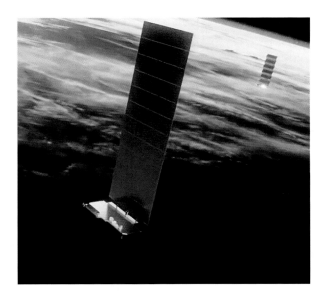

Elon Musk's SpaceX has launched a network of satellites called Starlink, illustrated here, and plans to launch thousands more, with the aim of providing high-speed internet access across the globe.

by more sunlight. Sometimes a satellite can give off bursts or flashes as sunlight hits reflective parts of the spacecraft body. Decommissioned, dead satellites and rocket boosters left in decaying orbits at times will emit a series of eye-catching light pulses as they tumble end over end.

Satellites can also sometimes be seen traveling in groups. The now defunct U.S. Navy's Naval Ocean Surveillance System (NOSS) was a set of three spacecraft that could be glimpsed flying in a triangular formation.

But the most dramatic, popular, and controversial of all satellite-spotting experiences has to be the Starlink trains. In May 2019, the private space company SpaceX began to launch its Starlink internet communication satellites, with the ultimate plan of putting into low-earth orbit a massive constellation of more than 40,000 objects. Each launch carries dozens of satellites, and in the days after their launch they can appear as trains of up to 30 satellites moving in synchrony across the sky. As the weeks pass, the satellites spread out and become less impressive.

Astronomers have expressed concern that so many visible satellites might cause a significant new form of light pollution, interfering with views of the night sky. To SpaceX's credit, the company has started to work on mitigation techniques to reduce its satellites' brightness—but only time will tell how successful those will be.

Various satellite spotting tools exist on the web, including Heavens-Above, Stellarium, and NASA's Spot the Station.

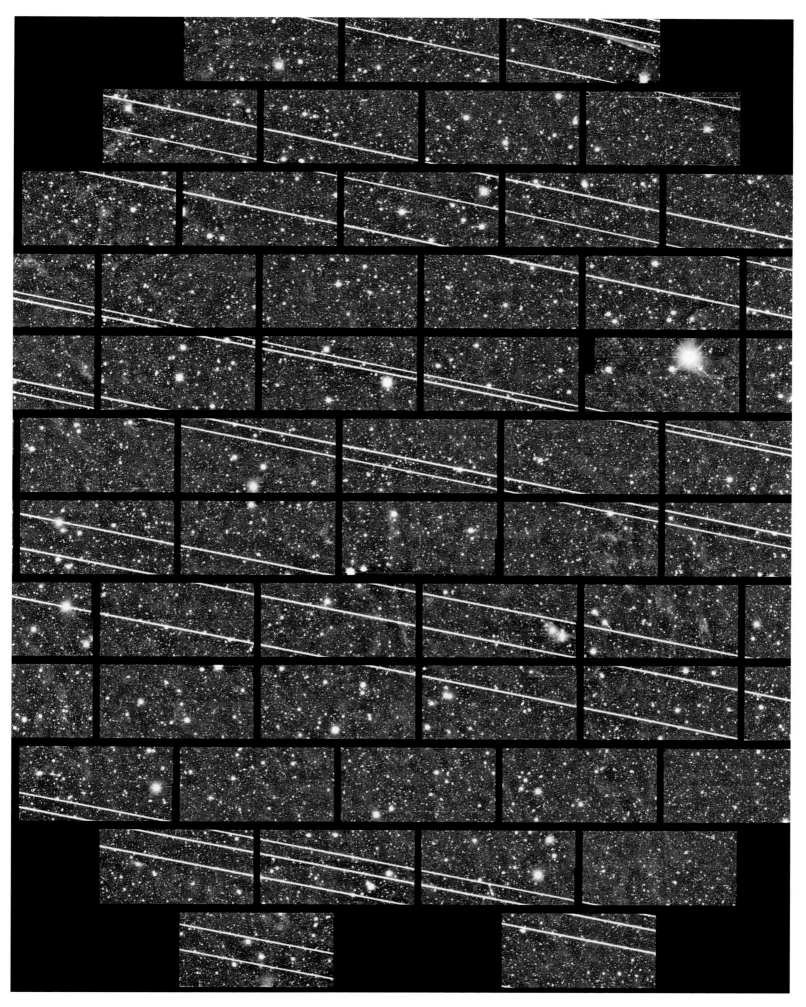

Satellites created streaks on images of the sky taken by the Cerro Tololo Inter-American Observatory in Chile.

GHOSTLY
SKY GLOWS

With their jaw-dropping visual appeal, auroras are awe-inspiring, must-see natural phenomena.

L ively, colorful displays of light naturally sparked by solar activity usually appear around Earth's polar regions and are the stuff of ancient folklore. But when strong geomagnetic storms move across our planet's magnetic field, they can unleash auroras producing a kaleidoscope of colors that can sweep across the night sky—even above heavily populated locations that don't usually get to experience them.

AURORAS

Considered by many to be one of the greatest natural sky glow shows, auroras—more popularly known as northern or southern lights depending on the hemisphere in which you observe them—come in an amazing variety of shapes, movements, and of course, colors.

Auroral displays can begin to appear very subtly, low in the sky. Over the course of a few minutes they will take shape near the horizon, appearing as patchy, indistinct glows. If you are lucky, the show intensifies, with the glows forming delicate streamers or rays of light running up the overhead sky. Particularly strong solar storms can create waving curtains of light for hundreds of miles, appearing from horizon to horizon.

Auroras occur when solar particles get funneled into the atmosphere, where they ride along Earth's magnetic fields and hit atoms, making them glow like a neon sign. Whether associated with weak or strong solar activity, by far the most commonly reported aurora color is green; this color is created by the ionization of oxygen atoms in the upper atmosphere. The red colors often seen come from the ionization of molecular oxygen and nitrogen. Auroral displays appear to be most frequent at the time of greatest sunspot activity, which peaks every 11 years. Expect the intensity and frequency of observable auroras to follow this general cycle.

While northern auroras tend be best visible near the far north, at or above the Arctic Circle, when solar activity is intense enough these eerie sky lights can be spotted across southern Canada and the northern United States. Viewers near the 40-degree latitude mark can witness auroras a few days each year on average. Around one percent of the time, during strong geomagnetic storms, auroras can overflow into more southerly locales, making them visible even in places like Texas and Florida.

Spring and fall, when Earth's protective magnetic bubble is weakened, tend to be the best times of the year to see the lights. One of the biggest obstacles to being able to catch auroras, however, is light pollution. The glow from cities beyond the curvature of Earth, along the horizon, has the ability to light up any scattered clouds, making it easy to mistake them for auroral glows. The darker your **observing site**, the better the chance to see even faint auroras. No need for binoculars or telescopes: The human eye is the perfect all-sky lens.

■ For more on viewing auroras, see pages 348–352.

ZODIACAL LIGHTS AND GEGENSCHEIN

Zodiacal lights are among the most challenging and ethereal of astronomical light shows. Visible as dim pyramid-shaped glows anchored across the horizon after sunset or before sunrise at certain times

(Opposite) In Lapland, Sweden, the northern lights (aurora borealis) do a colorful dance across the sky.

VALKYRIES AND DRAGONS

The mystical lights of auroras have been associated with ancient legends, folklore, and gods for cultures around the world. Vikings believed auroras were the reflections from the battle shields of the Valkyries. In northern China these dancing streaks of lights were said to be the exhalations of fire-breathing dragons.

Above the Laigu Glacier in Tibet, the gegenschein and the phenomenon known as airglow—the faint yellowish green luminescence of Earth's upper atmosphere—can be seen.

of the year, these astronomical phenomena are caused by countless tiny particles scattered across the plane of the solar system. As they reflect sunlight, they appear as a faint shine in Earth's skies. They are commonly mistaken for lights from a city just beyond the local horizon or for the pink-yellow light associated with the morning twilight. But they are caused mostly by dust from Mars being flung into space by the solar wind, and also in part by the original motes of dust left behind after the planets, moons, and asteroids formed some 4.6 billion years ago. Their moniker comes from the fact that this dust cloud appears densest along the zodiac or ecliptic—the path in the sky that the sun and planets follow. Therefore, observers across temperate latitudes, where the ecliptic appears, see zodiacal lights that appear tilted at an angle to the local horizon.

This interplanetary dust cloud becomes visible to the naked eye on very dark, moonless nights far away from light pollution. The best times to catch the zodiacal light are in the evening during the months of spring, looking toward the western horizon about two hours after local sunset, and then

again on fall mornings in the east, about two hours before sunrise. This is the same for both the Northern and Southern Hemispheres.

These dust clouds produce an even rarer celestial effect: the gegenschein (German for "counterglow"). Using averted vision, look for a faint brightening directly opposite the sun in the sky, only on the darkest of nights and near local midnight, when the gegenschein is at its highest. This extremely faint glow will be a few degrees across, equal to the width of 10 to 20 moon disks. The best time to catch the gegenschein is February and March and in October and November, when its glow is superimposed on dark, star-poor regions.

NOCTILUCENT CLOUDS

Every year from May through August, eye-catching displays of noctilucent clouds can grace the evening twilight across the Northern Hemisphere. These wispy clouds are found at the edge of space and are created by ice crystals forming around meteoritic dust particles falling into Earth's atmosphere.

Along with the moon (bottom, center) and Venus (top, center), zodiacal light can be seen above the Teide volcano in the Canary Islands, Spain.

In this long exposure, stars wheel above Casa Rinconada, a New Mexico archaeological site once used by ancestral Puebloans for rituals and community gatherings.

CHAPTER SIX

LET'S GO GAZING

TRAVEL THROUGH TIME

n myths and stories, humans walk among the stars, and the wonders of the heavens have inspired plenty of travel here on Earth. Ancient sites around the globe offer tantalizing clues to early astronomy. Dark-sky spots reflect our complicated relationship with the night in an age of overwhelming light. And at the spaceports that first sent humans beyond the reach of Earth's atmosphere, a new space race hints at what the future might hold.

The "overview effect," experienced by astronauts who have seen our planet from above, elicits feelings of awe and appreciation for the interconnectedness of life on Earth. Traveling has long been touted to produce a similar effect. To travel with a sense of the heavens, then, bridges these two perspectives—the scale and perspective afforded by thinking of the world as a pale blue dot fusing with the minute, specific delights we take in when visiting a new place for the first time. Some people travel to enjoy new cuisines, hear favorite musicians, or tour art galleries. Others head for the best stargazing spots or places where they can connect with the vast and varied history of our relationship with the night sky. The experiences we make when traveling help us connect with others and form meaningful memories that can influence our decisions. Even planning a future trip can be a source of present joy—conveniently, since travelers should always check ahead of time for sites' availability and potential restrictions.

And in a world where more people are traveling than ever before, travelers have more responsibility than ever. Many of the destinations attractive to those interested in the night sky are ones that require protection and careful management to preserve the features that enchant people in the first place. By limiting visitor numbers, encouraging activities with low environmental impact, and involving both locals and experts in policy decisions, we can keep night skies clear and historic places intact.

One day—maybe even one day soon—humans will journey to distant planets and chart their course under strange constellations. But for now, it's up to us to safeguard the places we already know and love on Earth.

At Chile's La Silla Observatory, in the Atacama Desert, the moon shines above a state-of-the-art telescope
and a rock marked with ancient petroglyphs.

ARCHAEOASTRONOMY

With their eyes on the sky, our ancestors built temples
and observatories we can still visit today.

There's a lot we don't know about how early humans lived; much of it we'll never know. But we can make a pretty good guess that they knew the night sky intimately.

Knowledge of the cosmos helped ancient peoples mark harvests, anticipate migrations, navigate vast distances, and plan rituals. Around the world, from petroglyphs to architecture, there's evidence that the heavens had a profound influence on how people lived.

Archaeoastronomy—a field, increasingly popular since the 1960s, that studies how past cultures interpreted celestial phenomena—enriches our understanding of archaeological sites. But it also encourages us to consider which parts of our ancestors' relationship with the night sky did not survive in the material record. What stories did they tell to interpret constellations? Did crowds journey far from home to watch auroras, eclipses, and meteor showers, as we do today?

Although these questions rarely have answers, many astronomy sites that ancient people left behind are open to visitors today—and for some, business is booming.

EUROPE

With a broad mix of prehistoric cultures and a well-established array of scientific institutions, Europe has been at the vanguard of archaeoastronomy since the discipline's inception. In fact, the oldest professional association of archaeoastronomers was formed in Europe in the early 1990s.

Sites of astronomical interest take a variety of forms across the continent. At Newgrange (Ireland) and Maeshowe (Scotland), sunlight shines down

tombs' narrow passageways on or near the winter solstice. Scholars have suggested celestial alignments at stone rings from Almendres (Portugal) to Karahunj (Armenia: geographically in Asia but politically in Europe). Ancient astronomy wasn't limited to architecture or engineering, either: At France's Lascaux Cave, for example, some scientists think the famous 17,000-year-old paintings might include early star charts.

At the end of the passageway into Newgrange, a 5,000-year-old Irish barrow, this spiral sits in year-round darkness until lit up by the winter solstice sun.

(Opposite) Interior of the passageways at Newgrange historic site in Ireland

There's much still to be discovered when it comes to clues left by Europe's ancient peoples about their relationship with the skies. But what we do know makes for fascinating, compelling visits to archaeoastronomy sites—and none is more compelling than Stonehenge.

■ For more on solstices, see page 87.

Stonehenge

Stonehenge is one of the world's most famous prehistoric sites, and its astronomical connection is among the most credible. A beloved symbol of British culture, recognized as a UNESCO World Heritage Site since 1986, the standing stones and their attendant earthworks have been studied for centuries, though much about their origin and purpose remains mysterious.

The complex consists of causeways, dozens of burial mounds (tumuli, or barrows), and the stone circle itself. The circle has three parts: an outer ring of 22-metric-ton (25-ton) megaliths quarried from nearby sandstone (called sarsen stones, after a local word meaning "pagan"); an inner arc of smaller bluestones likely dragged into place from Wales, 230 kilometers (143 mi) away; and a central horseshoe of sarsens. Construction began some 5,000 years ago with the circular earthwork, though another millennium would pass before the cromlech (stone ring) was raised.

The complex feat of engineering is a marvel in itself. But even more intriguing is the fact that the monument's central avenue aligns with sunrise on the summer solstice and sunset on the winter solstice. One feature of archaeoastronomical study is statistical analysis, used to determine whether such celestial alignments are accidental or intentional. But such analysis has its own imperfections. Some 19th- and 20th-century scientists used then-cutting-edge techniques to propose dozens of astronomical alignments—with the cycle of the moon, with certain stars and planets—only to be later discredited. And while Stonehenge's solstice alignment is easy enough to observe, it's still possible the original builders didn't intend that outcome, or that celestial alignments were secondary to the site's actual purpose.

Despite the mystery—or perhaps because of it—Stonehenge has captivated the public imagination. Typically prohibited from going within the ring in order to preserve the ancient site, tourists are allowed to gather there by the thousands on the summer solstice in an act that may mimic the ceremonies of the Neolithic and Bronze Age Britons who shaped this place over the centuries. (However, some scholars' recent work suggests that ancient Britons visited Stonehenge during the winter solstice, not the summer solstice.) More than a million tourists visit each year from around the world, bringing in about £23 million ($32 million) for English Heritage, the charity that manages Stonehenge and hundreds of other historic sites.

The megaliths these tourists come to see aren't

STAGES OF STONEHENGE

CIRCA 3000 B.C.
A circular ditch-and-bank monument, the "henge" in Stonehenge, is cut into the chalk of Salisbury Plain.

NEOLITHIC PERIOD
Timber posts are erected near the northeast entrance and across the center toward the southern entrance.

2500 B.C.
Circular or semicircular arrangements of stones appear, the earliest being pairs of bluestones. Also added: Station Stones, the Altar Stone, and the Heel Stone.

2500 B.C.
Stonehenge gains its iconic shape with the creation of the Sarsen Circle—30 worked stones topped by lintels.

2500 B.C.
Bluestones that had been cast aside are repositioned as a circle and a horseshoe within the Sarsen Circle, and a double ring of pits is dug.

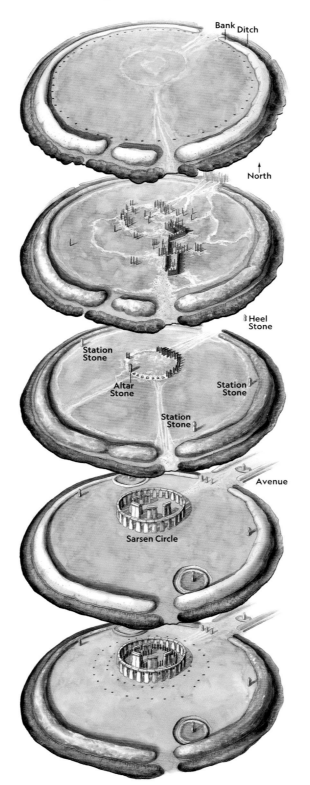

pristine or unchanging. Over the course of about a thousand years after the cromlech first was built, stones were rearranged or removed several times; modern restoration efforts have raised some stones that had fallen and supported others with concrete. Development—specifically a major highway that runs right within sight of the monument—provides its own challenges. But Stonehenge, having endured thousands of years with its symbolic power intact, is sure to remain a source of wonder for generations to come.

UNITED STATES

As in Europe, U.S. archaeoastronomy sites tend to be places where people designed architecture to align with celestial events, or where they left artistic renditions of phenomena such as supernovae, eclipses, or comets. But American sites have had to contend with half a millennium of colonization. Land has been stolen and Native sites have been disrupted or mismanaged, while Native knowledge is often ignored by scholars. Yet a rich heritage of historic Indigenous places remains.

One of the continent's most remarkable sites, Cahokia was a city of about 15,000 people—the largest urban center north of Mexico, and comparable to European cities of the time—that flourished in the 12th century in present-day Illinois. Organized around massive earthwork mounds (about 120 of them in its heyday), Cahokia also includes "Woodhenges": rings of towering red cedar posts planted in circles to serve as an astronomical observatory and gathering place.

Despite its fame, Stonehenge remains shrouded in mystery. Was it a meeting place, a center for healing, a monument to dead ancestors—or something else?

SEE FOR YOURSELF

STONEHENGE

Stonehenge lies just outside the city of Salisbury, a two-hour trip southwest from London. Because of the high volume of visitors, it's best to reserve tickets in advance. Begin at the visitors center, which offers a rotating exhibit of art and ephemera alongside multimedia displays of historical artifacts. Just outside, interpreters demonstrate cooking and crafting in replica Neolithic houses. Frequent shuttles span the mile-long road between the center and the stones. Visitors with general admission tickets are able to view the monument from a close distance, as a rope barrier blocks foot traffic within the stones themselves in order to preserve them. However, by paying extra for the Stone Circle Experience—an off-hours, hour-long, 30-person-maximum guided tour that tends to get booked months in advance—guests can see the sarsens up close. And though the circle itself is roped off, walkers can wander the surrounding landscape (much of it also administered by English Heritage) for the chance to spot grazing sheep, trace the remains of old earthworks, and explore ancient barrows (burial mounds).

MANHATTANHENGE

The New York City borough's gridded streets are a few degrees off from a true east-west axis. But for a couple of days a year, when the sun's azimuth lines up just right, it's possible to see the sun rise and set perfectly framed between skyscrapers. Neil deGrasse Tyson, director of the Hayden Planetarium, gave the phenomenon its nickname in the late 1990s. Today, an astrophysicist at the American Museum of Natural History (of which the planetarium is a part) calculates the dates in advance, since the streets aren't perfectly aligned to the solstice or equinox.

Farther east, in present-day Ohio, Serpent Mound was built around 1120 (although some scholars suggest it could have been built 1,400 years earlier). More than 400 meters (1,312 ft) long, the earthwork effigy depicts a twisting snake whose curves, some have suggested, align with the solstices and equinoxes.

Chaco Canyon

In the arid, echoing desert of present-day northwestern New Mexico, the ancestral Puebloan people built the complex known today as Chaco Culture National Historical Park. For the ancestral Puebloans, who lived in the American Southwest for more

than 2,000 years, Chaco Canyon was an important hub—for commerce and politics, for ceremony and culture. They built great houses with hundreds of interconnected rooms. They built kivas—circular, underground or semi-underground ritual spaces—of unique size and variety. And there's evidence they may have used the landscape and their own architecture to predict celestial events.

From the 800s to around 1250, Chaco Canyon flourished as an urban epicenter in the Southwest. The dozens of great houses—many aligned with the cardinal directions or with solar and lunar patterns—were built in different phases over the centuries. Larger houses, such as Pueblo Bonito (nearly 650 rooms) and Una Vida (about 175 rooms), tend to have been built earlier than smaller houses such as Wijiji and Pueblo Alto. Some of these great houses, such as Chetro Ketl, incorporated kivas within their symmetric floorplans. Freestanding kivas tended to be larger, and these great kivas were among the first public buildings in the region. The Chaco Canyon complex was also connected by a sophisticated road network to more than a hundred other great houses throughout the Southwest. Along with Aztec Ruins National Monument and five smaller federally managed archaeological sites, the 137-square-kilometer (53-sq-mi) Chaco Culture National Historical Park was recognized as a UNESCO World Heritage area in 1987.

Although Chaco Canyon's vast dwellings might have housed people year-round, descendants of the ancestral Puebloans say that the complex acted more as a gathering place for clans across the Four Corners region to meet for trade, celebration, and ceremony. Kivas are an important lifeline between modern Pueblo peoples and ancestral Puebloans, who probably shared beliefs that kivas represent the places where people emerged from the underworld into the current world. And while kivas were likely the foundation of spiritual practice at Chaco, the settlement's astronomical connections are found at a variety of sites: Fajada Butte, Wijiji, and—to some extent—Casa Rinconada.

Fajada Butte is a 2,000-meter (6,600-ft) promontory in the southeastern area of the canyon; at its peak are three stone slabs that arrange light and shadow on two spiral petroglyphs in a way that marks the solstices and equinoxes. This Sun Dagger site—named for the way a shard of sunlight bisects

Chaco Canyon's protected area shelters animals such as elk, bobcats, badgers, and porcupines.

Wijiji, one of the latest of the Chaccan great houses, is unique for its lack of great kivas and enclosed plazas.

At 19 meters (64 ft) in diameter, Casa Rinconada was one of Chaco's primary ceremonial sites. A road between it and Pueblo Bonito suggests the sites were ritually linked.

one of the spirals on those four days—is proof of the ancestral Puebloans' technological precision and cosmological knowledge. It's likely the spot had more of a ceremonial function than a daily calendrical one, given its remote perch on a sheer, towering butte. But it still indicates an important relationship between Chaco's inhabitants and the sun.

The Sun Dagger phenomenon captivated the public when it was publicized in the late 1970s. Tourism soon grew to unmanageable levels. In

1982, the National Park Service (NPS) closed access to Fajada Butte to protect the site. Then, in 1989, it was discovered that foot traffic had worsened the settling of one of the stone slabs, throwing off its alignment.

To the east of Chaco's visitors center lies Wijiji, one of the last great houses built by the ancestral Puebloans, completed around 1100. There are a number of solar observing spots around the house. But the real point of interest lies a little under a

PEOPLE OF THE FOUR CORNERS

From A.D. 100 to 1600, the Four Corners region—where present-day Utah, Colorado, Arizona, and New Mexico meet—was home to the ancestral Puebloans. (Non-Native scholars previously called this historical group the Anasazi, meaning "ancestors of the enemy" in Navajo; this term is no longer accepted.) Ancestral Puebloans, who were farmers, often came into conflict with more nomadic Apache and Navajo peoples. Conflicts increased as the Great Drought (1276–1299) strained food and water resources. By the early 1300s, ancestral Puebloans had abandoned their cliff palaces and great houses for more easily irrigable places to the south and east.

Modern Pueblo peoples (including the Zuni, Hopi, Laguna, and Acoma) retain their traditional languages and cultures. Their link to ancestral Puebloan sites is clear. But Chaco Canyon is also important to the nearby Navajo (right)—who trace several clans and ceremonies to the canyon—and to other Southwest nations, who consider it a spiritual place to be honored and protected.

mile farther east: a ledge marked with ancestral Puebloan petroglyphs (as well as later Navajo symbols) that was probably used to celebrate the sun's renewal on the winter solstice. In the 12th century, the solstice sunrise would have been visible from several vantages on the observer's ledge. The sunset, on the other hand, would have set precisely within a V-shaped notch in the rock—but only if viewed from a specific position.

Today, Casa Rinconada is perhaps the most famous example of archaeoastronomy at Chaco. One of the largest great kivas in the canyon, Casa Rinconada is oriented to the four cardinal directions; in fact, if its north axis were extended, it would connect with the great houses of Tsin Kletzin and Pueblo Alto, which are similarly oriented. The kiva is considered a calendrical marker of the summer solstice: For seven weeks leading up to the solstice day itself, sunlight enters an opening and tracks south along the wall until it illuminates a particular niche on the solstice. Park staff welcome visitors to the kiva to view this phenomenon, and guided tours describe it at other times of year. However, some scholars have pointed out that parts of Casa Rinconada were reconstructed in the 1930s in such a way that casts doubt on whether this solstice alignment was original to the ancestral Puebloans' version of the kiva.

MESOAMERICA AND SOUTH AMERICA

The famed Maya calendar has been widely misinterpreted by conspiracy theorists eager to press an ancient, still-thriving culture into service predicting the end of the world. Beneath the hokum, though, the astronomical accomplishments of the Maya—and of other cultures in Mesoamerica and South America—are a true marvel.

Since Western scholars became aware of it in 1911, Machu Picchu has captivated people from around the world. Now thought to have been a mountain retreat for Inca royals, the remote complex is remarkably well preserved—including the Intihuatana, or "hitching post of the sun," a complexly carved stone that aligns with the winter solstice and was likely used as an astronomical calendar.

Created centuries earlier, roughly 650 kilometers (404 mi) to the southwest, the Nasca lines are

SEE FOR YOURSELF

CHACO CANYON

This canyon lies in a remote, demanding landscape only accessible by dirt roads, so visitors should plan ahead. Check the official NPS site for driving directions from the major highways and call ahead for the latest road conditions, since some places may be impassable depending on weather. The park is a one- to three-hour drive from the nearest airports: Durango, Colorado, or Farmington, New Mexico, to the north, and Gallup or Albuquerque, New Mexico, to the south.

Although the park doesn't offer food or lodging, roughly 50 campgrounds are available by reservation. Park entrance tickets can be purchased in advance and allow seven days of admission. Start at the visitors center, open year-round (except for three holidays), to pick up self-guided trail info for sites along the 15-kilometer (9-mi) Canyon Loop Drive or to check the schedule of ranger-led tours. Backcountry trails offer dazzling vistas and glimpses of ancient petroglyphs, but permits are required. Overnight camping is prohibited.

The park is home to an observatory, and rangers offer dark-sky programming two nights a week from April through October (check for availability). During the equinoxes and solstices, visitors can observe sunrise alignments at Casa Rinconada or Kin Kletso.

another Peruvian site recognized by UNESCO for cultural and archaeological importance. Inheriting a tradition of petroglyph art from previous cultures, the Nasca people spent about a millennium carving lines into the plain, removing the oxidized surface material and exposing the paler rock beneath to create anthropomorphic figures and geometric patterns best viewed from above. Though some archaeologists staunchly argued for an astronomical interpretation of the Nasca lines, later scholars tend to favor the theory that the lines are linked to water rituals.

This 2,000-year-old earthwork in present-day Ohio was built by the Adena people to represent a giant snake holding an egg—hence its name, Serpent Mound.

The UNESCO-recognized Nasca lines—enormous, fragile, mysterious petroglyphs scored into Peru's coastal desert—include representations of animals, like this monkey, as well as abstract lines and shapes.

Chichén Itzá

On the tip of the Yucatán Peninsula, halfway between Mérida and Cancún, the magnificent stepped temples of Chichén Itzá rise from the forest. The Maya settled the area in the mid 400s, drawn by the easy water access provided by two cenotes (sinkholes that expose groundwater). These cenotes gave the place its name: chi (mouths), chen (wells), Itzá (the name of a Maya tribe who later resided there)—"at the mouth of the well of the Itzá." A slew of early buildings, including an observatory named El Caracol, preceded the grander architectural projects of the 10th century, like El Castillo, the 24-meter (80-ft) pyramid that has become perhaps Chichén Itzá's most famous spot. Construction had halted by the 13th century, and by the time of the Spanish conquest, the Maya of the Yucatán had abandoned cities like Chichén Itzá for smaller towns.

During its heyday, Chichén Itzá was the most prominent city among the northern Maya—a capital of sorts, overseeing the religious, political, and cultural lives of perhaps 35,000 people. Using advanced technology and mathematics, these people built massive, precise pyramids, ball courts, temples, colonnades, and mortuaries that have endured for centuries largely intact. Adorned with sculptures and brightly painted, these buildings also reflected the Maya's detailed understanding of calendars and astronomical events. For example, El Castillo's four sides face the cardinal directions, its 365 steps reflect the solar year, and its surface ripples with the shadow of an undulating snake on the equinoxes.

The architecture of El Caracol (The Snail) is similarly attuned to the mathematical patterns of the universe. Part of a cluster of buildings in the south-

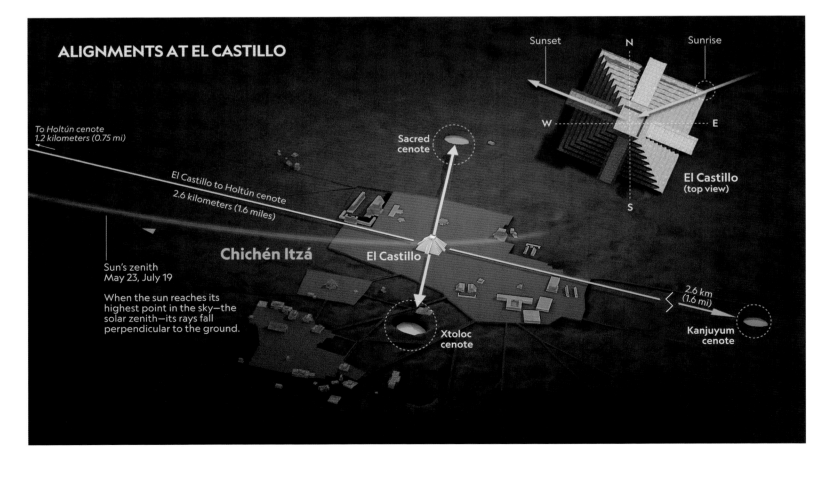

ALIGNMENTS AT EL CASTILLO

Sunset

Sunrise

N

W — — — E

S

El Castillo (top view)

To Holtún cenote
1.2 kilometers (0.75 mi)

El Castillo to Holtún cenote
2.6 kilometers (1.6 miles)

Sacred cenote

Chichén Itzá

El Castillo

Sun's zenith
May 23, July 19

When the sun reaches its highest point in the sky—the solar zenith—its rays fall perpendicular to the ground.

Xtoloc cenote

2.6 km (1.6 mi)

Kanjuyum cenote

ern part of the city, El Caracol—so named for the spiral staircase within its central tower—rises from two platforms that were designed to face important horizon events involving the sun and Venus, which bore special significance for the Maya. Dedicated to the powerful Mesoamerican god Quetzalcoatl (Kukulcán in Maya), the building likely served as an observatory, giving astronomers a vantage to track celestial bodies, since the flat Yucatán lowlands had few horizon features to act as reference points. The Maya's understanding of astronomy was so profound that they were able to predict solar eclipses.

AFRICA

One of the world's oldest complex civilizations arose in the long, sinuous valley of Egypt's Nile River. The ancient Egyptians we know from pyramids and hieroglyphs had a labyrinthine and evolving cosmology, which included a precise knowledge of the stars and seasons. But an even earlier civilization—nomadic cattle herders who roamed Nubia, traveling from one seasonal lake to another—might have

left a lasting influence on how the ancient Egyptians viewed the world. Some 1,100 kilometers (680 mi) south of the spot that would eventually host the Pyramids at Giza, this early civilization built a stone circle at Nabta Playa around 7,000 years ago, making it the world's oldest. The cromlech, buried in sand for thousands of years, was rediscovered in 1973. After years of studying the site, archaeologists

SEE FOR YOURSELF

CHICHÉN ITZÁ

Chichén Itzá is open to the public every day of the year and usually receives about 3,500 tourists a day—a number that doubles during peak season, especially on the equinoxes—meaning it faces the challenges of overtourism that also plague sites like Machu Picchu. Although it's necessary to tread lightly anywhere you visit, extra caution is required at Chichén Itzá to prevent degradation of the site itself. From international airports at Mérida and Cancún, visitors can take a bus service to the complex in about two hours. Those who find lodging in the area—for instance, in the nearby town of Pisté—have the chance to go see the site early, before day tours arrive from the cities, or later at night, when a light show is held.

Part of the complex of Chichén Itzá, the imposing Temple of the Warriors features four platforms and relief carvings of warriors, eagles, and jaguars.

The vast sweep of the Milky Way arcs above the Great Sphinx and the Great Pyramid of Khufu in Egypt.

began to believe it had served as an observatory or calendar: Around 4800 B.C., when the night sky was different, the brightest northern stars would have aligned with Nabta Playa's standing stones. Sadly, after information was published about the site in 1998, tourists found and damaged the site, and the Egyptian government moved the stones to a regional museum.

The area's desertification forced the people of Nabta Playa to migrate; some headed north, leading several scholars to suggest Egyptian civilization arose from these African peoples rather than spreading from Mesopotamia. This theory points to the resonance between Nabta Playa's cow worship and later Egyptians' reverence for Hathor, the cow goddess of fertility. However it arose, though, the civilization of ancient Egypt developed a unique cosmogony, making sense of their world through a pantheon of anthropomorphic deities and a devotion to preparing for the afterlife. Noticing the recurring patterns around them—the rising and setting of the sun, the annual flooding of the Nile, the wandering of the stars—the ancient Egyptians kept careful records and designed precise calendars.

For the /Xam San people of Africa, **the Milky Way** was created when a girl flung a handful of ashes into the sky.

Great Pyramid of Khufu

One of the Seven Wonders of the Ancient World— and the only one to survive to the modern age—the Great Pyramid of Khufu is best known as the extravagant burial complex of the Pharaoh Khufu. But the pyramid, along with its (slightly) smaller neighbor, the Pyramid of Khafre, was also engineered with careful attention to the paths of celestial bodies. As the site where the two pharaohs were sent into the afterlife with all the servants and possessions they'd need there, it was important that the pyramids were aligned with the cosmic patterns that reflected ancient Egyptians' understanding of the universe as a cyclical, ordered place. Some researchers have suggested that the Great Pyramid's inner shafts seemed to align with the two northern polar stars that would have been visible in that era, as well

For more on north stars, see page 227 (Draco).

In the Nubian Desert, the standing stones of 7,000-year-old Nabta Playa may once have been used as an astronomical calendar.

The falcon-headed creator god Ra traveled in a solar boat across the sky each day and fought a serpent in the underworld each night.

powerful signs: The circumpolar stars, always visible in the night sky, symbolized immortality, while the constellations represented the heavenly destination of pharaohs' souls.

The funerary complex at Giza includes the two larger pyramids, as well as a third, smaller pharaonic pyramid (that of Menkaure); the pyramids of several queens; workers' quarters and burial grounds; temples; boat pits; nobles' cemeteries and mastabas, or flat rectangular tombs; and the Great Sphinx. Built over the course of a century around 2500 B.C., the structures of this complex create additional astronomical alignments. Each of the three pyramids is built so that its four faces are perfectly oriented to the four cardinal directions. In fact, the Great Pyramid's north-south line is so precise that, even today, you can see how far your magnetic compass deviates from true north by placing it on the marble casing on the building's north face and observing the way the needle strays from the edge of the block. It's likely that this accuracy was calculated by observing the machinations of the heavens. The Great Sphinx, which lies on the eastern end of the complex, faces east itself, pointing toward the rising sun, Ra, the Egyptians' most important god. On the summer solstice, the sun sets at the midpoint between the pyramids of Khafre and Khufu when viewed from a certain point—re-creating the hieroglyph Akhet, which represents the horizon.

Some of the suggested astronomical alignments at Giza lean toward the dubious. In the 1980s, a popular theory began to circulate: What if the three great pyramids were deliberately built to mirror the three prominent stars of the constellation Orion? Duat, the entrance to the realm of the dead through which the sun god Ra made his way each night, was located within Orion; perhaps the pyramid builders sought to re-create this heavenly journey on Earth.

as with the constellation Orion and the star Sirius (although the shafts are blocked in places by other features of the pyramid). The degree to which the pyramids were built with the stars in mind remains a subject of hot debate. But if these alignments were deliberate, they connected these buildings with

SEE FOR YOURSELF

THE GIZA COMPLEX

An important source of revenue for Egypt, the Pyramids at Giza haven't returned to the levels of tourism they enjoyed before the Arab Spring reached the country in 2011. Several countries, including the U.S., have issued travel warnings or restrictions.

Those committed to visiting will likely fly into Cairo: The airport is barely 40 kilometers (25 mi) northeast of the pyramid complex, a distance easily traversed by taxi, rideshare app, bus, or metro. Different levels of admission are possible at two ticket booths: one near the main entrance and another by the Great Sphinx, which offers fewer crowds, especially in the morning. An extra charge paid at the main entrance lets viewers enter the pyramids themselves—two are open to the public while the third is being restored, a shift system that rotates every couple of years—though the interior passageways are often hot, cramped, and difficult to move around in. Smaller site features and the Giza Solar Boat Museum are oft-overlooked opportunities to avoid crowds.

Although a famous tourist rite of passage in years past, ascending the pyramids is now strictly forbidden, both for tourists' safety—several have died in the attempt—and to preserve the monuments themselves. Tours by horse or camel are another popular option, but be sure to check for reputable operators who take their animals' welfare seriously. Food and lodging are easily available near the complex itself.

Pyramid of Khafre

Pyramid of Khufu

Pyramid of Menkaure

Great Sphinx

However, this is considered a fringe theory. There's no evidence in the buildings' construction or in associated texts to confirm it; the pyramids would have to be flipped around to fit an Orion alignment; and the constellation didn't appear in the same place 5,000 years ago, anyway. It's true we don't have all the answers. Mysteries still surround the star-aligned shafts of the Great Pyramid, for example. But the allure of finding incredible connections between ancient sites and the stars sometimes leads to popular beliefs that are just that: incredible.

ASIA

Home to some of the world's oldest and most enduring cultures, the vast continent of Asia also enjoys a long legacy of astronomical discovery. For many modern-day Asian cultures, the stars and constellations have a profound and lasting influence on people's personalities and life prospects; these beliefs trace their roots to the region's historical cultures as well. Astronomy and astrology, which now face each other from opposite sides of the firmly drawn line of science, were once one and the same practice: The movements of celestial bodies advised rulers on matters of war and policy.

Such was the case at South Korea's Cheomseong-dae, which some claim to be the oldest astronomical observatory in East Asia. The nine-meter-tall (30-ft) stone tower was built in A.D. 647 during the reign of the first ruling queen of the Silla dynasty, which governed the Korean peninsula for nearly a millennium. The observatory's 365 granite blocks might represent the number of days in the lunar year, and its sole window lies between 12 layers of

At dusk, the Pyramids at Giza (above) are illuminated as part of an evening sound and light show. Seen next to human beings, the scale of the masonry blocks (below) becomes evident: The Great Pyramid is made of more than two million stones, each weighing on average almost 2.3 metric tons (2.5 tons).

Cheomseongdae Observatory is one of many monuments found in Gyeongju, a city in southeastern South Korea famous for its rich cultural history.

stones above and 12 layers below, which could symbolize the signs of the zodiac.

But whether or not these numerological coincidences were intentional, we do know that Cheomseongdae was used for rigorous scientific work. Making observations every hour of the day and every day of the year, official astronomers learned how to chart the paths of comets and eclipses. Along with temples, pagodas, royal tombs, and other monuments, the Cheomseongdae Observatory forms part of a historic area recognized by UNESCO for its "outstanding examples" of Korean Buddhist art. The stone tower even graces the back of North Korean coins.

Another, earlier observatory can be found at the archaeological site of Taosi, in northern China's Shanxi Province. An ancient provincial capital, the city flourished from 2300 to 1900 B.C. Its people built a palace complex, housing areas for nobles and commoners, ceremonial structures, craft districts, and cemeteries that—despite being almost totally destroyed by a rival regime—now form one of the richest historic areas in the country. During Taosi's Middle Period (2100–2000 B.C.), the citizens also built an observatory: an arc of rammed-earth blocks and a colonnaded platform that was used to

The world's **oldest surviving written record of a solar eclipse** was set down in northeastern China on July 17, 709 B.C.

make observations of the horizon and devise a calendar. This observatory site, which might have also functioned as a celestial altar, indicates the importance of Taosi at the time, linking knowledge of the heavens to the city-state's power on Earth.

Göbekli Tepe

In the low, sweeping hills of southeastern Turkey, not far from the Syrian border, rises the site of Göbekli Tepe, thought to be the oldest megalithic structure in the world. Between 9600 and 8200 B.C., Mesopotamian hunter-gatherers built rings of limestone megaliths, some more than five meters (16 ft) tall, some blank and some carved with the sinuous shapes of animals. Over centuries, the builders covered completed rings with dirt and began again atop the previous layer, resulting in a mound that gives the site its modern name—"belly hill" in Turkish.

Göbekli Tepe was first investigated in the 1960s

Visitors take in the excavation site of Taosi, a 4,200-year-old city that was called Pingyang during its time as the capital of legendary Chinese emperor Yao.

BUILDING GÖBEKLI TEPE

People must have gathered from far-flung settlements to erect the first known temples. Using flint tools, they carved pillars and shaped blocks for walls mortared with clay. When a new temple was completed, the old one was buried.

Human muscle moved the limestone pillars, weighing up to 16 tons, from quarries as far as a quarter mile away.

The inner ring had no door and may have been accessed with ladders. Animal pelts may have hung on the pillars as offerings.

Head
Arm
Belt
Hands
Animal-skin loincloth

Carvings mark the pillars as stylized human figures, but did they represent powerful people or supernatural beings?

Offerings

A pillar's shape was refined before being carved and placed.

by anthropologists who thought it nothing more than a medieval cemetery. But when a German-led archaeology team reexamined the site in the 1990s, they realized that tools used at the site were comparable to others at nearby sites that had been dated to 9000 B.C.; carbon testing later confirmed this timeline. Because Göbekli Tepe doesn't show much

evidence, if any, of permanent human settlement, it has thrown a wrench into scientists' understanding of societal evolution. Were monumental structures only possible after the development of sedentary, agricultural societies, or did the effort required to build monumental structures spur the development of agriculture?

Klaus Schmidt, who directed excavations at the site from 1995 until his death in 2014, believed Göbekli Tepe to be the world's oldest temple—perhaps a burial ground, with the bodies of ancient Mesopotamian hunters interred among a ritual landscape made holy by more than a thousand years of devotion. Other researchers think the pillars' T-shape represents a human body without a head, possibly suggesting a skull cult similar to those later found in the area. And some scientists have proposed Göbekli Tepe has astronomical connections—that it was deliberately oriented to certain stars, or that some of its carvings depict a comet impact that some think started an ice age called the Younger Dryas 13,000 years ago. Other archaeologists excavating the site object to such claims, however. They argue that even Göbekli Tepe's earliest phases date to the end of the Younger Dryas rather

SEE FOR YOURSELF

GÖBEKLI TEPE

When Göbekli Tepe was listed as a UNESCO site in 2018, that act paved the way for the increase in tourism typical of such international recognition. But political unrest in Turkey that year resulted in a U.S. State Department travel advisory, especially in the southeastern border province where Göbekli Tepe is located—and where ISIL terrorism activity was most dangerous. Even a national tourism campaign focusing on Göbekli Tepe the following year wasn't able to fully recoup the losses. As a result, Göbekli Tepe has received fewer visitors than would be expected for a site of such importance—just under half a million annual guests in its most successful year.

The spring, from March to May, is the best time to visit. Since much of the site is still unexcavated, head to the nearby Şanlıurfa Archaeology and Mosaic Museum for context—including artifacts and a replica of the temple—before visiting Göbekli Tepe itself. Şanlıurfa is the nearest city, about 17 kilometers (10 mi) west of Göbekli Tepe; a historic city in its own right, it offers food and lodging, as well as Byzantine ruins and a medieval mosque.

than its beginning. They also point out that it would be impossible to link specific constellations to specific structures, given that many columns' original positions are unclear due to alteration by later Neolithic construction—and given that it's tough to assume the same asterisms we know would mean anything to the people of Göbekli Tepe. Yet another group of scientists developed new archaeoastronomical methods to conclude that one of the site's enclosures is oriented to the star Deneb, another to the lunar standstill (when the moon hovers in the most extreme point of its transit in an 18.6-year cycle), and a third to the rising sun on a harvest festival halfway between the summer solstice and the autumn equinox.

But as Schmidt and others have pointed out, the enormous gulf of time separating us from the people of Göbekli Tepe frustrates any attempt to understand their symbols, intentions, or beliefs. Unlike many of the other archaeological sites discussed in this chapter, Göbekli Tepe offers no

NAVIGATING BY THE STARS

Cultures around the world have long used the stars as a navigation tool. But among the seafaring peoples of Polynesia—the area roughly triangulated by Hawaii, Easter Island, and New Zealand—celestial navigation is an art of astounding power. Using it, Maori sailors may have reached Antarctica in the seventh century, about 1,200 years before Europeans. In the Caroline Islands of neighboring Micronesia, navigators memorize complex star paths (the points on the horizon where clusters of stars rise and set over an island) in order to sail hundreds of miles, sometimes on routes they've never before traveled. Like many aspects of Indigenous knowledge, traditional celestial navigation has been obscured by colonialism. But as cultural revitalization movements take hold, knowledge keepers are finding those in the next generation who will sail by the stars.

written material or extant human lineage to help us parse its meaning. Its megaliths were carved six millennia before the invention of writing. Perhaps the 90-some percent of the site's area yet to be excavated will offer more answers. Until then, Göbekli Tepe's connection to the stars may be more a matter of belief than evidence.

Tall T-shaped pillars rise from the center of circular structures beneath a canopy and walkways installed to lessen tourists' impact on excavated areas at Göbekli Tepe.

ASTROTOURISM TODAY

From dark-sky sites to solar eclipses, our contemporary skies are a sight to behold.

What makes astrotourism unique? Plenty of traditional travel experiences revolve around experiencing the wonders of nature, whether it's watching a red desert sunrise, seeing the stars wheel above the mountains, or walking along the beach in the moonlight. Although it's unclear how the word was coined, "astrotourism" arose around the turn of the 21st century to describe a fast-growing subset of ecotourism that focuses on travel specifically geared toward space- and sky-related phenomena. If you're going stargazing, or touring observatories and planetariums, or chasing fleeting sights like eclipses and auroras, then congratulations—you're an astrotourist.

In such a new industry, it's difficult to quantify just how many people travel for astronomy, or exactly how much this kind of tourism contributes to destinations' coffers. But as parks, cities, and nations support more astronomical education programs, adopt more measures to control light pollution, and recognize the value of connecting visitors to our spectacular skies, it's certain that astrotourism will only continue to grow.

DARK-SKY SITES

With little need for specialized equipment or experience, stargazing should be one of the most accessible astrotourism activities—the night sky is available to everyone, after all, if you just go outside and look up. There's just one problem: Light pollution hides the stars from more people than ever.

By 2016, 99 percent of Europe and the continental U.S. had fallen under the glare of some kind of light pollution. There's no longer any spot on the planet

(Opposite) Maine's Katahdin Woods and Waters National Monument is recognized as an International Dark Sky Sanctuary.

from which astronomers can observe the stars free of light cast by space junk. A third of Earth's people can't even see the Milky Way.

Light pollution casts a haze that makes it hard for us to see the stars—a profound human loss we might not even be aware of. Stars have guided us through time and space, inspiring our stories and songs; they're deeply linked to so many aspects of culture and memory. To lose sight of them is to lose part of the richness of our life on Earth. Yet aside from that spiritual loss, light pollution—especially that from blue light—is also responsible for disrupting the very real physical rhythms that guide many animals' daily lives. It upsets the biochemical patterns related to the hormone production that governs our sleep and appetite, leading to long-term health risks. It disorients birds, causing off-kilter migrations and fatal collisions; it sends confused sea turtles wandering into traffic rather than the moonlit ocean. It throws off the delicate timing of coral spawns, reduces the habitat options for nocturnal animals, and decreases nighttime pollination.

Although the effects of light pollution are grave and widespread, the tide's not irreversible. Preserving dark-sky sites—and adopting policies to

■ For more on space junk, see page 311.

Many species migrate by the stars, but **dung beetles** are the only ones confirmed to navigate by the Milky Way.

Dawn breaks over Siding Spring Observatory and Warrumbungle National Park.

decrease light pollution in other places, in order to bring them closer to dark-sky status as well—can help. Tourism to these places can provide revenue that helps preserve them, as well as giving them the opportunity to educate visitors about how to reduce light pollution at home.

The International Dark-Sky Association (IDA) is one of the organizations on the vanguard of the fight against light pollution and for the protection of dark-sky spots. Since its International Dark Sky Places program launched in 2001, the Arizona-based nonprofit has used a tiered certification system to recognize places—urban enclaves, private and public parks, communities, nature reserves— that limit artificial light while educating residents

(and visitors) about the benefits of dark skies. So far, the IDA has noted more than 180 sites across the world, and applications quintupled between 2010 and 2018. The majority fall within the U.S. and Europe, though recognition is growing in South America, Africa, Asia, and Oceania. The effect is a powerful one: Many IDA-recognized sites see a spike in visitors after their new designation.

As dark-sky tourism increases, though, it's important to keep in mind the effects of overtourism. Dark skies are possible alongside human habitation—see the past couple million years of genus Homo's existence—but when it comes to tourism infrastructure, bigger crowds mean more lights. The IDA isn't an enforcement agency; it comes down to local governments to regulate how many people access dark-sky sites, in order to preserve the nature of what makes them special in the first place—and it comes down to tourists willing to be responsible in caring for the places they love.

Warrumbungle National Park, Australia

Australia, with its vast tracts of open land and 50,000 years of Aboriginal culture, is an astrotour-

ist's dream. Warrumbungle National Park is one of the continent's three IDA-recognized sites and its only International Dark Sky Park (IDSP). The 233-square-kilometer (90-sq-mi) park, about six hours' drive northwest of Sydney in New South Wales, has predictably clear skies—with views made even crisper by the low humidity and high altitude. Although it's a working research facility that's closed to the public at night, the nearby Siding Spring Observatory does welcome visitors during the day to wander its exhibition area or take a pre-booked tour of Australia's largest optical telescope. The observatory also hosts a StarFest event every October.

Warrumbungle's dramatic rock formations, including the aptly named Breadknife, make for excellent hiking and camping. The park experiences big swings in temperature and rainfall, making the more temperate spring the best season to visit (with the added bonus of blooming wildflowers).

Yeongyang Firefly Eco Park, South Korea

South Korea is one of the world's most population-dense countries, and more than 80 percent of its citizens live in urban areas; as a result, light pollution is a particularly pressing issue. Yet less than five hours outside Seoul, the country's bustling capital, lies the small Yeongyang Firefly Eco Park, which in 2015 became the first place in Asia to be recognized by the IDA as an IDSP. The 3.9-square-kilometer (1.5-sq-mi) spot is tucked within a river valley surrounded by mountains within a larger ecological reserve, and has been managed to protect its dark skies since 2005. Those who come to stargaze can also enjoy a luminous display that's much more down to Earth: the emergence of fireflies, which thrive in the park's dark nights.

Hortobágy National Park, Hungary

Hungary's first national park protects 75,000 hectares (185,300 acres) of ancient grasslands and

Dense star fields fill the skies over Hortobágy National Park.

wetlands, a fragment of the Eurasian Steppe that's seen little change since the last ice age. Enshrined by UNESCO for its cultural heritage and ecological value, Hortobágy National Park gained IDA recognition as an IDSP in 2011. Few people live within the park, but from April to October, traditional pastoralists graze their animals across the broad, flat meadows. This light human footprint makes the park a rich habitat for hundreds of bird species, as well as an essential rest stop for migratory birds such as cranes and geese. Hortobágy's light management plan minimizes effects on wildlife while also encouraging astrotourism. The park also offers night walks, educational programs, and a public astronomical observatory.

!Ae!Hai Kalahari Heritage Park, South Africa

In 2002, the South African government restored nearly 95,000 hectares (234,750 acres) along the Botswanan border to the Indigenous ‡Khomani San and Mier peoples, about 50,000 hectares (123,500 acres) of which now form the !Ae!Hai Kalahari Heritage Park. (Within some southern African languages, special characters like ‡ or ! represent different types of click consonants.) Jointly administered by South Africa's national park service and ‡Khomani San and Mier community members, the park preserves one of the world's darkest night skies in part by having only one lodge with permanent lighting, all retrofitted for less light pollution. Although the park's savannah calls to mind classic safari tourism, the area has grown increasingly interested in fostering astrotourism—in a way that respects and includes the sky's cultural importance to the ‡Khomani San and Mier.

Gabriela Mistral Dark Sky Sanctuary, Chile

On the southern edge of the Atacama Desert, in one of the world's premier astronomical research sites, lies the Gabriela Mistral Dark Sky Sanctuary. Where

■ For more on Chilean observatories, see page 379.

The Gabriela Mistral Dark Sky Sanctuary was named for the first Latin American laureate of the Nobel Prize in Literature, who grew up in the Elqui Valley.

Salt flats stud the Atacama, providing feeding grounds for flamingos, which get their bright pink coloring from their brine shrimp diet.

Craters of the Moon National Monument, Idaho

More than 10,000 years of volcanic eruptions—leftover ripples from the ancient Yellowstone hot spot—formed the stark, eerie landscape of Craters of the Moon on Idaho's Snake River Plain. The forbidding landscape discouraged settlement, and in 1924, the land was set aside as a national monument. Its apt name mirrors its long use as a research

IDA-ranked parks are spaces intended for tourists, IDA's "sanctuary" label describes places that are more remote and whose "conservation state is most fragile." At the center of this dark-sky sanctuary lies the Association of Universities for Research in Astronomy (AURA) Observatory, an international site backed by more than a billion dollars in investment, which supports four major facilities, including some of the world's most advanced telescopes.

The sanctuary doesn't just attract scientists. Tourists (planning well in advance) can book tours of the observatories. And the surrounding Elqui Valley, already famous for its lush vineyards and scenic views, is taking astrotourism seriously. Its arid air, high elevation, and scant cloud cover have encouraged owners of existing tourism infrastructure (such as glamping spots, inns, and tour companies) to lean into stargazing offerings.

spot for astronauts—the second crew to land on the lunar surface had studied the landscape of Craters of the Moon—and even today it is helping NASA scientists prepare for future journeys on extraterrestrial bodies.

Designated an International Dark Sky Park in 2017, Craters of the Moon holds Star Parties in the spring and fall, giving visitors the chance to use high-quality telescopes and learn from astronomical experts. Rangers lead full-moon hikes during the summer. And campgrounds provide an ideal spot for tourists to enjoy the sky for themselves with nothing but binoculars and a sense of wonder.

Dark Skies in U.S. National Parks

The IDA has supported the recognition of dark-sky spots around the world, but it has also inspired other bodies, from village councils to administrators of big-name parks, to adopt similar measures. In the

Craters of the Moon National Monument in Idaho sits above a rift zone, known as the Great Rift, which is the origin of its rugged, volcanic terrain.

reserves in the world. But it's the National Park Service that often takes the lead when it comes to night-sky preservation.

Among the 423 units the NPS oversees, dark-sky programs are becoming increasingly popular; about 100 parks now have astronomy-focused visitor programs. National parks such as Acadia (Maine), Shenandoah (Virginia), Rocky Mountain (Colorado), and Joshua Tree (California) host night-sky festivals with music, telescopes, youth programs, astronomy lectures, and constellation tours. Since 1999, an NPS team has monitored the quality of darkness in parks; since 2006, official agency policy describes how best to manage the lightscape in order to balance human safety with preserving darkness.

It's not all smooth sailing, though. NPS sites are among the most visited destinations in the world: In 2020, the 63 national parks saw 237 million visitors, a number made all the more shocking by the fact that it reflects a pandemic-caused 28 percent drop from the previous year's statistics. These numbers make parks especially vulnerable to the overtourism that can lead to increased light pollution. Add in the fact that the parks face a nearly $12 billion maintenance backlog, plus the fact that Congress doesn't actually require conservation efforts to include lightscape management, and it becomes

U.S., some state governments—especially western ones such as those of Nevada, Utah, and Colorado, which already have a strong reputation for ecotourism—are pushing for better outdoor lighting in the hopes of creating some of the largest IDA-approved

SKYLIST

HOW TO REDUCE LIGHT POLLUTION

IDA recognition requires destinations to have a commitment to dark-sky public education; many NPS night-sky programs have similar components. So what can we do to limit light pollution in our homes and neighborhoods?

- **WARM IT UP.** Scientists have shown that blue light creates more light pollution (in the form of ambient "sky glow") and has a greater effect on humans' and animals' physiological patterns. And though LED lights are often more energy- and cost-efficient, they're more likely to cast blue light. Instead, choose compact fluorescents or warmer-toned LEDs.

- **TURN IT DOWN.** When a light's not in use, turn it off; for outdoor lighting, use motion sensors or timers to keep lights dark until they're needed. Dimmers on both indoor and outdoor lights also help control light seepage.

- **KEEP IT FOCUSED.** Use shielding fixtures to aim light downward and prevent its glare from contributing to sky glow.

Cabins designed for stargazing are one of the ways tourism workers in Chile's Elqui Valley are adapting to the growing interest in astrotourism.

Using red light to diminish light pollution and preserve night vision, rangers educate visitors during the Night Sky Festival at Joshua Tree National Park.

apparent that the efforts of the NPS to protect dark skies are the result of individual people with a genuine passion for the stars.

TIME-SPECIFIC SIGHTS

Stars are ever present, but other phenomena appear more rarely, in certain seasons or certain centuries, captivating us with both their beauty and their brevity. Ancient astronomers often learned to track and predict them, and today's scientists have built on and refined this knowledge. Now, we can project eclipses, forecast auroras, and anticipate meteor showers down to the minute—which means we can travel to see them.

Solar Eclipses

In 1979, the writer Annie Dillard viewed a total solar eclipse with her husband in Yakima, Washington; an essay she wrote about her experience, published a few years later, is a widely read masterpiece. In "Total Eclipse," she describes the event's beginning: "The sky's blue was deepening, but there was no darkness. The sun was a wide crescent, like a segment of tangerine. . . . The sun was going, and the world was wrong. The grasses were wrong; they were platinum. Their every detail of stem, head, and blade shone lightless and artificially distinct as an art photographer's platinum print. This color has never been seen on earth."

For Dillard, the experience was as unnerving as it was fascinating, a wholly shocking sight that made her feel suspended from time, or from human life itself. And she's not alone. As long as people have been recording history, they have been writing accounts of the singular awe and terror that eclipses inspire. Dillard wrote, "Seeing a partial eclipse bears the same relation to seeing a total eclipse as kissing a man does to marrying him, or as flying in an airplane does to falling out of an airplane."

For many in the U.S., the thrill of 2017's total solar eclipse—the first to cross the lower 48 in nearly 40 years—opened the door to the wonders of astrotourism. A staggering 216 million people, or 88 percent of American adults, watched this eclipse in person or electronically; an estimated 21 million of these skywatchers traveled to a different spot for a better view. Eclipse boomtowns—places in the path of totality, many of them rural spots with a few hundred or few thousand residents—braced to welcome throngs of tourists by the tens of thousands. One Wyoming town (population 202) hosted an estimated 45,000 people.

Far from being a once-in-a-century flash in the pan, the public reaction to the 2017 eclipse is evidence of a burgeoning interest in astrotourism. One University of Michigan study found that adults who had participated in eclipse-viewing activities sought out scientific information at much higher rates

■ For more on solar eclipses, see page 89.

Although total solar eclipses happen about once every year and a half, any given location usually goes an average of 300 years between experiencing one itself. In fact, most eclipses cross only the ocean—after all, it makes up nearly three-quarters of Earth's surface. Within this century, North America will see two more cross-continental total eclipses running roughly parallel to the 2017 path: one from northern California to Florida in 2045, and another from British Columbia to North Carolina in 2099. North Africa and southern Europe will see one in 2027, and Australia in 2028.

Awesome a sight as it may be, a solar eclipse is still dangerous to look at except for the brief moments of totality, during which the sun is completely covered by the moon and the sun's feathery corona is visible. In every other moment of partial eclipse, look at the sun only if you're wearing special-purpose solar filters. Even then, don't look at the sun through binoculars, a camera, or a telescope, since the lenses will concentrate the solar rays, damage your solar filter, and injure your eye.

Auroras

If all the world's a stage, auroras are the curtains: shimmering, vibrant veils of reds, greens, purples, and blues that ripple in the night sky around the North and South Poles.

in the following months, suggesting a real interest in learning more about—and experiencing—astronomical phenomena. In 2019, another solar eclipse swept across South America and the Pacific Ocean; although its remote path of totality provided fewer viewing opportunities for landlubbers, Chile and Argentina saw significant eclipse-related tourism spikes. Many hotels and tourist destinations are already preparing for a solar eclipse in 2024 that will swing north from Mexico, through a large swath of the middle and eastern U.S., and over Canada.

SEE FOR YOURSELF

RISE AND SHINE

Maine's Acadia National Park, regularly among the top 10 most visited parks in the United States, is well known for its dark skies. Another claim to fame? It's the first American spot to see sunrise—at least for half of the year. From October to March, the 466-meter (1,530-ft) peak of Cadillac Mountain sees a couple hundred people huddled on the summit in the predawn dark, having made the trek into the remote island park in order to watch the sun's light spill over the rocky coastline. In the summer, as the sun moves north along the horizon, another Maine mountain (Mars Hill) takes the superlative. Not that Cadillac Mountain's visitors seem to mind.

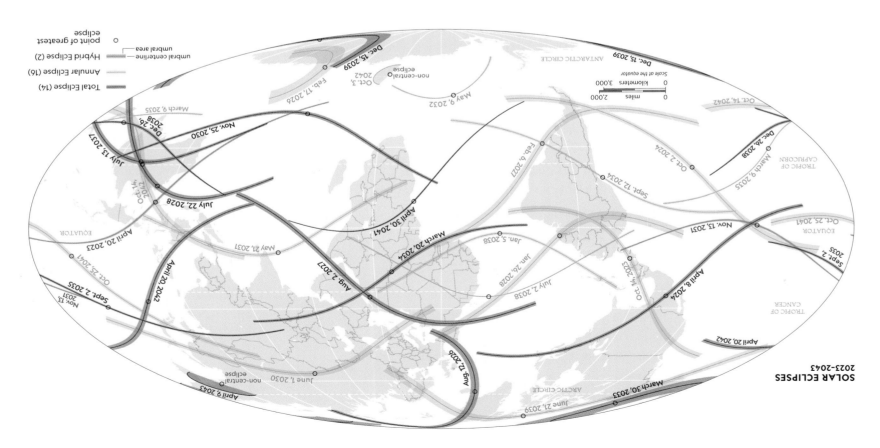

SOLAR ECLIPSES
2023-2043

Total Eclipse (14)
Annular Eclipse (16)
Hybrid Eclipse (2)
○ point of greatest eclipse
umbral area
umbral centerline

This composite, time-lapse image shows the 2019 total solar eclipse above La Silla Observatory in Chile over the course of two and a half hours.

Though rarer than its northern counterpart, the aurora australis (southern lights) is just as striking. Here it shimmers over Glenorchy, New Zealand.

Named for the Roman goddess of the dawn, these events are caused by charged solar particles hitting Earth's magnetosphere; some, caused by solar storms, can mess with satellites and the electrical grid, but they also cause more striking auroras that can be more widely seen. And if you've ever heard an eerie hissing or popping sound while viewing auroras, it's not your imagination. Scientists recently suggested that certain weather conditions can create a layer of warm air above cool air that acts like a lid between two different electrical charges, and when a geomagnetic storm breaks this barrier, the released charges create the mysterious sounds long dismissed as folklore.

Although it's possible for auroras to appear throughout the year in many places around the world—one 1958 sunstorm caused an aurora visible from Mexico City—they're far more common near the poles in spring and fall. And for some towns, aurora tourism is a big business.

the upper Midwest are becoming more popular, with citizen scientist groups keeping an eye out for the most favorable conditions in dark-sky spots, including Voyageurs National Park (Minnesota), Headlands International Dark Sky Park (Michigan), and Apostle Islands National Lakeshore (Wisconsin).

Although the southern lights (aurora australis) are caused by the same solar wind as their northern siblings, the two phenomena aren't perfect mirrors. The sun's magnetosphere is so powerful it actually distorts the symmetry of Earth's magnetosphere, flattening the field lines on the side of the planet facing the sun and creating a long tail on the other. As a result, aurora patterns that should be identical and simultaneous at both poles usually aren't.

The southern lights also tend to be more difficult for people to see: Antarctica, theoretically the best viewing spot because of its polar location and clear skies, is also famously remote and inhospitable. Some Southern Ocean cruises on the lookout for migrating whales happen to luck into an aurora sighting. But for more stable viewing sites—and all the accompanying amenities—most aurora-spotting hopefuls head to New Zealand and southern Australia, especially the island of Tasmania. In April and May, the best months for sightings, astrotourists should look for dark-sky spots such as New Zealand's Lake Tekapo, on the shores of which

The aurora borealis, also known as the northern lights, draws hundreds of thousands of visitors a year. Even as recently as a decade ago, places now ranked among the world's best aurora tourism spots saw little to no traffic; today, remote northern towns in Norway, Sweden, and Iceland welcome people from all over the world (especially from Asia). In North America, destinations like Fairbanks, Alaska, and Yellowknife, Canada, are increasingly marketing themselves to aurora-chasers. Even spots in

MEET STEVE

Around 2016, citizen scientists in Alberta, Canada, began reporting something strange: a long purple streak in the sky that looked like an aurora, but didn't seem to behave like one. Baffled, they dubbed it Steve—a reference to a 2006 kid's movie in which wild animals give that name to a manicured hedge because they don't know what else to call it. (Scientists later gave it the backronym Strong Thermal Emission Velocity Enhancement.)

Auroras form in high latitudes as a result of electron excitement; airglow, the other generally recognized category of night-sky optical phenomena, occurs around the world as atoms recombine and release energy as light. Some scientists now think that Steve, which includes a purple ribbon and often green "picket fence"–like markings, is a combination of aurora and airglow effects. But plenty of mystery remains. How do the two features relate to each other? How exactly do they form? Can we predict them? One thing that's certain is that citizen scientists will continue to play a key role in learning more about Steve.

rise Mount John and its observatory, or Tasmania's Tinderbox Nature Reserve, where vibrant birds call from the old-growth forest.

For now, Earth's two poles are the best places to enjoy the dazzling show that auroras present. But since our planet isn't the only body in the solar system with a magnetic field that's bombarded by the sun's charged particles, there's also the possibility that one day humans may witness auroras from Jupiter, Saturn, Uranus, or Neptune. Even Mars, with its weak magnetosphere, could provide an otherworldly light show: Remnants of its ancient mag-netic field still linger in some spots, reacting with solar wind to create faint, ghostly aurorae.

OBSERVATORIES AND PLANETARIUMS

In the Enlightenment era, the intellectual elite modeled the movement of the heavens with tabletop devices made of wires, gears, and globes. These devices—sometimes called orreries, also called planetariums—have evolved over the centuries into the highly advanced planetariums we recognize today, which project images onto a domed screen to immerse viewers in the universe itself. (Or to give that impression, at any rate.)

Though some working observatories are open to visitors, it's planetariums that are specifically designed to educate the public about space. They've inspired more than one curious kid to grow up into a career in astrophysics. And the buildings that house them are often architectural marvels in themselves.

L'Hemisfèric

One of the most prominent tourist destinations in Spain is Valencia's Ciutat de les Arts i les Ciències (City of Arts and Sciences), a grand complex of cultural and scientific buildings set in a scenic riverbed park. In 1998, the complex opened with the completion of L'Hemisfèric, a planetarium designed to look like a giant eye half submerged in water—the "eye of wisdom." The planetarium, which can seat 300 people under a 900-square-meter (9,700-sq-ft) screen, transports visitors to the farthest reaches of space, as well as to amazing sites on Earth, including the Amazon rainforest and CERN's Large Hadron Collider, a powerful instrument that studies nature's smallest particles. A scenic walkway named for Carl Sagan surrounds the theater, and the building also includes a space for rotating exhibitions.

Adler Planetarium

Chicago's Adler Planetarium was the first of its kind in the Western Hemisphere and has captivated audiences since its 1930 opening—in fact, one World's Fair exhibition drew more than a million people, roughly equal to a third of the city's entire population at the time. With its striking art deco exterior and copper dome, the building functions both as a planetarium and an astronomy museum: It holds three theaters, an observatory, and a world-

A FALLEN ICON

Made famous by appearances in *Goldeneye* and *The X Files*, Puerto Rico's Arecibo Observatory was the world's largest radio dish for half a century after its construction in the 1960s. Thanks to its supreme sensitivity—set into a natural sinkhole, its massive dish could focus faint radio waves to a number of receivers—Arecibo helped make dozens of groundbreaking discoveries. Scientists used it to detect the first binary pulsar and the first extrasolar planets, and to create detailed radar maps of our own planetary neighbors. Its antennae could help with study closer to home, too, by heating atmospheric plasma to create artificial aurorae.

But the observatory came to a tragic end. Already difficult to maintain due to the humid jungle environment and exposure to hurricane damage, the installation was marked for closure due to irreparable structural risk as early as 2006. Finally, between August and December 2020, a series of suspension cables snapped, eventually sending the receptor platform crashing through the dish and into the ground below. Arecibo's destruction was a tragic loss for science, and for the people of Puerto Rico, who took a fierce pride in the majestic telescope that had drawn scientists and visitors to the island for so many years.

renowned collection of historical instruments. A causeway connects the mainland to the small spit of land on which the Adler Planetarium was built, and this Lake Michigan vantage—complete with a glass-enclosed pavilion—gives Adler guests a spectacular view of the Chicago skyline. And though its urban locale makes it plenty accessible to in-person visitors, the planetarium also offers a host of online activities: citizen science projects, virtual field trips, and multimedia exhibitions.

Shanghai Astronomy Museum

Opened in July 2021 after years of construction, the world's largest museum dedicated solely to astronomy was designed to embody the celestial forces it studies. Like archaeoastronomy sites across the world, its gold-paneled oculus tracks the time and seasons, sending an arc of sunlight swooping across the court below to align with a black disk on the summer solstice. The sphere housing its planetarium is suspended in part of the building, so that it appears to float above an atrium like a planet caught in orbit. And its inverted glass dome, which hov-

ers over the entrance hall, provides a rooftop space meant for awe-inspiring views of the sky. These three main features also reference the "three-body" problem—the as-yet unsolved question of how to calculate three celestial objects' orbits in relation to each other. In addition to the planetarium, the Shanghai Astronomy Museum houses an observatory, a solar telescope taller than the Hollywood sign, and a series of permanent and rotating exhibitions.

L'Hemisfèric, part of an arts and sciences complex in Valencia, Spain, is a planetarium designed to resemble a giant human eye half submerged in a reflecting pool.

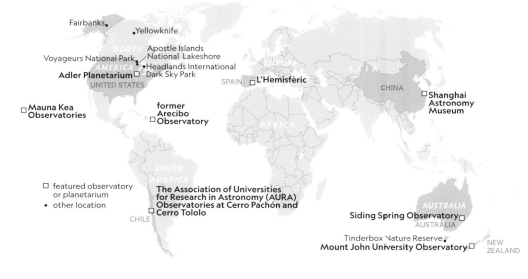

FUTURE FLIGHTS

In a new space age, even civilians may be able to travel beyond our planet.

Orbital space hotels. Zero-gravity space-planes. Weekend stays on the ISS. It may seem like the stuff of science fiction, but more likely than not the age of space tourism is about to begin.

Some might say it's already begun. In 2001, a seven-day stay on the ISS made American businessman Dennis Tito the world's first commercial space passenger. (Some object to the term "space tourist," given the rigorous demands of space travel.) In the two decades since then, just shy of a dozen others joined those ranks. Some paid tens of millions of dollars for the privilege—or were bankrolled by a wealthy patron.

Even from the beginning of tourism's evolution into the industry we recognize today, travel for leisure and pleasure's sake has largely been the domain of the wealthy elite—think 19th-century scions taking their Grand Tour of Europe. And though travel on Earth has certainly become more accessible as the middle class has grown, the exponential divide between the world's uber-rich and the other 99 percent means that space travel likely will stay out of reach for most of those without millions lying around.

As space travel increasingly includes civilians, the definition of "astronaut" gets blurrier. But some space historians have pointed out that the image of the astronaut is one deliberately built on an outdated idea of American masculinity: the white, straight, Christian family man who stood in opposition to Soviet values during the height of the Cold War. Even as we evolve our understanding of who has "the right stuff" to be an astronaut, tensions remain

about humans' role in space. In the 21st-century reprise of the space race, private corporations take center stage while geopolitical tensions continue to play out in the wings. The same questions of identity and purpose we grappled with in the 1960s have donned new costumes: How do we reckon with the environmental cost of space programs? Is it feasible to consider colonizing other bodies in the solar system as our own planet suffers the socioeconomic effects of centuries of colonization? As humans throw ourselves into the universe, maybe even one day to meet other intelligent life, who do we become in the process?

HIGH FLYING

So far, space travel has generally fallen into two categories: suborbital and orbital. Suborbital flights take passengers up to the edge of Earth's atmosphere, high enough to zoom through zero gravity for a few minutes. Such trips are available today from Blue Origin, the spaceflight company founded by Jeff Bezos, and from Richard Branson's Virgin Galactic (the cheaper of the two options, at nearly half a million dollars). Orbital flights, on the other hand, ferry passengers to destinations already in

(Opposite) The all-civilian crew of the Inspiration4 mission, launched in September 2021, included (top to bottom) Hayley Arceneaux (the youngest American to fly in space), Chris Sembroski, and Sian Proctor.

orbit—which at this point is pretty much the ISS, though one company has plans to build modular "space hotels" that can connect to each other, offering twice the square footage of the space station. The ISS itself reopened to tourism in 2020 at a rate of $35,000 a night, hoping to offset astronomical maintenance costs with private money as federal funds shift toward other programs.

Chartered voyages, offered by companies such as SpaceX and Axiom Space, are already being scheduled throughout the early 2020s. For non-billionaires, the route to orbital space travel requires being very, very lucky: Several planned expeditions have opened lotteries to bring civilians along. SpaceX, Elon Musk's aerospace company, has been resupplying the ISS since 2012 with cargo such as science experiments, equipment, and hardware,

■ For more on SpaceX, see page 312.

Each year, the Kennedy Space Center, on Florida's Merritt Island, hosts more than 1.5 million visitors.

and its development of reusable rockets could make it a leader in commercial space travel.

Although a seat on a spaceship may be out of reach for most, rocket launches and spaceport tours draw eager crowds around the U.S. Even as more spaceports are being built, most space launches take place at only a handful of spots, some of which aren't open to the public. To get a taste of both the storied history of American spaceflight and the energetic pace of today's innovations, there's really only one place to go.

Florida's Space Coast

The Kennedy Space Center is a sprawling installation on Merritt Island, near Cape Canaveral, Florida, at a midpoint on the state's Atlantic coast. As the spot that launched the first American crewed spaceflight and the world's first lunar landing, the Kennedy Space Center is the backbone of NASA's space program and remains one of the country's principal spaceports—and a prime tourist destination. At the Kennedy Space Center Visitor Complex, tourists can roam "mission zones" that interpret each of the U.S. space program's missions, from Mercury through Apollo and up to the current day. In kid-friendly play areas, future spacefarers can scramble through asteroid fields or launch themselves through a wormhole. Anyone feeling particularly resilient can test their mettle in the astronaut training experience. And if you find yourself tiring of the exhibitions, step outside to watch a rocket launch in person—a blaze of light, heat, and noise as about a million pounds of fuel are ignited to propel a payload into space. A daily admission ticket includes the ability to watch most smaller-profile launches from the main visitor complex. (High-profile launches typically require an additional event ticket.) The center includes three other launch observation spots, though not every spot will be available for every launch as a result of safety regulations. For launch times and special ticket sales—launch sites outside the main complex sell out quickly—check the Kennedy Space Center website ahead of time. Just down the coast from the Kennedy Space Center lies Cape Canaveral Space Force Station, a spaceport with an equally impressive résumé of missions. Although the base is a work site closed to visitors, its grounds include the Air Force Space & Missile Museum, where sched-

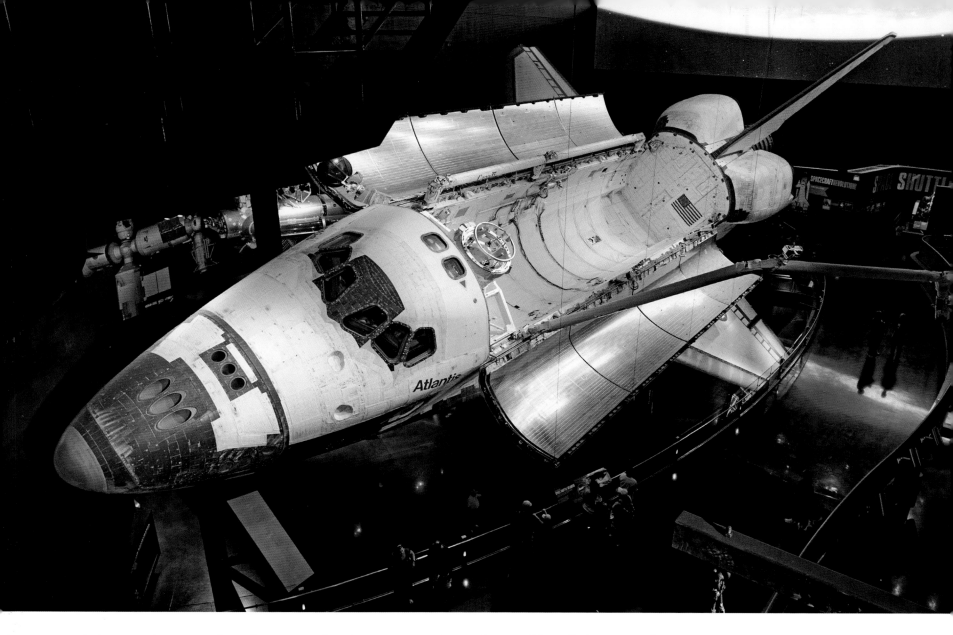

uled tours take guests through a maze of rocket parts and historical exhibits.

It's not all rockets, though. Most of the barrier island on which the Kennedy Space Center resides is part of the Merritt Island National Wildlife Refuge, which protects a majority of the 56,700 hectares (140,000 acres) that NASA acquired in 1962 for the space program as habitat for endangered species, including migratory birds and manatees. For a nominal day fee, visitors can explore by boat, on foot, or on a scenic drive; hunting and fishing are allowed as part of a wildlife management plan. Farther south, Canaveral National Seashore—a 39-kilometer (24-mi) stretch of development-free beach—offers the chance to view both nesting turtles and launching rockets.

SPACE ANALOGUES:
THE NEXT BEST THING

We're still a long way off from the casual tourist being able to pop to the moon or Mars for vacation—if that ever becomes reality at all.

In the meantime, though, earthbound tourists can visit the places astronauts and scientists use to approximate alien worlds, called space analogues. These spots help astronauts train for missions in forbidding environments that mimic conditions off-world, and they help scientists rethink what life can look like and how to search for it. They also captivate visitors with their stunning vistas and opportunities to learn more about distant planets, especially as space programs around the world gear up to send humans back to the moon and beyond.

(Above) *Atlantis*, a space shuttle retired in 2011, is displayed in the Kennedy Space Center Visitor Complex as if it had just undocked from the ISS. (Below) Paying passengers on an Airbus A330 aircraft enjoy the zero-g experience.

Visitors peer at the barren terrain of Lanzarote, easternmost of Spain's Canary Islands. The arid climate, cratered landscape, and lava flows make it an excellent space analogue.

Just off Route 66 near Winslow, Arizona, the massive 50,000-year-old Meteor Crater is uniquely suited for scientists researching how craters were formed.

Maybe the first person to step foot on Mars is even now climbing the craggy valleys of a place on Earth that, by some quirk of geology, can help us learn how to get there.

Moon Analogues

There's plenty we still don't know about our most important satellite. In the 1960s, Apollo 11 astronauts trained for their lunar mission in the barren terrain of Iceland. Since then, the number of moon analogues has only grown as we've learned more about the topographical and geological features future spacefarers are likely to face. Some of these analogues crop up in unexpected places. In the Canary Islands, for example—a Spanish archipelago off the coast of northwestern Africa—a violent seismic history has resulted in a jagged landscape perfect for simulating moonwalks, drawing astronauts from NASA and the European Space Agency (ESA). On the moon or Mars, lava tubes could make for an important shelter. On Lanzarote (one of the Canaries' seven main islands), enormous complexes of lava tubes stretch for miles, making the perfect site for creating pop-up labs and test-driving rovers. The archipelago has been a noted tourist destination since the mid-1950s, and much

of it is protected by UNESCO reserves and national parks that make these alien vistas accessible to the public.

The moon's famously pitted face is the result of hundreds of thousands of craters, scars left over from billions of years of meteoroid bombardment. Although craters erode and change shape much more quickly on Earth, our planet still has some remnants of impact craters that compare to the moon's. Arizona's Meteor Crater is one of these. The enormous basin is 1,200 meters (3,937 ft) across and 170 meters (558 ft) deep, revealing strata of ancient sedimentary rock, including limestone and sandstone from a collision 50,000 years ago. That's pretty young, geologically speaking, meaning that Meteor Crater is a well-preserved site particularly helpful for studying how such impacts work. Although Meteor Crater lies on private property still used for cattle ranching, a visitors center and other tourism infrastructure—including an easy one-hour drive from the nearest airport, in Flagstaff—makes it one of the more accessible moon analogues.

Mars

Our red, rugged neighbor is the next target of many space programs, both private and state-sponsored. This level of interest is reflected in the number of spots on Earth used as Mars analogues. Southern Utah is one of them: The region's isolation and geology are so strikingly Martian that thousands of people have used the area to rehearse potential missions at the Mars Desert Research Station. Airlocked habitats set within the red rock terrain allow crews to practice living in close quarters, communicating with other teams, and carrying out fieldwork in the laboratory. Luckily, landscapes like these are also open to the public: Federal land makes up nearly 65 percent of the state's acreage, and Utah is famous for its Mighty Five national parks (Arches, Bryce Canyon, Canyonlands, Capitol Reef, and Zion).

Even redder and remoter than Utah is Western Australia, with dunes and craters reminiscent of the red planet. Fossils play an important role here, too. Three-billion-year-old bacteria left traces in the area's ancient rock that look eerily similar to Martian meteorite samples. Whether or not these meteorite samples are proof of ancient life on Mars, scientists rely on this region—and other Mars analogues—to help hone the theories and

techniques that could help us perceive alien life off-planet.

Perhaps a more surprising Martian analogue than Utah or Australia is Iceland. Although not as red, Iceland is plenty rugged, and its high level of volcanic activity could give us clues about Mars's early days. Unlike Earth, Mars doesn't have an active plate tectonic system; the remains of its volcanoes lie quiet, but there's a possibility they could still erupt as ancient magma plumes finally burst through the crust. And studying the microbial life present in Iceland's volcanoes could help scientists detect traces of similar life that once thrived in those ancient hot spots—or that might still thrive there today, just under the surface.

I WANT TO BELIEVE

Is it retro to believe in aliens? Hardly. According to Gallup, in 2019, 30 percent of Americans thought UFOs were actually alien spacecraft; two years later, that number was four in 10. Although many scientists have dedicated themselves to searching for extraterrestrial intelligence with a rigor that is far from credulous, both skepticism and pseudoscience surround these efforts. And more than one destination has capitalized on tourists' interest—often with a healthy dose of kitsch. In Roswell, New Mexico, famous for a mysterious 1947 aircraft crash, little green men are the main draw: The town hosts an annual UFO festival and welcomes hundreds of people a day to the International UFO Museum and Research Center. Area 51, a highly classified military installation in the Nevada desert whose role in developing cutting-edge aircraft and surveillance techniques has made it an anchor for conspiracy theories and UFO sightings, is naturally closed to visitors. But that hasn't stopped tour guides from taking advantage of the hype to lead alien-lovers on a tour of the nearby "Extraterrestrial Highway," with stops at UFO sighting spots and campy bar/motel the Little A'le'Inn. And in the remote stretches of West Texas, the town of Marfa has attracted tourists from all over the world eager to see eerie glowing orbs, first glimpsed in 1883, as they weave and dash through its skies at night.

This hollowed-out cloud of gas and dust, known as the Jack-o-Lantern Nebula, was likely carved out by the outflows of a star 15 to 20 times more massive than the sun.

CHAPTER SEVEN

SEEING FARTHER

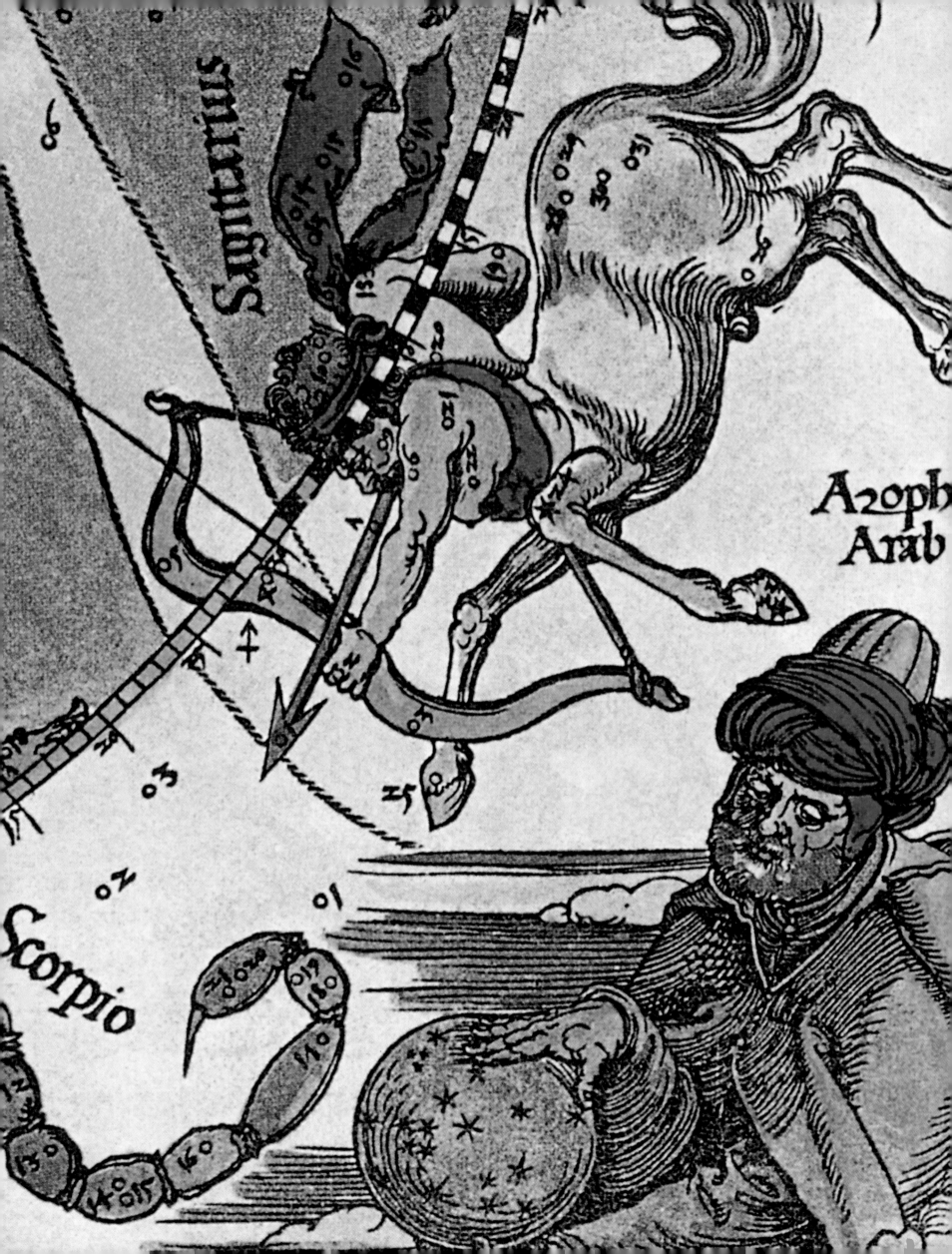

BEYOND
THE SOLAR SYSTEM

The Greek philosopher Democritus is said to have believed that the Milky Way's glow across the night sky came from "the coalition of many small bodies." More than 1,400 years later, in A.D. 964, the Persian astronomer Abd al-Rahman al-Sufi recorded the first formal observations of a galaxy outside the Milky Way. In his *Book of Fixed Stars,* al-Sufi noted a "little cloud" or "nebulous smear" near the midriff of the constellation Andromeda. This structure turned out to be the Andromeda galaxy, the Milky Way's neighboring spiral galaxy.

Only by the late 1920s did astronomers widely accept that the Milky Way wasn't the extent of the known universe. In other words, science's conception of a vast universe littered with galaxies is only about as old as sliced bread. Now, nearly a century after that vast revision, astronomers have established observatories on Earth and in space that span the electromagnetic spectrum, from radio waves all the way to gamma rays. And thanks to the new astronomy of gravitational waves, we even can detect the rumblings of space-time itself.

We find ourselves at a unique time in the history of our species. For decades, our robotic emissaries have journeyed to the farthest reaches of our solar system—and even past the boundaries that demarcate interstellar space. Five decades after the Apollo program's extraordinary visits to the lunar surface, humans have lived in orbit hundreds of miles above Earth's surface for more than 20 years straight. Networks of satellites circling Earth have beamed back data chronicling the exquisite beauty—and painful fragility—of our cosmic home. And as computers have gotten exponentially more powerful, and engineering advances enable bigger, more sophisticated telescopes and spacecraft, we can look out onto the universe—and even venture into it—as never before.

In this chapter, we'll take a look at a sample of what lies beyond our solar system, how we're seeing these objects, and what these near and distant bodies will add to our knowledge of the universe. We'll also see how future missions will fill out this vast cosmic picture—and the role that humans will play in our exploration of the cosmos.

Vigilant astronomer al-Sufi watches over the heavens, as imagined in this
1515 woodcut by master engraver Albrecht Durer.

OUR COSMIC NEIGHBORHOOD

Far-flying spacecraft are letting scientists explore the boundary between our solar system and interstellar space.

As British author Douglas Adams wrote about space in his famous novel *The Hitchhiker's Guide to the Galaxy,* "You just won't believe how vastly hugely mind-bogglingly big it is." Adams was right. The distances between planets in the solar system are enormous, not to mention the distances from here to other stars and galaxies. But if we're going to talk about looking and exploring beyond the solar system, we need to establish where our cosmic sidewalk ends. What is the outer bound of our solar system?

Exploring the solar system's outskirts is extremely difficult, and few spacecraft have tried. Only five spacecraft ever built have rocketed away from the sun with enough velocity to escape the sun's gravitational clutches and exit the solar system. Only three of these robots explored the worlds beyond Saturn.

Launched in 1977, NASA's twin spacecraft Voyager 1 and 2 performed flybys of Jupiter and Saturn—and Uranus and Neptune too, in Voyager 2's case—before passing through the solar system's functional edge. NASA's New Horizons spacecraft, which flew by Pluto in 2015 and the Kuiper belt object Arrokoth (see sidebar) in 2019, is now so far from Earth that from the spacecraft's perspective the location of some stars in the night sky has shifted. As we discussed in chapter 1, scientists can use this apparent shift—called parallax—to measure those stars' distances from the solar system.

What awaits these spacecraft beyond the sun's domain? As with so much else in our solar system, the answer is centered on our sun. As you saw in chapter 2, the sun throws off a breeze of electrically charged particles called the solar wind. This gust rushes out in all directions from the sun's surface at about 400 kilometers a second (nearly a million miles an hour), carrying the sun's magnetic field with it.

Four hundred kilometers a second might sound fast, but space is an enormous place. On average, it takes nearly four days for fresh solar wind to make it from the sun to Earth, even going that fast. In about a year's time, the solar wind makes it out to roughly 100 times the sun-Earth distance. By that point, the wind's makeup has changed, as it has picked up straggler particles along the way.

Eventually, this outwardly moving solar wind smashes into the interstellar medium: the diffuse, hot gas and dust that floats between the stars. Like oil and water, the solar wind and the interstellar medium don't perfectly mix, in part because of their different temperatures and mismatched magnetic

For more on parallax, see pages 18–19.

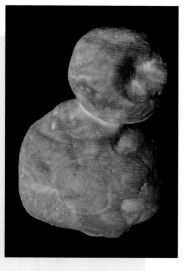

GOING THE DISTANCE

On New Year's Day 2019, New Horizons pulled off the most distant feat of exploration ever performed: the flyby of a small Kuiper belt object roughly 6.6 billion kilometers (4.1 billion mi) from Earth. That tiny, reddish world—now called Arrokoth, after the Powhatan word for "sky"—also made New Horizons the first spacecraft to visit an object discovered after the spacecraft's launch. Astronomers spotted Arrokoth with the Hubble Space Telescope in 2014, eight years after New Horizons began its journey toward Pluto.

Arrokoth is scientifically invaluable because its shape hints at how it formed, which tells us how other primordial pieces of the solar system likely came together. The object resembles a flattened red snowman some 35 kilometers (22 mi) long, with two squat lobes fused together. Scientists think that soon after the sun's ignition, innumerable tiny pebbles swirled out where Arrokoth orbits today, accreting as they went. Eventually, only two objects were left twirling around each other: Arrokoth's two lobes, which spiraled closer together until they touched at a mere walking pace. Arrokoth's two halves have stayed together for roughly 4.5 billion years.

(Opposite) In this artist's illustration, NASA's New Horizons spacecraft zooms through the outer solar system, the sun receding into the background.

As New Horizons journeys out of the solar system, its perspective on distant stars visibly shifts, as seen here in these superimposed images of Alpha Centauri.

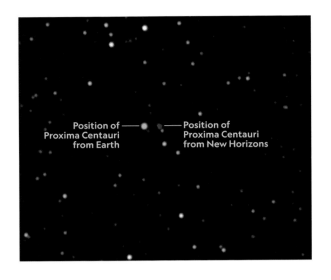

Position of ——— ● ——— Position of
Proxima Centauri   Proxima Centauri
from Earth   from New Horizons

fields. As a result, a magnetic bubble forms around the sun that is continually kept inflated by the solar wind. This bubble is called the heliosphere.

Gravitationally speaking, the solar system doesn't end where the heliosphere does. In fact, many billions of comets in the distant Oort cloud feebly orbit our home star at distances well past the solar wind's reach. But crossing the heliosphere's outermost boundary, a zone called the heliopause, still matters a great deal. Beyond that point, the electromagnetic environment isn't dictated by the sun—but rather by the vast, wild spaces between the stars.

Life on Earth reaps a huge benefit from the heliosphere, which shields everything inside it—including the pale blue dot we call home—from most of the galaxy's highest-energy radiation. That's one of the reasons why scientists are so eager to study the heliosphere. Understanding our own star's magnetic bubble could help us better under-

stand the radiation environments around other stars, which would refine what we know about the cosmic conditions associated with life.

EXPLORING THE HELIOSPHERE

The heliosphere isn't just a sphere, because the sun and solar system aren't just sitting still. Every 220 to 250 million years or so, the sun completes one circuit around the galaxy, at an average rotational velocity of 864,000 kilometers an hour (537,000 mph). As a result, the sun moves through the interstellar medium like a boat moving through water, which gives the heliosphere an extended wake.

The heliosphere's leading edge (the heliopause) is roughly 18 billion kilometers (11 billion mi) from the sun, or roughly four times farther from the sun than Neptune is. However, this distance isn't fixed; as the sun's activity ebbs and flows, so does the solar wind, which makes the heliosphere shrink and expand. At any rate, this boundary is extremely far from us—and no human spacecraft had ever pierced it until August 25, 2012, when Voyager 1 zoomed through the heliopause and beamed back data from the other side. In November 2018, Voyager 2 accomplished the same feat, sending back even more data than Voyager 1 (because it had more working instruments).

The data from Voyager 1 and 2 paint a fascinating picture of the boundary between our solar system and the vast interstellar space beyond. For instance, Voyager 2 found that as an object gets within 225 million kilometers (140 million mi) of the heliopause, the plasma surrounding it slows down, heats up, and gets more dense. On the other side of the boundary, the interstellar medium is at least 30,000°C (54,000°F), which is hotter than expected. However, this plasma is so thin and diffuse, the average temperature around the Voyager probes remains extremely cold.

Voyager 1 and 2 belong to an elite club. Though three other spacecraft are on their way out of the solar system, none will be working by the time they reach interstellar space. NASA's Pioneer 10 and 11 missions, launched in 1972 and 1973, respectively, are each zooming out of the solar system at more than 40,000 kilometers an hour (25,000 mph), but by 2003, Earth had lost contact with both of them. New Horizons is on its way out of the solar system, too, but it probably won't exit the

By the early 1990s, NASA's Pioneer and Voyager spacecraft had already cleared the orbit of Pluto.

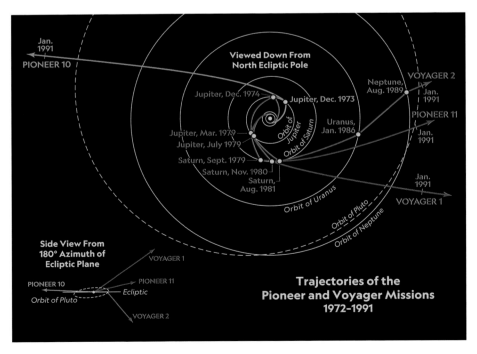

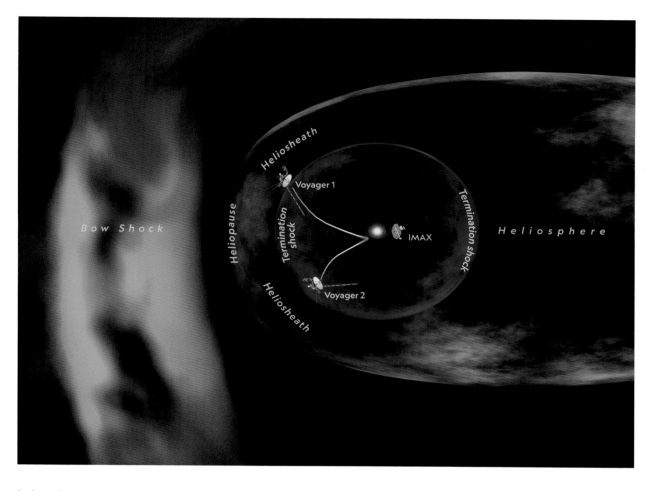

An artist's illustration depicts the heliosphere, a bubble in space created by the solar wind and the sun's magnetic field.

heliosphere until after its expected shutdown in the late 2030s, when it stops making enough electricity to keep itself warm.

Future missions are in the works, however. In 2021, China announced its goal to launch two Voyager-like interstellar probes: one that would fly through the heliosphere's leading edge, and another that would fly through its wake. The intent is to launch both spacecraft by 2024, so that by 2049— the 100th anniversary of the People's Republic of China—the two spacecraft will have each traveled 100 astronomical units, or roughly 15 billion kilometers (9.3 billion mi).

Likewise, U.S. scientists are sketching out plans for a proposed Interstellar Probe that, if launched between 2036 and 2041, could make it to the heliopause in 15 years, a full 20 years faster than Voyager 1—while hitting exit speeds of more than 119,000 kilometers an hour (74,000 mph). The Interstellar Probe is designed with Voyager-like endurance, with a baseline 50-year lifetime that would take it roughly 400 times farther from the sun than Earth. If the proposed mission were to outlast its design lifetime, as Voyager 1 and 2 have,

it could theoretically travel 1,000 times farther from the sun than Earth. As a result, the probe's designers have had to think through how to staff a space mission across multiple generations of scientists and engineers.

Other researchers are looking even farther afield and considering what it would take to pierce through the heliosphere at unprecedented speed— and fly all the way to the sun's neighboring stars.

WHAT'S THE HELIOSPHERE'S SHAPE?

It may sound surprising, but we don't yet know for sure what the sun's trailing wake looks like. Astronomers have seen distant stars with a variety of "astrosphere" shapes, from long, cometlike tails to squat globs that look like huge croissants. Using data from NASA's Cassini, New Horizons, and Voyager spacecraft, one 2020 study argued that the heliosphere is shaped like a squashed croissant. But is that really the case? Only time and more data will tell.

Future missions will help refine what we know about the heliosphere's shape, including NASA's upcoming Interstellar Mapping and Acceleration Probe (IMAP), a space telescope set to launch in 2025.

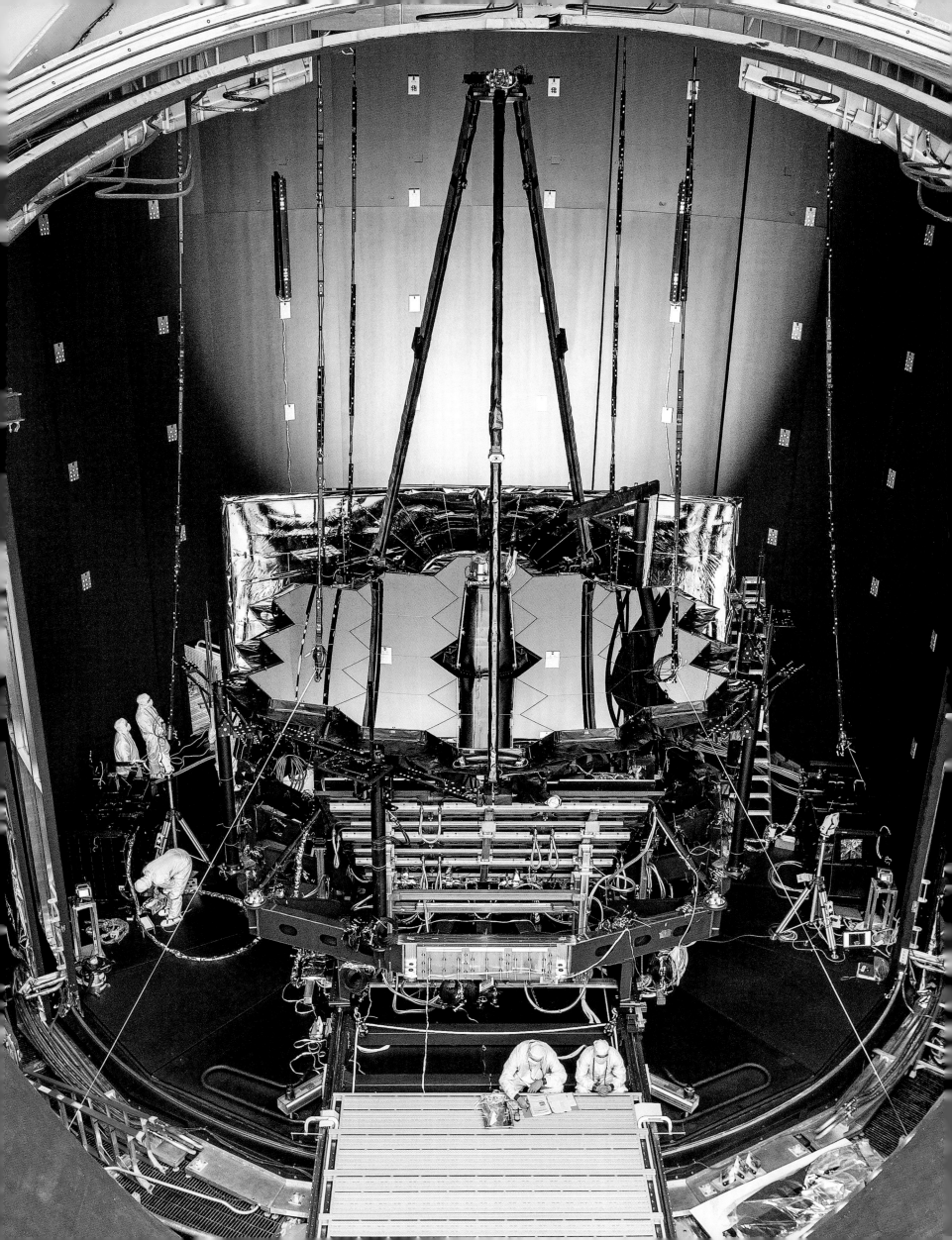

ADVANCES IN ASTRONOMY

Current and upcoming telescopes promise to reveal the many faces of the cosmos.

Armed with the best telescopes on Earth and in space, astronomers are exploring the skies as never before. There's so much happening in astronomy that summarizing the field as a whole would make for quite a long read, far beyond the scope of this chapter. Let's explore a few of the most exciting themes and developments, with an eye to how new space missions and telescopes will lead to even more change.

LOOKING BACK TO A BABY UNIVERSE

In a sense, telescopes are the closest thing we'll ever have to time machines. The farther our telescopes can see, the deeper into the universe's past our gaze can reach. So to see how the early universe spawned the galaxies we see today, astronomers are keenly interested in seeing distant objects.

We now can see distant galaxies that formed a mere 420 million years after the universe's birth, when the cosmos was just 3 percent of its current age. In 2020, researchers using NASA's Hubble Space Telescope managed to spot a galaxy that old, an object called GN-z11. NASA's James Webb Space Telescope, launched in December 2021, is a power-house of an infrared telescope designed to help with this observational time travel. It will take pictures at four times the resolution of the Hubble Space Telescope, with an infrared gaze that will let it see faint objects that, like GN-z11, formed in the universe's infancy.

Afterglow of the Big Bang

Some signals from the early universe date back even farther. Some 380,000 years after the big bang, the

hot, dense universe finally cooled down enough to let light shine through—creating a flash of light that we can still see today, as a faint afterglow of the universe's youth called the cosmic microwave background, or CMB for short.

The CMB's glow is almost exactly constant in all directions, with temperature fluctuations of just about one part in 100,000 that trace back to slight variations in the early universe's density. It is thought that these subtle differences helped sow the seeds for the universe's current large-scale structure. Some overdense parts of the cosmos are now chock-full of galaxies. Less dense parts are now voids with few stars.

Mapping this glow and statistically measuring its properties provides a snapshot of the baby universe's inventory of matter and energy. The CMB's properties have helped make the case that the universe's visible matter is about 5 percent of the

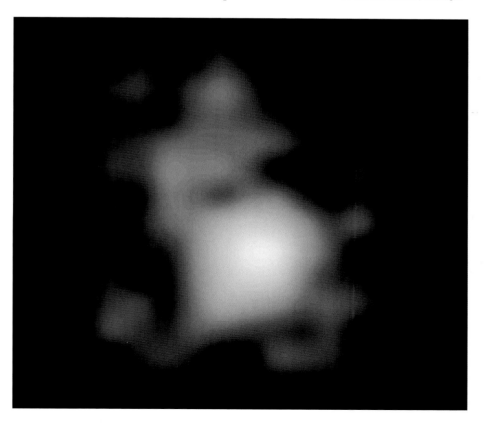

This reddish blob, as seen by NASA's Hubble Space Telescope, is an image of the infant galaxy GN-z11 as it looked 13.4 billion years ago.

(Opposite) In this test being set up inside Chamber A at NASA's Johnson Space Center, the James Webb Space Telescope had to withstand space-like cold: temperatures as low as -236°C (-393°F).

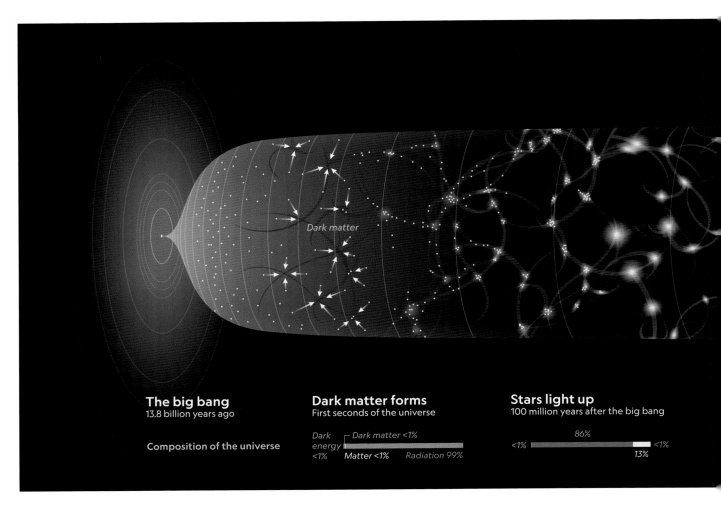

The big bang
13.8 billion years ago

Composition of the universe

Dark matter forms
First seconds of the universe

Dark energy <1%   ┌ Dark matter <1%
<1%   **Matter <1%**   Radiation 99%

Stars light up
100 million years after the big bang

86%
<1%   <1%
13%

matter and energy in the cosmos. For the CMB's structure to make sense as we understand it today, the universe must contain extra components: a strangely transparent, inert dark matter, and a dark energy that's accelerating the universe's rate of expansion.

The better that maps of the CMB get, the more refined our accounting of dark matter and dark

energy becomes. We also get better at measuring the universe's other fundamental properties, such as the nature of its curvature. The strides we've made in the last 30 years to measure the CMB have been astounding: Just compare the first map of its kind, made by NASA's COBE mission from 1989 to 1993, with the latest effort, produced by the European Space Agency's Planck space telescope. While these

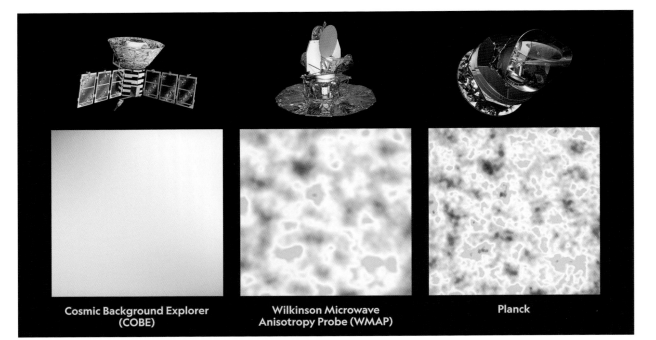

Each new mission to measure the cosmic microwave background has sharpened our view, from 1989's COBE spacecraft (left) to 2001's WMAP spacecraft (middle) and 2009's Planck spacecraft (right).

Cosmic Background Explorer (COBE)

Wilkinson Microwave Anisotropy Probe (WMAP)

Planck

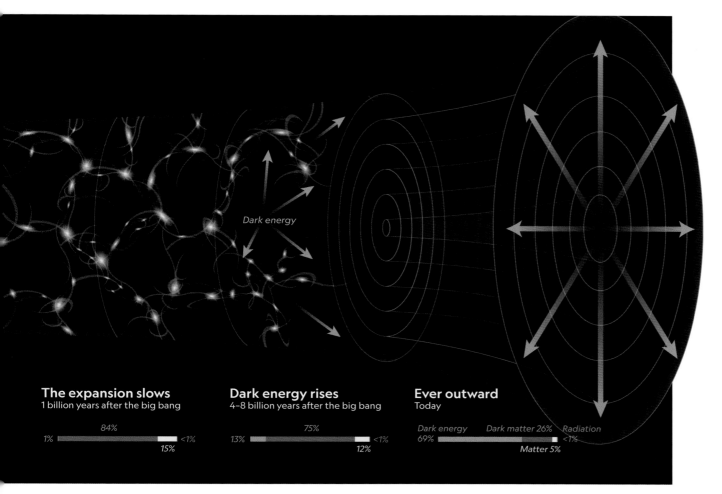

The expansion slows
1 billion years after the big bang

84%
1%   <1%
        15%

Dark energy

Dark energy rises
4–8 billion years after the big bang

75%
13%   <1%
        12%

Ever outward
Today

Dark energy   Dark matter 26%   Radiation
69%   <1%
        Matter 5%

A graphic illustrates cosmologists' current accounting of the universe's matter and energy. Within seconds of the big bang, dark matter formed, eventually laying down a gravitational scaffold for the galaxies we see today. In recent eons, a dark energy has accelerated the universe's expansion.

measurements have helped resolve some questions about the nature of the universe, they've also opened up others.

The Hubble Tension

As we learned in chapter 1, scientists in the early 20th century pieced together evidence that the universe is expanding. The telltale clue: The farther away an astronomical object is from us, the faster it appears to be receding from us (and us from it). The rate at which this occurs, called the Hubble constant, is a key measure of the universe's age and expansion rate.

However, new evidence suggests that the Hubble constant might not be constant after all. Measurements of the Hubble constant from the CMB are slightly smaller than estimates made using far younger stars, such as those in our Milky Way. This discrepancy—called the Hubble tension—exists even after accounting for other cosmic forces, such as the dark energy that's causing the expansion of the universe to speed up.

Measuring the Hubble constant is extremely hard, so it's possible that there's no Hubble tension at all and that the current gap comes from subtle errors somewhere in the calculations. But if the tension is confirmed, it means that the universe's basic ingredients are more complicated than we once thought, in a way we didn't see coming.

FILLING OUT THE MILKY WAY

It is no exaggeration to say that in the last decade, we've seen a revolution in our understanding of the Milky Way. This sea change traces back to the European Space Agency's Gaia space telescope, which since 2013 has precisely pinpointed the locations of 1.7 billion stars within our galaxy. For more than 1.3 billion of these stars, Gaia has measured the velocity of their motion across the night sky, as well as their parallax, the apparent shift in a star's location as Earth moves through its annual orbit. And for more than 100 million stars in the Milky Way, Gaia has measured those stars' distances to Earth. By contrast, the earlier Hipparcos spacecraft mentioned in chapter 1 collected distance data for 118,000 stars.

When complete, this data set will essentially let astronomers create a 3D movie of the Milky Way in motion, which will help scientists decipher the deep history of the galaxy. Already, the spacecraft's

For more on the Hubble constant, see page 28.

THREE VIEWS OF THE MILKY WAY

This view of the Milky Way reveals the brightness and color of nearly 1.7 billion stars within our home galaxy, courtesy of the European Space Agency's Gaia satellite.

This Gaia-based map of the Milky Way details star density, with lighter colors corresponding to more crowded regions.

Gaia also can reveal the galaxy's density of interstellar dust, from higher-density patches (lighter colors) to more diffuse areas (darker colors).

growing data sets are yielding major surprises. In 2018, scientists using Gaia found evidence that an ancient galaxy bigger than the Small Magellanic Cloud, called Gaia-Enceladus, merged with the Milky Way some 10 billion years ago. These stars now make up most of our galaxy's inner halo, and they can be found across the night sky.

FILLING OUT THE SOLAR SYSTEM

Today's telescopes can see more small, faint objects than ever before, letting astronomers survey the skies and fill in the wildly diverse population that surrounds the planets. In 2000, humans knew of roughly a hundred thousand solar system bodies; by late 2021, we'd discovered slightly more than a million. Remarkably, the pace of discovery is only going to accelerate.

Solar System Objects

The coming flood of newfound solar system objects will flow from the foothills of the Chilean Andes, the home of the U.S.-funded Vera C. Rubin Observatory. Once the facility begins operations in 2023, it will spend a decade mapping most of the southern night sky 825 times over. Each picture it takes will be so huge that displaying just one at full resolution would require a grid of 1,500 HD televisions.

If an obsessive film editor stitched all these pictures together into a cosmic time-lapse, the resulting video could run in full-color HD for 11 months straight.

This huge inventory of the heavens will number in the tens of billions of objects strewn throughout the cosmos, providing a treasure trove of data to astronomers and physicists of all stripes. And because the Rubin Observatory's survey will cover one patch of sky many times over, it'll be able to track short-term changes across the universe, on timescales ranging from years down to less than a minute. As a result, the survey is expected to turn up many supernovae and other transient events—

The Wide-field Infrared Survey Explorer (WISE) space telescope, taken out of hibernation and operating in a new mission as NEOWISE starting in 2013, has found more than 370 near-Earth asteroids and comets since 2009.

Located in the Cerro Pachón ridge in north-central Chile, the Vera Rubin Observatory will make a vast survey of the southern sky.

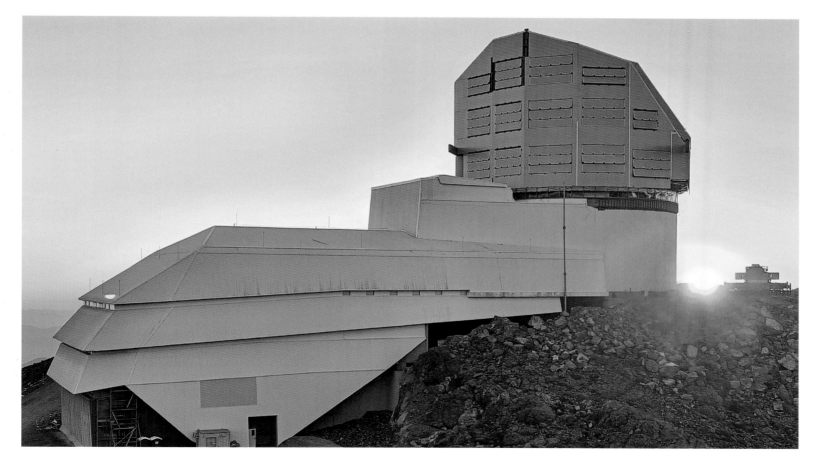

More than 90 percent of all known objects in the solar system **have been discovered since 2000.** Millions more will be found in the next decade.

■ For more on asteroids, see pages 159–161.

including types we've never found before, in all likelihood.

This massive data set also will make the Rubin Observatory an incredibly prolific mapper of the solar system. By the early 2030s, the Rubin Observatory is expected to add more than six million newfound objects to our map of the solar system, including another five million asteroids between Mars and Jupiter, 40,000 objects beyond Neptune, 10,000 comets, and between 10 and 100 interstellar objects passing through the solar system from other stars (as we'll talk about soon).

Near-Earth Objects

The Rubin Observatory is expected to find another 100,000 near-Earth objects, bodies whose orbits take them within 194 million kilometers (121 million mi) of the sun, close enough to Earth to merit extra caution. This haul should help astronomers make progress toward NASA's goal of finding 90

percent of all the near-Earth asteroids more than 140 meters (460 ft) wide.

So far, we've found pretty much all the asteroids big enough to cause cataclysms like the non-avian dinosaurs' extinction 66 million years ago, and none of these asteroids pose any threat to Earth for at least the next few centuries. Overall, the threat of a dangerous asteroid hitting Earth is very small. That said, unlikely catastrophes can and do happen, and the world needs to be prepared for them. Smaller asteroids in the 150-meter (almost 500-ft) range could still devastate an area the size of a state or province. Researchers estimate there may be 25,000 near-Earth objects more than 140 meters (460 ft) wide, but models suggest we've found only 40 percent of them.

Ongoing asteroid searches from ground-based telescopes, such as the Pan-STARRS survey in Hawaii, are helping fill in those details, as are long-running space missions such as NASA's NEOWISE infrared space telescope. The U.S.'s planned NEO Surveyor, an infrared space telescope that would launch no earlier than 2026, is expected to find thousands of near-Earth objects and fill out our knowledge of this portion of our cosmic backyard by the late 2030s.

Interstellar Objects

Telescope surveys have gotten so good that they

This graphic includes an artist's impression of 'Oumuamua (left), its path through the solar system (center), and the trajectory deviations that suggest 'Oumuamua was releasing gas (right).

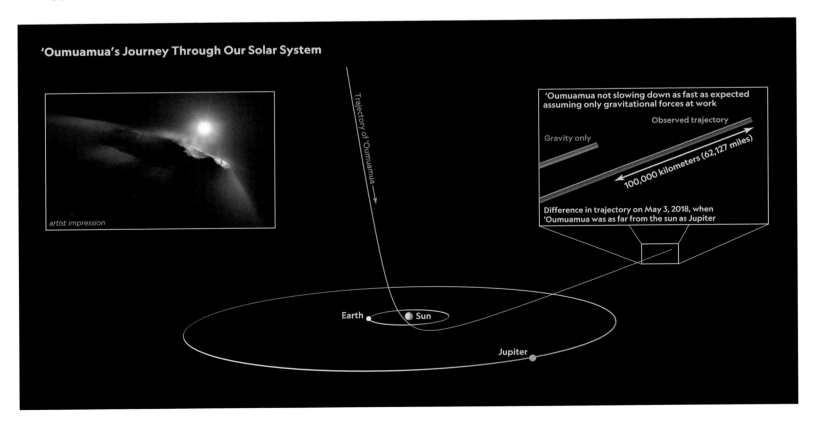

'Oumuamua's Journey Through Our Solar System

artist impression

Trajectory of 'Oumuamua

Earth   Sun

Jupiter

'Oumuamua not slowing down as fast as expected assuming only gravitational forces at work

Observed trajectory

Gravity only

100,000 kilometers (62,127 miles)

Difference in trajectory on May 3, 2018, when 'Oumuamua was as far from the sun as Jupiter

A typical near-Earth asteroid (blue) keeps its distance from Earth's orbit (green), but some potentially hazardous asteroids (orange) venture within eight million kilometers (5 million mi) of Earth's path around the sun, meriting extra caution.

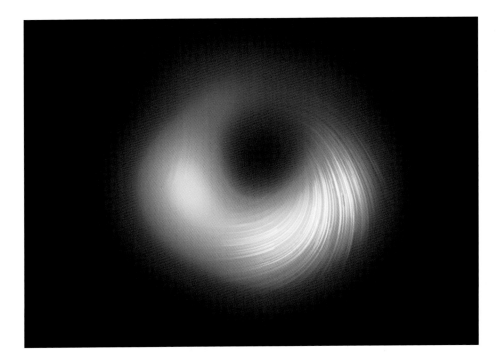

The silhouette of Pōwehi, M87's central supermassive black hole, looms large in the center of this Event Horizon Telescope image.

Other than the fact it was clearly interstellar, Borisov looked a lot like the comets we have, throwing off a veil of dust and gas rich with carbon monoxide as it neared our sun. Some astronomers think that Borisov is probably the most pristine comet ever seen, likely heated up for the first time when it zoomed past the sun in 2019. 'Oumuamua, by contrast, was far stranger than Borisov: an oblong, reddish lump that somehow picked up extra acceleration as it departed the inner solar system, likely through the release of unseen gases.

Thanks to 'Oumuamua and Borisov, researchers now know that star systems, including our own, act as huge hostels for interstellar drifters. At this very moment, thousands of interstellar objects the size of 'Oumuamua are thought to be passing within 4.5 billion kilometers (2.8 billion mi) of the sun, Neptune's average orbital distance. But even with Earth's best telescopes trained on these fleeting visitors, we still don't know much about them. Where did they come from? How did they form? What do they look like up close?

If we're lucky, we'll be able to answer these questions with a future spacecraft. In 2029, the European Space Agency will launch its Comet Interceptor mission, which will spend years "hovering" in a gravitational sweet spot 1.5 million kilometers (almost a million mi) behind Earth from the sun's point of view. The spacecraft's main target? Nobody

can spot faint objects moving so quickly through our solar system that they can't be orbiting the sun. In fact, these objects were jettisoned long ago from alien star systems and are merely passing through our neighborhood. Astronomers have found two such interstellar interlopers. In 2017, the Hawaii-based asteroid survey Pan-STARRS picked up the first, an object called 'Oumuamua, after the Hawaiian for "a messenger from afar arriving first." In 2019, Crimean amateur astronomer Gennady Borisov found the second, a comet now called Borisov.

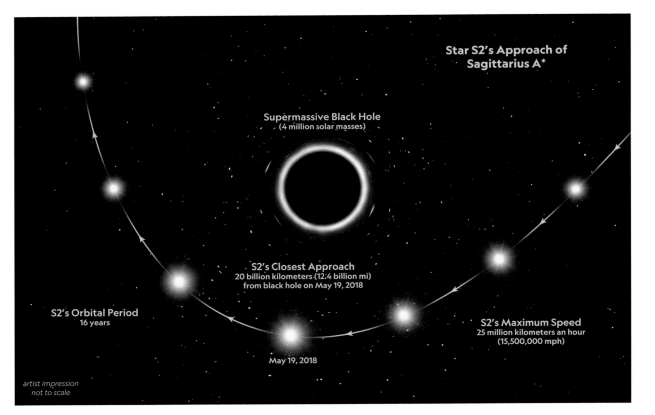

An artist's impression depicts the star S2 as it passes by Sagittarius A*, the Milky Way's central black hole. During its close approach, relativistic effects make S2 look redder than it would otherwise.

Star S2's Approach of Sagittarius A*

Supermassive Black Hole
(4 million solar masses)

S2's Closest Approach
20 billion kilometers (12.4 billion mi)
from black hole on May 19, 2018

S2's Orbital Period
16 years

S2's Maximum Speed
25 million kilometers an hour
(15,500,000 mph)

May 19, 2018

artist impression
not to scale

knows yet. The plan is to pick a newfound object the spacecraft can reach and perform a reconnaissance flyby of it.

The mission's main goal is to study a pristine comet from the solar system's farthest reaches as it approaches the sun for the first time. But if an interstellar object happens to pass within Comet Interceptor's reach, the spacecraft could give us our first look at alien terrain—or at least pieces of it.

LOOKING OUT FOR BLACK HOLES AND COSMIC FLASHES

The general theory of relativity predicts that certain massive objects can warp space-time so much that not even light can escape their clutches. Strange though they may sound, black holes are real, undeniable denizens of the cosmos. Many—if not most—galaxies have supermassive black holes at their heart, with each one millions to billions times more massive than our sun.

We've detected dozens of black holes across the universe at this point, by the intense light that superheated gases around them can give off or by the ripples in space-time they can produce (gravitational waves). We even have an image of a black hole's silhouette, thanks to the work of the Event Horizon Telescope (EHT) collaboration, which combined many telescopes to peer into the center of M87, a galaxy some 53.5 million light-years away in the constellation Virgo. This silhouette—an inky circle against a backdrop of superheated, glowing gases—also showed how the light around M87's central black hole was polarized, which revealed the vortex-like inward spirals of the black hole's magnetic fields.

The EHT also has turned its gaze toward the center of the Milky Way galaxy, where a powerful radio source called Sagittarius A* lurks. In all likelihood, the EHT will see a black hole there, too: Years of

An illustration captures the scale of the Fermi bubbles that extend from the Milky Way's center, as seen in x-rays (blue) and gamma rays (magenta).

As shown in this illustration, sometimes when the core of a giant star collapses into a black hole, the star's outermost layers get blown apart—releasing powerful gamma rays and triggering what is called a Type 1c supernova.

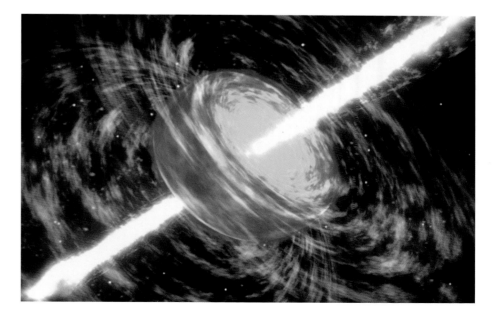

The European Southern Observatory's Extremely Large Telescope (above) will sit within the world's biggest telescope dome. The Giant Magellan Telescope (below) will include seven of the biggest mirrors ever built.

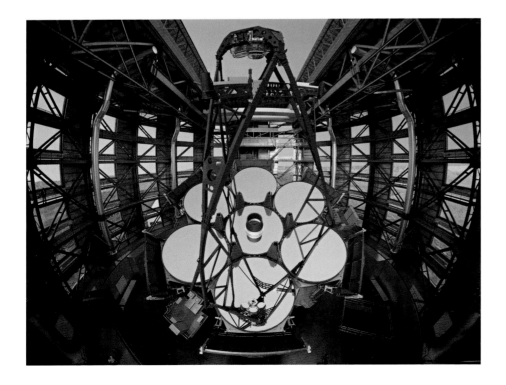

ground-based observations make an overwhelming case that Sagittarius A* is a black hole, an effort recognized by the 2020 Nobel Prize in Physics. We already know that Sagittarius A* corresponds with a compact object millions of times more massive than the sun. We've even watched a star orbit this object at speeds up to 2.5 percent of the speed of light!

Space-based telescopes also provide hints of how much energy our galaxy's center can give off. NASA's Fermi Gamma-ray Space Telescope, launched in 2008, has a knack for picking up some of the universe's most energetic flashes of light, including gamma-ray bursts given off by quasars and blazars: extremely bright galactic centers that can shoot out jets of light and high-energy particles, thanks to their supermassive black holes.

Fermi also has revealed previously hidden features of the Milky Way's architecture. In 2010, researchers using Fermi data showed that two vast balloons of gamma rays extended 25,000 light-years out of the Milky Way's disk, as if the galaxy's center were blowing highly energetic bubbles. These mysterious structures—called Fermi bubbles after the telescope that found them—are thought to be related to Sagittarius A*. An instrument on Russia's Spektr-RG space telescope has found that the Milky Way's center is giving off x-ray bubbles, too.

Another way to study black holes involves keeping watch for short, transient cosmic events. Case in point: AT 2018cow, a bizarre explosion nicknamed the Cow, whose light reached Earth on June 16, 2018. Thanks to dozens of telescopes, researchers know the Cow's basic structure: a "central engine" that glowed far brighter in x-rays than normal supernovae do, partially obscured by a blob of dust and gas. Some think the flash may show a blue supergiant star that collapsed and became a black hole.

Of course, black holes aren't the only extreme systems that can make sudden flashes of light. The Canadian Hydrogen Intensity Mapping Experiment (CHIME) telescope has spotted hundreds of some of the universe's weirdest phenomena: powerful "fast radio bursts" that last only milliseconds at a time and occasionally repeat. Many of these bizarre radio sources lie billions of light-years away, though at least one resides within the Milky Way. Researchers suspect that at least some of the signals come from short-lived, highly magnetized stellar corpses called magnetars. But it's still unclear how exactly magnetars produce the bursts—or even if magnetars are just one of several source types.

THE NEXT GENERATION OF EARTH-BASED TELESCOPES

Today's Earth-based telescopes are extraordinary pieces of scientific equipment, but in the next 20 years, a new generation of "extremely large telescopes" (that's the technical term) will transform our view of the cosmos. These versatile facilities will be able to see fainter, more distant objects than even our best space telescopes, while giving us sharper views of closer, more familiar objects. They represent one of astronomy's biggest priorities, and the science they enable will no doubt shake up what we know about the universe.

The Giant Magellan Telescope, under construction in Chile's Atacama Desert, will see the stars with a flower-shaped array of seven mirrors. Each mirror is more than 8.4 meters (27 ft) across, among the biggest ever built. With this profound ability to collect light, the GMT's resolution will reach seven to 10 times the resolution of the Hubble Space Telescope. It's scheduled to come online in 2029.

The Extremely Large Telescope (ELT), also being built in the Atacama, bills itself as the "world's biggest eye on the sky." It's not hard to see why: With a segmented mirror 39 meters (128 ft) across, the ELT will turn its gaze toward the faintest, earliest, most distant objects in the universe. Its goal is also to monitor Earthlike planets around distant stars for signs of life (more on this to come). The ELT is on track for completion in 2027.

As the name of the Thirty Meter Telescope (TMT) suggests, its 492-segment primary mirror will stretch 30 meters (98.4 ft) across, giving it a light-collecting area bigger than a basketball court. The telescope's resolution will be 12 times the Hubble Space Telescope's. It's currently slated for construction atop Hawaii's Mauna Kea volcano (though see sidebar).

GRAVITATIONAL WAVES

In one of the biggest astronomy developments in decades, scientists now can catch wind of cosmic phenomena by detecting waves totally unrelated to light—signals that could let us reach back to the farthest, deepest reaches of the universe's past.

These waves are a direct result of Einstein's vision of the cosmos, the general theory of relativity. According to this theory, matter warps space-time, and the geometry of space-time dictates how matter moves. To see how, imagine setting a barbell onto a taut trampoline. The trampoline flexes and warps, and the barbell's mass forms a divot in the trampoline's surface. Roll a marble across the warped trampoline, and its path now curves when it would have once moved in a straight line.

Now imagine you suddenly twisted the barbell.

PROTESTS AT THE TMT

Because of its altitude, clear skies, and favorable weather, the peak of Hawaii's Mauna Kea is among the most important astronomy sites in the world, home to 13 cutting-edge telescopes. Plans are underway to demolish some of the oldest telescopes there and replace them with the Thirty Meter Telescope (TMT). But science never exists in a vacuum, and while the TMT enjoys majority support across the state of Hawaii, many in Hawaii's Indigenous community oppose the effort.

Among other critiques, many Indigenous Hawaiians see Mauna Kea as sacred and worry about the environmental impacts of future construction. More importantly, they view TMT as a symbol of injustice and broken promises going back more than a century to the overthrow of the independent Hawaiian monarchy in 1893 and the U.S. annexation of Hawaii in 1898. As a result, large protests in 2015 and 2019 called for the project's cancellation, delaying the TMT's construction.

Indigenous Hawaiians have diverse perspectives on TMT, and some support it as a unique opportunity for the state. As of late 2021, TMT was still slated for construction in Hawaii. If construction in Hawaii loses support, the facility may be built instead in Spain's Canary Islands.

GRAVITATIONAL WAVE OBSERVATORIES
LAND-BASED
- ☐ Operational
- ☐ Under development
- ☐ Planned or proposed

The trampoline's surface would rebound or flex, in the form of ripples coursing outward from the twisting barbell. In the universe, these ripples are what are known as gravitational waves: traveling disturbances in the geometry of space-time, created by accelerating masses. The ones we can detect are made by incredibly massive objects caught in incredibly violent scenarios, such as colliding black holes.

Albert Einstein first theorized about gravitational waves in 1916. While many of his peers accepted the reality of these waves, Einstein himself later waffled on their existence and even published the occasional paper that tried to refute them. Since the 1970s, astronomers have had strong indirect evidence of the waves' existence (see sidebar), but detecting them directly has proven extremely difficult.

For one, gravitational waves' amplitude is predicted to be extremely tiny. As one of these waves passes through an object, it's predicted to stretch and strain the object. But the gravitational waves that astronomers expected to find—such as those thrown off by colliding black holes—would stretch and strain Earth by less than a proton's width. Detecting this subtle change is like measuring the distance from here to Proxima Centauri, the closest neighboring star to the sun, to within the width of a pollen grain. Yet scientists have managed to pull it off.

LIGO and the First Detection

On September 14, 2015, two detectors in Louisiana and Washington state felt a bump in the cosmos. The two facilities combine to form the U.S. Laser Interferometer Gravitational-Wave Observatory, or LIGO for short. That fall day, LIGO made history with the first confirmed direct detection of gravitational waves.

Each facility is shaped like an L, with each arm of the L some four kilometers (2.5 mi) long. By very precisely bouncing lasers back and forth down

INDIRECT EVIDENCE

In the 1970s, astronomers Joseph Taylor (pictured left) and Russell Hulse (right) found the first solid indirect evidence for gravitational waves in the form of extreme cosmic objects called pulsars: rapidly rotating, incredibly dense, magnetized objects formed from the remains of supernovae. Pulsars were first discovered by astronomer Jocelyn Bell Burnell in 1967, but the first ones were found floating by themselves. Hulse and Taylor's target—formally named PSR 1913+16—was the first binary pulsar ever discovered, a set of pulsars that orbited each other more than 13,000 light-years away from us.

PSR 1913+16 ended up being a perfect laboratory in which to test Einstein's theories of gravity. The longer Taylor and Hulse watched the system, the more they saw the pulsars spiral in toward each other—which meant the system had to be losing energy somehow. This energy loss exactly matched predictions for the energy the binary pulsars would be shedding in the form of gravitational waves. The discovery netted Hulse and Taylor the 1993 Nobel Prize in Physics.

In this simulation, two black holes spiral into each other and then merge. This kind of cosmic cataclysm warps the space-time nearby, forming ripples we can detect as gravitational waves.

the arms' length, each LIGO facility can measure its arms' lengths to astonishing levels of precision. If a gravitational wave passed through one of LIGO's facilities, it would stretch one of the arms and compress the other, enough to create a detectable signal. LIGO's two facilities are so sensitive that each can detect the rumble of cars on local roads or even the faint tremors of distant earthquakes. So to avoid false detections, both of LIGO's facilities must spot the same signal for it to be considered real.

On that fateful September day, both facilities saw the same signal within seven milliseconds of each other. Researchers think the signal was created during a merger of two black holes 1.3 billion light-years away, one 29 times more massive than the sun and the other 36 times more massive. The black hole that the merger created was 62 times more massive than our sun, three solar masses lighter than the two starter black holes. That missing mass hints at a truly enormous cataclysm. In the fraction of a second it took to form that bigger black hole, the collision gave off three suns' worth of energy in the form of gravitational waves. More than a billion years later, we managed to feel that aftershock.

LIGO announced this first detection in February 2016 and garnered 2017's Nobel Prize in Physics in the process. Thankfully for LIGO and scientists around the world, the event known as GW150914 wasn't a one-off. Now, Europe's Virgo observatory can combine with LIGO to detect more events and better place their points of origin on the night sky. As of early 2021, LIGO and Virgo had found 50 gravitational wave signals. Japan now has its own observatory, KAGRA, and more are on the way, including some in China and India. So too are more signals.

Now that we can spot gravitational waves, we can observe phenomena it would otherwise be impossible to see with light. Unless they're actively feeding and superheating surrounding gases in the process, black holes are just that: black. Without gravitational waves, events such as GW150914 couldn't be detected, which means gravitational waves give us a chance to see black holes in action and measure their key properties.

By now, we've spotted enough black holes to start to look at population trends, such as the overall proportions of different black hole masses. As more detections pile up, researchers also will be able to address vexing questions, such as how stars die and collapse into black holes in the first place. And gravitational waves will play a huge role in the coming age of "multi-messenger astronomy," in which astronomers will look at cosmic events in many different ways at once and learn more than they would have otherwise (see sidebar).

NEW DETECTORS

As productive as LIGO and Virgo have been, they're essentially the gravitational-wave equivalents of x-ray telescopes: designed to catch very high-frequency events. In the next two decades, astronomers will be able to find even lower frequencies, giving them access to vastly different cosmic phenomena. Here are some other efforts to detect gravitational waves, and the progress made on each.

LISA

For all the success that LIGO and Virgo have had, Earth is a noisy place compared to the vacuum

WHEN (NEUTRON) STARS COLLIDE

Arguably the most important gravitational wave signal found so far reached Earth on August 17, 2017. The LIGO and Virgo observatories detected a signal consistent with two neutron stars spiraling into each other and colliding. At about the same time, researchers saw a flash of light in that part of the night sky, which scientists soon realized was from the same event.

Because scientists recorded this event—named GW170817—in both light and gravitational waves, it has given scientists a huge amount of information, as if we had made the first talkie film of the universe after making only silent pictures of the stars.

Observations of the explosion suggest that collisions of neutron stars play an important role in generating many of the universe's heavier elements, including precious metals such as gold. And because the same event was seen in light and gravitational waves, researchers could compare how quickly each signal got to Earth. The two totally different signal types arrived within seconds of each other, an agreement that placed strict limits on theories of gravity that predicted otherwise.

of space. As a result, the European Space Agency is planning to launch a mission in the 2030s that will act like a much bigger, space-based version of a LIGO detector. The goal is to send three spacecraft into orbit around the sun and keep them flying in a triangular formation 2.5 million kilometers (1.6 million mi) long to a side. This array will detect lower-frequency gravitational waves than those LIGO can catch. In 2015, the ESA launched a proof-of-concept spacecraft called

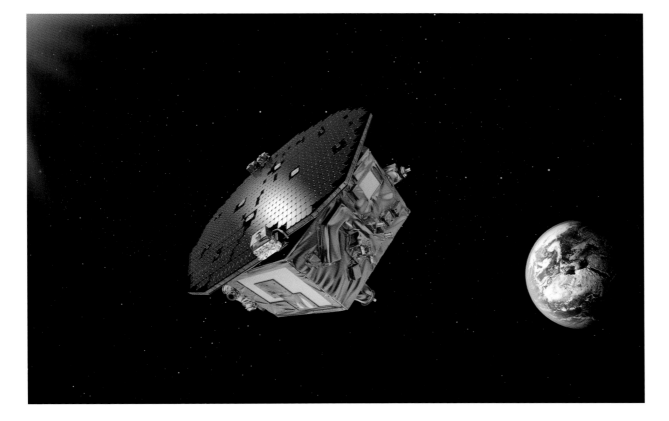

The LISA Pathfinder spacecraft orbited at a gravitational point between Earth and the sun.

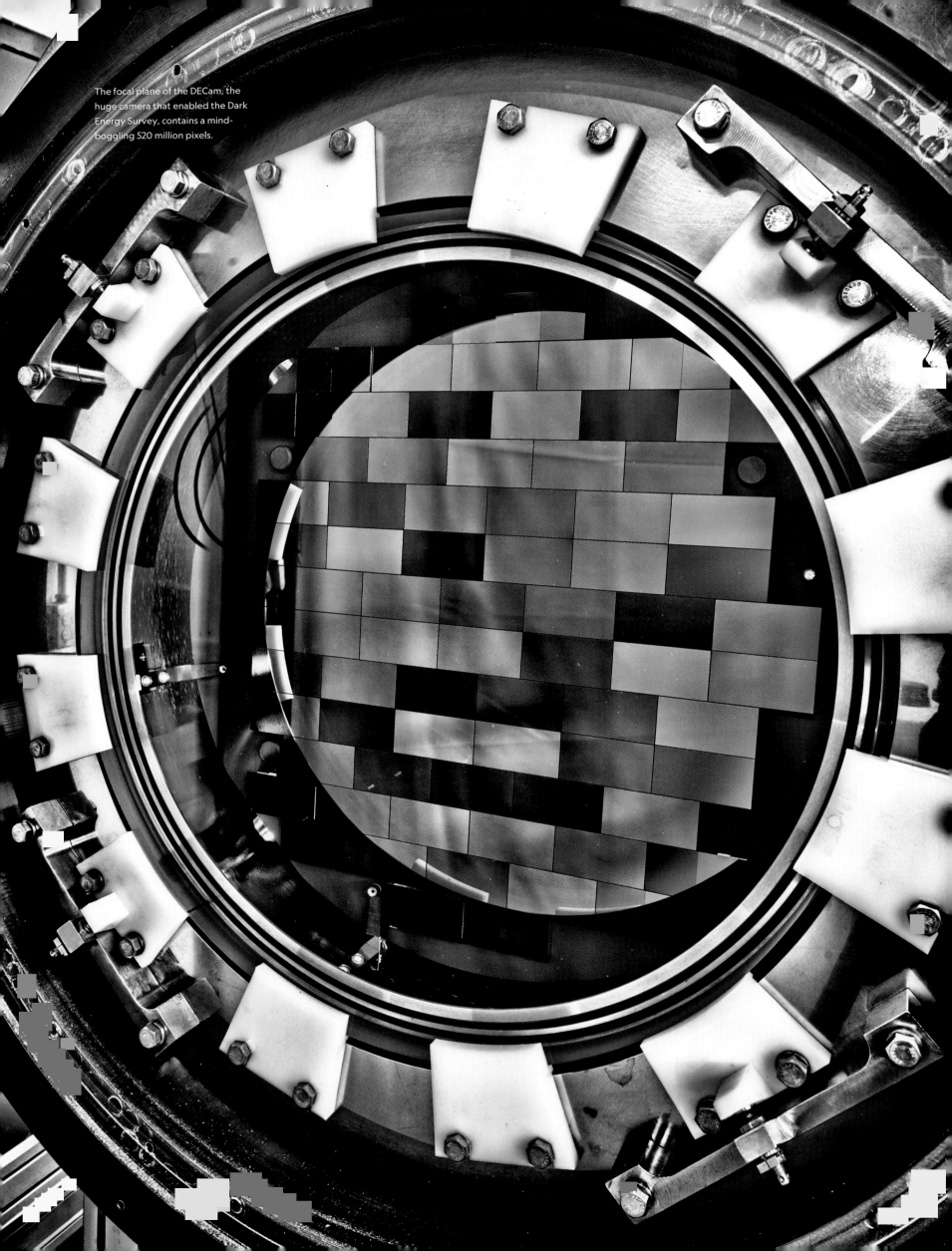

The focal plane of the DECam, the huge camera that enabled the Dark Energy Survey, contains a mind-boggling 520 million pixels.

LISA Pathfinder to test LISA's key tech. That mission worked, paving the way for the full mission.

NANOGrav

Galaxies collide, and many galaxies hold supermassive black holes at their hearts. It stands to reason, then, that somewhere in the cosmos, pairs of supermassive black holes are spiraling into each other and merging. The gravitational waves from these encounters are likely to have extremely low frequencies, however, which means it would take many years of observations to detect them.

A major U.S.-Canada collaboration called NANOGrav is trying to spot these low-rolling waves by tracking Earth's position relative to 47 distant pulsars. Like giant lighthouses, pulsars emit beacons of radiation that periodically pass Earth's way. Pulsars keep time so well that they can act as precise galactic clocks. If low-frequency gravitational waves are passing through Earth, as scientists think they are, they should compress and stretch the space between Earth and these pulsars. This slight stretch should throw off the pulsars' timing ever so slightly, in a telltale pattern that would vary depending on a pulsar's location in the night sky.

As of late 2020, NANOGrav researchers had identified a common set of noise across almost all the pulsars they are monitoring. If this signal holds with future data, astronomers could well be tuning into the universe's "stochastic gravitational wave background": the combined coffee-shop murmur of

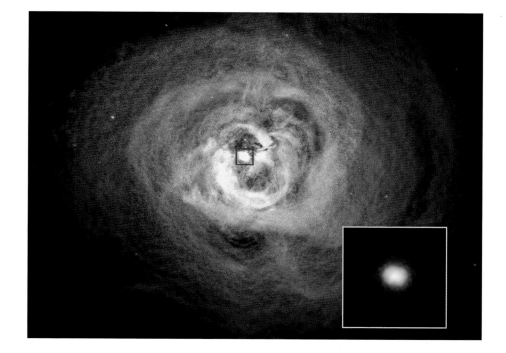

waves given off by many distant pairs of supermassive black holes. Confirming that signal will require more data to tease out the particular, sky-wide pattern that only gravitational waves could make.

DARK MATTER

As mentioned in chapter 1, scientists have lots of reasons to think that the stuff that makes up stars, planets, and people is about 5 percent of the universe's total matter and energy density. Some 26 percent of the universe is tied up in a mysterious substance called dark matter. Dark matter's name is a misnomer: It would be better if the stuff were called "clear matter," because it doesn't seem to

Extreme cosmic objects such as M87's central black hole, seen here in x-rays, can provide stringent tests of dark matter models. One 2020 study used M87's x-rays to rule out certain kinds of theoretical dark matter particles.

■ For more on pulsars, see page 47.

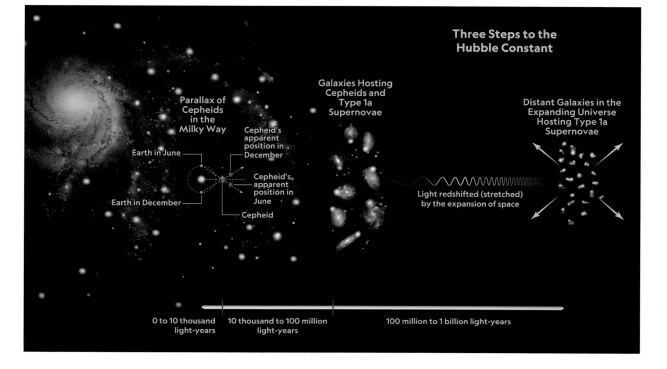

Three Steps to the Hubble Constant

Parallax of Cepheids in the Milky Way

Earth in June

Earth in December

Cepheid's apparent position in December

Cepheid's apparent position in June

Cepheid

Galaxies Hosting Cepheids and Type 1a Supernovae

Distant Galaxies in the Expanding Universe Hosting Type 1a Supernovae

Light redshifted (stretched) by the expansion of space

0 to 10 thousand light-years

10 thousand to 100 million light-years

100 million to 1 billion light-years

To measure the universe's expansion, astronomers build up a "cosmic distance ladder" by establishing our distance from nearby stars and supernovae and then building up to farther galaxies.

A theoretical particle called the axion (artist's conception at left) may be one ingredient of dark matter. Dark matter is thought to envelop many galaxies in spherical haloes, including the Small Magellanic Cloud (right). These haloes are thought to be densest toward their center and more diffuse at their periphery.

The four-meter-wide (13-ft) Victor M. Blanco Telescope is a cornerstone of the European Southern Observatory in Chile's Atacama Desert.

absorb or reflect light like the matter we're made of does. Instead, astronomers have to infer dark matter's presence from the signs of its gravitational field. We've also managed to suss out its presence from the cosmic microwave background.

Exactly what dark matter is, is still anybody's guess. Physicists haven't yet detected a dark matter particle in the lab. But if (when?) future detectors find dark matter, those instruments wouldn't just help define the vast clouds of dark matter we've inferred throughout the universe. In addition, detectors could act as telescopes in their own right, by letting us see and measure the vast spherical halo of dark matter that's thought to envelop and stabilize the Milky Way.

Many theories make predictions for particles that have the properties you'd want for dark matter, and efforts to find them go back decades. In the 1980s, many physicists had pinned their hopes on WIMPs, "weakly interacting massive particles" predicted by a theory called supersymmetry (which was

appealing for its own separate reasons). However, the detectors built to see these WIMPs—supercooled vats of liquid xenon buried deep underground—have come up empty so far.

If WIMPs are indeed real, this lack of a signal has clarified how much—or how little—they interact with everyday matter. This information is essential. When you're looking for a kind of particle nobody's ever seen, knowing how it doesn't behave is just as important as knowing how it does.

Hunting for the Axion

In the absence of WIMPs, some physicists have turned their focus to another dark matter candidate: the axion. The axion is a particle predicted by a theory that tries to explain a separate mystery of physics, the observed distribution of electrical charges within the neutron.

Conveniently for would-be axion hunters, the axion and related "axion-like" particles are predicted to convert into photons within a strong magnetic field, at some unknown (but likely very low) odds. If you construct an electromagnetically isolated, supercooled chamber, and you subject it to a strong magnetic field, you might force an axion to turn into a detectable photon within the chamber.

The frequency of these photons would depend on the axion's mass. But because nobody knows what the axion's mass is (or if the axion is even real), we don't know what frequency its photon spawn would have. As a result, axion hunters must sweep their devices across many possible frequencies, like turning the dial on your car radio to find music during a road trip. Several detectors, such as the University of Washington's Axion Dark Matter eXperiment (ADMX) haloscope, are actively turning their dials. Current theories allow for a fairly wide range of possible masses. If the axion exists, researchers might be scanning for a while.

WIMPs and axions are just two of the many types of possible dark matter that scientists are looking for. Other potential dark matter candidates abound. Could dark matter consist of smallish "primordial" black holes, so called because they formed in the universe's infancy? Ultralight "fuzzy" dark matter? Or are the signs of dark matter really a clue that our theory of gravity needs an overhaul? Only by keeping up the search will we ever know the answer.

DARK ENERGY

There's a lot of dark matter in the universe; it outmasses all the stuff we can see, from the tiniest ant to the biggest galaxy supercluster, by more than five to one. But even dark matter and regular matter combine to just 31 percent of the universe's matter and energy density. As you saw in chapter 1, the remaining 69 percent is made of a mysterious force called dark energy.

Huge ground-based telescopes have helped shed light on dark energy. They include the Dark Energy Survey, which from 2013 to 2019 used Chile's

Throughout the universe, dark matter outnumbers the stuff we're made of by more than five to one. Future detectors may reveal its identity.

Cerro Tololo Inter-American Observatory to map 300 million distant galaxies. Among other things, these images contained many distant supernovae, which let researchers measure galaxies' distances from the Milky Way (using the cosmic distance ladder in chapter 1) and infer some of the universe's key properties.

Future telescopes both on Earth and in space will enable new, better measurements of dark energy. For instance, the Rubin Observatory's massive survey of the southern night sky will be a vital source of data for dark energy hunters. The wide-field view on NASA's upcoming Nancy Grace Roman Space Telescope, expected to launch by 2027, will let astronomers more easily measure how the universe's expansion is accelerating.

The Roman Space Telescope's field of view (outer image) will dwarf that of Hubble (inner image), letting it take in cosmic phenomena at much larger scales.

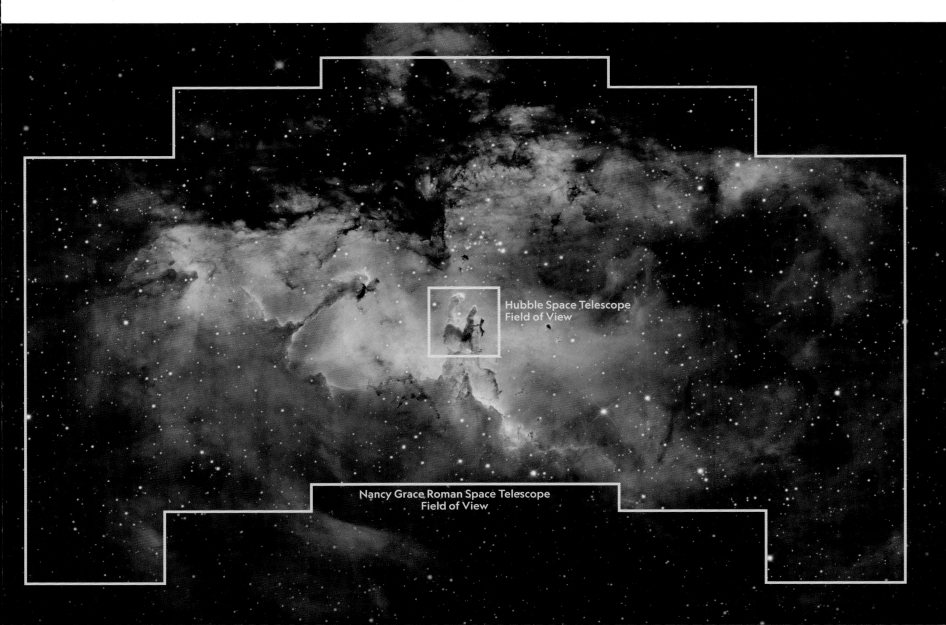

Hubble Space Telescope
Field of View

Nancy Grace Roman Space Telescope
Field of View

ALIEN PLANETS

Twins of Earth may be lurking around alien stars. So are strange worlds with no analogue in our solar system.

In one of the most profound scientific developments of the last 30 years, astronomers now can look out on the cosmos and find alien planets orbiting distant stars. So far, we've found more than 4,500 of these distant extrasolar planets, or exoplanets, including some around the stars next door.

The more exoplanets that scientists find, and the more data they collect on each, the more that they'll be able to find patterns in the galaxy's planet populations—and see how our solar system fits into the mix. For instance, some of the most common types of planets we've found so far are nowhere to be found in our solar system. Many other stars host giant "hot Jupiters" that orbit their home stars much more closely than Jupiter does our own sun. Some have "super-Earths" that fall in between the mass of Earth and Neptune—worlds for which we have no local analogue.

As these data sets grow, and as telescopes improve, we may get to a point where we can map distant planets' surfaces, figure out what's in their atmospheres, and even see if they could be home to alien life. But before we explore all these exciting possibilities, let's get one thing out of the way: How do astronomers find exoplanets in the first place?

THE RADIAL VELOCITY METHOD

If you've ever watched the Summer Olympics, you've probably seen the hammer throw event, in which athletes spin around with a heavy ball at the end of a thin wire and then let it go, sending the ball flying. You may have noticed that as the thrower spins, her balance isn't where it'd be if she were simply pirouetting in place. Instead, the thrower's center of mass shifts slightly toward the hammer, introducing a slight wobble to her spin.

The same is true for exoplanets. Think back to the Doppler shift you learned about in chapter 1: Stars receding from us look redder than they otherwise would, while stars moving toward us look bluer than they otherwise would. As planets move through their orbits, they cause their home stars to wobble toward and away from us, which imparts a repeating pattern of shifts in the star's light spectra. A star with an orbiting planet will appear slightly bluer as it approaches us, then slightly redder as it recedes, and then slightly bluer again.

In the right conditions, astronomers can see this wobble and use its particulars to learn about the orbiting planet. How much the star's spectra

As an exoplanet orbits its home star, it causes the star to wobble slightly. Sometimes this wobble shifts the star's light wavelengths enough to make the change detectable from Earth.

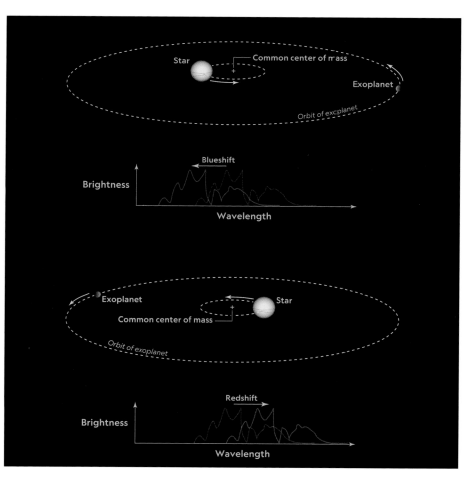

(Opposite) At 41 light-years from Earth, the planet 55 Cancri e lies far beyond human travel. This NASA travel poster imagines what it would be like to see the molten world up close.

The exoplanet Proxima b (seen here in an artist's conception) gets hit with such extreme UV radiation that it's not clear if the planet is habitable—or if it even has an atmosphere.

That planet, initially named 51 Pegasi b, came with a big surprise. Though it was approximately half the mass of Jupiter, the world orbited its home star hellishly close, keeping an average distance of just eight million kilometers (5 million mi)—seven times closer, on average, than Mercury gets to our own sun. The planet forced astronomers to think more broadly about how planets formed and migrated through newborn star systems. (Side note: In honor of its mass relative to Jupiter's, the planet is now called Dimidium, Latin for "half.")

Proxima b

Thanks to the radial velocity method, we also know that the red dwarf Proxima Centauri, the closest neighboring star to our solar system, has planets orbiting it. In 2016, a team called the Pale Red Dot announced that radial velocity measurements performed at the European Southern Observatory in Chile had picked up a planet at least 1.3 times more massive than Earth in orbit around Proxima Centauri.

shift, and how quickly this pattern repeats, depends on the planet's mass and orbital distance from its home star.

This technique—called the radial velocity method—was the first major planet-hunting method for sunlike stars. In 1995, astronomers Michel Mayor and Didier Queloz announced that they had used the radial velocity method to find an exoplanet orbiting 51 Pegasi, a sunlike star 50 light-years from us in the constellation Pegasus.

This planet—called Proxima b—likely spends most of its time in the star's habitable zone: the range of orbital distances in which a planet hypothetically could keep liquid water on its surface. Proxima b excites researchers because of its closeness to our solar system and possible tolerance of life. If habitable conditions can be found in the star system next door, how common could they be throughout the universe?

SKYLIST

CONSTELLATION STARS WITH EXOPLANETS

Most known exoplanets orbit stars that you can't readily see with the naked eye, especially if you live in an area with light pollution. But some bright stars in the night sky—even some in major constellations—have planets around them. All of these planets were found using the radial velocity method.

■ **TAURUS: AIN.** In 2006, astronomers discovered that the star Ain had a planetary companion: Epsilon Tauri b. The planet is a gas giant at least 7.6 times more massive than Jupiter and orbits Ain about once every 595 days. The planet also goes by Amateru, which refers to the sun goddess Amaterasu in Japanese mythology.

■ **TAURUS: ALDEBARAN.** In 2015, researchers found a gas giant in orbit around the red giant star Aldebaran. This world, at least 6.5 times more massive than Jupiter, is called Alpha Tauri b.

An artist's depiction of gas giant Alpha Tauri b imagines it in shades of orange and yellow.

■ **GEMINI: POLLUX.** In 2006, astronomers confirmed that Pollux is orbited by a gas giant at least twice as massive as Jupiter that goes around the star about every 585 days. This world, formally called Pollux b, was later named Thestias, after the mother of the Gemini twins Castor and Pollux in Greek mythology.

■ **OPHIUCHUS: NU OPHIUCHI.** A 2012 study using Japan's Okayama Astrophysical Observatory confirmed that Nu Ophiuchi is orbited by two brown dwarfs: bodies whose masses sit between the smallest stars and the biggest gas-giant planets. Both objects are more than 22 times Jupiter's mass. The closer of the two, Nu Ophiuchi b, orbits once every 530 days. The more distant, Nu Ophiuchi c, orbits once every 3,186 days.

■ **LEO: ALGIEBA.** In late 2009, astronomers announced that they had found a planet around one of the two stars in the binary star system of Algieba (Gamma Leonis). This planet, called Gamma 1 Leonis b, is a gas giant at least 8.8 times more massive than Jupiter.

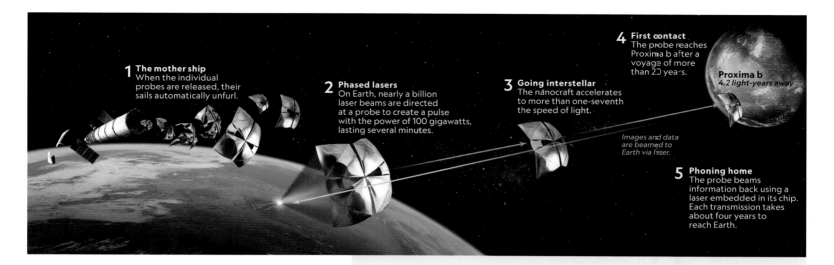

1 The mother ship
When the individual probes are released, their sails automatically unfurl.

2 Phased lasers
On Earth, nearly a billion laser beams are directed at a probe to create a pulse with the power of 100 gigawatts, lasting several minutes.

3 Going interstellar
The nanocraft accelerates to more than one-seventh the speed of light.

4 First contact
The probe reaches Proxima b after a voyage of more than 20 years.

Proxima b
4.2 light-years away

Images and data are beamed to Earth via laser.

5 Phoning home
The probe beams information back using a laser embedded in its chip. Each transmission takes about four years to reach Earth.

All that said, Proxima b's friendliness to life remains unclear. For one, Proxima Centauri has a nasty habit of emitting x-ray flares that are so intense that they run a risk of frying the planet's surface and sterilizing it. Nonetheless, in the future, Proxima b will almost certainly be a major focus of follow-up research. Future telescopes may even be able to detect its atmosphere (if it has one) or map its surface. One effort, the Breakthrough Starshot initiative, is looking into the technology required to send tiny spacecraft on reconnaissance flybys of the planet (see sidebar).

The Transit Method

In comic books, the police force of fictional Gotham City can call on the superhero Batman by projecting a circle of light on the night sky with a dark bat at its center. But how does this Bat-Signal work? In the comics, police have modified a standard spotlight by mounting a solid, bat-shaped plate in front of it. The plate blocks some of the spotlight's light while letting the light around the edges come through, which casts a bat-shaped shadow on the night sky.

Now, imagine a version of this scenario playing out at a cosmic scale, with a star playing the role of the spotlight and a planet playing the role of that bat-shaped plate. If star systems have planets, some fraction of them will have planets that, coincidentally, happen to pass between their home stars and our vantage point on Earth. If so, these alien worlds will pass between their home star and us once per orbit, temporarily blocking a small fraction of that star's light. Watch enough stars for long enough, and you'll eventually catch one in that fortunate alignment—and spot the universe's natural equivalent to the Bat-Signal.

BREAKTHROUGH STARSHOT

Billionaire tech investor Yuri Milner has poured millions of dollars into an astronomy and space exploration nonprofit called the Breakthrough Initiatives. One of its programs, the Breakthrough Starshot project, is trying to come up with ways to do a robotic flyby mission of the Alpha Centauri star system, in hopes of seeing planets such as Proxima b up close.

The plan calls for building many lightweight "nanocraft" weighing no more than a few grams each that would have large, highly reflective "lightsails." A ground-based array of powerful lasers would then direct its beams at the lightsails, which would give the spacecraft a slow, steady push toward Proxima Centauri. After years of constant acceleration, each spacecraft could get up to speeds of 161 million kilometers an hour (100 million mph), more than a seventh of the speed of light. At those mind-bending speeds, a flyby reconnaissance mission of Alpha Centauri would take approximately 20 years from launch.

The engineering challenges for such tiny spacecraft—and such powerful Earth-based lasers—are enormous. In all likelihood, it will take decades to make good on Starshot's vision of flying spacecraft through Alpha Centauri. But just imagine those pictures!

This exoplanet-hunting method, called transit photometry, has dominated the count of newfound exoplanets for the past 15 years, in large part thanks to NASA's Kepler Space Telescope, which by itself discovered more than half of all known planets beyond our sun.

Launched in 2009, Kepler was aimed at a fixed patch of sky about twice the size of the Big Dipper's scoop, monitoring the approximately 150,000 main-sequence stars there for any dips in brightness from transiting planets. But after the telescope suffered hardware failures in 2012 and 2013, scientists were forced to adjust Kepler's mission plan to sweep across wider swaths of sky with a less precise gaze. To date, data from Kepler have revealed more than 2,800 confirmed exoplanets and another 3,200 candidate exoplanets.

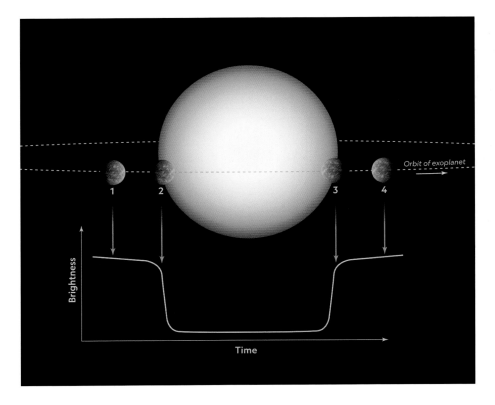

Transit photometry works by tracking a star's brightness over time. Regular dips in the star's brightness reveal planets whose orbits happen to pass between the star and Earth.

Though Kepler's the most prolific and famous planet-hunting telescope using the transit method, it's not the only game in town. NASA's TESS mission, a successor to Kepler, has found more than 107 planets since its 2018 launch. The European Space Agency's Cheops satellite, launched in 2019, specifically aims to do follow-up observations on already confirmed exoplanets, to better pin down these alien worlds' sizes.

Thanks to Cheops, scientists have worked out that as the planetary system TOI-178 formed 205 light-years away, five of its six planets' orbits settled into resonance with one another. For every 12 orbits the second innermost planet completes, the more distant planets complete nine, six, four and two orbits, respectively—a harmonic mini-sonata composed by gravity and time.

Ground-based telescopes also have had success with the transit method. In fact, the first exoplanet ever found via the transit method, the Jupiter-like OGLE-TR-56b, was found in 2002 using telescopes

Astronomers have found more than **4,500 alien planets** so far around distant stars. Some of them may one day be found to harbor life.

in Arizona, Chile, and Hawaii. In 2016, astronomers announced that telescopes in Chile had discovered TRAPPIST-1, a system that turns out to have seven Earth-size worlds, some of which could be in the temperature range to keep liquid water on their surface.

Transits don't just reveal whether an exoplanet exists. They also can provide helpful hints on the planet's properties. The bigger the transiting planet, the more starlight it blocks. The transit's length encodes how far away the exoplanet is passing from its home star. The bigger the planet's orbit, the longer the planet takes to complete the transit.

A planet's size also hints at its possible range of masses, based on everything we know about how planets form. If an exoplanet is as big as or bigger than Jupiter, its density probably resembles that of our gas giants, Jupiter and Saturn. If it's the size of Earth, it is probably formed from denser elements such as iron and silicon and resembles rocky planets like Earth or Venus.

In the best-case scenario, transits don't just reveal an exoplanet's size. They can even reveal something about an exoplanet's atmosphere, if it has one—and possibly whether it has moons.

While exoplanets themselves are opaque, their atmospheres can let light through, and depending on their composition, atmospheres absorb different specific wavelengths of light. In the right circumstances, astronomers can disentangle this faint atmospheric imprint and digitally sniff out the components of alien atmospheres from trillions of miles away.

As you might imagine, getting this kind of atmospheric measurement today requires conditions to be just right. So far, only a small fraction of exoplanets have had their atmospheres studied, and they skew toward nearby systems with bright stars, large planets, and scorching, diffuse atmospheres that extend far from the planet. (The taller the atmosphere, the more starlight can pass through it—which means Earthlings and their telescopes receive a bigger, cleaner signal.)

So far, no breathable alien atmospheres have been found—and many of the ones we've found so far are downright nightmarish. Consider HD 189733 b, a gas giant exoplanet some 64.5 light-years away from the solar system in the constellation Vulpecula. Though it's about 15 percent bigger

An artist's impression imagines
the surface of Proxima b. Proxima
Centauri looms largest in the sky;
the double star Alpha Centauri AB
is also visible.

than Jupiter, it orbits more than 30 times closer to its star than we do to our sun, completing its orbit once every 53 hours and 15 minutes, at a blistering average orbital speed of 549,000 kilometers an hour (341,000 mph).

HD 189733 b is just about perfect for atmo-

spheric study, making it one of the best understood exoplanets found so far. Astronomers know, for instance, that the atmosphere's temperature varies between 700 and 940°C (1,292 to 1,724°F)—hot enough to melt aluminum—and contains water vapor, carbon monoxide, methane, carbon dioxide, and sodium. This scorching world is the first exoplanet whose color we've learned: a deep, dark blue, thanks to a truly hellish sort of rain. The blue color appears as light gets scattered by silicate particles—aka tiny shards of glass—in HD 189733 b's high-altitude clouds. Ensnared in eastward winds that blow from the planet's day to night side, these particles whip around the planet at more than 8,690 kilometers an hour (5,400 mph), seven times the speed of sound in Earth's atmosphere.

On some exoplanets, astronomers have even seen alien atmospheres shift and change on the timescales of weather here on Earth. Recent observations of Kepler-434 b—a warm Jupiter-like exoplanet discovered in 2015—have revealed clouds of potassium chloride and sodium chloride (aka table salt) that evaporate into rising gas, condense, rain back down, and evaporate again, all in a few hours' time.

As telescopes get more sensitive, astronomers may soon gain the ability to tease out the tiny additions that exomoons contribute to transiting exoplanets' shadows. For now, no definitive exomoons have been found, but hints of them are starting to appear. In 2018, astronomers Alex Teachey and David Kipping suggested that the transit data of Kepler-1625 b, a Jupiter-like gas planet, could be better explained if the planet had a gargantuan moon about the size of Neptune. Two years later, other astronomers found that timing variations in the transits of six exoplanets might be explained by orbiting moons' tiny tugs. Future work will surely turn up clear-cut evidence for exomoons. The quest for them is only just beginning.

OTHER PLANET-FINDING TECHNIQUES

While the transit and radial velocity methods are the primary planet-finding tools today, there are a few more worth mentioning, both for their historical importance and their sheer coolness.

Pulsars

The first exoplanet ever discovered, found in 1992, wasn't orbiting a sunlike star. Instead, researchers located it by picking up changes in the timing of PSR B1257+12, a pulsar about 1,960 light-years from us. In the years since, astronomers have looked for cyclical shifts in the pulsar's timing and found two more planets in orbit around the dense little husk of an exploded star.

As far as planetary systems go, PSR B1257+12 is fairly creepy. Because the pulsar was formed from the aftermath of a supernova, one could call it "undead," and the planets are likely inhospitable due to the pulsar's radiation. Fittingly, PSR B1257+12 is now called Lich, after a powerful undead entity in fiction. Its three planets have the spooky names Poltergeist, Phobetor, and Draugr.

Microlensing

Stars are so massive, they can warp the space-time around them enough to bend incoming light. If a star happens to pass between Earth and a distant light source, it can temporarily magnify that distant light from our point of view and make a flash. Sometimes, planets orbiting such a "microlensing" star can cause this flash to vary enough that astronomers can infer the planets' presence. So far, more than 120 exoplanets have been found using this method.

Gaia

As you saw earlier in this chapter, the European Space Agency spacecraft Gaia is making the best ever map of the Milky Way, which also makes Gaia a powerful planet-finding tool. As a distant star moves across Gaia's field of view, any orbiting planets could cause the star to wobble enough for Gaia

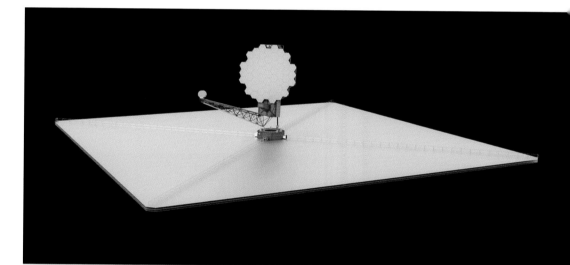

SPACE TELESCOPES AIMING AT EXOPLANETS

In November 2021, the U.S. National Academies of Sciences, Engineering, and Medicine released a major report that outlined a plan for U.S. astronomy through the early 2030s. One of the report's biggest priorities: a next-generation space telescope with amazing planet-spotting capabilities.

The plan calls for the construction of a roughly six-meter-wide (20-ft) space telescope that can detect infrared, visible, and ultraviolet light, like a much bigger version of the 2.4-meter (7.9-ft) Hubble Space Telescope. Thanks to its coronagraph, the proposed telescope would be powerful enough to discern the atmospheres around roughly 25 distant, potentially habitable planets—and possibly sense the traces of life.

Launching no sooner than the early 2040s, the proposed telescope is a mashup of two NASA mission concepts. One, HabEx, called for a telescope with a coronagraph and a foldable starshade 52 meters (170 ft) wide—nearly twice as long as a basketball court—that would fly 76,000 kilometers (47,000 mi) in front of the telescope. The telescope's other parent concept, LUVOIR, envisioned a telescope with an enormous folding mirror at least eight meters (26.2 ft) wide.

Astronomers can use stars' spirals through the Milky Way to infer the presence of planets.

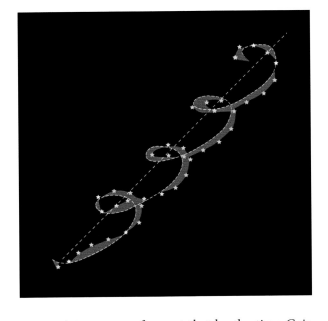

to see. Astronomers forecast that by the time Gaia is done, it could find thousands of new exoplanets around stars up to 1,600 light-years from the sun.

DIRECT IMAGING

All the planet-finding techniques we've mentioned so far offer indirect hints of a planet's presence. But for a small handful of the more than 4,500 known exoplanets, astronomers can pick the plan-

This detailed image from ALMA reveals the protoplanetary disk surrounding the young star HL Tauri. Dark patches within the disk may correspond to newly forming planets.

ets out directly—letting them see the outlines of alien worlds.

How can we possibly see faraway planets? Above all else, astronomers need the right conditions. The planets with the best shot of being seen are the ones that are young and hot enough to glow bright in the infrared. To increase their odds of being seen, planets also need to orbit far from their home stars: at least as far as Neptune from our sun, if not tens to hundreds of times farther. If a planet is too close to its home star, the star's brightness simply overwhelms the planet's fainter glow.

Even in the best conditions, spotting a planet is still a technical challenge, like trying to read a far-distant sign on a bright summer's day. But just as you would squint your eyes or hold up a hand to your face to get a better look, astronomers who want to see planets use telescope add-ons called coronagraphs to block the light of the planets' home star. That way, the much brighter star doesn't overwhelm the telescope, and the planets' dimmer glows shine through.

So far, the best glimpses we've gotten of exoplanets have been fuzzy dots. But even without seeing a planet's finer features, astronomers have learned a great deal. Take HR 8799, a star system about 130 light-years from us in the constellation Pegasus, appearing roughly halfway between the stars Markab and Scheat. In 2008, researchers announced that after blocking out HR 8799's light with a coronagraph, they had identified three planets in orbit around the star. By 2010, they had found a fourth using the same technique.

Since the planets' discovery, efforts to look at their atmospheres have found that some of the planets—all of which appear to be gas giants bigger than Jupiter—don't chemically match anything in our solar system. The closest orbiting of the four, HR 8799 e, absorbs light as if it were a redder version of Saturn. But even more remarkably, researchers have been watching the HR 8799 system long enough to stitch together a "movie" of its four planets moving in orbit. It'll take a while yet, though, to watch the planets complete a full circuit around their home star: The closest-in planet, HR 8799 e, takes about 57 years to complete an orbit. The farthest one out takes more than 400.

Direct imaging techniques may have revealed another planet closer to home. In February 2021,

astronomers announced that they had spotted something intriguing around Alpha Centauri A, which at 4.24 light-years away is one of the closest stars to our sun. When researchers blotted out the star's light, they saw a bright, glowing spot next to it. They don't yet know for sure whether it's an unknown artifact of their measurements, a glowing cloud of dust, or a planet roughly between Saturn and Neptune in size.

New advances also have given scientists the chance to see where and how planets are born. For instance, the Atacama Large Millimeter Array (ALMA), which helped take the first images of a black hole's silhouette, also has imaged the debris disks around distant stars. Gaps in these disks might be carved out by baby planets in the process of forming. These disks are not only beautiful; they're informative. More than 4.5 billion years ago, our solar system probably looked something like one of these dust rings. ALMA has even spotted what may be the makings of alien moons. In July 2021, astronomers announced that ALMA had seen a debris disk around one of the gas giants orbiting PDS 70, a young star some 370 light-years from Earth in the constellation Centaurus. So far, no exomoon has been found around the planet, but its "circumplanetary" disk is exactly the sort of structure in which moons could form.

In 1992, astronomer Aleksander Wolszczan found the first exoplanets around the pulsar PSR B1257+12, now called Lich. The three planets orbiting Lich probably get hit with charged particles, likely giving rise to alien aurorae.

■ For more on Alpha Centauri A, see page 213.

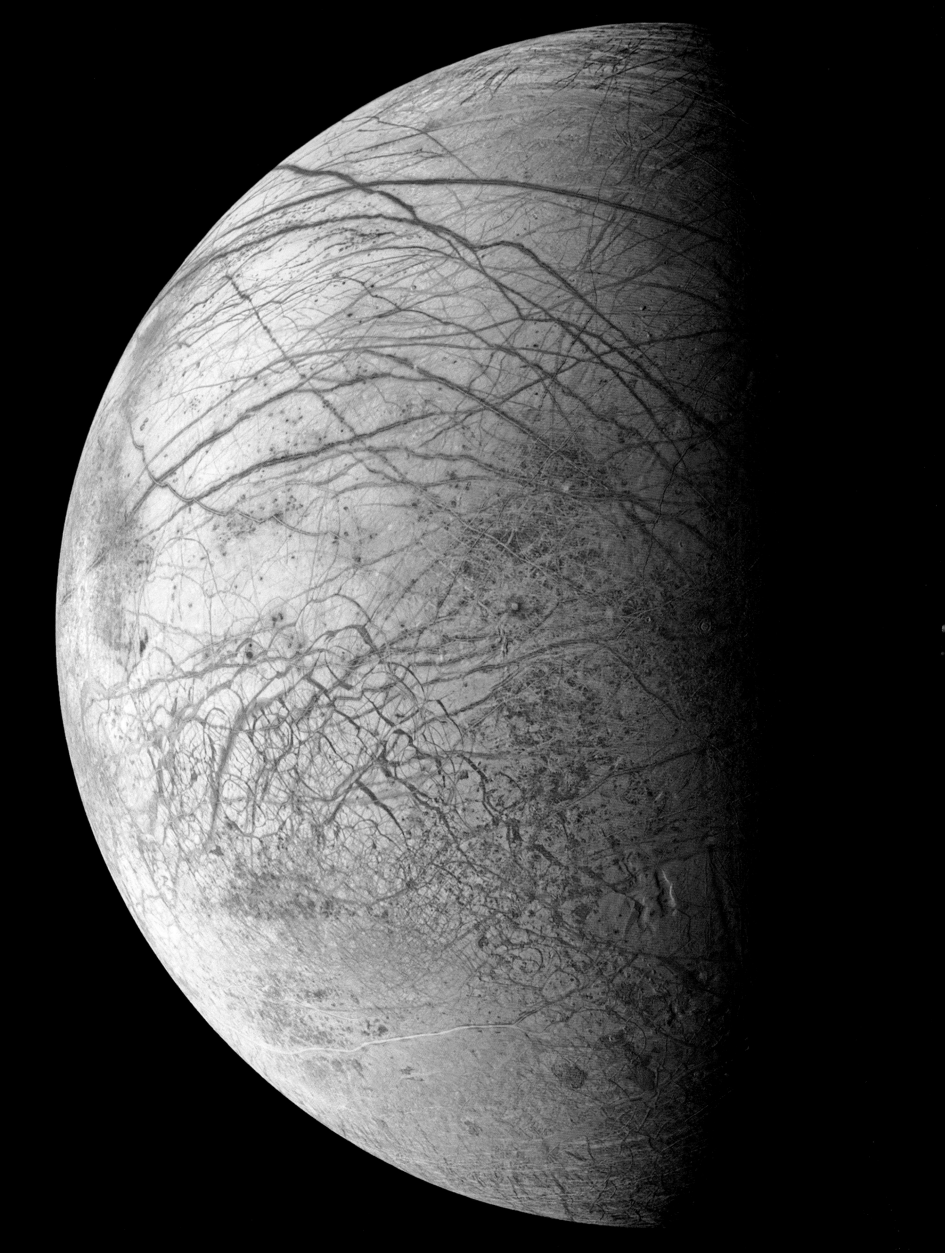

SEARCHING FOR LIFE

New techniques may one day detect chemical hints of alien life—or even its technological signals.

t's one thing to look for planets; it's another thing to test whether any of them harbors the one thing currently unique to Earth, the confirmed presence of life. The search for these hints of life, called biosignatures, is playing out across the solar system and throughout the Milky Way.

Thanks in large part to the Kepler telescope, astronomers now suspect that at least a third of all sunlike stars in the Milky Way, if not more, harbor Earthlike worlds of our planet's mass within their habitable zones. Our galaxy alone, then, could hold millions of Earthlike worlds. Scaled up across the billions and billions of galaxies in the observable universe, unfathomable numbers of alien worlds with masses similar to Earth's could hypothetically keep liquid water on their surfaces. As the vast differences between Earth and Venus illustrate, however, mass and location aren't the only considerations for habitability. Could we ever test an alien atmosphere to see if life as we know it could thrive there? What if we even detected signs of life itself?

It's not at all clear what the first signs of alien life will be, if we ever manage to find them. Signs of life could come from a distant planet's atmosphere, or perhaps take the form of fossils from as close as the planet next door. What is clear: In all likelihood, the first signals of alien life will not be clean-cut. Science often moves in increments from years to decades. Nothing about the search for life beyond Earth will be exempt from this slow, steady approach.

MARS

The idea of life—even intelligent life—on the red planet goes back centuries, though we've had to hit a hard reset on those dreams since pictures of

Mars's desiccated red surface beamed back to Earth in the 1960s. Whether life lives there now remains to be seen. (If it does, it's likely underground, in the form of hardy soil microbes beneath the planet's hostile surface.)

But as you saw in chapter 3, Mars was likely far more hospitable for life as we know it billions of years ago, when the red planet had a thicker, warmer atmosphere and enough liquid water on the surface to form lakes, carve out riverbeds, and feed hydrothermal systems.

For several decades, space agencies around the world have "followed the water" on Mars, tracing where it could be present today and where it flowed in the distant past. These locations may well harbor signs of fossils of past life from Mars's early history, if not the last survivors of native Martian life.

Efforts are underway to hunt for Martian fossils. In February 2021, NASA's Perseverance rover landed in Mars's Jezero crater, with a mission to find and cache rock samples with the best chance of preserving fossils of ancient, long-dead Martian microbes. Consider just how amazing a moment

▤ For more on Mars, see pages 121–127.

In this artist's depiction, ESA's Rosalind Franklin Mars rover (foreground) leaves its accompanying lander (background) to explore—and eventually drill into—the red planet's surface.

(Opposite) Its icy surface crisscrossed with reddish cracks and ridges, Jupiter's moon Europa is thought to have a subsurface ocean of liquid water beneath its outer shell—one that might be suitable for life as we know it.

On March 4, 2021, NASA's Perseverance Mars rover took its first drive, leaving tire marks across the surface of another planet.

we live in right now: We have a nuclear-powered robotic paleontologist trundling across the surface of another planet.

Perseverance comes with patience—or a need for it, at least. The NASA rover is just the first step in a multistage effort to bring samples from Mars back to scientists' tricked-out labs on Earth. In several years' time, a follow-on mission from NASA and the European Space Agency will send a rover to pick up after Perseverance and launch its samples into orbit around the red planet. That sample-ferrying rocket will rendezvous with a waiting

orbiter, which will then ferry the precious samples from Mars back to Earth. If this elaborate mission works, scientists will have successfully retrieved pristine samples from the surface of another planet for the first time. Whatever these rocks contain, they will make history.

Perseverance is far from the only new effort to study the surface of Mars. In the mid-2020s, the European Space Agency and the Russian space agency Roscosmos hope to launch the Rosalind Franklin rover, part of the two agencies' joint ExoMars initiative. This rover will drill up to two meters (6.6 ft) beneath the punishing Martian surface and analyze those buried soils in an on-board laboratory, in search of potential "biomarkers"—chemical hints of life.

OCEAN WORLDS

Mars is not the only target in the search for alien life. In fact, it might not even be the best one in our own backyard. Many of the solar system's icy bodies, including moons and dwarf planets, appear to have hidden subsurface oceans of liquid water. At least one—Saturn's moon Enceladus—has enough internal energy to fuel vast amounts of hydrothermal activity, possibly of the sort that could fuel a version of deep-sea life.

Here's an overview of some recent and future missions to some candidate ocean worlds.

Moons of Jupiter

Data from the Voyager and Galileo missions have long suggested that Jupiter's moon Europa might have a global ocean of liquid water beneath an icy shell (see chapter 3). In 2024, NASA intends to launch a spacecraft called Europa Clipper to explore this moon in more detail. Once it arrives in 2030, Europa Clipper will perform about 45 flybys of Europa's surface and come within 10 kilometers (16 mi) of the moon. Along the way, it will take measurements that should help determine the outer ice shell's thickness, as well as how deep and salty Europa's oceans really are.

Europa Clipper won't be alone. The European Space Agency's upcoming JUICE mission will tour three of Jupiter's largest moons, with a special focus on the

FLIGHT ON OTHER CELESTIAL BODIES

For decades, space agencies have sent landers and rovers to celestial bodies, but fliers through alien atmospheres are few and far between. The first flight on another planet occurred on Venus in June 1985, when the Soviet spacecraft Vega 1 and Vega 2 each released a balloon probe that floated some 55 kilometers (34 mi) above the planet's surface. Each battery-powered balloon lasted about 47 hours and flew more than 11,000 kilometers (more than 6,800 mi) before falling out of contact with Earth.

On April 19, 2021, the NASA helicopter Ingenuity performed the first controlled, heavier-than-air flight on another planet—a true Kitty Hawk moment. Fittingly, the spot where Ingenuity first hovered is now called Wright Brothers Field. In all likelihood, Ingenuity will be just the first of many Mars helicopters to come. The future for flying space explorers is bright. In 2027, NASA will launch a rotorcraft called Dragonfly to Saturn's moon Titan, whose dense atmosphere and lower gravity makes flying around a relative breeze. Every 16 Earth days, the craft will hop from place to place and study the moon's surface chemistry. Over its planned two-year mission, Dragonfly will cover several hundred miles.

biggest moon in the solar system, Ganymede (which is bigger than Pluto and Mercury). Like Europa, Ganymede's magnetic field implies that salty fluids are coursing through the moon's interior, suggesting the presence of a subsurface ocean that holds more water than all of Earth's oceans combined.

Titan

In some respects, Saturn's moon Titan is like a bizarre, extremely cold version of Earth. It's the only other place in the solar system with a thick atmosphere, stable liquid on its surface, and a strangely Earthlike weather cycle: storms, rainfall, the whole shebang. However, Titan's weather is based on liquid hydrocarbons, not water, and its surface liquid is mostly methane.

If life somehow has found a home on the surface of this small, frigid world, it seems likely that it plays by very different biochemical rules than life on Earth does and uses liquid methane where Earth life uses water. As weird as that mode of life sounds, it might be chemically possible. In labs on Earth, researchers have shown that certain molecules that could be present on Titan can form structures that behave like our cell membranes, which would be essential for any Titanian life. However, lab experiments haven't yet found a good Titanian analogue for DNA.

In 2005, the European Space Agency's Huygens lander plunged through Titan's atmosphere and landed on its surface, the most distant landing on a celestial body performed to date. In 2027, NASA will send a follow-up mission to the moon: a nuclear-powered rotorcraft called Dragonfly (see "Flight on Other Worlds" sidebar).

VENUS

Venus? Life? Its present-day surface conditions are far from hospitable, but some models of Venus's evolution suggest that it's possible the planet was habitable in its youth for one or two billion years, with mild temperatures suitable for liquid water. If so, perhaps fossils of Venusian life lie buried

This spacecraft "selfie," taken by NASA's Perseverance rover on April 6, 2021, shows the rover next to its high-flying companion, the experimental Mars helicopter Ingenuity.

ESA's JUICE mission will explore the icy moons of Jupiter, with a special focus on the solar system's largest moon, Ganymede (bottom).

An illustration depicts ESA's Huygens probe after its landing on Saturn's moon Titan. Data from Huygens suggest the probe landed on a surface resembling sand made of ice grains.

beneath the planet's scorched and scoured surface.

Surviving on Venus's surface would be a nightmare today. Temperatures are an average of 471°C (880°F), and the CO_2-choked atmosphere is so thick, standing on Venus's surface would subject your body to the same pressure that you'd feel under more than 900 meters (more than 3,000 ft) of water

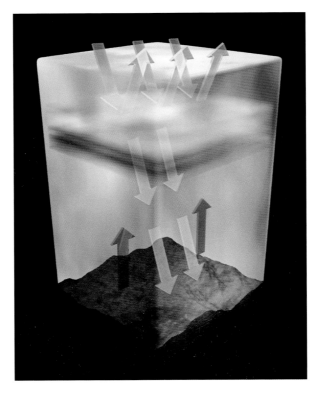

Venus's atmosphere traps heat from the sun (yellow arrows), causing a runaway greenhouse effect. The ground also radiates heat (red arrows), reaching surface temperatures over 400°C (752°F).

on Earth. As if that weren't punishment enough, the atmosphere is full of corrosive sulfuric acid. Small wonder that, to date, Venus landers haven't lasted longer than 127 minutes. Future technologies may well let us build more Venus-proof robotic explorers (see sidebar).

EXTRATERRESTRIAL INTELLIGENCE?

While we're on the subject of extraterrestrial life (E.T.), we might as well be searching for all forms of it—including those creatures that, like humans, have achieved enough intelligence to leave their mark on the heavens. In addition to biosignatures, researchers are on the hunt for "technosignatures": any observable hint of intelligent life in the cosmos. You've likely heard of SETI: the search for extraterrestrial intelligence.

Finding a technosignature might sound less likely than finding a biosignature, but keeping our eyes peeled brings enormous upside. For one, if we ever did receive a message that was clearly from E.T., it would be an unequivocal sign of life among the stars—and far less debatable than the chemical makeup of a distant planet's atmosphere. Likewise, if technosignatures exist, they could be abundant.

In theory, an advanced enough extraterrestrial intelligence could fan out across its home star system and even beyond, which would increase its total footprint—and possibly our odds of detecting it.

Though humankind has dreamed of intelligent life elsewhere for a long time, the systematic, scientific search for E.T. began in 1960 at the National Radio Astronomy Observatory in Green Bank, West Virginia. It was there that young astronomer Frank Drake turned a 26-meter (85-ft) radio telescope to the sky and scanned for any artificial pulses sent by distant civilizations. Drake called his experiment Project Ozma after the queen of Oz, a land in L. Frank Baum's books that Drake described as "very far away, difficult to reach, and populated by strange and exotic beings."

In the decades since, scientists studying technosignatures have kept up the search for radio transmissions. They've also broadened their gaze to more exotic possibilities. For a moment, imagine that intelligent, faraway extraterrestrials built huge laser arrays that they used to propel spacecraft away from their home planet. In theory, if one of these laser arrays happened to point in our direction, we might be able to pick it up. Likewise, these aliens may be such adept builders that they could craft massive, energy-harvesting shells around entire stars. In theory, these hypothetical structures—called "Dyson spheres" after physicist Freeman Dyson, who first proposed them—would glow with waste heat, perhaps brightly enough to spot from a distance.

As fun as it is to speculate on far-future technologies, though, they represent just one type of potential technosignature. In fact, experts say that the best kinds of technosignatures to look for would be those that don't assume technologies any more advanced than we had on Earth in the year 2000. For instance, if we ever mapped an exoplanet's surface in detail, it's possible we could detect the heat of alien cities or the glow of artificial lights.

Even if we don't see anything, we'll learn plenty along the way. In fact, scientists who design searches for technosignatures craft their experiments so that even if no trace of E.T. is found, the resulting data are still useful. Case in point: a bizarre star that, for a time, had the nickname "WTF" (for "where's the flux?").

In 2015, astronomers and citizen scientists noticed that in Kepler data, a main-sequence star in the constellation Cygnus called KIC 8462852 had dimmed erratically, as if oddly shaped objects were passing between the star and Earth. Soon, the odd star—nicknamed Tabby's Star after astronomer Tabetha Boyajian, who led the study describing it—attracted a lot of attention, after researchers flagged the remote possibility that a vast "megastructure" could be causing the dimming, while stressing that natural causes were far likelier.

Follow-up analyses showed that in some frequencies of light, Tabby's Star wasn't that obscured, which pointed away from a big, opaque infrastructure project hovering between us and the star. Instead, the star seemed to be obscured by large dust clouds, which may have formed as a one-time moon orbiting the star steadily evaporated in the star's heat. Though Tabby's Star wasn't E.T. hiding in plain sight, putting it to the test yielded a lot of good science.

On November 16, 1974, the Arecibo radio telescope beamed this signal toward a star cluster 25,000 light-years away—the first time humans had ever sent a signal designed as a message for E.T.

BUILDING A VENUS-PROOF ROBOT

Building spacecraft that can survive on Venus's surface requires rethinking electronics from the ground up. To make a Venus-proof computer chip, some researchers are looking into swapping out silicon for a different building material: silicon carbide, which is used on Earth to make high-performance brake disks. To date, test circuits made using silicon carbide have been run for more than a year at 500°C (932°F) and in simulated Venus conditions for upwards of 60 days.

Other designs try to minimize the need for electronics at all. The AREE rover concept, published by NASA's Jet Propulsion Lab in 2017, does as much as possible to sense obstacles and move with mechanical, clockwork-type mechanisms. The rover's power source: a small wind turbine mounted on its back, which would wind its main spring.

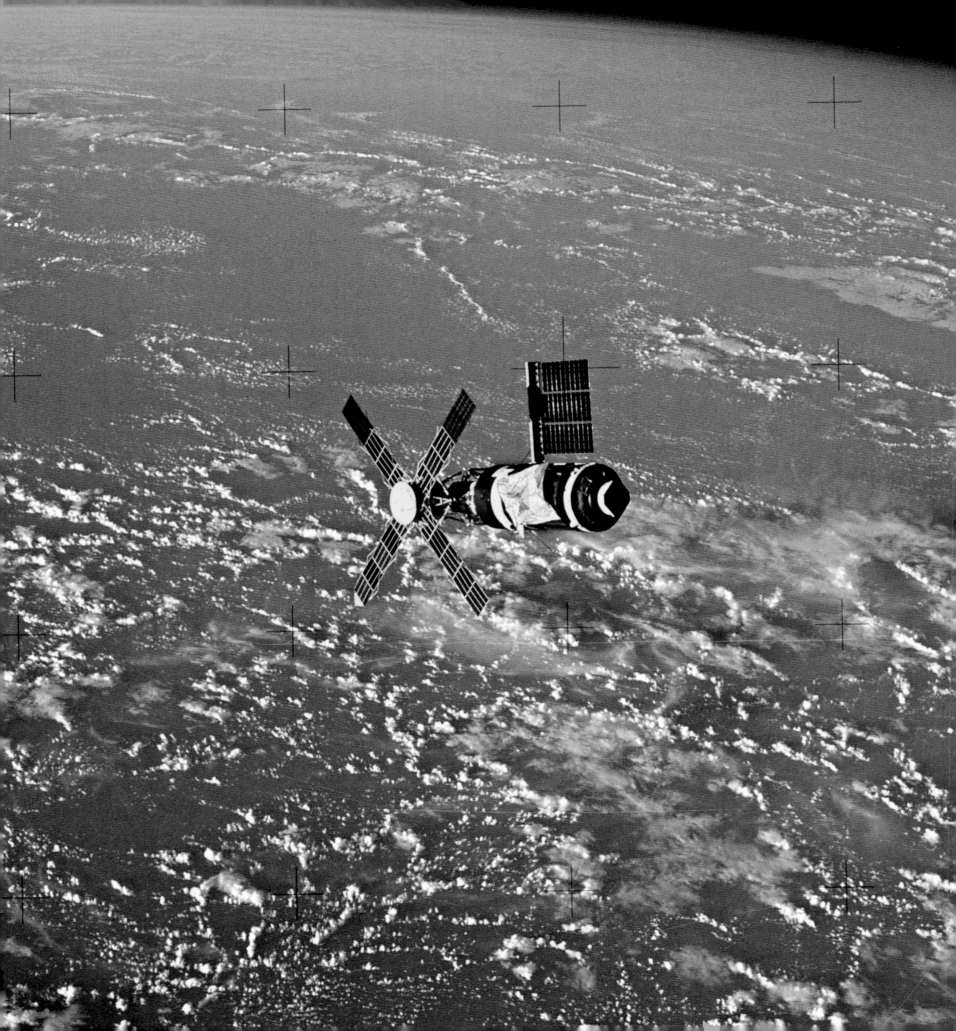

WHERE DO HUMANS FIT IN?

Humankind may now be able to return to the moon and venture to Mars and beyond—this time, to stay.

So far in this chapter, we've talked a lot about space and Earth-based telescopes and robotic missions to other places—but we haven't yet talked about humans' future voyages deeper into the solar system. What could the future of crewed exploration look like?

As they have for decades, robots will be taking the lead on exploring much of the solar system. In fact, some argue that robots offer many advantages over sending humans. For one, they are long-lasting and durable in every way that a living thing can't be in the vacuum of space. And as technology improves, robotic spacecraft can operate autonomously with increasing skill—a critical step to overcoming the communications delays between Earth and far-flung destinations. But advocates for human spaceflight argue that even now, humans offer a curious intelligence—and a capacity to inspire—that robotic explorers can't quite match.

As we saw in earlier chapters, dreams of humans venturing into the cosmos go back centuries, if not millennia. But humans' first steps toward the stars became reality in 1961, when Soviet cosmonaut Yuri Gagarin launched from the steppes of what's now Kazakhstan into a single orbit around Earth. That launch not only marked a major step for humankind, but also a major escalation in the space race between the U.S. and the Soviet Union. That geopolitical contest culminated in the late 1960s and 1970s with the U.S. Apollo program, which sent six crewed missions to the moon's surface and back again.

After Apollo and its massive budgets ended, the short-term future of human spaceflight veered away from the moon and toward low-Earth orbit.

Starting in the 1970s, the U.S. and Soviet Union began launching space stations. The U.S. sent crews to its Skylab station from 1973 to 1974, with the station falling to Earth in 1981. The Soviets launched a series of shorter-term Salyut stations through the 1970s and 1980s until starting construction on a bigger station, called Mir, in 1986. Mir was deorbited in 2001.

The Soviet Union, and later Russia, stayed consistent with its ride to space, relying on refinements to its stalwart Soyuz rocket and crew capsule. With an eye to cheap, reliable access to space, NASA worked through the 1970s on the space shuttle: a reusable crewed spacecraft designed for low-Earth orbit. Two solid rocket boosters and a massive fuel tank gave the space shuttle a boost into orbit, and once the spacecraft had finished its mission, the shuttle reentered Earth's atmosphere on its heat-resistant belly and then glided to a runway landing.

From its first launch in 1981, the space shuttle cemented itself as a technological marvel.

On April 12, 1961, cosmonaut Yuri Gagarin launched into orbit on Vostok 1, becoming the first human ever to venture into space.

(Opposite) The U.S. space station Skylab drifts above Earth in this 1973 image taken by its first crew. Skylab helped pave the way for today's human presence in low-Earth orbit.

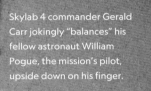

Skylab 4 commander Gerald Carr jokingly "balances" his fellow astronaut William Pogue, the mission's pilot, upside down on his finger.

The fleet of space shuttles flew 133 successful missions, including key missions to repair the Hubble Space Telescope and assemble the International Space Station (more below). However, designers' dreams of a cheap flight to space didn't exactly come true. Technical hurdles put the shuttle years behind schedule, and the vehicle turned out to be expensive to launch and refurbish. The program also suffered two major tragedies: the 1986 loss of the space shuttle *Challenger* during launch, and the 2003 breakup of the space shuttle *Columbia* during reentry. The disasters killed 14 and shook the program to its core.

U.S. officials decided in the mid-2000s to retire the space shuttle and finalized its retirement in 2011. Nearly a decade would pass until the U.S. regained its own ride to low-Earth orbit: Crew Dragon, a capsule built and operated by the private U.S. company SpaceX under a NASA contract. (Under the same NASA program, Boeing's CST-100 Starliner spacecraft is expected to send astronauts into low-Earth orbit as well.)

HUMAN SPACEFLIGHT TODAY

As of 2021, Russia's Soyuz rocket and the SpaceX spacecraft under NASA's Commercial Crew program were humankind's most consistent rides into orbit. Their destination: the International Space Station. Called the ISS for short, this space station is among the most expensive and technologically complex objects ever built: a $150-billion pressurized habitat as long as a football field, whizzing

Since 2000, not all humans have lived on Earth's surface at the same time, thanks to the ever orbiting **International Space Station.**

400 kilometers (248 mi) above Earth's surface at more than 28,000 kilometers an hour (about 17,500 mph). Over the decades, more than 240 women and men from around the world have temporarily called the space station home, some for nearly a full year at a time.

Like Earth's most enduring structures, the ISS was decades in the making. Born out of the U.S. concept for "Space Station Freedom" in 1984, the project gradually evolved into a 15-nation pact among the U.S., Canada, Japan, Russia, and the 11 member states of the European Space Agency. The first pieces of the ISS started arriving in orbit in 1998, and Expedition 1 crew members climbed aboard the newborn station on November 2, 2000.

Since then, a steady stream of follow-on crews have visited the station and kept it occupied for more than two decades straight, performing scientific experiments and testing the conditions that

■ For more on the International Space Station, see pages 311–312.

Space suits have come a long way, from cotton-rubber pressure suits in the 1930s (left) to the EVA suits worn by Apollo astronauts (right). Modern suits (second from right) have even more features.

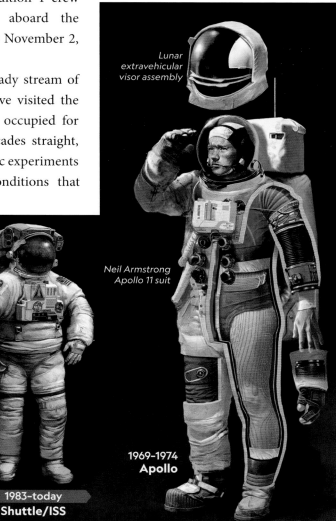

Lunar extravehicular visor assembly

Neil Armstrong Apollo 11 suit

1969–1974 Apollo

1934–1935 (years worn)	1959–1968	1961–1963	1965–1966	1983–today
Winnie Mae	**X-15**	**Mercury**	**Gemini**	**Shuttle/ISS**

future spacefarers would face on extended trips deeper into the solar system. For a moment, consider just how astounding that is: For more than 20 years now, not all humans have lived on Earth's surface at the same time.

To keep the station running and its inhabitants alive, crew members and global support teams must work together, as if the station were a miniature United Nations. More than anything, this diplomatic coordination is the real triumph of the ISS. No matter what tensions are playing out on terra firma, world governments have worked together to keep their citizens in orbit alive and healthy.

With more than 120,000 orbits and more than five billion kilometers (more than 3 billion mi) traveled above Earth's surface, the ISS is still going strong—and it's more of a global effort than ever, with visitors from 19 different countries.

The station is currently slated to run until 2030, and much of its hardware is certified to operate safely until at least 2028, if not longer for its younger components. For its part, NASA is trying to boost commercial companies' use of the station, including welcoming tourists and commercial research-

ers. In 2020, NASA awarded a docking module in the station's U.S. section to the startup Axiom Space, which plans to add a commercial section to the ISS—including an enormous, many-windowed cupola. NASA isn't alone in its creative use of the ISS: In October 2021, Roscosmos launched a Russian film director and actress to the station, where they worked for nearly two weeks on the first full-length movie ever shot and directed in space.

THE FUTURE OF HUMANS IN LOW-EARTH ORBIT

Even if commercial interests take over the U.S. stake in the ISS, nothing lasts forever. In early 2021, Russia signaled that it wanted to wind down its contributions to the ISS as soon as 2025 and build a new Russian station. Whether that station will proceed, and when, is unclear. In addition, teams of private companies have started making plans for commercial space stations of their own, envisioning these craft as shared, research-focused "office parks" in orbit.

Meanwhile, China, which is not one of the ISS member countries, has been steadily building up its own crewed spaceflight program, having launched

Multiple spacecraft can dock with the International Space Station at once. In this 2016 image, an Orbital ATK Cygnus cargo craft (left) sits next to a Russian Soyuz crew vehicle (middle) and a Russian Progress cargo craft (right).

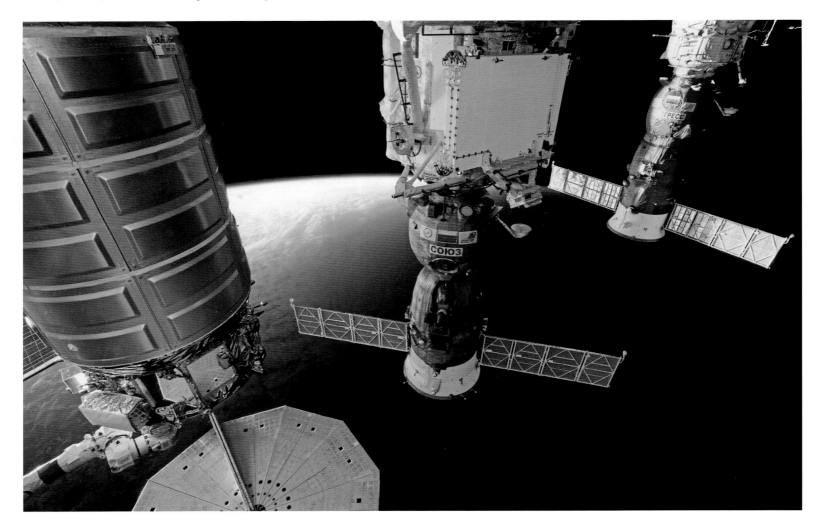

multiple crews into orbit since 2003. The country is at work constructing its own space station: Tiangong, Chinese for "heaven's palace." In April 2021, China launched the station's first module, called Tianhe.

It's also possible to go to space without fully going into orbit, and many private companies are working on space tourism ventures to send the lucky—or the fabulously wealthy—toward the stars for a few minutes at a time. In the summer of 2021, Virgin Airlines mogul Richard Branson and Amazon founder Jeff Bezos had both taken brief flights more than 80 kilometers (50 mi) above Earth's surface—the U.S. definition of the edge of space—aboard spacecraft operated by their respective companies, Virgin Galactic and Blue Origin.

You might be noticing a theme: private companies getting involved with spaceflight. While commercial firms have had a hand in space for decades—think satellite TV and radio, or telecommunications—space travel has been an exclusive club. However, as launch costs have steadily declined in recent decades, more countries and private entities, including for-profit companies, can operate satellites in low-Earth orbit. Their efforts include satellite internet and continuously generated maps of Earth from space, which can yield useful data for weather, agriculture, and many other pursuits. Going into the 2020s, the U.S.-based companies SpaceX and Planet Labs operate the world's two biggest satellite fleets—a statement that would have been unbelievable two decades ago.

As the number of satellites increases, though, so too do the changes to the night sky we all share, in

In 2020, SpaceX launched Demo-2, the first crewed voyage of its Crew Dragon craft (above). Its two passengers (below): NASA astronauts Bob Behnken (left) and Doug Hurley

In this artist's concept, the Gateway station orbiting the moon (left) prepares to receive supplies from a SpaceX Dragon XL spacecraft.

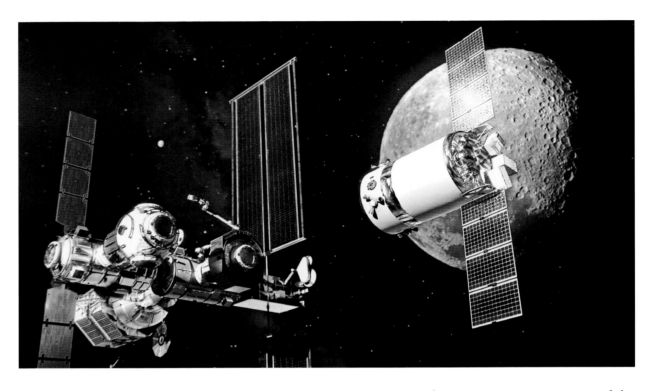

a way that may transform our common experience of the heavens. For instance, private companies are racing to launch "megaconstellations" of satellites that will beam internet access to vast swaths of Earth's midlatitudes. Astronomers are already sounding the alarm, partially out of concern that these satellite swarms, such as SpaceX's Starlink constellation, will streak across their fields of view and disrupt their measurements of the cosmos.

HUMANS' RETURN TO THE MOON

The idea of traveling to and living on the moon has been with us for centuries—but successfully pulling off more than the occasional visit won't be easy.

A SPACE STATION AROUND THE MOON?

As part of its longer-term Artemis plans, NASA wants to build an outpost in orbit around the moon. This spacecraft, called the Gateway, will act as a sort of way station between spacecraft coming in from Earth and spacecraft specifically designed to land on the moon's surface.

The Gateway will occupy a special type of orbit called a near-rectilinear halo orbit. Every seven days, the spacecraft will come in as close as 3,500 kilometers (2,200 mi) to the moon's surface and then move more than 71,000 kilometers (44,000 mi) away. The orbit will take very little energy to maintain, and the Gateway will also be able to reposition itself and change which part of the moon lies beneath it during its close approaches. If all goes to plan, the Gateway should make it easier to get people from Earth to anywhere on the moon's surface.

This station is already being built, and its first two sections are set for launch no sooner than late 2024. It's also being designed with international collaboration in mind. As of 2021, Canada, Italy, Japan, and other countries have committed money and hardware in support of Gateway.

In recent years, however, space agencies around the world have refocused on exploring and even sending humans once again to the moon.

NASA is the furthest along in realizing its lunar ambitions. In the last few years, the agency has moved aggressively to return U.S. astronauts to the lunar surface for the first time since the Apollo 17 mission in 1972. Its program—named Artemis, after the Greek goddess of the moon—has racked up support from public and private partners around the world. Though the program's future is uncertain, it has ambitious goals.

First, the Artemis program wants to return people to the moon, including the first woman and the first person of color. Next, NASA wants to go back to the moon to stay, with longer-term plans to regularly revisit the lunar surface, run unprecedented scientific missions, and test some of the tech needed to send humans to Mars.

How Artemis Will Work

Parts of the Artemis architecture were put into place years before the program formally started. As the U.S.'s long-term human spaceflight strategy remained flexible—or, less charitably, formless—in the 2010s, U.S. legislators funded technologies they thought would be essential to any future plan. As a result, the U.S. committed to building a capsule that could keep astronauts alive in deep space for weeks at a time, called Orion, and a massive heavy-lift rocket, called Space Launch System (SLS).

On January 20, 2020, NASA astronauts Jessica Meir (top) and Christina Koch conducted a spacewalk to finalize upgrades to part of the International Space Station's power system.

In this 2019 photo, NASA astronaut Christina Koch (right) poses for a picture with fellow NASA astronaut Jessica Meir as Meir prepares for a spacewalk.

The current plan calls for Artemis astronauts to pack into Orion and launch on SLS. This successor to Apollo's Saturn V rocket draws on space shuttle technology, including some of the same rocket engines built and used during the space shuttle program. SLS is the most powerful rocket NASA has ever built. It's capable of generating four million kilograms (8.8 million lbs) of thrust during lift-off—15 percent more oomph than the Apollo program's Saturn V mustered.

The first flight of Orion and SLS will be part of an uncrewed test mission to and from the moon's vicinity, called Artemis I. Next on the docket: Artemis II, which will follow no sooner than 2024. Like 1968's Apollo 8 mission, Artemis II will take a crew around the moon and back to Earth.

The Artemis III mission—which probably won't launch sooner than 2025—will mark NASA's first attempt in more than 50 years to land people on the moon. But because of the added requirement of landing on the surface, the mission is slightly more complicated. While SLS has enough thrust to get Orion near the moon, it doesn't pack enough punch

to get Orion and its crew all the way to the lunar surface. NASA's plan is to insert Orion into orbit around the moon, where the capsule will meet up with a prestaged "Human Landing System" (HLS), which will take the Artemis III crew to and from the lunar surface. After staying on the moon's surface for up to a week, the Artemis III astronauts will ride the HLS back into lunar orbit, where they will once again board Orion and return to Earth.

What to Do on the Moon?

According to NASA, the goal of Artemis isn't merely to put U.S. boots back on the moon and call it a day. The space agency's plans for the program call for international collaboration and the maintenance of a "sustainable" human presence in the moon's backyard, with multiple missions to the same spot near the lunar south pole. The agency imagines a more robust infrastructure on the moon, including a communications "LunaNet" that will act much like the internet on Earth.

There's also quite a bit of science to do on the lunar surface—including astronomy. Because the lunar far side consistently points away from Earth's surface, it's isolated from the radio signals that Earth's talkative technology constantly gives off. In theory, a large radio telescope could be installed within a lunar crater and collect exquisite data on far-off, ancient cosmic phenomena.

THE WORLD'S JOURNEY TO THE MOON

Beyond sending humans, space agencies around the world are targeting renewed robotic exploration of the moon, which remains crucial to our understanding of the solar system—and, in turn, the cosmos. A lot of what we know about the age and evolution of the solar system traces back to estimates of the ages of the moon's craters, estimates made possible by lunar rock samples from the Apollo missions.

China has already brought lunar samples to Earth. NASA's recent Commercial Lunar Payload Services (CLPS) program has signed contracts with private companies to send NASA instruments to the moon. One will even deliver a NASA rover, VIPER, to the lunar south pole.

The world's renewed interest in lunar exploration also pushes the boundaries of the laws that currently govern who can do what in space. As humans begin to increase our presence off-world, the framework that has governed human affairs in space for more than 50 years risks becoming outdated. Under what framework would lunar areas of scientific or cultural interest, such as the landing sites of past missions, be protected, if at all?

PUTTING OUR BEST FOOT FORWARD

Though humankind's near-term spaceflight focus is the moon, Mars is clearly the long-term target. The challenges of living there—with a thin, unbreathable atmosphere, brutally cold temperatures, punishing radiation, and utter isolation from Earth—will be extraordinary. At minimum, it seems reasonable that future generations may try to establish a space station in Martian orbit, or perhaps an Antarctica-style research base on the red planet's surface. Some dream of a grandiose multiplanetary future, including domed cities on Mars made possible with fully reusable rockets (see sidebar).

STARSHIP

A silvery steel rocket that looks plucked out of 1950s sci-fi, Starship is a heavy-lift rocket developed by the U.S. space company SpaceX. The company's promise is that, like commercial airplanes, Starship could be launched, landed, refueled, and re-flown within a matter of hours. SpaceX even hopes to design Starship so that it can be refueled in orbit, which would let the spacecraft land large payloads of more than 100 metric tons (more than 110 tons) on the surface of Mars.

The full rocket will consist of two pieces: a 70-meter-tall (230-ft) booster called Super Heavy, and a 50-meter-tall (165-ft) upper stage called Starship. (Together, the booster and upper stage of the rocket are also known as Starship.) Building on SpaceX's past rockets, whose first-stage boosters can land themselves, Starship is designed to be fully reusable. It's also slated to be extremely powerful. If Starship can be fully reused as often as SpaceX hopes, the cost of launch into low-Earth orbit and beyond could decrease dramatically, which would let more entities—both public and private—launch bigger and more complex payloads into space than ever before.

Versions of Starship also could hold people. In 2021, NASA signed a contract with SpaceX to have a modified form of the Starship's upper stage ferry astronauts from lunar orbit to the moon's surface. SpaceX has an agreement to send Japanese fashion billionaire Yusaku Maezawa and a group of artists around the moon and back no sooner than 2023. SpaceX's founder and CEO Elon Musk has even bigger goals for Starship: He would like to use the craft to build and populate a city on the surface of Mars.

If humankind decides to explore the solar system in person, we should do so with a deep humility for the vast, wild cosmos into which we'll be voyaging. How can we be the best visitors to other worlds we can possibly be? And at the same time, how can we be the best stewards for that extraordinary oasis for life: Earth itself?

A visitor at China's National Museum in Beijing looks at a display of lunar samples brought back from the moon by the country's Chang'e-5 lunar lander.

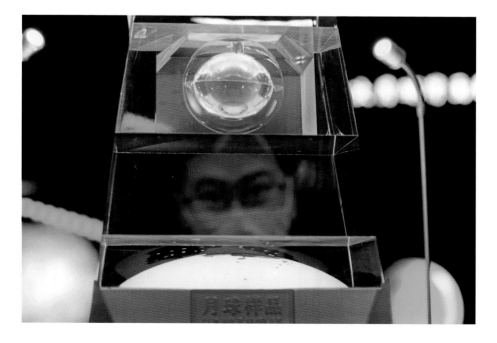

In 2015, NASA astronaut Terry Virts took this selfie during a series of spacewalks he conducted at the International Space Station.

GLOSSARY

aphelion: the point in an object's orbit when it is farthest from the sun

archaeoastronomy: the study of astronomical knowledge and practice in ancient cultures

asterism: a distinct pattern of stars that forms a shape in the night sky, but is not a constellation

asteroid: small rocky body revolving around the sun. The majority of asteroids orbit between Mars and Jupiter, in the asteroid belt.

astronomical unit (AU): the average distance from Earth to the sun, 149,597,870 kilometers (92,955,730 mi). Employed as a standard unit of astronomical measurement

atmosphere: the gaseous layer surrounding a planet or star that is retained as a result of the gravitational field and temperature around the object

aurora: luminous colored lights in the upper atmosphere, caused by solar particles traveling along Earth's magnetic field lines

big bang: the widely accepted theory of the origin and evolution of the universe that states that the universe began in an infinitely compact state and is now expanding

binary star: a pair of stars in orbit around each other. An optical binary is a pair of stars that appear to be next to each other in the view from Earth but are not gravitationally connected.

black hole: an object or region of space whose gravitational pull is so strong that nothing, not even light, can escape from it

celestial sphere: the imaginary surface surrounding Earth upon which all celestial objects appear and which is used to describe the positions of objects in the sky

comet: an icy body orbiting in the solar system that typically develops a long tail as it approaches the sun

conjunction: an alignment of two solar system bodies as seen from Earth

constellation: a shape assigned to an arbitrary arrangement of stars that serves as a tool to differentiate between the stars. There are 88 different recognized constellations.

core: the dense, innermost part of a planet, large moon, asteroid, or star

corona: the sun's outermost atmosphere; the visible light from the sun, seen around the disk of the moon as it covers the sun during a total eclipse

crust: the solid, outermost layer of a planet or moon

dark energy: a property of space that propels the acceleration of the expanding universe

dark matter: unseen matter in space that exerts a gravitational pull within galaxies

density: the ratio of the mass of an object to its volume, commonly measured in grams per cubic centimeters (g/cm^3) or grams per liter (g/L)

Doppler shift: the observed change in the wavelength of light or sound caused by motion in the observer, the object, or both

dwarf planet: a body orbiting the sun that has sufficient mass to be nearly round in shape, but which has not cleared its path of other bodies of comparable size

eclipse: an occasion when the light from one celestial body is cut off by the presence of another celestial body

ecliptic: the apparent path of the sun on the celestial sphere; also, the plane of Earth's orbit around the sun

exomoon: a natural satellite orbiting an exoplanet

exoplanet: an extrasolar planet; a planet orbiting a star other than the sun

fusion: a nuclear reaction that occurs when light nuclei join to produce a heavier nucleus; nuclear fusion

galaxy: a large collection of stars bound by mutual gravitational attraction

gas giant: a large, low-density planet composed primarily of hydrogen and helium; in our solar system, Jupiter, Saturn, Uranus, and Neptune are considered gas giants, with Uranus and Neptune also called ice giants

gegenschein: a faint atmospheric glow, probably caused by sunlight reflecting off dust, visible in the night sky at a point opposite from the sun

globular cluster: a densely packed, roughly spherical cluster of thousands or millions of stars

gravitational wave: a ripple in space-time caused by the acceleration of massive objects

habitable zone: the orbital distance within which liquid water exists on a planet's surface

Hertzsprung-Russell (H-R) diagram: the graph on which the luminosity and temperatures of stars are plotted

heliosphere: the region in space affected by the sun and the solar wind

impact crater: the depression formed as a result of a high-speed solid hitting a rigid surface, such as the circular craters on the surface of the moon

infrared: having a wavelength longer than the red end of the visible light spectrum and shorter than that of microwaves

inner planets: the four small, rocky planets nearest to the sun; also called terrestrial planets

Kuiper belt: (pronounced KI-per) a disk-shaped region located 30 to 50 AU from the sun that is filled with icy bodies

light-year: the distance light travels in a vacuum in one year; equivalent to about 9.5 trillion kilometers (6 trillion mi)

luminosity: the total amount of energy given off by a star

magnitude: the brightness of a star or other astronomical object. Apparent magnitude measures brightness as seen from Earth; absolute magnitude measures intrinsic brightness.

mantle: the layer between the crust and the core of a large moon or planet

meteor: the bright streak of light produced by a meteoroid, a rocky particle, entering Earth's atmosphere; a shooting star

meteorite: the fragment of a meteoroid that reaches Earth's surface

meteoroid: a small piece of rock or dust traveling through space

Milky Way: the large spiral galaxy that is home to the sun; also, the central plane of that galaxy seen in the night sky

nebula: a cloud of gas and dust in interstellar space

neutron star: a small, extremely dense star composed almost entirely of neutrons

nova: a white dwarf in a binary star system that pulls matter from its companion star, periodically brightening in a brilliant burst

Oort cloud: an enormous spherical cloud surrounding our solar system that contains billions of icy objects

parallax: the apparent change in the position of an object when seen from different vantage points

planet: a body in orbit around a star, possessing sufficient mass to form a nearly round shape, and gravitationally dominant, meaning it will have cleared its path of other bodies of comparable size

planetary nebula: a luminous shell of gas surrounding a star late in its evolution

planetesimal: small primordial body formed by accretion in the early solar system

plasma: a gas made up of charged particles

precession: the motion of the rotation axis of a spinning body as it traces out a conelike shape

pulsar: a rapidly rotating neutron star that emits regular pulses of radio waves

radiation: energy emitted in the form of waves or particles

redshift: the increase in the wavelength of radiation sent out by a celestial body as a result of the Doppler effect, gravitation, or cosmological expansion

retrograde: moving east to west, or backward, in rotation or in the sky

solar nebula: the spinning gaseous cloud that condensed to form the solar system

standard candle: an object whose absolute magnitude (innate brightness) is well known, and so can serve as an indicator of distance

supergiant star: a member of a class of the largest and most luminous stars known, with radii between 100 and 1,000 times that of the sun

supernova: an enormous stellar explosion; the death of a massive star

technosignature: a detectable signal of technology from another planet

tidal force: the varying gravitational force one massive object exerts on another

variable star: a star whose brightness changes, regularly or irregularly

NOTABLE DEEP-SKY OBJECTS

Astronomers know the universe beyond our solar system as the "deep sky": a treasure house of stars, star clusters, galaxies, nebulae, and more. In the 18th century, French astronomer Charles Messier, in order to distinguish comets from other fuzzy items in the sky, created a catalog of the most notable deep-sky objects. This Messier catalog, whose members are designated by "M" numbers, includes many of the most beautiful objects visible today. Later catalogs have added to the lists, notably the *The New General Catalogue of Nebulae and Clusters of Stars* (NGC), which was compiled beginning in the 1880s. Dedicated astronomers know them well. Amateurs with good backyard telescopes can see many of them, though not in great detail. More can be viewed in all their beauty online at such websites as the Hubble Telescope's HubbleSite (http://hubblesite.org/gallery/), the Messier Catalog at SEDS (http://messier.seds.org), or the NGC Catalog Online (http://spider.seds.org/ngc/ngc.html). To find them in our book, check the index on pages 426–431.

GALAXIES

Spiral Galaxies

M31: The Andromeda Galaxy
M33: The Triangulum Galaxy
M51: The Whirlpool Galaxy
M58
M61
M63: The Sunflower Galaxy
M64: The Black Eye Galaxy
M65
M66
M74
M77
M81: Bode's Galaxy
M83: The Southern Pinwheel Galaxy
M88
M90
M91
M94
M95
M96
M98
M99
M100
M101: The Pinwheel Galaxy
M104: The Sombrero Galaxy

M106
M108
M109
NGC 253
NGC 891
NGC 1055
NGC 2403
NGC 2903
NGC 3628
NGC 4565
NGC 4571
NGC 4631: The Whale Galaxy
NGC 4656
NGC 5907
NGC 6946
NGC 7331
NGC 7479

Lenticular (SO) Galaxies

M84
M85
M86
M102
NGC 5866: The Spindle Galaxy

Elliptical Galaxies

M32
M49
M59
M60
M87
M89
M105
M110
Leo I: The Regulus Galaxy
SagDEG
Canis Major Dwarf

Irregular Galaxies

M82: The Cigar Galaxy
NGC 2976
NGC 3077
NGC 5128
NGC 5195
NGC 6822: Barnard's Galaxy
IC 10
IC 5152
The Large Magellanic Cloud
The Small Magellanic Cloud

Supernova Remnants

M1: The Crab Nebula

Barnard's Loop

Planetary Nebulae

M27: The Dumbbell Nebula

M57: The Ring Nebula

M76: The Little Dumbbell Nebula

M97: The Owl Nebula

NGC 2438

NGC 6543: The Cat's Eye Nebula

NGC 7009: The Saturn Nebula

NGC 7293: The Helix Nebula

Star-Forming Nebulae

M8: The Lagoon Nebula

M16: The Eagle Nebula

M17: The Omega Nebula

M20: The Trifid Nebula

M42: The Orion Nebula

M43: De Mairan's Nebula

M78

NGC 2023

NGC 2237, 2238, 2239, 2246: The Rosette Nebula

NGC 2264: The Cone Nebula

NGC 3372: The Carina Nebula

NGC 7000: The North America Nebula

Dark Nebulae

Barnard 33: The Horsehead Nebula

Caldwell 99: The Coalsack Nebula

Open Clusters

M6: The Butterfly Cluster

M7: Ptolemy's Cluster

M11: The Wild Duck Cluster

M18

M21

M23

M25

M26

M29

M34

M35

M36

M37

M38

M39

M41

M44: Praesepe (The Beehive Cluster)

M45: The Pleiades (The Seven Sisters)

M46

M47

M48

M50

M52

M67

M93

M103

Globular Clusters

M2

M3

M4

M5

M9

M10

M12

M13: The Hercules Cluster

M14

M15

M19

M22

M28

M30

M53

M54

M55

M56

M62

M68

M69

M70

M71

M72

M75

M79

M80

M92

M107

SKYWATCHER'S CALENDAR

2022

October:
10/25: New moon
10/25: Partial solar eclipse (best seen in parts of Russia and Kazakhstan)

November:
11/4, 5: Peak of Taurids meteor shower
11/8: Total lunar eclipse (visible throughout eastern Russia, Japan, Australia, the Pacific Ocean, and parts of western and central North America)
11/17, 18: Peak of Leonids meteor shower
11/23: New moon

December:
12/8: Full moon
12/13, 14: Peak of Geminids meteor shower
12/21: December solstice (first day of winter solstice in the Northern Hemisphere and summer solstice in the Southern Hemisphere)
12/21, 22: Peak of Ursids meteor shower
12/23: New moon

2023

January:
1/3, 4: Peak of Quadrantids meteor shower
1/6: Full moon
1/21: New moon

February:
2/5: Full moon
2/20: New moon

March:
3/7: Full moon
3/20: March equinox (first day of spring, or vernal equinox, in the Northern Hemisphere and fall, or autumnal equinox, in the Southern Hemisphere)
3/21: New moon

April:
4/6: Full moon
4/20: New moon
4/20: Hybrid solar eclipse (the eclipse path will begin in the southern Indian Ocean and move across parts of western Australia and southern Indonesia)
4/22, 23: Peak of Lyrids meteor shower

May:
5/5: Full moon
5/5: Penumbral lunar eclipse (will be visible throughout all of Asia and Australia and parts of eastern Europe and eastern Africa)
5/6, 7: Peak of Eta Aquarids meteor shower
5/19: New moon

June:
6/4: Full moon
6/18: New moon
6/21: June solstice (first day of summer, or summer solstice, in the Northern Hemisphere and winter, or winter solstice, in the Southern Hemisphere)

July:
7/3: Full moon, supermoon
7/17: New moon
7/28, 29: Peak of Delta Aquarids meteor shower

August:
8/1: Full moon, supermoon
8/12, 13: Peak of Perseids meteor shower
8/16: New moon
8/31: Full moon, supermoon, blue moon

September:
9/15: New moon
9/23: September equinox (first day of fall, or autumnal equinox, in the Northern Hemisphere and spring, or vernal equinox in the Southern Hemisphere)
9/29: Full moon, supermoon

October:
10/7: Peak of Draconids meteor shower
10/14: New moon
10/14: Annular solar eclipse (the eclipse path will begin in the Pacific Ocean off the coast of southern Canada and move across the southwestern United States and Central America, Colombia, and Brazil)
10/21, 22: Peak of Orionids meteor shower
10/28: Full moon
10/28: Partial lunar eclipse (the eclipse will be visible throughout all of Europe, Asia, Africa, and western Australia)

November:
11/4, 5: Peak of Taurids meteor shower
11/13: New moon
11/17, 18: Peak of Leonids meteor shower
11/27: Full moon

December:
12/12: New moon
12/13, 14: Peak of Geminids meteor shower
12/21, 22: Peak of Ursids meteor shower
12/22: December solstice (the first day of winter, or winter solstice, in the Northern Hemisphere and summer, or summer solstice, in the Southern Hemisphere)
12/27: Full moon

2024

January:
1/3, 4: Peak of Quadrantids meteor shower
1/11: New moon
1/25: Full moon

February:
2/9: New moon
2/24: Full moon

March:
3/10: New moon
3/20: March equinox (first day of spring, or vernal equinox, in the Northern Hemisphere and fall, or autumnal equinox, in the Southern Hemisphere)
3/25: Full moon
3/25: Penumbral lunar eclipse (eclipse will be visible throughout all of North America, Mexico, Central America, and South America)

April:
4/8: New moon
4/8: Total solar eclipse (the path of totality will begin in the Pacific Ocean and move across parts of Mexico and the eastern United States and Nova Scotia)
4/22, 23: Peak of Lyrids meteor shower
4/23: Full moon

May:
5/6, 7: Peak of Eta Aquarids meteor shower
5/8: New moon
5/23: Full moon

June:
6/6: New moon
6/20: June solstice (first day of summer, or summer solstice, in the Northern Hemisphere and winter, or winter solstice, in the Southern Hemisphere)
6/22: Full moon

July:
7/5: New moon
7/21: Full moon
7/28, 29: Peak of Delta Aquarids meteor shower

August:
8/4: New moon
8/12, 13: Peak of Perseids meteor shower
8/19: Full moon, blue moon

September:
9/3: New moon
9/18: Full moon, supermoon
9/18: Partial lunar eclipse (eclipse will be visible throughout most of North America, Mexico, Central America, South America, the Atlantic Ocean, and most of Europe and Africa)
9/22: September equinox (first day of fall, or autumnal equinox, in the Northern Hemisphere and spring, or vernal equinox, in the Southern Hemisphere)

October:
10/2: New moon
10/2: Annular solar eclipse (the eclipse path will begin in the Pacific Ocean off the coast of South America and move across parts of southern Chile and Argentina)

10/7: Peak of Draconids meteor shower
10/17: Full moon, supermoon
10/21, 22: Peak of Orionids meteor shower

November:
11/1: New moon
11/4, 5: Peak of Taurids meteor shower
11/15: Full moon, supermoon
11/17, 18: Peak of Leonids meteor shower

December:
12/1: New moon
12/13, 14: Peak of Geminids meteor shower
12/15: Full moon
12/21: December solstice (first day of winter, or winter solstice, in the Northern Hemisphere and summer, or summer solstice, in the Southern Hemisphere)
12/21, 22: Peak of Ursids meteor shower
12/30: New moon

2025

January:
1/3, 4: Peak of Quadrantids meteor shower
1/13: Full moon
1/29: New moon

February:
2/12: Full moon
2/28: New moon

March:
3/14: Full moon
3/14: Total lunar eclipse (eclipse will be visible throughout all of North America, Mexico, Central America, and South America)
3/20: March equinox (first day of spring, or vernal equinox, in the Northern

Hemisphere and fall, or autumnal equinox, in the Southern Hemisphere)
3/29: New moon
3/29: Partial solar eclipse (will be visible throughout Greenland and most of northern Europe and northern Russia)

April:
4/13: Full moon
4/22, 23: Peak of Lyrids meteor shower
4/27: New moon

May:
5/6, 7: Peak of Era Aquarids meteor shower
5/12: Full moon
5/27: New moon

June:
6/11: Full moon
6/21: June solstice (first day of summer, or summer solstice, in the Northern Hemisphere and winter, or winter solstice, in the Southern Hemisphere)
6/25: New moon

July:
7/10: Full moon
7/24: New moon
7/28, 29: Peak of Delta Aquarids meteor shower

August:
8/9: Full moon
8/12, 13: Peak of Perseids meteor shower
8/23: New moon

September:
9/7: Full moon
9/7: Total lunar eclipse (will be visible throughout all of Asia and Australia and the central and eastern parts of Europe and Africa)
9/21: New moon
9/21: Partial solar eclipse

(will only be visible in New Zealand, Antarctica, and the southern Pacific Ocean)
9/22: September equinox (first day of fall, or autumnal equinox, in the Northern Hemisphere and spring, or vernal equinox, in the Southern Hemisphere)

October:
10/7: Full moon, supermoon
10/7: Peak of Draconids meteor shower
10/21: New moon
10/21, 22: Peak of Orionids meteor shower

November:
11/4, 5: Peak of Taurids meteor shower
11/5: Full moon, supermoon
11/17, 18: Peak of Leonids meteor shower
11/20: New moon

December:
12/4: Full moon, supermoon
12/23, 24: Peak of Geminids meteor shower
12/20: New moon
12/21: December solstice (first day of winter, or winter solstice, in the Northern Hemisphere and summer, or summer solstice, in the Southern Hemisphere)
12/21, 22: Peak of Ursids meteor shower

ABOUT THE AUTHORS

ANDREW FAZEKAS (foreword, chapters 4 and 5), also known as The Night Sky Guy, is a science writer, speaker, and broadcaster who shares his passion for the wonders of the universe through all media. He writes a popular weekly stargazing column for National Geographic, hosts a weekly Night Sky livestream event on Facebook and Twitter, and is the author of the National Geographic books *Backyard Guide to the Night Sky* and *Star Trek: The Official Guide to Our Universe.* Andrew is a syndicated correspondent for television and radio broadcast networks, the communications manager for Astronomers Without Borders, and an active skywatching member of the Royal Astronomical Society of Canada since 1983. Co-creator of the world's first open-air augmented-reality planetarium experience, Andrew lives in Montreal, Canada, with his wife and two daughters. As he likes to say, he has never met a clear sky he didn't like.

JAMES TREFIL (chapters 1 and 2), Clarence J. Robinson Professor of Physics at George Mason University, is internationally recognized as an expert in making complex scientific ideas comprehensible. He is the author of numerous magazine articles and books on science for the general public, including National Geographic's *Space Atlas, The Story of Innovation,* and, coauthored with Neil deGrasse Tyson, *Cosmic Queries.* He is a fellow of the American Physical Society, the American Association of the Advancement of Science, and the World Economic Forum. He lives with his wife in Fairfax, Virginia.

MAYA WEI-HAAS (chapter 3) is a staff writer at National Geographic who has a particular affection for rocks and reactions. Her fascination with the processes that shape the planets led her to pursue geology as an undergraduate at Smith College and earn a Ph.D. from The Ohio State University in earth science, working in both Antarctica and Alaska. Maya made the jump to journalism in 2015 as an AAAS Mass Media Fellow. Before starting at National Geographic, she was an assistant editor for Smithsonian.com. When not writing, Maya can be found adventuring in the forest with her husband, Travis, and dog, Jasper, near their home in Virginia.

RACHEL BROWN (chapter 6) was a staff member at National Geographic for five years, where she worked as a producer, writer, and editor on the travel desk. She is a writer of nonfiction, fiction, and poetry. A Chesapeake native, Rachel attended college in Maryland, studying anthropology, Spanish, and creative writing. A childhood experience of the Leonid meteor shower sparked her interest in the stars, and a love of the outdoors cemented it. Rachel enjoys the clear night skies of her home in Tucson, Arizona—headquarters of the International Dark-Sky Association—where she lives with her two partners, three cats, and a three-legged dog.

MICHAEL GRESHKO (chapter 7) has never forgotten the awe he felt as a child when he first saw Jupiter and Saturn through a backyard telescope. Now a staff writer at National Geographic, he has written many magazine articles and online stories about advances in planetary science, physics, and human spaceflight, including the September 2021 cover story "Mysteries of the Solar System." Closer to Earth, he also writes about paleontology, climate change, and human origins. He lives with his wife, Jaclyn, and dog, Luna, in Washington, D.C., and in his spare time enjoys singing, magic, and a good book.

ACKNOWLEDGMENTS

We would like to thank Melissa Hulbert from the Sydney Observatory, part of the Museum of Applied Arts and Sciences, in Millers Point, Australia, for contributing her expertise and knowledge of the southern sky. We would also like to thank James Hedberg of the City College of New York Planetarium and Physics Department for his constellation illustrations, which appear throughout chapter 4. Thank you to John Steele at Brown University for his help regarding the MUL.APIN tablets and to Jessie Christiansen of the California Institute of Technology for her insights on exoplanets.

ILLUSTRATIONS CREDITS

All illustrations appearing in the constellation maps on pages 196–276 were created by James Hedberg.

Cover, Babak Tafreshi/National Geographic Image Collection; back cover: (UP) credit: Alan Dyer/VWPics/Science Source; (LO LE), illustration by James Hedberg; (LO CTR), Gareth Davies/Alamy Stock Photo; (LO RT), ESO; end-papers, Library of Congress Geography and Map Division (G3190 1810 .C7); 2–3, Jeff Dai/Science Source; 4, NASA, ESA, A. Sarajedini (Florida Atlantic University), and G. Piotto (Università degli Studi di Padova). Processing: Gladys Kober (NASA/Catholic University of America); 6, Babak Tafreshi/Science Source; 10–11, Alan Dyer/VWPics/Science Source; 12–13, Tunç Tezel; 14, NASA, ESA, and T. Brown and S. Casertano (STScI); CC BY 4.0. Acknowledgment: NASA, ESA, and J. Anderson (STScI); 16, NASA, ESA, and M. Montes (University of New South Wales); 17, NASA/Ames/JPL-Caltech. Source: ESA/ATG medialab; 18, NASA, ESA, K. Luhman (Penn State University), and M. Robberto (STScI); 19 (UP). Based on image by Tim Brown/Science Source, with photos from NASA and SOHO (ESA & NASA); 19 (LO), WENN/Alamy Stock Photo; 21, Eckhard Slawik/Science Source; 22, NASA; 23, INTERFOTO/History/Alamy Stock Photo; 24 (UP), NASA/JPL-Caltech; 24 (LO), NASA Goddard; 25 (UP), The History Collection/Alamy Stock Photo; 25 (LO), NASA/JPL-Caltech; 26, NASA, ESA, J. Dalcanton (University of Washington, U.S.A.), B. F. Williams (University of Washington, U.S.A.), L. C. Johnson (University of Washington, U.S.A.), the PHAT team, and R. Gendler; 26–7, ESO, ALMA (ESO/NAOJ/NRAO)/A. Schruba, VLA (NRAO)/Y. Bagetakos/Little THINGS; 28, David Parker/Science Source; 29, Mark Garlick/Science Source; 30, NASA, ESA, D. Lennon and E. Sabbi (ESA/STScI), J. Anderson, S. E. de Mink, R. van der Marel, T. Sohn, and N. Walborn (STScI), N. Bastian (Excellence Cluster, Munich), L. Bedin (INAF, Padua), E. Bressert (ESO), P. Crowther (University of Sheffield), A. de Koter (University of Amsterdam), C. Evans (UKATC/STFC, Edinburgh), A. Herrero (IAC, Tenerife), N. Langer (AifA, Bonn), I. Platais (JHU), and H. Sana (University of Amsterdam); 31, Thomas Pesquet/ESA/NASA; 32, NASA; 33, NLM/Science Source; 34 (UP), Lucas Vieira; 34 (LO), Mark Garlick/Science Source; 35 (LE), Science History Images/Alamy Stock Photo; 35 (RT), Smithsonian Institution Libraries/Science Source; 36 (UP), N.A. Sharp/NOAO/AURA/NSF/Science Source; 36 (LO), NASA and STScI; 37, HUPSF Observatory (45), olvwork360664. Harvard University Archives; 38, Mark Garlick/Science Source; 39, ESO/Science Source; 40, Sheila Terry/Science Source; 41, NASA, ESA, and the Hubble Heritage Team (STScI/AURA), A. Nota (ESA/STScI), and the Westerlund 2 Science Team; 42, ESO/S. Steinhöfel; 43, NASA, ESA, and the Hubble Heritage Team (STScI/AURA). Acknowledgment: R. Sahai and J. Trauger (Jet Propulsion Laboratory); 44 (UP), ESO, P. Kervella, Digitized Sky Survey 2, and A. Fujii; 44 (LO), anttoniart/Shutterstock; 45, ESO/L. Calçada; 46, X-ray: NASA/CXC/U.Texas/S.Post et al. Infrared: 2MASS/UMass/IPAC-Caltech/NASA/NSF; 47 (UP LE), NASA/JPL-Caltech; 47 (UP RT), NASA/STScI; 47 (LO), NASA/JPL-Caltech; 48–9, NASA, ESA, and S. Beckwith (STScI) and the HUDF Team; 50, NASA's Goddard Space Flight Center; 51, X-ray: NASA/CXC/CfA/M.Markevitch. Optical and lensing map: NASA/STScI, Magellan/U.Arizona/D.Clowe. Lensing map: ESO WFI; 52–3, NASA/JPL-Caltech/GSFC; 54, Alpha Stock/Alamy Stock Photo; 56, Samuel Baylis/Alamy Stock Photo; 57, Source: Miles Hatfield, NASA Goddard Space Flight Center; 58 (LE), NSO/AURA/NSF; 58 (RT), Library of Congress/Science Source; 59, David Chenette, Joseph B. Gurman, Loren W. Acton; 60, NSO/AURA/NSF; 61, Sources: W. Dean Pesnell, NASA Solar Dynamics Observatory; Miles Hatfield, NASA Goddard Space Flight Center. Data: AIA and HMI science teams, NASA/SDO; 62, Solar Dynamics Observatory, NASA; 63 (UP), Image taken by the Solar Dynamics Observatory on February 24, 2014; 63 (LO), Image taken by the Solar Dynamics Observatory on June 18, 2015; 64 (LE). Source: NASA/GSFC; 64 (RT), Richard Treptow/Science Source; 65, Royal Astronomical Society/Science Source; 66 (UP), U.S. Department of Energy; 66 (LO), Goronwy Tudor Jones, University of Birmingham/Science Source; 67, Mark Garlick/Science Source; 68, Sergei Krasnoukhov/TASS via Getty Images; 69, Werner Forman/Universal Images Group/Getty Images; 70, Mikkel Juul Jensen/Science Source; 71, Dana Berry/National Geographic Image Collection; 72–3, Global Mosiac: Lunar Reconnaisance Orbiter, NASA, Arizona State University; 74 (UP), NASA/Eugene Cernan; 74 (LO), NASA/John W. Young; 75 (UP), NASA/Ames/JPL-Caltech; SOHO (ESA & NASA); 75 (LO), Paul Harris/Getty Images; 76, NASA, Lunar-Reconnaissance Orbiter (LRO); 77, NASA; 78, NASA/Cory Huston; 79 (UP), RIA Novosti/Science Source; 79 (LO), From *The First Men in the Moon*, by H.G. Wells, 1901, George Newnes, London, via Wikimedia Commons; 80 (UP), China National Space Administration/AFP via Getty Images; 80 (LO), D. Ducros/European Space Agency/Science Source; 81, Gregoire Cirade/Science Source; 82, ESA—G. Porter; 83, Y. Beletsky (LCO)/ESO; 84, ESO/A. Ghizzi Panizza; 85, Andrew Brookes, National Physical Laboratory/Science Source; 86 (UP), György Soponyai; 86 (LO), Based on graphic by Hong Kong Space Museum, with photo from SOHO (ESA & NASA); 87 (UP), Sheila Terry/Science Source; 87 (LO), Aizar Raldes/AFP via Getty Images; 88, Mark Garlick/Science Source; 88–9, NASA/Aubrey Gemignani; 89, Chris Dillmann/The Daily via AP; 90, X-ray: NASA/CXC/PSU/L.Townsley et al. Optical: NASA/STScI. Infrared: NASA/JPL/PSU/L.Townsley et al.; 91 (UP), NASA/WMAP Science Team; 91 (LO), NASA Goddard; 92–3, United Launch Alliance; 94, Zolt Levay; 96, NASA/JPL-Caltech; 97, Mark Garlick/Science Source; 98–9, Mark Garlick/Science Source; 99, Dana Berry/National Geographic Image Collection; 100–1, 3D art and render: Antoine Collignon. Diagrams: Matt Twombly. Final art: Manuel Canales. Sources: David Jewitt, UCLA; Michael K. Shepard, Bloomsburg University; Scott S. Sheppard, Carnegie Institute for Science; Nick Moskovitz, Lowell Observatory, Francesca E. Demeo, Massachusetts Institute of Technology; David Nesvorny, Southwest Research Institute; John J. Loiacono, Arctic Slope Regional Corporation; NASA; 102, NASA/JPL/USGS; 103, Detlev van Ravensswaay/Science Source; 104–5, Global Mosaic: MESSENGER (MEcury Surface Space ENvironment, GEochemistry, and Ranging), NASA, Johns Hopkins University Applied Physics Laboratory, Carnegie Institute of Washington; 106 (UP), NASA/Johns Hopkins University Applied Physics Laboratory/Carnegie Institution of Washington; 106 (LO), © The Trustees of the British Museum/Art Resource, NY; 107, NASA/Johns Hopkins University Applied Physics Laboratory/Carnegie Institution of Washington; 108, NASA/JPL; 109, JAXA/NASA; 110–1, Global Mosaic: Magellan Synthetic Aperture Radar Mosaics, NASA, JPL (Jet Propulsion Laboratory, California Institute of Technology); 112 (UP), Miguel Claro/Science Source; 112 (LO), RIA Novosti/Science Source; 113, NASA/JPL; 114, NASA/Terry Virts; 115, blickwinkel/F. Neukirchen/Alamy Stock Photo; 116–7, Surface Satellite Mosaic: NASA Blue Marble, NASA's Earth Observatory; 118, The COMET Program and Chapel Design & Marketing; 119 (UP), Chris Johns/National Geographic Image Collection; 119 (LO), Chuck Carter, Eagre Games, Inc.; 120, NASA/JPL-Caltech/Univ. of Arizona; 121, NASA/JPL-Caltech; 122–3, Global Mosaic: NASA Mars Global Surveyor; National Geographic Society; 124, NASA/JPL-Caltech/MSSS; 125 (UP), Detlev van Ravensswaay/Science Source; 125 (LO), NASA/JPL/USGS; 126 (UP), ILC Dover, LP; 126 (LO), NASA/JPL-Caltech/MSSS; 127, Manuel Canales and Matthew W. Chwastyk, NGM Staff; Alexander Stegmaier. Art: Antoine Collignon. Sources: Ashley Palumbo, Brown University; Robin Wordsworth, Harvard University; NASA; 128, NASA/JPL-Caltech/SwRI/MSSS/Gerald Eichstädt/Seán Doran; 129, NASA/ESA/NOIRLab/NSF/AURA/M.H. Wong and I. de Pater (UC Berkeley) et al. Acknowledgment: M. Zamani; 130, Global Mosaic: NASA Cassini Spacecraft, NASA, JPL (Jet Propulsion Laboratory, California Institute of Technology), Space Science Institute; 131 (UP), NASA/JPL-Caltech; 131 (LO), SPL/Science Source; 132–3, All Global Mosaics: NASA Galileo Orbiter, NASA, JPL (Jet Propulsion Laboratory, California Institute of Technology), University of Arizona; 135 (UP), Enhanced image by Betsy Asher Hall and Gervasio Robles based on images provided courtesy of NASA/JPL-Caltech/SwRI/MSS; 135 (LO), Mark R. Showalter, SETI Institute; 136, NASA/JPL-Caltech/Space Science Institute; 137, NASA/JPL-Caltech; 138 (UP), NASA/JPL-Caltech/Space Science Institute; 138 (LO), NASA Cassini Spacecraft, NASA, JPL (Jet Propulsion Laboratory, California Institute of Technology), Space Science Institute; 139–141, All Global Mosaics: NASA Cassini Spacecraft, NASA, JPL (Jet Propulsion Laboratory, California Institute of Technology), Space Science Institute; 142, Royal Astronomical Society/Science Source; 143, NASA-JPL/Caltech; 144, Claus Lunau/Science Source; 145, Tim Brown/Science Source; 146, NASA/JPL-Caltech; 147–8, Global Imagery: NASA Voyager II, NASA, JPL (Jet Propusion Laboratory, California Institute of Technology); 149 (UP), NASA, ESA, and L. Lamy (Observatory of Paris, CNRS, CNES); 149 (LO), NSA Digital Archive/Getty Images; 150, NASA, ESA, and M. Showalter (SETI Institute), CC BY 4.0; 151, NASA/JPL; 152–3, Global Imagery: NASA Voyager I, NASA, JPL (Jet Propulsion Laboratory, California Institute of Technology); 154, NASA/Johns Hopkins University Applied Physics Laboratory/Southwest Research Institute; 155, NASA/JPL-Caltech/UCLAMPS/DLR/IDA; 156, Mark Garlick/Science Source; 157 (UP), NASA/JFL-Caltech/UCLA/MPS/DLR/IDA, Ceres Dawn FC Global Mosaic 140m v1; 157 (LO), Global Imagery: NASA, Johns Hopkins University Applied Physics Laboratory, Southwest Research Institute, Lunar and Planetary Institute; 158, Mark Garlick/Science Source; 159, NASA/Goddard/University of Arizona; 160 (UP LE), ESA/Rosetta/MPS for OSIRIS Team MPS/UPD/LAM/IAA/SSO/INTA/UPM/DASP/IDA; 160 (UP RT), NASA/JPL/JHUAPL; 160 (LO), Tim Brown/Science Source; 161, NASA/ESA/STScI; 162–3, Images: NASA Goddard Space Flight Center; NASA/JPL-Caltech; Johns Hopkins University Applied Physics Laboratory; Carnegie Institution of Washington; ESA; JAXA; 164–5, Y. Beletsky (LCO)/ESO; 166, Royal Astronomical Society/Science Source; 168, Alan Dyer/VWPics via AP Images; 169, Celestron; 170 (UP), Courtesy Nikon, Inc.. 170 (LO), Babak Tafreshi/Science Source; 171 (UP), Alan Dyer/VWPics via AP Images; 171 (LO), Orion Telescopes and Binoculars; 172, John R. Foster/Science Source; 174–89, Mapping Specialists, Ltd.; 192–3 (UP), Mapping Specialists, Ltd.; 194, Album/British Library/Alamy Stock Photo; 196 (LO), NASA, ESA, Digitized Sky Survey 2 (Acknowledgment: Davide De Martin); 197 (LO), Adam Block; 198 (UP), Mapping Specialists, Ltd.; 198 (LO), ESA/Hubble & NASA; 199 (LO), NASA, ESA, C.R. O'Dell (Vanderbilt University), and M. Meixner, P. McCullough, and G. Bacon (Space Telescope Science Institute); 200 (LO), Mapping Specialists, Ltd.; 201 (LO), Göran Nilsson & The Liverpool Telescope; 202 (LO), Adam Block/Mount Lemmon SkyCenter/University of Arizona; 203 (LO), Mapping Specialists, Ltd.; 204 (LO), Advanced Camera Data: NASA, ESA, A. Aloisi (STScI/ESA), J. Mack and A. Grocholski (STScI), M. Sirianni (STScI/ESA), R. van der Marel (STScI), L. Angeretti, D. Romano, and M. Tosi (INAF-OAB), and F. Annibali, L. Greggio, and E. Held (INAF-OAP). Wide Field Planetary Camera 2 Data: NASA, ESA, P. Shopbell (California Institute of Technology), R. Dufour (Rice University), D. Walter (South Carolina State University, Orangeburg), and A.

Wilson (University of Maryland, College Park); 205 (LO), NOIRLab/NSF/AURA; 206 (LO), ESA/Hubble & NASA, G. Piotto et al.; 207, NASA, ESA, S. Beckwith (STScI), and the Hubble Heritage Team (STScI/AURA); 208 (LO), ESO; 209 (LO), ESO; 210 (LO), The Hubble Heritage Team (AURA/STScI/NASA/ESA); 211 (UP), Mapping Specialists, Ltd.; 211 (LO), ESO/G. Beccari; 212 (UP), NASA/JPL-Caltech/UCLA; 213 (UP), Mapping Specialists, Ltd.; 213 (LO), ESO; 214 (LO), Canada-France-Hawaii Telescope/J.-C. Cuillandre/Coelum, and the Digitized Sky Survey; 215 (LO), Mapping Specialists, Ltd.; 216 (UP), Mapping Specialists, Ltd.; 217 (UP), Mapping Specialists, Ltd.; 217 (LO), NASA and the Hubble Heritage Team (AURA/STScI). Acknowledgment: S. Smartt (Institute of Astronomy) and D. Richstone (U. Michigan); 218 (UP), Mapping Specialists, Ltd.; 218 (LO), ESO; 219 (LO), Adam Block/Mount Lemmon SkyCenter/University of Arizona; 220 (LO), NASA, ESA, and the Hubble Heritage Team (STScI/AURA)-ESA/Hubble Collaboration. Acknowledgment: B. Whitmore (Space Telescope Science Institute) and James Long (ESA/Hubble); 221 (LO), ESA/Hubble & NASA, P. Erwin et al.; 222 (UP), Mapping Specialists, Ltd.; 222 (LO), ESO/S. Brunier; 223, ESO; 224 (UP), NASA, ESA, Digitized Sky Survey 2. Acknowledgment: Davide De Martin; 225 (LO), ESA/Hubble & NASA; 226 (UP), Mapping Specialists, Ltd.; 226 (LO), ESA/NASA, ESO, and Danny LaCrue; 228 (LO), Mapping Specialists, Ltd.; 229 (LO), NASA, ESA, Andrew Fruchter (STScI), and the ERO team (STScI + ST-ECF); 230 (LO), ESO; 231 (LO), ESA/Hubble & NASA. Acknowledgment: Gilles Chapdelaine; 232 (UP), Mapping Specialists, Ltd.; 232 (LO), NASA, ESA, Dan Maoz (Tel-Aviv University, Israel, and Columbia University, U.S.A.); 233 (LO), NASA, ESA, and the Hubble Heritage Team (STScI/AURA). Acknowledgment: W. Blair (STScI/JHU), Carnegie Institution of Washington (Las Campanas Observatory), and NOAO; 234, X-ray: NASA/CXC/Univ. Potsdam/L. Oskinova et al. Optical: ESA, NASA/STScI. Infrared: NASA/JPL-Caltech; 235 (UP), Mapping Specialists, Ltd.; 235 (LO), ESA/Hubble & NASA; 236 (UP), Mapping Specialists, Ltd.; 237 (LO), ESO/INAF-VST/OmegaCAM. Acknowledgment: OmegaCen/Astro-WISE/Kapteyn Institute; 239 (LO), Sloan Digital Sky Survey; 240 (UP), Mapping Specialists, Ltd.; 240 (LO), Radio: NRAO/AUI/NSF/GBT/VLA/Dyer, Maddalena & Cornwell. X-ray: Chandra X-ray Observatory; NASA/CXC/Rutgers/G. Cassam-Chenaï, J. Hughes et al. Visible light: 0.9-metre Curtis Schmidt optical telescope. NOAO/AURA/NSF/CTIO/Middlebury College/F. Winkler and Digitized Sky Survey; 241 (LO), ESA/Hubble & NASA, J. Walsh; 242 (LO), NASA & ESA. Acknowledgment: Gilles Chapdelaine; 243, Hubble data: NASA, ESA, C. Robert O'Dell (Vanderbilt University), and David Thompson (LBTO); 244 (BOTH), Mapping Specialists, Ltd.; 245 (LO), ESO/S. Guisard; 246 (UP), Mapping Specialists, Ltd.; 246 (LO), Raghvendra Sahai and John Trauger (JPL), the WFPC2 science team, and NASA/ESA; 247 (UP), Mapping Specialists, Ltd.; 247 (LO), NASA/Space Telescope Science Institute; 248 (UP), Mapping Specialists, Ltd.; 248 (LO), ESO; 249 (LO), ESO; 250 (LO), SPECULOOS Team/E. Jehin/ESO; 251 (UP), Mapping Specialists, Ltd.; 251 (LO), ESA/Hubble & NASA; 252 (UP), ESA/Hubble & NASA/D. Milisavljevic (Purdue University); 253 (LO), N.A.Sharp/NOIRLab/NSF/AURA; 254 (UP), ESO; 254 (LO), Mapping Specialists, Ltd.; 255 (UP), Mapping Specialists, Ltd.; 255 (LO), X-ray: NASA/CXC/Univ of Hertfordshire/M.Hardcastle et al. Radio: CSIRO/ATNF/ATCA; 256 (UP), NASA, ESA, the Hubble Heritage Team (STScI/AURA)-ESA/Hubble Collaboration, and B. Whitmore (STScI); 258 (UP), Mapping Specialists, Ltd.; 258 (LO), NASA, ESA, and the Hubble Heritage Team (STScI/AURA); 259 (UP), Mapping Specialists, Ltd.; 260 (LO), ESO; 261 (UP), NASA, ESA, and J. Kastner (RIT); 262 (UP), Mapping Specialists, Ltd.; 262 (LO), ESO/M. Wittkowski; 263, X-ray: NASA/CXC; Optical: NASA/STScI; 264 (LO), ESA/Hubble & NASA, L. Stanghellini; 265 (UP), NASA, ESA/Hubble, and the Hubble Heritage Team; 266 (LO), X-ray: NASA/CXC/Univ. of Alabama/K.Wong et al. Optical: ESO/VLT; 267 (UP), NASA, ESA, and Allison Loll/Jeff Hester (Arizona State University). Acknowledgment: Davide De Martin (ESA/Hubble); 268 (UP), Mapping Specialists, Ltd.; 268 (LO), Data from Hubble/NASA with tweaks by Mike Peel via Wikimedia Commons; 269 (LO), ESO/Digitized Sky Survey 2. Acknowledgment: Davide De Martin; 270 (UP), Mapping Specialists, Ltd.; 270 (LO), ESA, ESO, and NASA; 271 (UP), Mapping Specialists, Ltd.; 271 (LO), NASA, ESA, and the Hubble Heritage (STScI/AURA)-ESA/Hubble Collaboration. Acknowledgment: J. Mack (STScI) and G. Piotto (University of Padova, Italy); 272 (UP), Robert Williams and the Hubble Deep Field Team (STScI) and NASA/ESA; 273 (LO), NASA, ESA, and the Hubble SM4 ERO Team; 274 (UP), The Hubble Heritage Team (STScI/AURA/NASA); 274 (LO), Mapping Specialists, Ltd.; 275 (LO), NASA, ESA, and the Hubble Heritage Team (STScI/AURA); 276 (UP), Mapping Specialists, Ltd.; 277, T.A. Rector (University of Alaska Anchorage) and H. Schweiker (WIYN and NOIRLab/NSF/AURA); 278–9, Brad Goldpaint; 280, funkyfood London—Paul Williams/Alamy Stock Photo; 282, Petr Horálek; 283, Abdullah Çelik/Anadolu Agency/Getty Images; 284 (UP), Péter Mocsonoky/Alamy Stock Photo; 284 (LO), © Canadian Space Agency, 2015; 285, Mark Garlick/Science Photo Library/Alamy Stock Photo; 286 (LO), © AccuWeather; 287, Pekka Parviainen/Science Source; 288, Composite image by Tom Buckley-Houston and created with photos by Stephen Rahn and NASA; 289, A. Feild (STScI); 290, NASA, ESA, and AURA/Caltech; 291, NASA, ESA, J. Hester, and A. Loll (Arizona State University); 292, ESA/Hubble & NASA; 293-4, Mapping Specialists, Ltd.; 295 (UP), X-ray: NASA/CXC/CfA/S.Wolk et al. Optical: NSF/AURA/WIYN/Univ. of Alaska/T.A.Rector; 295 (LO), Bray Falls; 296 (UP), Giuseppe Donatiello; 296 (LO), NASA & ESA; 297 (UP), Mapping Specialists, Ltd.; 297 (LO), NASA/JPL-Caltech/University of Wisconsin; 298, Mapping Specialists, Ltd.; 299 (UP), NASA/JPL-Caltech and the Hubble Heritage Team (STScI/AURA); 299 (LO), ESA/Hubble & NASA, L. Ho et al.; 300, NASA/JPL-Caltech; 301 (UP), NASA/JPL-Caltech/P.S. Teixeira (Center for Astrophysics); 301 (LO), Mapping Specialists, Ltd.; 302, ESO/G. Beccari; 303 (UP), Mapping Specialists, Ltd.; 303 (LO), ESO; 304, Chandra: NASA/CXC/RIKEN/T. Sato et al.; NuSTAR: NASA/NuSTAR; Hubble: NASA/STScI; 305, NASA, ESA, and P. Challis (Harvard-Smithsonian Center for Astrophysics); 306, Adam Block/Steward Observatory/University of Arizona; 307 (UP), Robert Hynes; 307 (LO), X-ray: NASA/CXC. Ultraviolet/Optical: NASA/STScI. Combined image: NASA/ESA/N. Smith (University of Arizona), J. Morse (BoldlyGo Institute), and A. Pagan; 308, G.Hüdepohl (atacamaphoto)/ESO; 310, Peter Komka/MTI via AP; 311, NASA; 312 (BOTH), SpaceX; 313, CTIO/NOIRLab/NSF/AURA/DECam DELVE Survey; 314, Babak Tafreshi/Science Source; 316 (UP), Jeff Dai/Science Source; 316 (LO), Pekka Parviainen/Science Source; 317, Juan Carlos Casado (starryearth)/Science Source; 318–9, Bob Sacha/National Geographic Image Collection; 320, Babak Tafreshi/Science Source; 322, nevio/Shutterstock; 323 (LO), Charles Walker Collection/Alamy Stock Photo; 324, Art by Oliver Uberti, NGM Staff. Sources: Mike Parker Pearson, University of Sheffield; Science and Stonehenge, Barry Cunliffe and Colin Renfrew, eds.; 325, The Walker/Getty Images; 326, Gary Hershorn/Getty Images; 327, Janzig/USA/Alamy Stock Photo; 328 (UP), Mladen Antonov/AFP via Getty Images; 328 (LO), Universal Images Group via Getty Images; 329, MPI/Getty Images; 330 (UP), Stefano Ravera/Alamy Stock Photo; 330 (LO), Art by Hernán Cañellas. Sources: Guillermo de Anda, University of Yucatán; Arturo Montero, University of Tepeyac; 331, Jordi CAMÍ/Alamy Stock Photo; 332, Alfredo Garcia Saz/Alamy Stock Photo; 333, Mike P Shepherd/Alamy Stock Photo; 334 (UP), Helene Rogers/Art Directors & TRIP/Alamy Stock Photo; 334 (LO), Pierre Mion; 335 (UP), David Degner/Getty Images; 335 (LO), Greg Balfour Evans/Alamy Stock Photo; 336, Zoonar/Insung Choi/Alamy Stock Photo; 337, Imaginechina/Alamy Stock Photo; 338, Art by Fernando G. Baptista, NGM Staff; research by Patricia Healy. Sources: Klaus Schmidt, Jens Notroff, and Oliver Dietrich, German Archaeological Institute; Ian Kuijt, University of Notre Dame; 339, Chris McGrath/Getty Images; 340, John T. Meader, © 2019; 342, Babak Tafreshi/Science Source; 343, Kiss Gergo/EyeEm/Alamy Stock Photo; 344, NOAO/AURA/NSF; 345, Janice and Nolan Braud/Alamy Stock Photo; 346, Babak Tafreshi/Science Source; 347, Natural History Library/Alamy Stock Photo; 348 (UP), Data provided by Xavier Jubier; 348 (LO), Pat & Chuck Blackley/Alamy Stock Photo; 349, ESO/P. Horálek; 350–1, Gareth Davies/Alamy Stock Photo; 351, Vanexus Photography; 352, Ricardo Arduengo/AFP via Getty Images; 353, Stefano Politi Markovina/Alamy Stock Photo; 354, Inspiration4/John Krauss/Cover Images via AP Images; 355, Source: Thomas G. Roberts, CSIS Aerospace Security Project; 356, Songquan Deng/Alamy Stock Photo; 357 (UP), Atlantis display at the Kennedy Space Center/Alamy Stock Photo; 357 (LO), Novespace/ABACAPRESS/Alamy Stock Photo; 358 (UP), Kumar Sriskandan/Alamy Stock Photo; 358 (LO), Witold Skrypczak/Alamy Stock Photo; 359, Robin Loznak/ZUMA Press/Alamy Stock Photo; 360–1, NASA/JPL-Caltech; 362, Photo Researchers/Science History Images/Alamy Stock Photo; 364, NASA/JHUAPL/SwRI; 365, NASA/Johns Hopkins University Applied Physics Laboratory/Southwest Research Institute/Roman Tkachenko; 366 (UP), NASA; 367 (UP), NASA; 367 (LO), Prof. Merav Opher, Boston University; 368, NASA/Chris Gunn; 369, NASA, ESA, P. Oesch (Yale University), G. Brammer (STScI), P. van Dokkum (Yale University), and G. Illingworth (University of California, Santa Cruz); 370, NASA/JPL-Caltech/ESA; 370–1, Art by Jason Treat, NGM Staff. Source: Tom Abel; 372, Acknowledgment: Gaia Data Processing and Analysis Consortium (DPAC). Top and middle: A. Moitinho/A. F. Silva/M. Barros/C. Barata, University of Lisbon, Portugal; H. Savietto, Fork Research, Portugal. Bottom: Gaia Coordination Unit 8; M. Fouesneau/C. Bailer-Jones, Max Planck Institute for Astronomy, Heidelberg, Germany; 373 (UP), NASA/JPL-Caltech; 373 (LO), Rubin Obs/NSF/AURA; 374, ESA. Artist impression: ESA/Hubble, NASA, ESO, M. Kornmesser; 375, NASA/JPL-Caltech; 376 (UP), EHT Collaboration; 376 (LO), ESO/M. Kornmesser; 377 (UP), NASA/Goddard Space Flight Center; 377 (LO), NASA/SkyWorks Digital; 378 (UP), ESO; 378 (LO), Giant Magellan Telescope—GMTO Corporation; 379, Gabriele Holtermann-Gorden/Sipa USA via AP Images; 380 (LO), Department of Special Collections and University Archives, W.E.B. Du Bois Library, University of Massachusetts Amherst; 381, The SXS (Simulating eXtreme Spacetimes) Project; 382, The Virgo Collaboration/CCO 1.0; 383, ESA - C. Carreau; 384, DOE/FNAL/DECam/R. Hahn/CTIO/NOIRLab/NSF/AURA; 385 (UP), NASA/CXC/Cambridge Univ./C.S. Reynolds; 385 (LO), NASA, ESA, A. Feild (STScI), and A. Riess (STScI/JHU); 386 (UP LE), Ramon Andrade 3Dciencia/Science Source; 386 (UP RT), Dark matter, R. Caputo et al. 2016. Background, Axel Mellinger, Central Michigan University; 386 (LO), CTIO/NOIRLab/NSF/AURA/P. Marenfeld; 387, GSFC; 388, NASA/JPL; 389, Based on graphic by ESA; 390 (UP), NASA Visualization Technology Applications and Development (VTAD); 390 (LO), Exoplanet Exploration Program and the Jet Propulsion Laboratory for NASA's Astrophysics Division; 391, Sources: Zac Manchester, Stanford University; BREAKTHROUGH Initiatives; 392, Based on graphic by ESA; 393, ESO/M. Kornmesser; 394 (UP), ESO/M. Kornmesser; 394 (LO), NASA/JPL-Caltech/T. Pyle (SSC); 395, NASA/GSFC; 396 (UP), ESA; 396 (LO), ALMA (ESO/NAOJ/NRAO); 397, NASA/JPL-Caltech; 398, NASA/JPL-Caltech/SETI Institute; 399, ESA/ATG medialab; 400 (UP), NASA/JPL-Caltech; 400 (LO), Richard Bizley/Science Source; 401 (UP), NASA/JPL-Caltech/MSSS; 401 (LO), Spacecraft: ESA/ATG medialab. Jupiter: NASA/ESA/J. Nichols (University of Leicester). Ganymede: NASA/JPL. Io: NASA/JPL/University of Arizona. Callisto and Europa: NASA/JPL/DLR; 402 (UP), NASA/JPL/ESA; 402 (LO), Carlos Clarivan/Science Source; 403 (UP), SPL/Science Source; 403 (LO), NASA/JPL-Caltech; 404, NASA; 405, RIA Novosti/Science Source; 406, NASA; 407, Art: Fernando G. Baptista (NGM Staff), Jose Daniel Cabrera Peña. Research: Kaya Berne. Sources: Kenneth S. Thomas; NASA; Smithsonian's National Air and Space Museum; Richard D. Watson; Amy J. Ross; 408, NASA; 409 (BOTH), SpaceX; 410, NASA/Johnson; 411, NASA; 412, NASA; 413 (UP), SpaceX; 413 (LO), AP Photo/Ng Han Guan; 414–5, NASA.

INDEX

Boldface indicate
illustrations.

A

Acadia National Park, Maine
346, 348, **348**
Achernar (star) 192, 228, 254
Adams, Douglas 365
Adams, John Couch 151
Adler Planetarium, Chicago
352–353
!Ae!Hai Kalahari Heritage
Park, South Africa 344
Africa
archaeoastronomy
331–335, **332, 333, 334**
astrotourism 344
Ain (star) 390
Airbus A330 aircraft **357**
Airglow **316**
Akatsuki spacecraft 112–113
Al-Sufi, Abd al-Rahman **362,**
363
Aldebaran (star) 184, 191, 192,
267, 390
Algieba (star) 237, 390
Alien life, search for 399–403
belief in 359
Jupiter's moons **398,**
400–401, 401
Mars 125, **399,** 399–400,
400, 401
Saturn's moons 143, 401,
402
Venus 401–402, **402, 403**
Alien planets see Exoplanets
Alignments 309
ALMA radio telescope
164–165, 396, 397
Alnair (star) 193, 230
Alpha Centauri (star) 24,
184, 192, 193, 213, 216, 240,
366, 397
Alphard (star) 174, 190, 233
Alpheratz (star) 178, 196,
252
Ancestral Puebloans **318–319,**
326–329, **327, 328**
Andromeda (constellation)
178, **178–179, 182–183,** 191,
196, **196**
Andromeda galaxy (M31)
26, **26,** 178, 196, **196, 289,**
291, 363
Ant Nebula (Mz3) 247, **247**
Antares (star) **34,** 176, 186,
188, 190, 193, 240, 261
Antlia (constellation) **182–183,**
197, **197**
Apollo missions 72, **74, 77,**
79, 99, 162, 212, 274, 358,
405, **407**
Apus (constellation) **188–189,**
198, **198**
Aquarius (constellation) 178,
178–179, 182, **182–183,** 199,
199

Aquila (constellation) **166,**
176, **176–177, 188–189,** 190,
200, **200**
Ara (constellation) **188–189,**
200, **200**
Arceneaux, Hayley **354**
Archaeoastronomy 322–339
Africa 331–335, **332, 333,**
334, 335
Asia 335–339, **336, 337, 338,**
339
celestial navigation 339
constellations 195
defined 416
Europe **322, 323,** 323–325,
324, 325
map 323
Mesoamerica 329–331, **330,**
331
Milky Way beliefs 15
MUL.APIN tablets 106, **106**
South America 222, 329,
330
sun mythology **54,** 55
United States **318–319,**
325–329, **326, 327, 328,**
329
Archytas of Tarentum 28
Arcturus (star) 176, 188, 190,
193, 203
Area 51, Nevada 359
Arecibo Observatory, Puerto
Rico 352, **352,** 403
AREE rover 403
Aries (constellation) **178–179,**
182, **182–183,** 191, 201, **201**
Aristarchus of Samos 95
Armstrong, Neil 79, 99, **407**
Arrokoth (2014 MU69) 97, **101,**
158, 163, **163,** 365, **365**
Artemis program **78,** 80, 410,
412–413
Asia
archaeoastronomy 69,
335–339, **336, 337, 338,**
339
astrotourism 343, 353
Asteroid belt 98, 100, 156,
159–161
Asteroids
Bennu **159,** 163, **163**
collision with Earth 160,**285**
collision with Mercury
106–107
composition 159–160
defined 416
Eros **160,** 163, **163**
exploration 163, **163**
field guide 284, **284**
near-Earth **160,** 374, **375**
Pan-STARRS survey 374,
376
solar system in action 100,
100, 101
Trojans 101, 163
Vesta **155,** 163, **163**

Astrology 17, 106
Astronomical units (AUs)
23–24, **24,** 169–170, 416
Astronomy, advances in
368–387
axion, hunting for **386,**
386–387
big bang afterglow
369–371, **370–371**
black holes **376, 377,**
377–379, **381,** 382–383,
385, **385**
cosmic flashes 377–379
cosmic microwave
background **370**
dark energy **370–371, 384,**
387
dark matter **370–371, 385,**
385–387, **386**
Earth-based telescopes
378, 379, **386**
gravitational waves
379–383, **381, 382,** 385
Hubble constant 371, **385**
interstellar objects 374,
374, 376–377
Milky Way 371, **372,** 373
near-Earth objects 100,
374, **375**
neutron star collisions 383
observatories **373, 378,**
382, 386
solar system 373–377
space telescopes **368, 369,**
373, 387
Astrotourism 340–353
about 341
dark-sky sites 341–347
maps 341, 348, 353
observatories and
planetariums **352,**
352–353, 353
reducing light pollution
346, **347**
rocket launches and
spaceport tours 355,
356–357
solar eclipses **65,** 89, **89,**
347–348, **349**
Steve 351, **351**
time-specific sights
347–352
UFOs and aliens 359, **359**
U.S. national parks **340,**
345, 345–347, **347,** 348,
348
see also Space travel
AT 2018cow explosion 378
Atacama Large Millimeter
Array (ALMA) **164–165,**
396, 397
Atlantis (space shuttle) **357**
Atlas V rockets **92–93**
Atmosphere
defined 416
Earth 31, 33, 118, **118**
exoplanets 392, 394

Jupiter 129
Mars 121
Neptune **151,** 151–152
Uranus 145, 146, 151
Venus 109, 112, 402, **402**
Atomic clock 85, **85**
Auriga (constellation)
178–181, 180, 184, 191, 202,
202
Auroras 58, 62, **314,** 315, 348,
350–351, 350–352, 416
AUs (astronomical units)
23–24, **24,** 169–170, 416
Australia
astrotourism **342,**
342–343,351–352
Mars analogues 359
Axiom Space 356, 408
Axion, hunting for **386,**
386–387
Aymara people **87**
Aztec **54,** 55, **69**

B

Barnard, E. E. 300
Barnard's galaxy **26–27**
Barnard's Loop 300
Barnard's Star 249
Bayer, Johann 235, 236
Bean, Alan 77
Becquerel, Henri 32
Behnken, Bob **409**
Bennu (asteroid) **159,** 163,
163
BepiColombo spacecraft
107, 162
Bessel, Friedrich 24, 28
Beta Centauri (star) 184, 192,
193, 213, 216, 240
Betelgeuse (supergiant star)
34, 44, 45, 180, 192, 209,
250, 305–306, **306**
Bethe, Hans 43
Bezos, Jeff 355, 409
Big bang 28, **29,** 369–371,
370–371, 416
Big Dipper asterism 170, 174,
190, 191, **191,** 272, 293
Binary star systems 40, **40,**
47–48, 416
Binoculars 169, 170, **170,** 171
Black holes
advances in astronomy
376, 377, 377–379, **381,**
382–383, 385, **385**
closest to Earth 268
defined 416
emitting plasma **47**
formation of 47
in Monoceros constellation
245
Blood moon 83, **83**
Blue moon 83
Blue Origin (company) 355,
409
Blue Planetary Nebula (NGC
3918) 302

Boeing's CST-100 Starliner
407
Boötes (constellation)
174–177, 176, 188, 190, 203,
203, 206
Borisov, Gennady 376
Boyajian, Tabetha 403
Branson, Richard 355, 409
Breakthrough Starshot 391,
391
Brown dwarf stars 39
Bullet Cluster **51**
Bunsen, Robert 33–34
Burnell, Jocelyn Bell 380

C

Cacciatore, Niccolò 225
Caelum (constellation)
184–185, 203, **203**
Cahokia mounds, Illinois 325
Caldwell 1 (NGC 188) 214, 293
Caldwell 84 (NGC 5286) **4,** 303
Calendars 287, 420–421
California Nebula (NGC 1499)
253, 295, **295**
Calisto (Jupiter's moon) 131,
133, 163
Camelopardalis
(constellation) **180–181,**
204, **204**
Canada
astrotourism **10–11,** 168,
351, **351**
space program 379, 410
Canary Islands 358, **358**
Cancer (constellation)
174–175, 186, **186–187,** 205,
205
Canes Venatici (constellation)
182–185, 206, **206,** 292, **292**
Canis Major (constellation)
180, **180–181,** 184, **184–185,**
192, 208, **208,** 250
Canis Minor (constellation)
180, **180–181, 184–185,** 192,
209, **209,** 250
Cannon, Annie Jump **35,**
35–36
Canopus (star) 184, 186, 192,
203, 211, 302
Cape Canaveral, Florida
92–93, 356–357
Capella (star) 180, 184, 191,
202
Capricornus (constellation)
180–181, 184–185, 192, 210,
210
Carina (constellation)
about 211, **211**
Eta Carinae 211, 306–307,
307
in False Cross 193
Gem Cluster (NGC 3293)
302, **302**
sky maps **182–189,** 184, 186
star-hopping 192
Carr, Gerald **406**

Carrington, Richard 62–63
Cartwheel Galaxy 262, **263**
Cassini, Giovanni 60
Cassini spacecraft 137, **137**, 138, 142, 143, 162–163, **162–163**
Cassiopeia (constellation) **174–181**, 178, 191, 212, **212**, **294**, 294–295, **295**
Cassiopeia A supernova remnant **304**
Castor (star) 180, 191, 192, 209, 229
Cat's Eye Galaxy (M94) 206, 292, **292**
Cat's Eye Nebula (NGC 6543) 293
Celestial navigation 339
Centaurs (unstable bodies) 101
Centaurus (constellation)
    about 213, **213**
    deep-sky treasures 302–303, **303**
    exomoon 397
    sky maps 182, 184, **184–187**, 186, 188
    see also Alpha Centauri
Cepheid variables 25–26, 48
Cepheus (constellation) **178–179**, 191, **214**, 214, 293, **294**, 294–295
Ceres (dwarf planet) 155, 156, **157**, 163, **163**
Cerro Tololo Inter-American Observatory, Chile 313, 387
Cetus (constellation) 178, **178–179**, 182, **182–183**, 191, 192, 215, **215**
Chaco Canyon, New Mexico **318–319**, 326–329, **327**, **328**
Chaffee, Roger 274
Challenger space shuttle 73, 407
Chamaeleon (constellation) **182–189**, 215, **215**
Chandra X-ray Observatory 257, 264, 266, **304**
Chandrasekhar, Subrahmanyan 48
Chandrasekhar limit 48
Chandrayaan orbiters 72, 80
Chang'e spacecraft 72, 73, 80, 162, **413**
Charon (Pluto's moon) 101, **101**, 156
Cheomseongdae, South Korea 335, **336**, 337
Cheops satellite 392
Chichén Itzá **330**, 330–331, **331**
Chicxulub impact 160
Chile
    astrotourism **83**, **344**, 344–345, **346**, 349
    observatories **44**, **83**, 84, **164–165**, 308, 313, **320**, 345, **349**, **373**, 373–374, **378**, 379, **386**, 387, 390
    sun god 55
CHIME telescope 379
China
    archaeoastronomy **194**, 315, 337, **337**

astrotourism 353
crewed spaceflight program 408–409
interstellar probes 367
moon exploration 72, 73, 80, **80**, 413, 413
solar system missions 162–163, **162–163**
Zhurong rover 126, 163
Christmas Tree Cluster (NGC 2264) 245, 300, **301**
Circinus (constellation) **182–189**, 216, **216**
Clement XI, Pope 87
CLPS (Commercial Lunar Payload Services) 413
Clusters 294–297, **294–297**, **300**, 300–301, **301**, 419; see also Galaxy clusters; Globular clusters; Star clusters
CMB (cosmic microwave background) 369–371, **370**
CMEs (coronal mass ejections) 59, 60–64, **61**, **62**, **63**
Coathanger asterism (Collinder 399) 276, **296**, 297
COBE (Cosmic Background Explorer) 370, **370**
Columba (constellation) **180–181**, **184–185**, 216, **216**
Columbia space shuttle 407
Coma Berenices (constellation) **174–175**, **186–187**, 217, **217**, 299
Comet Interceptor mission (ESA) 376–377
Comets
    about 286–287, 416
    exploration 163, **163**
    field guide 284, **284**
    Halley's comet 163, **280**
    interstellar 376–377
    Jupiter collision 135, **135**
    Kuiper belt **160**
    long-period comets 161
    NEOWISE **282**
    short-period comets 159
    solar system in action 101
    Swift-Tuttle 253
    Ursid meteor shower 273
Commercial Lunar Payload Services (CLPS) 413
Comte, Auguste 33, 34
Conjunctions **308**, 309, 416
Constellations
    about 195
    alphabetical list 196–276
    defined 416
    exoplanets 197, 252, 274, 390, **390**
    guide to 167
    meteor showers named for 283–284, **286–287**
    navigating the night sky 169–171
    sky maps 173–193
Copernicus, Nicolaus 20, 55, 85, 95, 100
Corona **57**, 58, **61**, 416

Corona Australis (constellation) **176–177**, **188–189**, 218, **218**
Corona Borealis (constellation) 176, **176–177**, 188, 219, **219**
Coronal mass ejections (CMEs) 59, 60–64, **61**, **62**, **63**
Corvus (constellation) **174–175**, **186–187**, 193, **220**, 220, 221
Cosmic Background Explorer (COBE) 370, **370**
Cosmic flashes 377–379
Cosmic microwave background (CMB) 369–371, **370**
Cow (AT 2018cow explosion) 378
Crab Nebula (M1) 191, 267, **267**, **291**
Crater (constellation) 174–175, 186–187, 221, **221**
Craters of the Moon National Monument, Idaho 345, **345**
Crew Dragon 407, **409**
Crux (Southern Cross) (constellation)
    about 222, **222**
    deep-sky treasures 302–303
    sky maps 182, **182–189**, 184, 186, 188
    star-hopping 192, 193
CST-100 Starliner 407
Curiosity rover 124, 126, **126**, 163
Cygnus (constellation)
    about 224, **224**
    deep-sky treasures 296–297, **297**
    Lacerta and 236
    61 Cygni (star) 24
    sky maps 176, **176–179**, 182
    star-hopping 190
    Tabby's Star 403

D

Daphnis (Saturn's moon) **138**, **143**
Dark energy 50, **50**, 370, **370–371**, **384**, 387, 416
Dark matter 50, **50**, 370, **370–371**, **385**, 385–387, **386**, 416
Dark-sky sites **168**, **340**, 341–347
DART (Double Asteroid Redirection Test) 161, 163
Davis, Raymond, Jr. 66–67
Dawn spacecraft **155**, 156, 163
DECam **384**
Deep-sky objects 288–304
    about 289
    catalogs 290–291, **291**
    clusters 419
    galaxies **288**, **289**, 418
    nebulae 419
    tours 292–303, **292–303**
    viewing tips 289–290

Deep space missions 162–163, **162–163**
Deep Space Network **22**
Deimos (Mars's moon) 125
Delphinus (constellation) **166**, **176–179**, 225, **225**
Democritus (philosopher) 363
Deneb (star) 176, 182, 188, 190, 224, 339
Denebola (star) 190, 193, 237
Differentiation 70, 98
Dillard, Annie 347
Dinosaur extinction 160, 285
Diphda (star) 191, 192, 215
Distances
    distance ladder 17, **25**, 25, 48, 50, **385**
    measurement tips 193, **193**
    stellar distances 24–25, 48, 191, 193
    see also Parallax; Standard candles
Doppler, Christian 28
Doppler effect 28, **28**
Doppler shift 389, 416
Dorado (constellation) **184–185**, 226, **226**
Double Asteroid Redirection Test (DART) 161, 163
Double star systems 40, **40**, 47–48, 416
Draco (constellation) **174–181**, 176, 227, **227**, 293
Dragon XL spacecraft **410**
Dragonfly rotorcraft 400, 401
Drake, Frank 403
Dreyer, Johan 291
Duke, Charles 74
Dumbbell Nebula (M27) 276, **277**
Durer, Albrecht, engraving by **362**
Dwarf planets 101, 154–157, **156**, 159, 416; see also specific dwarf planets
Dyson, Freeman 403

E

Eagle Nebula 265, **265**
Earth 84–91, 114–119
    about 54, 85, 115, 118
    asteroid collisions 160, **285**
    atmosphere 31, 33, 118, **118**
    axis 86–89, **88**
    CME collisions 62–64
    diameter 19
    equinoxes 88–89
    exploration 162, **162**
    facts 115
    formation of 70
    geology 115, **115**, **118**, 118–119
    grip on moon 76, **76**
    land tides 76
    Late Heavy Bombardment **99**
    life **119**
    maps **116**, **117**
    measuring 19–20
    meteorites 160
    orbiting observatories 90–91

relative size and orbit **98**, 99
    rotation 85–86
    seasons 88
    shape 23
    solar system in action 100, **100**
    sun's red giant phase 67, **67**
    supernova dangers 306
    tectonic plates 115, 119, **119**
    viewed from ISS **31**, **114**
Eclipses **65**, 89–90, 416, 420, 421; see also Lunar eclipses; Solar eclipses
Ecliptic 86, 89, 416
Eddington, Sir Arthur 36
Egypt 19, **19**, 20, 331–335, **332**, **333**, **334**
EHT (Event Horizon Telescope) **376**, 377
Eight-Burst Nebula (NGC 3132) 274, **274**
Einstein, Albert 31, 32, 42, 377, 379–380
Electromagnetic spectrum 31–36, **32**, **33**, **36**
Elephant's Trunk Nebula (IC 1396) **215**, 294
ELT (Extremely Large Telescope) **378**, 379
Enceladus (Saturn's moon) **140**, 143, 400
Equinoxes 88–89, 326, 330, 420, 421
Equuleus (constellation) **176–179**, 227, **227**
Eratosthenes of Cyrene 19, 20
Eridanus (constellation) 182, **184–185**, 192, 228, **228**, **300**, 300–301
Eris (dwarf planet) 155, **156**, 159
Eros (asteroid) 160, 163, **163**
ESA see European Space Agency
Eta Aquarid meteor showers 287, 420, 421
Eta Carinae (double star system) 211, 306–307, **307**
Eudoxus of Cnidus 20, 23
Europa (Jupiter's moon) 115, 131, **131**, **132**, 163, 398, 400
Europa Clipper 135, 400
Europe
    archaeoastronomy 15, **322**, **323**, 323–325, **324**, **325**
    astrotourism **343**, 343–344, 351, 352, **353**
    viewed from ISS **31**
European Southern Observatory, Chile **44**, 84, **378**, 379, **386**, 390
European Space Agency (ESA) missions **162–163**, 162–163; see also specific missions, objects, and planets
Event Horizon Telescope (EHT) **376**, 377
Exomoons 397, 416
Exoplanets 388–397
    atmospheres 392, 394

Breakthrough Starshot 391, **391**
constellation stars 197, 252, 274, 390, **390**
defined 416
direct imaging **396,** 396–397
HD 189733 b 392, 394, **394**
planet-finding techniques **389,** 389–396, **392, 395, 397**
Proxima b **390,** 390–391, **391, 393**
travel poster **388**
Expanding universe 26, 28, 44, 48, 50, **385**
Extinctions 160, 285
Extraterrestrial intelligence 359, 402–403; see also Alien life
Extremely Large Telescope (ELT) **378,** 379

F

False Cross asterism 193, 274
Faraday, Michael 59
Fermi, Enrico 65
Fermi bubbles **377,** 378
Fermi Gamma-ray Space Telescope 378
55 Cancri e (exoplanet) **388**
Finé, Oronce **23**
Florida's space coast **78,** **92–93, 356,** 356–357, **357**
Fomalhaut (star) 178, 191, 192, 230, 257
Fornax (constellation) **182–185,** 228, **228**
Four Corners region, U.S. 328, **328**
Fusion 42, 43, 57, 65–66, 416

G

G299 (supernova) **46**
Gabriela Mistral Dark Sky Sanctuary, Chile **344,** 344–345, **346**
Gagarin, Yuri 99, 405, **405**
Gaia spacecraft 371–373, 395–396
Galaxies
deep-sky **292,** 292–293, **293,** 298–299, 418
defined 416
gravity 48
Hubble Space Telescope images 26, **48–49**
star-forming clouds **26–27**
types of **289**
see also Andromeda galaxy; Milky Way
Galaxy clusters **16, 51**
Galilei, Galileo 23, 58, 131, 142, 151
Galileo missions 162, 163, 400
Galle, Johann Gottfried 151
Gamma Crucis (star) 193
Gamma rays 33, 57, 62–63, **377,** 378

Ganymede (Jupiter's moon) 131, **133, 163,** 401, **401**
Gateway space station (proposed) **80,** 410, **410**
Gegenschein 315–316, **316,** 416
Gem Cluster (NGC 3293) 302, **302**
Gemini (constellation)
about 229, **229**
Canis Minor and 209
exoplanets 390
sky maps **174–175,** 180, **180–181,** 186
star-hopping 190, 191, 192
Geminid meteor showers 283, 286, 287, 420, 421
Geminos (Greek writer) 218
General theory of relativity 377, 379–380
George III, King 149
Giant Magellan Telescope (GMT) **378,** 379
Giza pyramids 332, 333–335, **334, 335**
Globular clusters **4, 14,** 303, 416, 419
Glossary 416–417
GMT (Giant Magellan Telescope) **378,** 379
GN-z11 galaxy 369, **369**
Göbekli Tepe, Turkey 337–339, **338, 339**
Gravitational waves 377, 379–385, **381, 382,** 416
Gravity
birth of stars 40, 41–42
black holes 47
Earth's 76, **76**
galaxies 48
health effects of low gravity environment 81–82
moon 71, 81–82
Neptune 159
stars 15, 43, 44, 47
Great Pyramid of Khufu **332,** 333–335, 335
Great Sphinx, Egypt **332,** 334
Great Square of Pegasus asterism 178, 182, 191, 192, 196, 199, 230, 252
Greek cosmology 20, **23,** 95
Greenhouse gases 119
Greenwich Observatory, England 59–60
Grissom, Virgil "Gus" 212, 274
Grus (constellation) **182–183, 188–189,** 193, 230, **230**
Gum 29 (stellar nursery) **41**

H

H-R diagram **39,** 39–40, 43, 44, 417
Halley, Edmond 206
Halley's comet 163, **280**
Harvard University 25, 35, **37**
Harvest moon 83
Haumea (dwarf planet) 155, **156,** 159
Hawaii **58, 60,** 374, 376, 379, **379**

Hawking, Stephen 245
HD 189733 b (exoplanet) 392, 394, **394**
HD 209458 b (exoplanet) **394**
Heliocentric universe 20, 23, 55, 95, 100
Helios (Greek sun god) 55
Heliosphere 366–367, **367,** 417
Helium **64,** 64–65
Helix Nebula **47,** 199, **199**
L'Hemisfèric planetarium, Spain 352, **353**
Henderson, Thomas 24
Hercules (constellation) **174–177,** 176, 188, 190, **231,** 231
Herschel, Caroline 149
Herschel, John 59
Herschel, William 145, 149, 245, 296
Hertz, Heinrich 32
Hertzsprung, Ejnar 36, 39–40
Hertzsprung-Russell (H-R) diagram **39,** 39–40, 43, 44, 417
Hevelius, Johannes 206, 236, 241, 264, 266
High Precision PARallax COllecting (Hipparcos) Satellite 25, 371
Hind's Variable Nebula (NGC 1555) 301
Hinode mission 109
Hipparchus (Greek astronomer) 20, 25, 169, 290
Hipparcos (satellite) 25, 371
Hippocamp (Neptune's moon) **150,** 152
The Hitchhiker's Guide to the Galaxy (D. Adams) 365
Horologium (constellation) **182–185,** 232, **232**
Hortobágy National Park, Hungary **343,** 343–344
Houtman, Frederick de 198, 230
Hubble, Edwin 26, 28
Hubble constant 28, 371, **385**
Hubble expansion 26, 28, 44, 48, 50, **385**
Hubble Space Telescope
Andromeda galaxy **26**
Ant Nebula 247
Arrokoth 365
Caldwell 84 **4**
capabilities 169
Cartwheel Galaxy 262
Cat's Eye Nebula 293
ESO 69-6 270
Eta Carinae star 211
field of view **387**
galaxy cluster **17**
GN-z11 galaxy 369, **369**
Henize dwarf galaxy 257
Hippocamp **150**
Hubble Deep Field South 271
Jupiter **129**
M56 (globular cluster) 297
Modest Galaxy (UGC 3855) 241

NGC 1512 232
orbit 90, **91**
repairs 407
Stingray Nebula 200
10,000 galaxies **48–49**
website 418
Hubble tension 371
Hulse, Russell 380, **380**
Hungary 15, **343,** 343–344
Hurley, Doug **409**
Huygens, Christiaan 142, 232
Huygens lander 138, 163, 401, **402**
Hyades star cluster 180, 267
Hydra (constellation) 174, **174–177, 182–183,** 190, 233, **233**
Hydrus (constellation) 235, **235**

I

Iapetus (Saturn's moon) **141**
IAU (International Astronomical Union) 155, 195, 259
IC 1396 (Elephant's Trunk Nebula) **215,** 294
IC 2118 (Witch Head Nebula) **300,** 300–301
IC 2944 (Running Chicken Nebula) 213, **223**
Iceland 358, 359
IDA (International Dark-Sky Association) 342–346
IMAP (Interstellar Mapping and Acceleration Probe) 367
Impact craters **284, 358,** 359, 417
Inca 329
Index Catalogue 291
India, space missions 72, 80, 162–163, **162–163**
Induced currents 63
Indus (constellation) **182–183,** 236, **236**
Infrared 33, 417
Ingenuity helicopter **121,** 400, **401**
Inouye Solar Telescope, Hawaii **58,** 60
Inspiration4 mission **354**
International Astronomical Union (IAU) 155, 195, 259
International Dark-Sky Association (IDA) 342–346
International Space Station (ISS)
about 407–408
assembly 407
commercial passengers 355, 356, 408
Earth viewed from **31, 114**
space suit **407**
spacecraft docking **408**
spacewalks **411, 414–415**
tracking **311,** 311–312
Interstellar Mapping and Acceleration Probe (IMAP) 367

Interstellar objects 374, **374,** 376–377
Interstellar Probe 367
Io (Jupiter's moon) 131, **132, 163**
Ireland 322, 323, **323**
Israel, space missions 72, 162–163, **162–163**
ISS see International Space Station
Italy, space program 382, *382, 383,* 410

J

Jack-o-Lantern Nebula **360–361**
James Webb Space Telescope 91, **91, 368,** 369
Japan Aerospace Exploration Agency (JAXA)
Akatsuki 112–113
BepiColombo 107, 162
Gateway space station 410
Hinode mission 109
KAGRA observatory 382
Mars's moons 125
moon exploration 72, 73, 80, 82
solar system missions 162–163, **162–163**
JUICE (JUpiter ICy moons Explorer) 135, 400–401, **401**
Juno spacecraft 129, 131, 135, 162, **162–163,** 163
Jupiter 128–135
atmosphere 129
comet collision 135, **135**
conjunctions **308**
exploration 135, 163, **163**
facts 129
Galileo's observations 131, **131**
Great Red Spot 129
Hubble Space Telescope image **129**
location 98
map **130**
moons 115, 131, *131, 132–133,* 134, 163, **163, 398,** 400–401, **401**
planetary slingshot 134
relative size and orbit **98,** 99
rings 134–135
solar system in action 100, **100,** 101
storms **128,** 129, 135, **135**
strange center 129, 131
track chart **134–135**
Trojans (asteroids) 101, 163

K

KAGRA observatory 382
Kapteyn's Star 255
KBOs (Kuiper belt objects) 159; see also Arrokoth; Dwarf planets
Kennedy, John F. 79
Kennedy Space Center, Florida **78, 356,** 356–357, **357**

Kepler, Johannes 23
Kepler Space Telescope 391–392, 399
Keyser, Pieter Dirkszoon 198, 230
Keystone asterism 176, 190, 231
ǂKhomani San people 344
KIC 8462852 (Tabby's Star) 403
Kipping, David 395
Kirchhoff, Gustav 33–34
Kirkwood gaps 100, **100**
Koch, Christina **411, 412**
Kohoutek 4-55 (star) **43**
Kuiper, Gerard 149
Kuiper belt 98, 101, 152, 159, **160,** 417; *see also* Arrokoth; Dwarf planets

L

La Silla Observatory, Chile **320, 349**
Lacaille, Nicolas-Louis de 203, 211, 228, 247, 262
Lacerta (constellation) 176–179, 236, **236**
Lagrange, Joseph-Louis 91
Lagrange points 91, **91**
Laika (dog) 99
Lanzarote, Canary Islands 358, **358**
Laplace, Pierre-Simon, Marquis de **40,** 40–41
Large Magellanic Cloud 184, 226, **226, 235,** 244, 305, **305;** *see also* Tarantula Nebula
Laser Interferometer Gravitational-Wave Observatory (LIGO) 380, 382–383
Lassell, William 149, 152
Late Heavy Bombardment **99**
Le Verrier, Urbain Jean Joseph 151
Leavitt, Henrietta Swan 25–26
Leo (constellation)
   about 237, **237**
   Coma Berenices and 217
   deep-sky treasures **298,** 298–299, **299**
   exoplanets 390
   sky maps 174, **174–175,** 186, **186–187**
   star-hopping 190, 191
Leo Minor (constellation) **174–175,** 238, **238**
Leonid meteor showers 283, 287, 420, 421
Lepus (constellation) **180–181, 184–185,** 238, **238**
L'Hemisfèric planetarium, Spain 352, **353**
Libra (constellation) **176–177, 188–189,** 239, **239**
Lich (planetary system) 395, **397**
Life *see* Alien life; Extraterrestrial intelligence

Light pollution **168,** 289–290, 312, **340,** 341–347, **347**
Light wavelengths 31–32, 33
Light-years 170, 417
LIGO 380, 382–383
Linnaeus, Carolus 35
LISA 383, **383,** 385
Little Dipper asterism 273
Local Group of galaxies 269
Lockyer, Joseph Norman 64–65, **65**
Long-period comets 161
Low-Earth orbit 311, 312, 405, 407, 408–410
Lowell, Percival 125
Lowell Observatory, Arizona 125, 155
Luminosity 36, **39,** 39–40, 417
Luna spacecraft 72, 80, 81, 162
Lunar *see* Moon (Earth's)
Lunar eclipses **83,** 89–90, 420, 421
Lunokhod 1 lunar rover **79**
Lupus (constellation) **174–175, 186–187,** 240, **240,** 302–303
Lynx (constellation) **174–175, 180–181,** 241, **241**
Lyra (constellation) **176–177,** 193, 242, **242, 296,** 296–297; *see also* Vega
Lyrid meteor showers 287, 420, 421

M

M Centauri (star) 303
M1 (Crab Nebula) 191, 267, **267, 291**
M11 (Wild Duck Cluster) 264
M27 (Dumbbell Nebula) 276, **277**
M31 (Andromeda galaxy) 26, **26,** 178, 196, **196, 289,** 291, 363
M33 (Triangulum galaxy) 191, 269, **269**
M34 (open star cluster) 253, 295
M42 (Orion Nebula) **18,** 41, 180, 216, 250, 291
M45 (Pleiades star cluster) **6,** 180, 192, 267, **290,** 291
M49 (elliptical galaxy) 275, 299
M51 (Whirlpool Galaxy) 206, **207,** 305
M52 (open star cluster) 212, 294–295
M56 (globular cluster) 242, **242, 296,** 296–297
M57 (Ring Nebula) 242, **243**
M63 (Sunflower Galaxy) 206, 292–293
M85 (lenticular galaxy) 299, 305
M87 (supergiant elliptical galaxy) 275, **376,** 377, **385**
M94 (Cat's Eye Galaxy) 206, 292, **292**
M101 (Pinwheel Galaxy) **289,** 293, 305

M104 (Sombrero Galaxy) 275, **275,** 298, **299,** 305
M106 (spiral galaxy) 206, 292
Machu Picchu, Peru 329
Macondo (exoplanet) 197
Maezawa, Yusaku 413
Magellan, Ferdinand 226
Magellan spacecraft 109, 112
Magnetars 379
Main asteroid belt *see* Asteroid belt
Main sequence stars 39, **42,** 43
Maine, astrotourism **340,** 346, 348, **348**
Makemake (dwarf planet) **101,** 155, **156,** 159
Manhattanhenge 326, **326**
Maori 69, 198, 222, 339
Maps
   archaeoastronomy 323
   astrotourism 341, 348, 353
   Ceres **157**
   Earth **116, 117**
   gravitational-wave observatories 380
   Jupiter **130**
   Jupiter's moons **132, 133**
   launches by spaceport 355
   Mars **122, 123, 125, 127**
   Mercury **104, 105**
   moon **72, 73**
   Neptune **153**
   observatories and planetariums 353, 380
   Pluto **157**
   Saturn **139**
   solar eclipses 348
   Triton (Neptune's moon) **152**
   Uranus **147**
   Venus **110, 111**
   see also Sky maps; Track charts
Marconi, Guglielmo 32
Mariner spacecraft 103, 106, 112, 126, 162, 163
Mars 120–127
   analogues 359
   atmosphere 121
   canals 125, **125**
   conjunctions **308**
   exploration 92–93, 121, 124, 126, **126,** 163, **163,** 400, **401**
   facts 121
   geology **120,** 121, 124, **124,** 126
   human travel to 413
   life, search for 125, **399,** 399–400
   maps **122, 123, 125, 127**
   moons 125
   relative size and orbit 98
   rovers 121, 124, 126, **126, 399,** 399–400, **400, 401**
   solar system in action **100**
   track chart **124–125**
   watery past 124–125, **127,** 399
Maskelyne, Nevil 24
Mauna Kea, Hawaii 379, **379**

Maunder, Edward Walter 59–60
Maunder Minimum 59–60
Maya 329, **330,** 330–331, **331**
Mayor, Michel 390
Meir, Jessica **411, 412**
Melotte, P. J. 217
Mensa (constellation) **182–189,** 244, **244**
Mercury 102–107
   archaeoastronomy 106, **106**
   conjunctions **308**
   exploration 103, 106, 107, **107,** 162, **162**
   facts 103
   geology **102, 106,** 106–107
   maps **104, 105**
   observations 103, 106
   orbit 98, 103
   relative size 98
   solar system in action **100**
   sunrise and sunset 103, **103**
   sun's red giant phase 67
   viewing tips 309
Mesoamerican archaeo-astronomy 329–331, **330, 331**
MESSENGER spacecraft 106, **107,** 162
Messier, Charles 191, 290, 418
Messier catalog 290–291, 418
Messier objects 174, 191, 233
Meteor Crater, Arizona **284, 358,** 359
Meteor showers
   about 283
   calendar 287, 420, 421
   colors 286, **286**
   named for constellations 283–284, **286–287**
   Perseids **10–11,** 253, **278–279,** 283–284, 287, 420, 421
   Ursids 273, 420, 421
   viewing tips 283–284, 286
Meteorites 70, 160, **283,** 284, **284,** 417
Meteoroids 284, **284,** 417
Meteors 283–284, 284, 286, **286, 287, 358,** 359, 417
Mexico, archaeoastronomy **330,** 330–331, **331**
Microlensing 395
Microscopium (constellation) **182–183,** 244, **244**
Mier people 344
Milky Way
   advances in astronomy 371–373, **372**
   archaeoastronomy 15
   defined 417
   diameter 26
   Great Rift 259
   in infrared light 297
   south galactic pole (SGP) 262
   star-hopping 190, 192, 193
   stargazing **12–13, 94, 172,** 176, 182, 188, 296, **332**
Milner, Yuri 391

Maunder, Edward Walter 59–60
Mimas (Saturn's moon) **140**
Mir space station 405
Miranda (Uranus's moon) **148,** 149
Mistral, Gabriela 344
Monoceros (constellation) **180–181, 184–185,** 245, **245,** 300, **301**
Moon (Earth's) 68–83
   about 54, 69
   analogues 358–359
   blood moon 83
   blue moon 83
   distance from Earth 20
   Earth's gravity and 76, **76**
   exploration **72, 73, 74, 77,** 78–81, **79, 80,** 99, 162, **162,** 358, 405, 407, 413, **413**
   formation of 69–71, **71,** 118
   geology 71, 74–75, 118
   gravity 71, 81–82
   harvest moon 83
   human settlements on **81,** 81–82, 83
   humans' return to **80, 410,** 410–414
   international interest in **79,** 79–81, **80,** 413
   lunar eclipses **83,** 89–90
   maps **72, 73**
   phases 69, 70, **70,** 420, 421
   radiation 82
   religion and mythology 69, **69,** 71
   resources 82–83
   solar eclipses 89
   supermoon **68,** 83
   temperature swings 78
   tidal locking 76, 78
   tides and **75,** 75–76
   wolf moon 83
Moons
   dwarf planets **156**
   Jupiter 115, 131, **131, 132–133,** 134, 163, **163, 398,** 400–401, **401**
   Mars 125
   Neptune 101, **150,** 152, **152**
   Pluto 101, **101,** 156, **156**
   Saturn 138, **138, 140–141,** 142–143, **143,** 163, **163,** 400, 401, **402**
   Uranus **148,** 149
MUL.APIN tablets 106, **106**
Multiple star systems 40, **40**
Musca (constellation) **182–189,** 246, 246
Musk, Elon 312, 356, 413; *see also* SpaceX

N

Nabta Playa, Egypt 331, 333, **333**
Nancy Grace Roman Space Telescope 387, **387**
NANOGrav 385
NASA missions 162–163, **162–163;** *see also specific missions, objects, and planets*

Nasca lines, Peru 15, 329, **330**
National Park Service, U.S. 345–347, **347**
Navajo 328, **328**
Navigation, celestial 339
Near-Earth objects 100, **160, 374, 375**
Nebulae
  birth of stars 41
  deep-sky 294–297, **294–297, 300**, 300–301, **301**, 419
  defined 41, 417
  evolution **38**
  redshift 26, 28
  see also specific nebulae
Nebular hypothesis 41, 70
NEO Surveyor (infrared space telescope) 374
NEOWISE (comet) **282**
NEOWISE (infrared space telescope) **373**, 374
Neptune 150–153
  atmosphere **151**, 151–152
  exploration 163, **163**
  facts 151
  formation of 145
  gravity 159
  as ice giant 145
  map **153**
  moons 101, **150**, 152, **152**
  relative size and orbit 99, **99**
  rings 152
  solar system in action 101, **101**
Neutrinos 65–67, **66**
Neutron stars 47, 383, 417
The New General Catalogue of Nebulae and Clusters of Stars (NGC) 291, 418
New Horizons spacecraft 97, 134–135, **154**, 155, 162–163, **162–163, 364**, 365, **366**, 367
New Mexico **318–319**, 326–329, **327, 328**, 359
New York City 326, **326**
New Zealand
  astrotourism **350–351**, 351–352
  Maori 69, 198, 222, 339
Newgrange, Ireland **322**, 323, **323**
NGC 188 (open star cluster) 214, 293
NGC 281 (Pacman Nebula) 295, **295**
NGC 891 (Outer Limits Galaxy) 196, 295
NGC 1499 (California Nebula) 253, 295, **295**
NGC 1555 (Hind's Variable Nebula) 301
NGC 2264 (Christmas Tree Cluster) 245, 300, **301**
NGC 2903 (spiral galaxy) 237, 299, **299**
NGC 2997 (spiral galaxy) 197, **197**

NGC 3132 (Eight-Burst Nebula) 274, **274**
NGC 3293 (Gem Cluster) 302, **302**
NGC 3766 (Pearl Cluster) 213, 302, **303**
NGC 3918 (Blue Planetary Nebula) 302
NGC 4636 (elliptical galaxy) 298–299
NGC 4945 (spiral galaxy) 213, 302–303
NGC 5286 (Caldwell 84) **4**, 303
NGC 6543 (Cat's Eye Nebula) 293
Noctilucent clouds 316, **316**
Norma (constellation) **182–183**, 247, **247**
North Celestial Pole 273
North Star **88**, 88–89, 273
Northern Cross see Cygnus
Northern lights (aurora borealis) 58, **314**, 351
Nu Ophiuchi (star) 390
Nuclear fusion 42, 43, 57, 65–66, 416

O
Oberon (Uranus's moon) 149
Observatories
  advances in astronomy **373, 378, 382, 386**
  astrotourism **342**, 343, **352**, 352–353
  maps 353, 380
  orbiting 90–91
  see also Space telescopes
Octans (constellation) **182–189**, 193, 248, **248**
Oort, Jan 161
Oort cloud 98–99, **160**, 161, 417
Ophiuchus (constellation) **176–177, 188–189**, 249, **249**, 265, 390
Opportunity rover 121, 126, 163
Orbital ATK Cygnus cargo craft **408**
Orion (capsule) 410, 412
Orion (constellation)
  about 250, **250**
  archaeoastronomy 334–335
  Canis Major and 208, 250
  Canis Minor and 209, 250
  deep-sky treasures **300**, 300–301
  Scorpius and 250, 261
  sky maps 180, **180–181**, 184, **184–185**
  star-hopping 192
  stargazing **6, 21**
  Very Large Telescope images **44**
  see also Betelgeuse; Rigel
Orion Nebula (M42) **18**, 41, 180, 216, 250, 291
Orionid meteor showers 287, 420, 421
OSIRIS-REx spacecraft 159, 163
'Oumuamua (interstellar object) **161, 374**, 376

Outer Limits Galaxy (NGC 891) 196, 295
Overview effect 321
Ozma project 403

P
Pacman Nebula (NGC 281) 295, **295**
Pale Red Dot (team) 390
Parallax 18–20, **19**, 23, 25, 36, 48, 170, 417
Parsec system 170
Pavo (constellation) **182–189**, 186, 192, 251, 251
Peacock (star) 192, 251
Pearl Cluster (NGC 3766) 213, 302, **303**
Pegasus (constellation)
  about 252, **252**
  Andromeda and 196
  exoplanets 390, 396
  sky maps 178, **178–179**, 182, **182–183**
  star-hopping 191, 192
Perseid meteor showers **10–11**, 253, **278–279**, 283–284, 287, 420, 421
Perseus (constellation)
  about 253, **253**
  deep-sky treasures **294**, 294–295, 295
  sky maps 178, **178–179**, **182–183**
  star-hopping 191
Perseverance rover **92–93**, 126, 163, 399–400, **400, 401**
Peru, archaeoastronomy 15, 55, 222, 329, **330**
Petroglyphs 15, **320**, 329, 330
Phaethon (son of Helios) 55
Phobos (Mars's moon) 125
Phoebe (Saturn's moon) 142–143
Phoenix (constellation) **182–183**, 254, **254**
Photons 31–32, 33, 57, 386
Pictor (constellation) **186–187**, 255, **255**
Pillars of Creation 265, **265**
Pinwheel Galaxy (M101) **289**, 293, 305
Pioneer missions 135, 138, 162–163, **162–163**, 366, **366**
Pisces (constellation) 178, **178–179**, 256, **256**
Piscis Austrinus (constellation) **178–179, 182–183**, 191, 192, 257, **257**
Plancius, Petrus 204, 245
Planck space telescope 370, **370**
Planetariums **352**, 352–353
Planetary nebulae 44, 417, 419
Planetesimals 70, 97, **97**, 100, **100**, 417
Planets see Exoplanets; specific planets by name
Plasma 42, 47, 57, 58–59, **61**, 63, 417
Plate tectonics see Tectonic plates

Pleiades star cluster (M45) **6**, 180, 192, 267, **290**, 291
Pluto (dwarf planet)
  exploration 155, 163, **163**
  geology **154**, 155–156
  location 98
  map **157**
  moons 101, **101**, 156, **156**
  orbit 152, 159
  planetary status 95, 155
  solar system in action 101, 101
Pogue, William **406**
Pointers see Alpha Centauri; Beta Centauri
Polaris (North Star) **88**, 88–89, 273
Pollux (star) 180, 191, 209, 229, 390
Pōwehi (black hole) **376**
Proctor, Sian **354**
Procyon (star) 180, 191, 192, 209
Progress cargo craft **408**
Project Artemis **78**, 80
Project Ozma 403
Protoplanetary disk 40, 41–42, 97, 98, **396**
Protoplanets 70, 163, **163**
Proxima b (exoplanet) **390**, 390–391, **391, 393**
Proxima Centauri (star) 390–391, **393**
PSR 1913+16 (binary pulsar) 380
Psyche (asteroid) 160
Ptolemy, Claudius 20, 195, 200, 219
Puebloans **318–319**, 326–329, **327, 328**
Puerto Rico 352, **352**, 403
Pulsars 47, 380, 385, 395, **397**, 417
Puppis (constellation) **180–183**, 257, **257**
Pyramids **330**, 330–331, **332**, 333–335, **335**
Pyxis (constellation) **180–181, 184–185**, 258, **258**

Q
Quadrantid meteor showers 287, 420, 421
Quadruple star systems 40, **40**, 242
Queloz, Didier 390

R
Radial velocity method 389–395
Radio telescopes **164–165**, 352, **352**, 396, 397, 403, 413
Radio waves 32, 33
Ramsay, William 65
Red dwarf stars 39, **39**
Red giant stars 36, 39, **39, 42**, 43–44, 67, **67**
Red stars 36, 39
Redshift 26, 28, 417
Regulus (star) 186, 190, 237

Relativity, general theory of 377, 379–380
Reticulum (constellation) **182–189**, 259, **259**
Retrograde 103, 106, 109, 134, 146, 152, 210, 417
Rigel (star) **34**, 180, 191, 192, 250
Ring Nebula (M57) 242, **243**
Roman Space Telescope 387, **387**
Rosalind Franklin Mars rover **399**, 400
Roscosmos see Russia; Soviet Union
Roswell, New Mexico 359
Rubin, Vera 50
Rubin Observatory, Chile **373**, 373–374, 387
Running Chicken Nebula (IC 2944) 213, **223**
Russell, Henry Norris 39–40
Russia
  ExoMars initiative 400
  International Space Station 408
  moon exploration 81
  Progress cargo craft **408**
  Rosalind Franklin Mars rover **399**, 400
  solar system missions 162–163, **162–163**
  Soyuz spacecraft 407, **408**
  Spektr-RG space telescope 378
  supermoon 68
  see also Soviet Union

S
Sagitta (constellation) **176–177, 188–189**, 259, **259**, 307
Sagittarius (constellation) 176, **176–177**, 188, **188–189**, 190, 192, 193, 260, **260**
Sagittarius A* (black hole) **376**, 377–378
Saiph (star) 192
Salyut space stations 405
Satellites
  how to spot satellites 311–312
  orbits 91, **91**
  Sputnik 1 79, 99, 311
  tracking **310**, 310–313, **311, 312, 313**
  see also specific satellites
Saturn 136–143
  exploration 137, **137**, 138, 142, 163, **163**
  facts 137
  jet streams 137–138
  map **139**
  moons 138, **138, 140–141**, 142–143, **143**, 163, **163**, 400, 401, 402
  planetary slingshot 134
  relative size and orbit 98, 137
  rings **136**, 137, 138, **138**, 142, **142**

solar system in action 100, **100–101**
track chart **142–143**
Saucepan asterism 192
Schiaparelli, Giovanni 125
Schmidt, Klaus 338–339
Schmitt, Harrison **74**
Schwabe, Heinrich 59
Scorpius (constellation)
about 261, **261**
Libra and 239
Orion and 250, 261
Sagittarius and 260
sky maps 176, **176–177,** 186, 188, **188–189**
star-hopping 190, 192, 193
see also Antares
Sculptor (constellation) **178–179, 182–183,** 262, **262**
Scutum (constellation) **176–177,** 190, 264, **264**
SELENE spacecraft 72, 80, 82
Sembroski, Chris **354**
Serpens (constellation) **176–177, 188–189,** 249, 265, **265**
Serpent Mound, Ohio 326, **329**
Seven Sisters *see* Pleiades
Sextans (constellation) **174–175, 186–187,** 266, **266**
Shanghai Astronomy Museum, China 353
Shoemaker-Levy 9 comet 135, **135**
Shooting stars *see* Meteors
Short-period comets 159
Sidereal day 86, **86**
La Silla Observatory, Chile **320,** 349
Sirius (star) 180, 184, 186, 191, 192, 208, 209
61 Cygni (star) 24
Skidmore, Owings, & Merrill 83
Sky maps 173–193
Northern Hemisphere **174–181, 190–191**
Southern Hemisphere **182–189, 192–193**
Tang dynasty scroll map **194**
using 173, **173**
see also Track charts
Skylab space station **404,** 405, **406**
Skywatching *see* Stargazing
SLS (Space Launch System) 410, 412
Small Magellanic Cloud (SMC) 184, 186, **234,** 271, **386**
SNEWS (SuperNova Early Warning System) 305
Sojourner rover 126, 163
Solar cycle 59, **59**
Solar day 86, **86**
Solar eclipses
about 89
astrotourism 89, **89,** 347–348, **349**
calendar 420, 421

map 348
phases **88–89,** 89
scientific expeditions **65**
Solar flares 60–64, **61, 63**
Solar neutrino problem 65–67, **66**
Solar system 92–163
in action 100–101, **100–101**
advances in astronomy 373–377, **374, 375**
asteroid belt 98, 100, 156, 159–161
Copernicus's model 55
exploration missions 162–163, **162–163**
gas giants 98
ice giants 98
interstellar objects 374, **374,** 376–377
near-Earth objects 100, **160,** 374, **375**
objects **24,** 373–374
Oort cloud 98–99, **160,** 161, 417
origins 40, 70, 85, 86, **97,** 97–99, 100, **100**
planet sizes and orbits **98–99**
planetary family 95
planets, formation of 97–99, 100, **100**
terrestrial planets 98, 100
see also Dwarf planets; *specific planets*
Solar wind 58, 67, 365–366
Solstices
archaeoastronomy 323, 324, 326, 329, 334
calendar 420, 421
traditions 87, 87–88, 260
SOM (Skidmore, Owings, & Merrill) 83
Sombrero Galaxy (M104) 275, 275, 298, **299,** 305
South Africa, astrotourism 344
South America
archaeoastronomy 15, 55, 222, 329–331, **330**
astrotourism **344,** 344–345, **346**
South Celestial Pole 193, 248
South Korea 335, **336,** 337, 343
Southern Cross *see* Crux
Southern lights (aurora australis) 58, **350,** 351–352
Soviet Union
Luna spacecraft 72, 80, 81, 162
Mir space station 405
moon exploration **79,** 80
Salyut space stations 405
solar system missions 162–163, **162–163**
Sputnik 1 79, 99, 311
Vega spacecraft 163, 400, **400**
Venera probes 112, **112,** 162
Vostok 1 **405**
see also Russia
Soyuz spacecraft 407, **408**

Space Coast, Florida **78, 92–93, 356,** 356–357, **357**
Space Launch System (SLS) 410, 412
Space races 24, 79, 355, 405, **405**
Space shuttles 73, 357, 405, 407, 407
Space stations *see* International Space Station; Mir; Salyut; Skylab; Tiangong
Space suits **407**
Space telescopes
advances in astronomy **368, 369, 373,** 374, 377–378, **387**
aiming at exoplanets 395, **395**
James Webb 91, **91, 368,** 369
Kepler 391–392, 399
Nancy Grace Roman 387, **387**
see also Hubble Space Telescope
Space travel
flights on other celestial bodies 400, **400**
Florida's Space Coast **356,** 356–357, 357
future flights **354,** 354–359, 409
future human voyages 405, 407
future of humans in low Earth orbit 408–410
human spaceflight today 407–408
humans' return to moon 410–414
launches by spaceport 355
Mars **126,** 359
moon analogues 358–359
orbital flights 355–356
space analogues **357,** 357–359, **358**
suborbital flights 355
zero-g experience **357**
see also Astrotourism
SpaceX (company) 312, **312,** 356, 407, **409,** 409–410, **410,** 413, **413**
Spain 352, **353,** 358, **358**
Spectroscopy 33, 34, **36,** 64–65
Spectrum 31–36, **32, 33, 36**
Speed of light 32–33
Spektr-RG space telescope 378
Spica (star) 174, 186, 193, 275
Spirit rover 126, 163
Spring Triangle asterism 190, 193
Sputnik 1 (satellite) 79, 99, 311
Stade (unit of length) 19, 20
Standard candles **24,** 25–26, 48, 50, 417
Star charts *see* Sky maps
Star clouds 294–297, **294–297**
Star clusters **302,** 302–303,

303; *see also* Globular clusters
Star-hopping **190–193,** 190–193
Stargazing 318–359
calendars 287, 420–421
joys of 7
measuring distances 169–170, **193**
tools 170–171 (*see also* Binoculars; Telescopes)
travel through time 321
see also Archaeo-astronomy; Astrotourism
Starlink satellites 312, **312,** 409–410
Stars 12–51
astronomical unit 23–24
atomic fingerprints 33–35
birth **26–27, 30,** 40–42, **41**
catalogs 290
chemical composition 33–34
classification **34,** 35–36, **37, 39,** 39–40, 169
death 31, **42,** 43–44
distances 24–25, 48, 191, 193
Doppler effect 28
energy 43
expanding universe 26, 28
gravity and 15, 43, 44, 47
hydrogen burning 43
life cycle **38,** 38–51
light from 33–34
long train journey 43
luminosity 36, 39, **39**
magnitudes 169, 191, 193
main sequence 39, **42,** 43
measuring Earth 19–20
motion 17, **17, 18**
multiple star systems 40, **40**
nuclear fusion 42, 43
parallax 18–19, **19,** 20, 23, 24, 417
perfect spheres 20, 23
spectrum 32–36, **36**
standard candles 25–26
staying alive 43
surface temperature 39, **39**
universe end 48, 50
wavelengths 31–32
what are they? 30–37
where are they? 17–29
you are starstuff 44, **44**
see also Cepheid variables; Supernovae
Starship rocket 413, **413**
Steve (purple streak) 351, **351**
Stingray Nebula 200
Stonehenge, England **324,** 324–325, **325**
Struve, Friedrich Georg Wilhelm von 24
Sufi, Abd al-Rahman al- **362,** 363
Summer Triangle asterism 176, 190, 200, 224, 242
Sun 52–67
about 55, 57
core 57, **57**

corona **57,** 58, **61,** 416
coronal mass ejections 59, 60–64, **61, 62, 63**
energy 39, 57, 60, 67
exploration 162, **162**
future 67, **67**
as G-type star 35
heliocentric universe 20, 23, 55, 95, 100
heliosphere 366–367, **367,** 417
helium 64, 64–65
hydrogen burning 43, 67
interior **57,** 65–67
magnetic field 59, 60, 62, **63**
Maunder Minimum 59–60
nuclear fusion 57, 65–66
origins 40, **96,** 97, 100, 100
path through sky 86, 87–88
red giant phase 67, **67**
religion and mythology **54,** 55, 69
rotation 59, 366
size 99
solar cycle **59,** 59
solar flares 60–64, **61, 63**
solar neutrino 65–67, **66**
solar wind 58, 67, 365–366
sunspots 57, **58,** 58–60, 61
surface 57–58, **60, 61, 63**
surface temperature 33, 39, 59
tracking from Earth 87, **87**
x-rays **52–53**
see also Solar eclipses
Sunflower Galaxy (M63) 206, 292–293
Supermoon **68,** 83, 420, 421
Supernovae
1987 supernova 226, 305, **305**
about 305, 417
candidates for 304–307, 306, 307
Cassiopeia A **304**
debris field 46
early warning system 305
evolution **38**
life cycle 44, 47
remnants 419
Type Ia 47–48, 50
Type Ic 377
Type II 47
Supersymmetry theory 386

T
Tabby's Star (KIC 8462852) 403
Tang dynasty sky map **194**
Taosi, China 337, **337**
Tarantula Nebula **30, 90,** 226, **226**
Taurus (constellation)
about 267, **267**
deep-sky treasures 300–301
exoplanets 390, **390**
sky maps 180, **180–181,** 184
star-hopping 191, 192
Taylor, Joseph 380, **380**

Teachey, Alex 395
Teapot asterism 190, 192, 193
Technosignatures 402–403, 417
Tectonic plates 115, 119, **119**, 131
Telescopes
  advances in astronomy **368**, 368–387, **378**, 379, **386**
  Earth-based **378**, 379, **386**
  Herschel's 149, **149**
  for stargazing **169**, 170, **170**, **171**, 171, 289–290
  see also Space telescopes
Telescopium (constellation) **188–189**, 268, **268**
Temperature scales 57
TESS mission 392
Tethys (Saturn's moon) **141**, **163**
Thirty Meter Telescope (TMT), Hawaii 379, **379**
Tiangong space station 409
Tides 75, 75–76, 85, 417
Titan (Saturn's moon) 138, 143, 163, **163**, 400, 401, **402**
Titania (Uranus's moon) **148**, 149
Tito, Dennis 355
TMT (Thirty Meter Telescope), Hawaii 379, **379**
Tombaugh, Clyde 155
Tonatiuh (Aztec sun god) **54**, 55
Track charts
  Jupiter **134–135**
  Mars **124–125**
  Saturn **142–143**
Transit photometry 391–392, **392**, 394–395
Travel see Astrotourism; Space travel
Triangulum (constellation) **178–179**, 191, 269, **269**
Triangulum Australe (constellation) **182–189**, 270, **270**
Triangulum galaxy (M33) 191, 269, **269**
Triton (Neptune's moon) 101, 152, **152**

Trojans (asteroids) 101, 163
Tucana (constellation) **182–189**, 271, 271
Turkey, archaeoastronomy 337–339, **338**, **339**
Tyson, Neil deGrasse 326

U
UFOs 359; see also Alien life; Extraterrestrial intelligence
Umbriel (Uranus's moon) **148**, 149
United Arab Emirates, space missions 162–163, **162–163**
United States
  archaeoastronomy 15, **318–319**, 325–329, **326**, **327, 328, 329**
  astrotourism **340**, 345, 345–347, **347**, 348, **348**, 351
  see also NASA; specific space missions
Universe
  age 28
  big bang 28, **29**
  big bang afterglow 369–371, **370–371**
  determining size of 17–18
  end of 48, 50
  Hubble expansion 26, 28, 44, 48, 50, **385**
  Hubble tension 371
  pie chart **50**
  size 24
  see also Dark energy; Dark matter
Uranus 144–149
  atmosphere 145, 146, 151
  aurorae **149**
  exploration 163, **163**
  facts 145
  formation of 145
  Herschel studies 145, 149
  as ice giant 145
  magnetosphere **145**, 146
  map **147**
  moons **148**, 149
  as not so plain 145–146
  relative size and orbit **98**
  rings **144**, **145**, 149

rotation 86–87, **144**, **145**, 146
  seasons 149
  solar system in action 100, **101**
Ursa Major (constellation) **174–181**, 190, 191, **272**, 272, 292–293, **293**; see also Big Dipper asterism
Ursa Minor (constellation) **174–181**, 273, **273**, 292–293, **293**
Ursid meteor shower 273, 420, 421
Utah, as Mars analogue 359

V
V Sagittae (binary star system) 307, **307**
Variable stars **35**, 417
Vega (star) 24, **88**, 88–89, 176, 188, 193, 242
Vega spacecraft 163, 400, **400**
Veil Nebula Complex 224, **224**, 296
Vela (constellation) **182–189**, 193, 274, **274**
Venera probes 112, **112**, 162
Venus 108–113
  alien life, search for 113, 401–402
  atmosphere 109, 112, 402, **402**
  conjunctions **308**
  exploration 106, **112**, 112–113, 162, **162**, 403, **403**
  facts 109
  geology **108**, 113, **113**
  maps **110**, **111**
  relative size and orbit **98**
  robots 403, **403**
  rotation 87, 109
  solar system in action **100**
  sun's red giant phase 67
  transit **109**
  Vega balloon probes 400, **400**
  viewed from Earth **112**, **317**
  viewing tips 309
  volcanoes 113, **113**

Vera C. Rubin Observatory, Chile **373**, 373–374, 387
Verne, Jules 78
Vespucci, Amerigo 270
Vesta (asteroid) **155**, 163, **163**
Victor M. Blanco Telescope, Chile **386**
Vikings 315
Virgin Galactic (spaceflight company) 355, 409
Virgo (constellation)
  about 275, **275**
  Coma Berenices and 217
  deep-sky treasures **298**, 298–299, **299**
  sky maps 174, **174–175**, 186, **186–187**
  star-hopping 193
Virgo observatory, Italy 382, **382**, 383
Virts, Terry 115, **414–415**
Volans (constellation) **186–187**, 276, **276**
Volcanoes, Venus 113, **113**
Vostok 1 **405**
Voyager missions
  beyond solar system 365, 366
  to Europa 400
  flight paths 162–163, **162–163**, 366
  to Jupiter 134, 135
  to Neptune **151**, 152, 163
  Oort cloud 161
  planetary slingshot 134
  to Saturn 134, 135, 138, 163
  to Uranus 146, **146**, 149, 163
Vulpecula (constellation)
  about 276, **276**, **277**
  Coathanger asterism 276, **296**, 297
  HD 189733 b 392, 394, **394**
  sky maps **176–177**
VY Canis Majoris (star) **34**

W
Warrumbungle National Park, Australia **342**, 342–343
WASP-66 (exoplanet) 197
Wavelengths 28, 31–32
Weakly interacting massive particles (WIMPs) 386

Wells, H. G. 78, **79**
Whirlpool Galaxy (M51) 206, **207**, 305
White dwarf stars 39, **43**, 44, 47–48
Wide-field Infrared Survey Explorer (WISE) space telescope **373**
Wild Duck Cluster (M11) 264
Wilkins, John 78
Wilkinson Microwave Anisotropy Probe (WMAP) 370
WIMPs (weakly interacting massive particles) 386
Winter Hexagon asterism 191
WISE (Wide-field Infrared Survey Explorer) space telescope 373
Witch Head Nebula (IC 2118) **300**, 300–301
WMAP (Wilkinson Microwave Anisotropy Probe) **370**
Wolf moon 83
Wolszczan, Aleksander 397

X
X-rays 32, 33, **52–53**, 57, 62–63, **377**, 378

Y
Yeongyang Firefly Eco Park, South Korea 343
Yutu-2 lunar rover **80**

Z
Zero-g experience **357**
Zhurong rover 126, 163
Zodiacal lights **2–3**, 315–316, **317**
Zubenelgenubi (star) 193, 239
Zubeneschamali (star) 193, 239

Since 1888, the National Geographic Society has funded more than 14,000 research, conservation, education, and storytelling projects around the world. National Geographic Partners distributes a portion of the funds it receives from your purchase to National Geographic Society to support programs including the conservation of animals and their habitats.

Get closer to National Geographic Explorers and photographers, and connect with our global community. Join us today at nationalgeographic.com/join

For rights or permissions inquiries, please contact National Geographic Books Subsidiary Rights: bookrights@natgeo.com

ISBN: 978-1-4262-2220-7

Printed in South Korea

22/SPSK/1

EXPLORE THE NEW FRONTIER

NEIL DEGRASSE TYSON with JAMES TREFIL

COSMIC QUERIES
StarTalk's Guide to Who We Are, How We Got Here, and Where We're Going

NATIONAL GEOGRAPHIC

BACKYARD GUIDE TO THE NIGHT SKY
SECOND EDITION

ANDREW FAZEKAS

NATIONAL GEOGRAPHIC

SECOND EDITION

SPACE ATLAS
| MAPPING THE UNIVERSE AND BEYOND |

INCLUDES "THE MOON FIFTY YEARS LATER" BY BUZZ ALDRIN

JAMES TREFIL

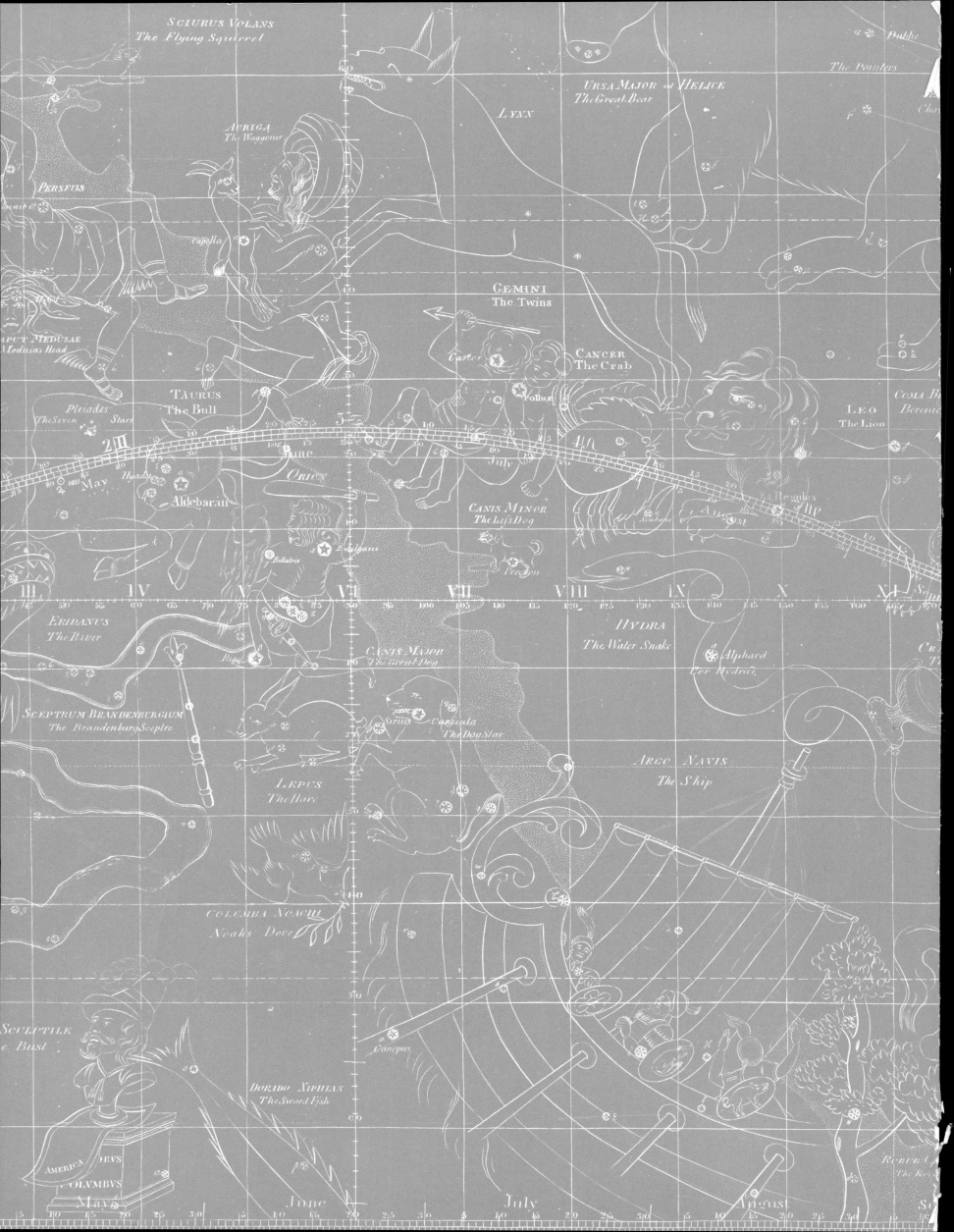